# PERCEPTUAL DIGITAL IMAGING

## METHODS AND APPLICATIONS

# Digital Imaging and Computer Vision Series

**Series Editor**

Rastislav Lukac

*Foveon, Inc./Sigma Corporation*
*San Jose, California, U.S.A.*

# PERCEPTUAL DIGITAL IMAGING

## METHODS AND APPLICATIONS

EDITED BY
## RASTISLAV LUKAC

CRC Press
Taylor & Francis Group
Boca Raton London New York

CRC Press is an imprint of the
Taylor & Francis Group, an **informa** business

CRC Press
Taylor & Francis Group
6000 Broken Sound Parkway NW, Suite 300
Boca Raton, FL 33487-2742

First issued in paperback 2017

© 2013 by Taylor & Francis Group, LLC
CRC Press is an imprint of Taylor & Francis Group, an Informa business

No claim to original U.S. Government works
Version Date: 20120822

ISBN 13: 978-1-138-07740-9 (pbk)
ISBN 13: 978-1-4398-6856-0 (hbk)

**Visit the Taylor & Francis Web site at**
**http://www.taylorandfrancis.com**

**and the CRC Press Web site at**
**http://www.crcpress.com**

*All our knowledge has its origins in our perceptions.*

—*Leonardo da Vinci (1452–1519)*

# Dedication

*To my supporters and friends*

# Contents

# *Preface*

Visual perception is a complex process requiring interaction between the receptors in the eye that sense the stimulus and the neural system and the brain that are responsible for communicating and interpreting the sensed visual information. This process involves several physical, neural, and cognitive phenomena whose understanding is essential to design effective and computationally efficient imaging solutions. Building on the research advances in computer vision, image and video processing, neuroscience, and information engineering, *perceptual digital imaging* has become an important and rapidly developing research field. It greatly enhances the capabilities of traditional imaging methods, and numerous commercial products capitalizing on its principles have already appeared in divergent market applications, including emerging digital photography, visual communication, multimedia, and digital entertainment applications.

The purpose of this book is to fill the existing gap in the literature and comprehensively cover the system design, implementation, and application aspects of perceptual digital imaging. Because of the rapid developments in specialized imaging areas, the book is a contributed volume where well-known experts are dealing with specific research and application problems. It presents the state-of-the-art as well as the most recent trends in image acquisition, processing, storage, display, and visual quality evaluation. The book serves the needs of different readers at different levels; it can be used as textbook in support of graduate courses in computer vision, digital imaging, visual data processing, computer graphics, and visual communication, or as stand-alone reference for graduate students, researchers, and practitioners.

This book provides a strong, fundamental understanding of theory and methods, and a foundation on which solutions for many of today's most interesting and challenging imaging problems can be built. It details recent advances in the field and explores human visual system-driven approaches across a broad spectrum of applications, including image quality and aesthetics assessment, digital camera imaging, white balancing and color enhancement, thumbnail generation, image restoration, super-resolution imaging, digital halftoning and dithering, color feature extraction, semantic image analysis and multimedia, video shot characterization, image and video encryption, display quality enhancement, and more.

The book begins by focusing on human visual perception. The human visual system can be subdivided into two major components, that is, the eyes, which capture light and convert it into signals that can be understood by the nervous system, and the visual pathways in the brain, along which these signals are transmitted and processed. Chapter 1 discusses *characteristics of human vision*, focusing on the anatomy and physiology of the above components as well as a number of phenomena of visual perception that are of particular relevance to digital imaging.

As motion is ubiquitous in normal viewing conditions, it is essential to analyze the effects of various sources of movement on the retinotopic representation of the environment. Chapter 2 deals with *an analysis of human visual perception based on real-time constraints of ecological vision*, considering two inter-related problems of motion blur and moving ghosts. A model of retino-cortical dynamics is described in order to provide a mathematical framework for dealing with motion blur in human vision.

Chapter 3 addresses important issues of perceptual *image and video quality assessment*. Built on the knowledge on perception of images and videos by humans and refined computational models of visual processing, a number of assessment methods capable of producing the quality scores can be designed. Although human qualitative opinion represents the palatability of visual signals, subjective quality assessment is usually time consuming and impractical. Thus, a more efficient approach is to design the algorithms that can objectively evaluate visual quality by automatically generating the quality scores that correlate well with subjective opinion.

Chapter 4 focuses on *visual aesthetic quality assessment of digital images*. Computational aesthetics is concerned with exploring techniques to predict an emotional response to a visual stimulus and with developing methods to create and enhance pleasing impressions. Among various modules in the aesthetic algorithm design, such as data collection and human study, feature extraction, and machine learning, constructing and extracting the features using the knowledge and experience in visual psychology, photography, and art is essential to overcome the gap between low-level image properties and high-level human perception of aesthetics.

The human visual system characteristics are also widely considered in the digital imaging technology design. As discussed in Chapter 5, digital camera designers largely rely on *perceptually based image processing* to ensure that a captured image mimics the scene and is visually pleasing. Perceptual considerations affect the decisions made by the automatic camera control algorithms that adjust the exposure, focus, and white balance settings of the camera. Various camera image processing steps, such as demosaicking, noise reduction, color rendering, edge enhancement, and compression, are similarly influenced in order to execute quickly without sacrificing perceptual quality.

Chapter 6 presents the framework that addresses the problem of *joint white balancing and color enhancement*. The framework takes advantage of pixel-adaptive processing that combines the local and global spectral characteristics of the captured visual data in order to produce the image with the desired color appearance. Various example solutions can be constructed within this framework by following simple but yet powerful spectral modeling and combinatorial principles. The presented design methodology is efficient, highly flexible, and leads to visually pleasing color images.

Taking advantage of their small size, thumbnails are commonly used in preview, organization, and retrieval of digital images. *Perceptual thumbnail generation*, explored in Chapter 7, aims to provide a faithful impression about the image content and quality. Unlike the conventional thumbnail, its perceptual counterpart displays both global composition and important visual features, such as noise and blur, of the original image. This allows the user to efficiently judge the image quality by viewing the low-resolution thumbnail instead of inspecting the original full-resolution image.

Chapter 8 reviews the principles of *patch-based image models* and explores their possible scientific connections with human vision models. The evolution from first-generation patch models, which relate to dictionary construction and learning, to second-generation patch models, which include structural clustering and sparsity optimization, offers insights on how locality and convexity have served in mathematical modeling of photographic images. The potential of patch-based image models is demonstrated in various image processing applications, such as denoising, compression artifact removal, and inverse halftoning.

Super-resolution imaging aims at producing a high-resolution image or a sequence of high-resolution images from a set of low-resolution images. The process requires an image acquisition model that relates a high-resolution image to multiple low-resolution images and involves solving the resulting inverse problem. Chapter 9 surveys existing relevant methods, with a focus on efficient *perceptually driven super-resolution techniques*. Such techniques utilize various models of the human visual system and can automatically adapt to local characteristics that are perceptually most relevant, thus producing the desired image quality and simultaneously reducing the computational complexity of processing.

Digital halftoning refers to the process of converting a continuous-tone image or photograph into a binary pattern of black and white pixels for display on binary devices, such as ink-jet printers. Similar to dithering used in computer graphics, this process creates the illusion of depth when outputting an image on a device with a limited palette. Chapter 10 discusses the *methods of dither array construction employing models of visual perception*, including the extension of the stochastic dither arrays to nonzero screen angles and the challenging problem of lenticular printing.

Color features are widely used in content analysis and retrieval. However, most of them show severe limitations due to their poor connection to the color perception mechanism of the human visual system and their inability to characterize all the properties of the color composition in a visual scenery. To overcome these drawbacks, Chapter 11 focuses on *perceptual color descriptors* that reflect all major properties of prominent colors. Extracted global and spatial properties using these refined descriptors can be combined further to form the final descriptor that is unbiased and robust to non-perceivable color elements in both spatial and color domains.

Exploiting information in the sense of visual semantics, context, and implicit or explicit knowledge not only allows for better scene understanding by bridging the semantic and conceptual gap that exists between humans and computers but also enhances content-based multimedia analysis and retrieval performance. To address this problem, Chapter 12 deals with *concept-based multimedia processing using semantic and contextual knowledge*. Such high-level concepts can be efficiently detected when an image is represented by a model vector with the aid of a visual thesaurus and visual context, where the latter can be interpreted by utilizing an ontology-based fuzzy representation of knowledge.

Chapter 13 presents *perceptually driven video shot characterization*, employing an unsupervised approach to identify meaningful components that influence the semantics of the scene through their behavioral and perceptual attributes. This is done by using the perceptual grouping and prominence principles. Namely, the former takes advantage of an organizational model that encapsulates the grouping criteria based on spatiotemporal consistency exhibited by emergent clusters of grouping primitives. The latter models the cognitive saliency of the subjects based on attributes that commonly influence human judgment. The video shot is categorized based on the observations that direct visual attention of a human observer across the visualization space.

With the proliferation of digital imaging devices, protecting sensitive visual information from unauthorized access and misuse becomes crucial. Given the extensive size of visual data, full encryption of digital images and videos may not be necessary or economical in some applications. Chapter 14 discusses *perceptual encryption of digital images and videos* that can be implemented by selectively encrypting part of the bitstream representing the visual data. Of particular interest are attacks on perceptual encryption schemes for popular image and video formats based on the discrete cosine transform.

Finally, Chapter 15 explores perceptual effects to *exceed physical limitations of display devices*. By considering various characteristics of human visual perception, display qualities can be significantly enhanced, for instance, in terms of perceived contrast and disparity, brightness, motion smoothness, color, and resolution. Similar enhancement could often be achieved only by improving physical parameters of displays, which might be impossible without fundamental design changes in the existing display technology and clearly may lead to overall higher display costs.

As the above overview suggests, this book is a unique up-to-date reference that should be found useful in the design and implementation of various digital imaging-related tasks. Moreover, each chapter offers a broad survey of the relevant literature, thus providing a good basis for further exploration of the presented topics. The book includes numerous examples and illustrations of perceptual digital imaging results, as well as tables summarizing the results of quantitative analysis studies. Complementary material for further reading is available online at *http://www.colorimageprocessing.org*.

I would like to thank the contributors for their effort, valuable time, and motivation to enhance the profession by providing material for a wide audience while still offering their individual research insights and opinions. I am very grateful for their enthusiastic support, timely response, and willingness to incorporate suggestions from me to improve the quality of contributions. Finally, a word of appreciation for CRC Press / Taylor & Francis for giving me the opportunity to edit a book on perceptual digital imaging. In particular, I would like to thank Nora Konopka for supporting this project, Jessica Vakili for coordinating the manuscript preparation, Jim McGovern for handling the final production, Andre Barnett for proofreading the book, and John Gandour for designing the book cover.

**Rastislav Lukac**
Foveon, Inc. / Sigma Corp., San Jose, CA, USA
*E-mail: lukacr@colorimageprocessing.com*
*Web: www.colorimageprocessing.com*

# The Editor

**Rastislav Lukac** (www.colorimageprocessing.com) received M.S. (Ing.) and Ph.D. degrees in telecommunications from the Technical University of Kosice, Slovak Republic, in 1998 and 2001, respectively. From February 2001 to August 2002, he was an assistant professor with the Department of Electronics and Multimedia Communications at the Technical University of Kosice. From August 2002 to July 2003, he was a researcher with the Slovak Image Processing Center in Dobsina, Slovak Republic. From January 2003 to March 2003, he was a postdoctoral fellow with the Artificial Intelligence and Information Analysis Labora-tory, Aristotle University of Thessaloniki, Thessaloniki, Greece. From May 2003 to August 2006, he was a postdoctoral fellow with the Edward S. Rogers Sr. Department of Electrical and Computer Engineering, University of Toronto, Toronto, Ontario, Canada. From September 2006 to May 2009, he was a senior image processing scientist at Epson Canada Ltd., Toronto, Ontario, Canada. In June 2009, he was a visiting researcher with the Intelligent Systems Laboratory, University of Amsterdam, Amsterdam, the Netherlands. Since August 2009, he has been a senior digital imaging scientist at Foveon, Inc. / Sigma Corp., San Jose, California, USA. Dr. Lukac is the author of five books and four textbooks, a contributor to twelve books and three textbooks, and he has published more than 200 scholarly research papers in the areas of digital camera image processing, color image and video processing, multimedia security, and microarray image processing. He holds 12 patents and has authored 25 additional patent-pending inventions in the areas of digital color imaging and pattern recognition. He has been cited more than 700 times in peer-review journals covered by the *Science Citation Index* (SCI).

Dr. Lukac is a senior member of the Institute of Electrical and Electronics Engineers (IEEE), where he belongs to the Circuits and Systems, Consumer Electronics, and Signal Processing societies. He is an editor of the books *Perceptual Digital Imaging: Methods and Applications* (October 2012), *Computational Photography: Methods and Applications* (October 2010), *Single-Sensor Imaging: Methods and Applications for Digital Cameras* (September 2008), and *Color Image Processing: Methods and Applications* (October 2006), all published by CRC Press / Taylor & Francis. He is a guest editor of *Real-Time Imaging*, Special Issue on Multi-Dimensional Image Processing, *Computer Vision and Image Understanding*, Special Issue on Color Image Processing, *International Journal of Imaging Systems and Technology*, Special Issue on Applied Color Image Processing, and *International Journal of Pattern Recognition and Artificial Intelligence*, Special Issue on Facial Image Processing and Analysis. He is an associate editor for the *IEEE Transactions on Circuits and Systems for Video Technology* and the *Journal of Real-Time Image*

*Processing*. He is an editorial board member for *Encyclopedia of Multimedia* (2nd Edition, Springer, September 2008). He is a *Digital Imaging and Computer Vision* book series founder and editor for CRC Press / Taylor & Francis. He serves as a technical reviewer for various scientific journals, and participates as a member of numerous international conference committees. He is the recipient of the 2003 North Atlantic Treaty Organization / National Sciences and Engineering Research Council of Canada (NATO/NSERC) Science Award, the Most Cited Paper Award for the *Journal of Visual Communication and Image Representation* for the years 2005–2007, the 2010 Best Associate Editor Award of the *IEEE Transactions on Circuits and Systems for Video Technology*, and the author of the #1 article in the ScienceDirect Top 25 Hottest Articles in *Signal Processing* for April–June 2008.

# Contributors

**James E. Adams, Jr.**  Eastman Kodak Company, Rochester, New York, USA

**Gonzalo R. Arce**  University of Delaware, Newark, Delaware, USA

**Murat Birinci**  Tampere University of Technology, Tampere, Finland

**Alan C. Bovik**  The University of Texas at Austin, Austin, Texas, USA

**Santanu Chaudhury**  Indian Institute of Technology Delhi, New Delhi, India

**Tsuhan Chen**  Cornell University, Ithaca, New York, USA

**Aaron T. Deever**  Eastman Kodak Company, Rochester, New York, USA

**Piotr Didyk**  MPI Informatik, Saarbrücken, Germany

**Elmar Eisemann**  Telecom ParisTech (ENS) – CNRS (LTCI), Paris, France

**Wei Feng**  Tianjin University, Tianjin, P. R. China

**Moncef Gabbouj**  Tampere University of Technology, Tampere, Finland

**Gonzalo J. Garateguy**  University of Delaware, Newark, Delaware, USA

**Gaurav Harit**  Indian Institute of Technology Rajasthan, India

**Lina Karam**  Arizona State University, Tempe, Arizona, USA

**Serkan Kiranyaz**  Tampere University of Technology, Tampere, Finland

**Stefanos Kollias**  National Technical University of Athens, Athens, Greece

**Daniel L. Lau**  University of Kentucky, Lexington, Kentucky, USA

**Congcong Li**  Cornell University, Ithaca, New York, USA

**Shujun Li**  University of Surrey, Surrey, UK

**Xin Li**  West Virginia University, Morgantown, West Virginia, USA

**Zhouchen Lin**  Microsoft Research Asia, Beijing, P. R. China

**Zhi-Qiang Liu**  City University of Hong Kong, Hong Kong, P. R. China

**Rastislav Lukac**  Foveon, Inc. / Sigma Corp., San Jose, California, USA

**Anush K. Moorthy**  The University of Texas at Austin, Austin, Texas, USA

**Efraín O. Morales**   Eastman Kodak Company, Rochester, New York, USA

**Phivos Mylonas**   National Technical University of Athens, Athens, Greece

**Karol Myszkowski**   MPI Informatik, Saarbrücken, Germany

**Haluk Öğmen**   University of Houston, Houston, Texas, USA

**Bruce H. Pillman**   Eastman Kodak Company, Rochester, New York, USA

**Tobias Ritschel**   Telecom ParisTech (ENS) – CNRS (LTCI), Paris, France

**Nabil Sadaka**   Arizona State University, Tempe, Arizona, USA

**Kalpana Seshadrinathan**   Intel Corporation, Santa Clara, California, USA

**Evaggelos Spyrou**   National Technical University of Athens, Athens, Greece

**Liang Wan**   Tianjin University, Tianjin, P. R. China

**Stefan Winkler**   Advanced Digital Sciences Center, Singapore

**Tien-Tsin Wong**   The Chinese University of Hong Kong, Hong Kong, P. R. China

# 1

## *Characteristics of Human Vision*

**Stefan Winkler**

## 1.1 Introduction

Vision is perhaps the most essential of human senses. A large part of human brain is devoted to vision, which explains the enormous complexity of the human visual system. The human visual system can be subdivided into two major components: the eyes, which capture light and convert it into signals that can be understood by the nervous system, and the visual pathways in the brain, along which these signals are transmitted and processed. This chapter discusses the anatomy and physiology of these components as well as a number of phenomena of visual perception that are of particular relevance to digital imaging.

The chapter is organized as follows. Section 1.2 presents the optics and mechanics of the eye. Section 1.3 discusses the properties and the functionality of the receptors and neurons in the retina. Section 1.4 explains the visual pathways in the brain and a number of components along the way. Section 1.5 reviews human sensitivity to light and various related mathematical models. Section 1.6 discusses the processes of masking and adaptation. Section 1.7 describes the representation of color in the visual system and other useful color spaces. Section 1.8 briefly outlines the basics of depth perception. Section 1.9 provides conclusions and pointers for further reading.

## 1.2 Eye

### 1.2.1 Physical Principles

From an optical point of view, the eye is the equivalent of a photographic camera. It comprises a system of lenses and a variable aperture to focus images on the light-sensitive retina. This section summarizes the optical principles of image formation.

The optics of the eye rely on the physical principles of refraction. Refraction is the bending of light rays at the angulated interface of two transparent media with different refractive indices. The refractive index $n$ of a material is the ratio of the speed of light in vacuum $c_0$ to the speed of light in this material $c$, that is, $n = c_0/c$. The degree of refraction depends on the ratio of the refractive indices of the two media as well as the angle $\phi$ between the incident light ray and the interface normal, resulting in $n_1 \sin\phi_1 = n_2 \sin\phi_2$. This is known as Snell's law.

Lenses exploit refraction to converge or diverge light, depending on their shape. Parallel rays of light are bent outward when passing through a concave lens and inward when passing through a convex lens. These focusing properties of a convex lens can be used for image formation. Because of the nature of the projection, the image produced by the lens is rotated $180°$ about the optical axis.

Objects at different distances from a convex lens are focused at different distances behind the lens. In a first approximation, this is described by the Gaussian lens formula:

$$\frac{1}{d_s} + \frac{1}{d_i} = \frac{1}{f},$$

$$(1.1)$$

where $d_s$ is the distance between the source and the lens, $d_i$ is the distance between the image and the lens, and $f$ is the focal length of the lens. An infinitely distant object is focused at focal length, resulting in $d_i = f$. The reciprocal of the focal length is a measure of the optical power of a lens, that is, how strongly incoming rays are bent. The optical power is defined as $1\text{m}/f$ and is specified in diopters.

Most optical imaging systems comprise a variable aperture, which allows them to adapt to different light levels. Apart from limiting the amount of light entering the system, the aperture size also influences the depth of field, that is, the range of distances over which objects will appear in focus on the imaging plane. A small aperture produces images with a large depth of field and vice versa. Another side effect of an aperture is diffraction, which is the scattering of light that occurs when the extent of a light wave is limited. The result is a blurred image. The amount of blurring depends on the dimensions of the aperture in relation to the wavelength of the light.

Distance-independent specifications are often used in optics. The visual angle $\alpha = 2\arctan(s/2D)$ measures the extent covered by an image of size $s$ at distance $D$ from the eye. Likewise, resolution or spatial frequency are measured in cycles per degree (cpd) of visual angle.

### 1.2.2 Optics of the Eye

Attempts to make general statements about the eye's optical characteristics are complicated by the fact that there are considerable variations between individuals. Furthermore, its components undergo continuous changes throughout life. Therefore, the figures given in the following can only be approximations.

The optical system of the human eye is composed of the cornea, the aqueous humor, the lens, and the vitreous humor, as illustrated in Figure 1.1. The refractive indices of these four components are 1.38, 1.33, 1.40, and 1.34, respectively [1]. The total optical power of the eye is approximately 60 diopters. Most of it is provided by the air-cornea transition, where the largest difference in refractive indices occurs (the refractive index of air is close

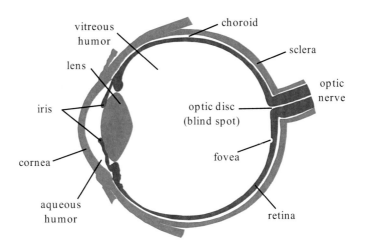

**FIGURE 1.1**

The human eye (transverse section of the left eye).

to 1). The lens itself provides only a third of the total refractive power due to the optically similar characteristics of the surrounding elements.

The lens is important because its curvature and thus its optical power can be voluntarily increased by contracting muscles attached to it. This process is called accommodation. Accommodation is essential to bringing objects at different distances into focus on the retina. In young children, the optical power of the lens can extend from 20 to 34 diopters. However, this accommodation ability decreases gradually with age until it is lost almost completely, a condition known as presbyopia.

Just before entering the lens, the light passes the pupil, the eye's aperture. The pupil is the circular opening inside the iris, a set of muscles that control its size and thus the amount of light entering the eye depending on the exterior light levels. Incidentally, the pigmentation of the iris is also responsible for the color of the eyes. The diameter of the pupillary aperture can be varied between 1.5 and 8 mm, corresponding to a thirtyfold change of the quantity of light entering the eye. The pupil is thus one of the mechanisms of the human visual system for light adaptation, which is discussed in Section 1.5.1.

### 1.2.3   Optical Quality

The physical principles described in Section 1.2.1 pertain to an ideal optical system, whose resolution is only limited by diffraction. While the parameters of an individual healthy eye are usually correlated in such a way that the eye can produce a sharp image of a distant object on the retina, imperfections in the lens system can introduce additional distortions that affect image quality. In general, the optical quality of the eye deteriorates with increasing distance from the optical axis. This is not a severe problem, however, because visual acuity also decreases there, as will be discussed in Section 1.3.

The blurring introduced by the eye's optics can be measured [2] and quantified by the point spread function (PSF) or line spread function of the eye, which represent the retinal images of a point or thin line, respectively; their Fourier transform is the modulation transfer function. A simple approximation of the foveal PSF of the human eye according to Reference [3] is shown in Figure 1.2 for a pupil diameter of 4 mm. The amount of blurring depends on the pupil size. Namely, for small pupil diameters up to 3 or 4 mm, the optical

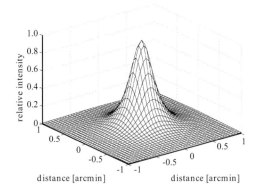

**FIGURE 1.2**

Point spread function of the human eye as a function of visual angle [3].

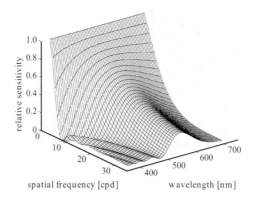

**FIGURE 1.3**

Variation of the modulation transfer function of a human eye model with wavelength [5].

blurring is close to the diffraction limit; as the pupil diameter increases (for lower ambient light intensities), the width of the PSF increases as well, because the distortions due to cornea and lens imperfections become large compared to diffraction effects [4]. The pupil size also determines the depth of field.

Because the cornea is not perfectly symmetric, the optical properties of the eye are orientation dependent. Therefore, it is impossible to perfectly focus stimuli of all orientations simultaneously, a condition known as astigmatism. This results in a point spread function that is not circularly symmetric. Astigmatism can be severe enough to interfere with perception, in which case it has to be corrected by compensatory glasses.

The properties of the eye's optics, most important the refractive indices of the optical elements, also vary with wavelength. This means that it is impossible to focus all wavelengths simultaneously, an effect known as chromatic aberration. The point spread function thus changes with wavelength. Chromatic aberration can be quantified by determining the modulation transfer function of the human eye for different wavelengths. This is shown in Figure 1.3 for a human eye model with a pupil diameter of 3 mm and in focus at 580 nm [5]. It is evident that the retinal image contains only poor spatial detail at wavelengths far from the in-focus wavelength (note the sharp cutoff going down to a few cycles per degree at short wavelengths). This tendency toward monochromacy becomes even more pronounced with increasing pupil aperture.

### 1.2.4   Eye Movements

The eye is attached to the head by three pairs of muscles that provide for rotation around its three axes. Several different types of eye movements can be distinguished [6]. Fixation movements are perhaps the most important. The voluntary fixation mechanism allows to direct the eyes toward an object of interest. This is achieved by means of saccades, high-speed movements steering the eyes to the new position. Saccades occur at a rate of two to three per second and are also used to keep scanning the entire scene by fixating on one highlight after the other. One is unaware of these movements because the visual image is suppressed during saccades. The involuntary fixation mechanism locks the eyes on the object of interest once it has been found. It involves so-called micro-saccades that counter the tremor and slow drift of the eye muscles. The same mechanism also compensates for head movements or vibrations.

Additionally, the eyes can track an object that is moving across the scene. These so-called pursuit movements can adapt to object trajectories with great accuracy. Smooth pursuit works well even for high velocities, but it is impeded by large accelerations and unpredictable motion.

Understanding what drives the eye movements, or in other words, why people look at certain areas in an image, has been an intriguing problem in vision research for a long time. It is important for perceptual imaging applications since visual acuity of the human eye is not uniform across the entire visual field. In general, visual acuity is highest only in a relatively small cone around the optical axis (the direction of gaze) and decreases with distance from the center. This is due to the deterioration of the optical quality of the eye toward the periphery (see above) as well as the layout of the retina (see Section 1.3).

Experiments presented in Reference [7] demonstrated that the saccadic patterns depend on the visual scene as well as the cognitive task to be performed. The direction of gaze is not completely idiosyncratic to individual viewers; however, a significant number of viewers will focus on the same regions of a scene [8], [9]. These experiments have given rise to various theories regarding the pattern of eye movements. Salient points attracting attention is a popular hypothesis [10], which is appealing in passive viewing conditions, such as when watching television. Salient locations of the image are based on local image characteristics, such as color, intensity, contrast, orientation, motion, etc. However, because this hypothesis is purely stimulus driven, it has limited applicability in real life, where semantic content rather than visual saliency drives eye movements during visual search [11]. There are also information-theoretic models that attempt to explain the pattern of eye movements [12].

## 1.3   Retina

The optics of the eye project images of the outside world onto the retina, the neural tissue at the back of the eye. The functional components of the retina are illustrated in Figure 1.4. Light entering the retina has to traverse several layers of neurons before it reaches the light-sensitive layer of photoreceptors and is finally absorbed in the pigment layer. The anatomy and physiology of the photoreceptors and the retinal neurons is discussed in more detail below.

### 1.3.1   Photoreceptors

The photoreceptors are specialized neurons that make use of light-sensitive photochemicals to convert the incident light energy into signals that can be interpreted by the brain. There are two different types of photoreceptors, namely, rods and cones. The names are derived from the physical appearance of their light-sensitive outer segments (Figure 1.4). Rods are responsible for scotopic vision at low light levels, while cones are responsible for photopic vision at high light levels.

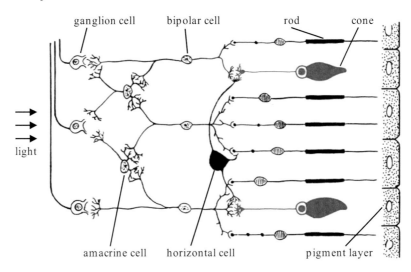

**FIGURE 1.4**

Anatomy of the retina. © 1991 W.B. Saunders

Rods are very sensitive light detectors. With the help of the photochemical rhodopsin, they can generate a photocurrent response from the absorption of only a single photon [13]. However, visual acuity under scotopic conditions is poor, even though rods sample the retina very finely. This is because signals from many rods converge onto a single neuron, which improves sensitivity but reduces resolution.

The opposite is true for the cones. Several neurons encode the signal from each cone, which already suggests that cones are important components of visual processing. There are three different types of cones, which can be classified according to the spectral sensitivity of their photochemicals. These three types are referred to as L-, M-, and S-cones, corresponding to their sensitivity to long, medium, and short wavelengths, respectively. Therefore, sometimes cones are also referred to as red, green, and blue cones, respectively. Estimates of the absorption spectra of the three cone types are shown in Figure 1.5 [14],

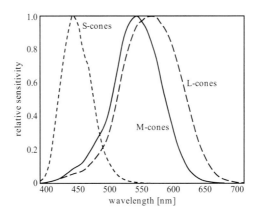

**FIGURE 1.5**

Normalized absorption spectra of L-, M-, and S-cones [15], [16].

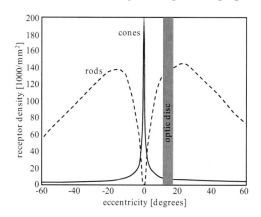

**FIGURE 1.6**

The distribution of photoreceptors on the retina [17]. Cones are concentrated in the fovea at the center of the retina, whereas rods dominate in the periphery. The gap around 15 degrees eccentricity represents the blind spot around the optic disc, where no receptors are present.

[15], [16]. The peak sensitivities occur around 570 nm, 570 nm, and 440 nm, respectively. As can be seen, the absorption spectra of the L- and M-cones are very similar, whereas the S-cones exhibit a significantly different sensitivity curve. The cones form the basis of color perception. The overlap of the spectra is essential to fine color discrimination (color perception is discussed in more detail in Section 1.7).

There are approximately 5 million cones and 100 million rods in each eye. Their density varies greatly across the retina, as is evident from Figure 1.6 [17]. There is also a large variability between individuals. Cones are concentrated in the fovea, a small area near the center of the retina, where they can reach a peak density of up to $300,000/mm^2$ [18]. Throughout the retina, L- and M-cones are in the majority; S-cones are much more sparse and account for less than 10% of the total number of cones [19]. Rods dominate outside of the fovea, which explains why it is easier to see very dim objects (for example, stars) when they are in the peripheral field of vision than when looking straight at them. The central fovea contains no rods at all. The highest rod densities (up to $200,000/mm^2$) are found along an elliptical ring near the eccentricity of the optic disc. The blind spot around the optic disc, where the optic nerve exits the eye, is completely void of photoreceptors.

The spatial sampling of the retina by the photoreceptors is illustrated in Figure 1.7. In the fovea, the cones are tightly packed and form a hexagonal sampling array. In the periphery, the sampling grid becomes more irregular; the separation between the cones grows, and rods fill in the spaces. Also note the size differences: the cones in the fovea have a diameter of 1 to 3 μm; in the periphery, their diameter increases to 5 to 10 μm. The diameter of the rods varies between 1 and 5 μm.

The size and spacing of the photoreceptors determine the maximum spatial resolution of the human visual system. Assuming an optical power of 60 diopters and thus a focal length of approximately 17 mm for the eye, distances on the retina can be expressed in terms of visual angle using simple trigonometry. The entire fovea covers approximately 2 degrees of visual angle. The L- and M-cones in the fovea are spaced approximately 2.5 μm apart, which corresponds to 30 arc seconds of visual angle. The maximum resolution of around

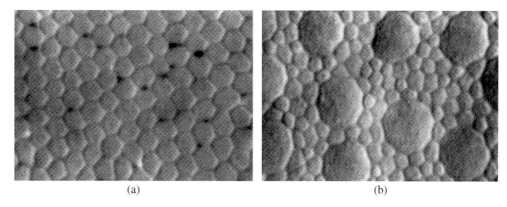

(a)                                                                                          (b)

**FIGURE 1.7**

The photoreceptor mosaic on the retina [17]. (a) In the fovea, the cones are densely packed on a hexagonal sampling array. (b) In the periphery, their size and separation grows, and rods fill in the spaces. Each image shows an area of $35 \times 25$ μm$^2$. ©1990 John Wiley & Sons

60 cpd attained here is high enough to capture all of the spatial variation after the blurring by the eye's optics. S-cones are spaced approximately 50 μm or 10 minutes of arc apart on average, resulting in a maximum resolution of only 3 cpd [19]. This is consistent with the strong defocus of short-wavelength light due to the axial chromatic aberration of the eye's optics (see Figure 1.3). Thus, the properties of different components of the visual system fit together nicely, as can be expected from an evolutionary system. The optics of the eye set limits on the maximum visual acuity, and the arrangements of the mosaic of the S-cones as well as the L- and M-cones can be understood as a consequence of the optical limitations (and vice versa).

### 1.3.2   Retinal Neurons

The retinal neurons process the photoreceptor signals. The anatomical connections and neural specializations within the retina combine to communicate different types of information about the visual input to the brain. As shown in Figure 1.4, a variety of different neurons can be distinguished in the retina:

- Horizontal cells connect the synaptic nodes of neighboring rods and cones. They have an inhibitory effect on bipolar cells.

- Bipolar cells connect horizontal cells, rods, and cones with ganglion cells. Bipolar cells can have either excitatory or inhibitory outputs.

- Amacrine cells transmit signals from bipolar cells to ganglion cells or laterally between different neurons. About 30 types of amacrine cells with different functions have been identified.

- Ganglion cells collect information from bipolar and amacrine cells. There are about 1.6 million ganglion cells in the retina. Their axons form the optic nerve that leaves the eye through the optic disc and carries the output signal of the retina to other processing centers in the brain (see Section 1.4).

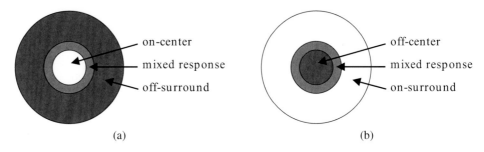

(a)                                                                          (b)

**FIGURE 1.8**

Center-surround organization of the receptive field of retinal ganglion cells: (a) on-center, off-surround, and (b) off-center, on-surround.

The interconnections between these cells give rise to an important concept in visual perception, the receptive field. The visual receptive field of a neuron is defined as the retinal area in which light influences the neuron's response. It is not limited to cells in the retina; many neurons in later stages of the visual pathways can also be described by means of their receptive fields (see Sections 1.4.1 and 1.4.2).

The ganglion cells in the retina have a characteristic center-surround receptive field, which is nearly circularly symmetric as shown in Figure 1.8. Light falling directly on the center of a ganglion cell's receptive field may either excite or inhibit the cell. In the surrounding region, light has the opposite effect. Between center and surround, there is a small area with a mixed response. About half of the retinal ganglion cells have an on-center, off-surround receptive field; that is, they are excited by light on their center. The other half have an off-center, on-surround receptive field with the opposite reaction. This receptive field organization is mainly due to lateral inhibition from horizontal cells. The consequence is that excitatory and inhibitory signals basically neutralize each other when the stimulus is uniform. However, for example, when edges or corners come to lie over such a cell's receptive field, its response is amplified. In other words, retinal neurons implement a mechanism of contrast computation (see also Section 1.5.4).

Ganglion cells can be further classified in two main groups:

- P-cells constitute the large majority (nearly 90%) of ganglion cells. They have very small receptive fields; that is, they receive inputs only from a small area of the retina (only a single cone in the fovea) and can thus encode fine image details. Furthermore, P-cells encode most of the chromatic information as different P-cells respond to different colors.

- M-cells constitute only 5 to 10% of ganglion cells. At any given eccentricity, their receptive fields are several times larger than those of P-cells. They also have thicker axons, which means that their output signals travel at higher speeds. M-cells respond to motion or small differences in light level but are insensitive to color. They are responsible for rapidly alerting the visual system to changes in the image.

These two types of ganglion cells represent the origin of two separate visual streams in the brain, the so-called magnocellular and parvocellular pathways (see Section 1.4.1).

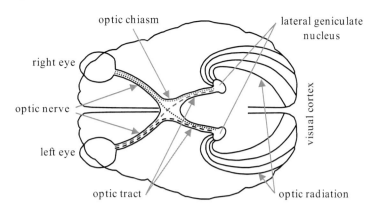

**FIGURE 1.9**

Visual pathways in the human brain. ©1991 W.B. Saunders

As becomes evident from this intricate arrangement of neurons, the retina is much more than a device to convert light to neural signals; the visual information is thoroughly pre-processed here before it is passed on to other parts of the brain.

## 1.4 Visual Pathways

The optic nerve leaves the eye to carry the visual information from the ganglion cells of the retina to various processing centers in the brain. These visual pathways are illustrated in Figure 1.9. The optic nerves from the two eyes meet at the optic chiasm, where the fibers are rearranged. All the fibers from the nasal halves of each retina cross to the opposite side, where they join the fibers from the temporal halves of the opposite retinas to form the optic tracts. Since the retinal images are reversed by the optics, the left visual field is processed in the right hemisphere, and the right visual field is processed in the left hemisphere.

Most of the fibers from each optic tract synapse in the lateral geniculate nucleus (see Section 1.4.1). From there fibers pass by way of the optic radiation to the visual cortex (see Section 1.4.2). Throughout these visual pathways, the neighborhood relations of the retina are preserved; that is, the input from a certain small part of the retina is processed in a particular area of the lateral geniculate nucleus and of the primary visual cortex. This property is known as retinotopic mapping.

There are a number of additional destinations for visual information in the brain apart from the major visual pathways listed above. These brain areas are responsible mainly for behavioral or reflex responses. One example is the superior colliculus, which seems to be involved in controlling eye movements in response to certain stimuli in the periphery.

### 1.4.1 Lateral Geniculate Nucleus

The lateral geniculate nucleus comprises around one million neurons in six layers. The two inner layers, the magnocellular layers, receive input almost exclusively from M-type ganglion cells. The four outer layers, the parvocellular layers, receive input mainly from P-type ganglion cells. As mentioned in Section 1.3.2, the M- and P-cells respond to different types of stimuli, namely, motion and spatial detail, respectively. This functional specialization continues in the lateral geniculate nucleus and the visual cortex, which suggests the existence of separate magnocellular and parvocellular pathways in the visual system.

The specialization of cells in the lateral geniculate nucleus is similar to the ganglion cells in the retina. The cells in the magnocellular layers are effectively color-blind and have larger receptive fields. They respond vigorously to moving contours. The cells in the parvocellular layers have rather small receptive fields and are differentially sensitive to color. They are excited if a particular color illuminates the center of their receptive field and inhibited if another color illuminates the surround. Only two color pairings are found, namely, red-green and blue-yellow. These opponent colors form the basis of color perception in the human visual system and will be discussed in more detail in Section 1.7.2.

The lateral geniculate nucleus not only serves as a relay station for signals from the retina to the visual cortex but also controls how much of the information is allowed to pass. This gating operation is controlled by extensive feedback signals from the primary visual cortex as well as input from the reticular activating system in the brain stem, which governs a general level of arousal.

### 1.4.2   Visual Cortex

The visual cortex is located at the back of the cerebral hemispheres (see Figure 1.9). It is responsible for all higher-level aspects of vision. The signals from the lateral geniculate nucleus arrive at an area called the primary visual cortex (also known as area V1, Brodmann area 17, or striate cortex), which makes up the largest part of the human visual system. In addition to the primary visual cortex, more than 20 other cortical areas receiving strong visual input have been discovered. Little is known about their exact functionalities, however.

There is an enormous variety of cells in the visual cortex. Neurons in the first stage of the primary visual cortex have center-surround receptive fields similar to cells in the retina and in the lateral geniculate nucleus (see above). A recurring property of many cells in the subsequent stages of the visual cortex is their selective sensitivity to certain types of information. A particular cell may respond strongly to patterns of a certain orientation or to motion in a certain direction. Similarly, there are cells tuned to particular frequencies, colors, velocities, etc. This neuronal selectivity is thought to be at the heart of the multichannel organization of human vision, which is discussed in Section 1.4.3.

The foundations of knowledge about cortical receptive fields were laid in References [20] and [21]. Based on physiological studies of cells in the primary visual cortex, several classes of neurons with different specializations were identified.

Simple cells behave in an approximately linear fashion; that is, their responses to complicated shapes can be predicted from their responses to small-spot stimuli. They have receptive fields composed of several parallel elongated excitatory and inhibitory regions, as illustrated in Figure 1.10. In fact, their receptive fields resemble Gabor patterns [22]. Hence, simple cells can be characterized by a particular spatial frequency, orientation, and phase. Serving as an oriented bandpass filter, a simple cell thus responds to a certain, limited range of spatial frequencies and orientations.

**FIGURE 1.10**

Idealized receptive field of a simple cell in the primary visual cortex.

Complex cells are the most common cells in the primary visual cortex. Like simple cells, they are also orientation-selective, but their receptive field does not exhibit the on- and off-regions of a simple cell; instead, they respond to a properly oriented stimulus anywhere in their receptive field.

A small percentage of complex cells respond well only when a stimulus (still with the proper orientation) moves across their receptive field in a certain direction. These direction-selective cells receive input mainly from the magnocellular pathway and probably play an important role in motion perception. Some cells respond only to oriented stimuli of a certain size. They are referred to as end-stopped cells. They are sensitive to corners, curvature, or sudden breaks in lines. Both simple and complex cells can also be end-stopped. Furthermore, the primary visual cortex is the first stage in the visual pathways where individual neurons have binocular receptive fields; that is, they receive inputs from both eyes, thereby forming the basis for stereopsis and depth perception.

### 1.4.3 Multichannel Organization

As mentioned above, many neurons in the visual system are tuned to certain types of visual information, such as color, frequency, and orientation. Data from experiments on pattern discrimination, masking, and adaptation (see Section 1.6) yield further evidence that these stimulus characteristics are processed in different channels in the human visual system. This empirical evidence motivated the multichannel theory of human vision, which provides an important framework for understanding and modeling pattern sensitivity.

A large number of neurons in the primary visual cortex have receptive fields that resemble Gabor patterns (Figure 1.10). Hence, they can be characterized by a particular spatial frequency and orientation, and essentially represent oriented bandpass filters. There is still a lot of discussion about the exact tuning shape and bandwidth, and different experiments have led to different results. For the achromatic visual pathways, most studies give estimates of approximately one to two octaves for the spatial frequency bandwidth and 20 to

60 degrees for the orientation bandwidth, varying with spatial frequency [23]. These results are confirmed by psychophysical evidence from studies of discrimination and interaction phenomena. Interestingly, these cell properties can also be related with and even derived from the statistics of natural images [24], [25].

Fewer empirical data are available for the chromatic pathways. They probably have similar spatial frequency bandwidths [26], [27], whereas their orientation bandwidths have been found to be significantly larger, ranging from 60 to 130 degrees [28].

Many different transforms and filters have been proposed as approximations to the multichannel representation of visual information in the human visual system. These include Gabor filters (Figure 1.10), the Cortex transform [29], a variety of wavelets, and the steerable pyramid [30]. While the specific filter shapes and designs are very different, they all decompose an image into a number of spatial frequency and orientation bands. With a sufficient number of appropriately tuned filters, all stimulus orientations and frequencies in the sensitivity range of the visual system can be covered.

## 1.5    Sensitivity to Light

### 1.5.1    Light Adaptation

The human visual system is capable of adapting to an enormous range of light intensities. Light adaptation allows to better discriminate relative luminance variations at every light level. Scotopic and photopic vision together cover twelve orders of luminance magnitude, from the detection of a few photons to vision in bright sunlight. However, at any given level of adaptation, humans only respond to an intensity range of two to three orders of magnitude. Three mechanisms for light adaptation can be distinguished in the human visual system:

- The mechanical variation of the pupillary aperture. As discussed in Section 1.2.2, it is controlled by the iris. The pupil diameter can be varied between 1.5 and 8 mm, which corresponds to a thirtyfold change of the quantity of light entering the eye. This adaptation mechanism responds in a matter of seconds.

- The chemical processes in the photoreceptors. This adaptation mechanism exists in both rods and cones. In bright light, the concentration of photochemicals in the receptors decreases, thereby reducing their sensitivity. On the other hand, when the light intensity is reduced, the production of photochemicals and thus the receptor sensitivity is increased. While this chemical adaptation mechanism is very powerful (it covers five to six orders of magnitude), it is rather slow; complete dark adaptation in particular can take up to an hour.

- Adaptation at the neural level. This mechanism involves neurons in all layers of the retina, which adapt to changing light intensities by increasing or decreasing their signal output accordingly. Neural adaptation is less powerful, but faster than the chemical adaptation in the photoreceptors.

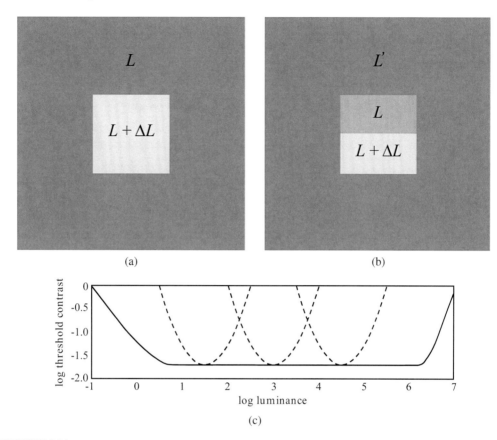

**FIGURE 1.11**

Illustration of the Weber-Fechner law. Using the basic stimulus shown in (a), threshold contrast is constant over a wide range – shown as a solid line in (c). When the adapting luminance $L'$ is different from $L$ as shown in (b), this contrast constancy is lost – shown as dotted lines for various $L'$ in (c).

### 1.5.2 Contrast Sensitivity

The response of the human visual system depends much less on the absolute luminance than on the relation of its local variations to the surrounding luminance. This property is known as the Weber-Fechner law. Contrast is a measure of this relative variation of luminance. Mathematically, Weber contrast can be expressed as

$$C^W = \frac{\Delta L}{L}. \tag{1.2}$$

This definition is most appropriate for patterns consisting of a single increment or decrement $\Delta L$ to an otherwise uniform background luminance.

The threshold contrast, which is the minimum contrast necessary for an observer to detect a change in intensity, is shown as a function of background luminance in Figure 1.11. As can be seen, it remains nearly constant over an important range of intensities (from faint lighting to daylight) due to the adaptation capabilities of the human visual system, that is, the Weber-Fechner law holds in this range. This is indeed the luminance range typically encountered in most image processing applications. Outside of this range, inten-

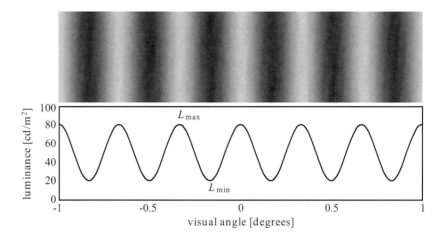

**FIGURE 1.12**

Sinusoidal grating with Michelson contrast of $C_M = 0.6$ and its luminance profile.

sity discrimination ability of the human eye deteriorates. Under optimal conditions, the threshold contrast can be less than 1%. The exact figure depends greatly on the stimulus characteristics, most important, its color and spatial frequency (see Section 1.5.3).

The Weber-Fechner law is only an approximation of the actual sensory perception. Most important, it presumes that the visual system is adapted to the background luminance $L$. This assumption is generally violated when looking at a natural image in print or on screen. If the adapting luminance $L'$ is different, as depicted in Figure 1.11b, the required threshold contrast can become much larger, depending on how far $L$ and $L'$ are apart. Even this latter scenario is a simplification of realistic image viewing situations, because the adaptation state is determined not only by the environment, but also by the image content itself. Besides, most images are composed of many more than just two colors, so the response of the visual system becomes much more complex.

### 1.5.3 Contrast Sensitivity Functions

The dependencies of threshold contrast on stimulus characteristics can be quantified using contrast sensitivity functions (CSFs); contrast sensitivity is simply the inverse of the contrast threshold. In these CSF measurements, the contrast of periodic (often sinusoidal) stimuli with varying frequencies is defined as the Michelson contrast:

$$C_M = \frac{L_{\max} - L_{\min}}{L_{\max} + L_{\min}}, \tag{1.3}$$

where $L_{\min}$ and $L_{\max}$ are the luminance extrema of the pattern (see Figure 1.12).

CSF approximations to measurements from Reference [31] are shown in Figure 1.13. Achromatic contrast sensitivity is generally higher than chromatic, especially at high spatial frequencies. This is the justification for chroma subsampling in image compression applications; for example, humans are relatively insensitive to a reduction of color detail. Achromatic sensitivity has a distinct maximum around 2 to 8 cpd (again depending

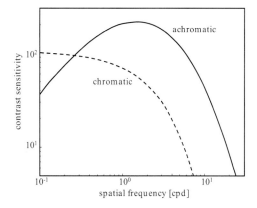

**FIGURE 1.13**

Approximations of achromatic and chromatic contrast sensitivity functions to data from Reference [31].

on stimulus characteristics) and decreases at low spatial frequencies,[1] whereas chromatic sensitivity does not. The chromatic CSFs for red-green and blue-yellow stimuli are very similar in shape; the blue-yellow sensitivity is slightly lower overall, and its high-frequency decline sets in a bit earlier.

Aside from the pattern and color of the stimulus, the exact shape of the contrast sensitivity function depends on many other factors. Among them are the retinal illuminance [33], which has a substantial effect on the location of the maximum, and the orientation of the stimulus [34] (the sensitivity is highest for horizontal and vertical patterns, whereas it is reduced for oblique stimuli).

Various CSF models have been proposed in the literature. A simple yet effective engineering model that can fit both achromatic and chromatic sensitivity measurements in different situations was suggested in Reference [35]:

$$CSF(f) = (a_0 + a_1 f)e^{a_2 f^{a_3}}, \tag{1.4}$$

where $a_i$ are design parameters. More elaborate models have explicit luminance and orientation parameters [36], [37].

### 1.5.4 Image Contrast

The two definitions of contrast in Equations 1.2 and 1.3 are not equivalent and do not even share a common range of values: Michelson contrast can range from 0 to 1, whereas Weber contrast can range from $-1$ to $\infty$. While they are good predictors of perceived contrast for simple stimuli, they fail when stimuli become more complex and cover a wider frequency range, for example, Gabor patches [38]. It is also evident that none of these simple global definitions are appropriate for measuring contrast in natural images, since a few very bright or very dark spots would determine the contrast of the whole image. Actual human contrast perception on the other hand varies with the local average luminance.

---

[1]This apparent attenuation of sensitivity toward low frequencies may be attributed to implicit masking, that is, masking by the spectrum of the window within which the test gratings are presented [32].

In order to address these issues, Reference [39] proposed a local band-limited contrast that measures incremental or decremental changes with respect to the local background:

$$C_j^P[x,y] = \frac{\psi_j * L[x,y]}{\phi_j * L[x,y]}, \tag{1.5}$$

where $L[x,y]$ is the luminance image, $\psi_j$ is a bandpass filter at level $j$ of a filter bank, and $\phi_j$ is the corresponding lowpass filter. This definition is analogous to the symmetric (in-phase) responses of vision mechanisms.

However, a complete description of contrast for complex stimuli has to include the anti-symmetric (quadrature) responses as well [40]. Analytic filters represent an elegant way to achieve this. The magnitude of the analytic filter response, which is the sum of the energy responses of in-phase and quadrature components, exhibits the desired behavior in that it gives a constant response to sinusoidal gratings.

While the implementation of analytic filters in the one-dimensional case is straightforward, the design of general two-dimensional analytic filters is less obvious because of the difficulties involved when extending the Hilbert transform to two dimensions. Oriented measures of contrast can still be computed, because the Hilbert transform is well defined for filters whose angular support is smaller than $\pi$. Such contrast measures are useful for many image processing tasks. They can implement a multichannel representation of low-level vision in accordance with the orientation selectivity of the human visual system (Section 1.4.3) and facilitate modeling aspects, such as contrast sensitivity and pattern masking. They have been used in many vision models and their applications, for example, in perceptual quality assessment of images and video [41]. Using analytic orientation-selective filters $\eta_k[x,y]$, this oriented contrast can be expressed as:

$$C_{jk}^O[x,y] = \frac{|\psi_j * \eta_k * L[x,y]|}{\phi_j * L[x,y]}. \tag{1.6}$$

The design of an isotropic contrast measure is more difficult. As pointed out before, the contrast definition from Equation 1.5 is not suitable because it lacks the quadrature component, and isotropic two-dimensional analytic filters as such do not exist. In order to circumvent this problem, a class of nonseparable filters can be used that generalize the properties of analytic functions in two dimensions. These filters are directional wavelets whose Fourier transform is strictly supported in a convex cone with the apex at the origin. For these filters to have a flat response to sinusoidal stimuli, the angular width of the cone must be strictly less than $\pi$. This means that at least three such filters are required to cover all possible orientations uniformly. Using a technique described in Reference [42], such filters can be designed in a very simple and straightforward way; it is even possible to obtain dyadic-oriented decompositions that can be implemented using a filter bank algorithm.

Essentially, this technique assumes $K$ directional wavelets with Fourier transform $\hat{\Psi}(r,\varphi)$ that satisfy the above requirements and

$$\sum_{k=0}^{K-1} |\hat{\Psi}(r,\varphi - 2\pi k/K)|^2 = |\hat{\psi}(r)|^2, \tag{1.7}$$

where $\hat{\psi}(r)$ is the Fourier transform of an isotropic dyadic wavelet. Using these oriented filters, one can define an isotropic contrast measure $C_j^I$ as the square root of the energy sum

(a)                                                    (b)

**FIGURE 1.14**

Isotropic local contrast: (a) an example image and (b) its isotropic local contrast $C_j^I[x,y]$ as given by Equation 1.8.

of the filter responses, normalized as before by a lowpass band:

$$C_j^I[x,y] = \frac{\sqrt{2\sum_k |\Psi_{jk} * L[x,y]|^2}}{\phi_j * L[x,y]}, \tag{1.8}$$

where $L[x,y]$ is the input image, and $\Psi_{jk}$ denotes the wavelet dilated by $2^{-j}$ and rotated by $2\pi k/K$. The term $C_j^I$ denotes an orientation- and phase-independent quantity.

Being defined by means of analytic filters it behaves as prescribed with respect to sinusoidal gratings (that is, $C_j^I[x,y] \equiv C^M$ in this case). This combination of analytic-oriented filters thus produces a meaningful phase-independent isotropic measure of contrast. The example shown in Figure 1.14 demonstrates that it is a very natural measure of local contrast in an image. Isotropy is particularly useful for applications where nondirectional signals in an image are considered [43].

### 1.5.5 Lightness Perception

Weber's law already indicates that human perception of brightness, or lightness, is not linear. More precisely, it suggests a logarithmic relationship between luminance and lightness. However, Weber's law is only based on threshold measurements, and it generally overestimates the sensitivity for higher luminance values. More extensive experiments with tone scales were carried out by Munsell to determine stimuli that are perceptually equidistant in luminance (and also in color). These experiments revealed that a power-law relationship with an exponent of $1/3$ is closer to actual perception of lightness. This was standardized by CIE as $L^*$ (see Section 1.7.3). A modified log characteristic that can be tuned for various situations with the help of a parameter was proposed in Reference [44]. The different relationships are compared in Figure 1.15.

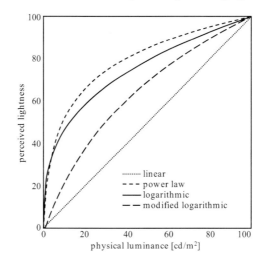

**FIGURE 1.15**

Perceived lightness as a function of luminance.

## 1.6 Masking and Adaptation

Masking and adaptation are very important phenomena in vision in general and in digital imaging in particular as they describe interactions between stimuli. Masking occurs when a stimulus that is visible by itself cannot be detected due to the presence of another. As demonstrated in Figure 1.16, the same distortion can be disturbing in certain regions

(a)                                              (b)

**FIGURE 1.16**

Demonstration of the masking effect. The same noise has been added to rectangular areas (a) at the top of the image, where it is clearly visible in the sky, and (b) at the bottom of the image, where it is much harder to see on the rocks and in the water due to masking by the heavily textured background.

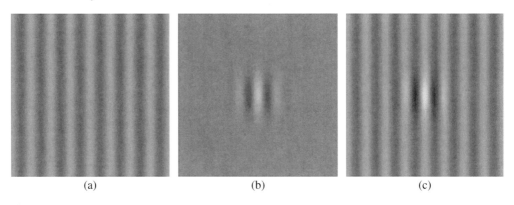

| (a) | (b) | (c) |

**FIGURE 1.17**

Typical contrast masking experiment: (a) a cosine masker, (b) a Gabor target, and (c) a superimposed result. The visibility of the target is highly dependent on the contrast of the masker.

of an image while it is hardly noticeable elsewhere. Thus, within the framework of digital imaging it is helpful to think of features of the processed image being masked by the original image. Results from masking and adaptation experiments were also the basis for developing a multichannel theory of vision (see Section 1.4.3).

### 1.6.1 Contrast Masking

Spatial masking effects are usually quantified by measuring the detection threshold for a target stimulus when it is superimposed on a masker with varying contrast. The stimuli (both maskers and targets) used in contrast masking experiments are typically sinusoidal gratings or Gabor patches, as shown in Figure 1.17.

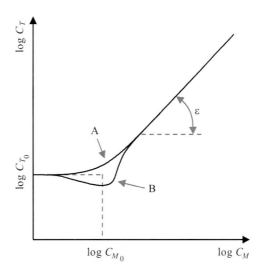

**FIGURE 1.18**

Illustration of typical masking curves. The detection threshold for the target is elevated by increasing the contrast of the masker. For stimuli with different characteristics, masking is the dominant effect (case A). Facilitation occurs for stimuli with similar characteristics (case B).

Figure 1.18 shows an example of curves approximating the data typically resulting from such experiments. The horizontal axis shows the log of the masker contrast $C_M$, and the vertical axis the log of the target contrast $C_T$ at detection threshold. The detection threshold for the target stimulus without any masker is indicated by $C_{T_0}$. For contrast values of the masker larger than $C_{M_0}$, the detection threshold grows exponentially with increasing masker contrast. As can be seen in this figure, case A shows a gradual transition from the threshold range to the masking range. This typically occurs when masker and target have different characteristics. In case B, the detection threshold for the target actually decreases when the masker contrast is close to $C_{M_0}$, which implies that the target is easier to perceive due to the presence of a (weak) masker in this contrast range. This effect is known as facilitation and occurs mainly when target and masker have very similar properties.

The exact masking behavior depends greatly on the types of target and masker stimuli and their relationship to each other (for example, relative frequency and orientation). Masking is strongest when the interacting stimuli have similar characteristics, such as frequencies, orientations, colors, etc. Masking also occurs between stimuli of different orientation [45], between stimuli of different spatial frequency [46], and between chromatic and achromatic stimuli [26], [47], [48], although it is generally weaker.

### 1.6.2 Pattern Masking

The stimuli (both maskers and targets) used in the experiments cited above are typically sinusoidal gratings or Gabor patches. Compared to natural scene content, they are artificial and quite simplistic. Therefore, masking models based on such experiments do not apply to more complex cases of masking, such as a compression artifact or a watermark in a natural image. Because the observer is not so familiar with the patterns, uncertainty effects become more important, and masking can be much larger.

Not many studies investigate masking in natural images. Reference [49] studied signal detection on complex backgrounds in the presence of noise for medical images and measured defect detectability as a function of signal contrast.

Reference [50] coined the term "entropy masking" to emphasize that the difference between contrast masking with regular structures, such as sinusoidal or Gabor masks on one hand and noise masking (also referred to as texture masking) with unstructured noise-like stimuli on the other, lies mainly in the familiarity of the observer with the stimuli. Natural images are somewhere in between these two extremes.

Reference [51] compared a number of masking models for the visibility of wavelet compression artifacts in natural images. This was one of the first comparative studies of masking in natural images, even if the experiments were quite specific to JPEG2000 encoding. The best results were achieved by a model that considered exclusively the local image activity instead of simple pointwise contrast masking.

Similarly, Reference [52] measured the RMS-based contrast of wavelet distortions at the threshold of visibility in the presence of natural image maskers, which were selected from a texture library. The results were used to parameterize a wavelet-based distortion metric, which was then applied to an associated watermarking algorithm.

Most recently, Reference [53] investigated the visibility of various noise types in natural images. The targets were Gaussian white noise and bandpass filtered noise of varying en-

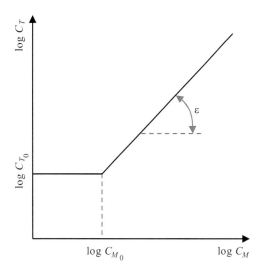

**FIGURE 1.19**

Simple model of masking given by Equation 1.9.

ergy. Psychophysical experiments were conducted to determine the detection threshold of these noise targets on 30 different images with varying content. The achieved results were consistent with data from other masking experiments in a number of ways; namely, detection ability was higher on a plain background, and a clear masking effect was observed. In particular, noise thresholds increase significantly with image activity.

### 1.6.3 Masking Models

Numerous masking models of different complexity have been proposed [51], [54]. Because the facilitation effect is usually limited to a rather small and low range of masker contrasts, and because in digital imaging applications, the masker (the image) and the target (the distortions) often have quite different characteristics, facilitation can usually be neglected. The most important effect is clearly masking, that is, the significant increase of the target's visibility threshold with increasing masker contrast $C_M$. Hence, a simple masking model, shown in Figure 1.19, can be formulated as follows:

$$C_T(C_M) = \begin{cases} C_{T_0} & \text{if } C_M < C_{M_0}, \\ C_{T_0}\,(C_M/C_{M_0})^\varepsilon & \text{otherwise.} \end{cases} \tag{1.9}$$

Other popular masking models in digital imaging are based on contrast gain control, which can explain a wide range of empirical contrast masking data. These models were inspired by analyses of the responses of single neurons in the visual cortex of the cat [55], [56], where contrast gain control serves as a mechanism to keep neural responses within the permissible dynamic range while at the same time retaining global pattern information.

Contrast gain control can be modeled by an excitatory nonlinearity that is inhibited divisively by a pool of responses from other neurons. Masking occurs through the inhibitory effect of the normalizing pool [45], [57]. A further generalization of these models [58] facilitates the integration of channel interactions. Namely, let $a = a[c, f, \varphi, x, y]$ be a coefficient of the perceptual decomposition in color channel $c$, frequency band $f$, orientation band $\varphi$, at location $[x,y]$. Then the corresponding sensor output $s = s[c,f,\varphi,x,y]$ is computed as

$$s = \alpha \frac{a^p}{\beta^2 + h * a^q}. \tag{1.10}$$

The excitatory path in the numerator consists of a power-law nonlinearity with exponent $p$. Its gain is controlled by the inhibitory path in the denominator, which comprises a nonlinearity with exponent $q$ and a saturation constant $\beta$ to prevent division by zero. The factor $\alpha$ is used to adjust the overall gain of the mechanism.

In the inhibitory path, filter responses are pooled over different channels by means of a convolution with the pooling function $h$. In its most general form, $h = h[c,f,\varphi,x,y]$, the pooling operation in the inhibitory path may combine coefficients from the dimensions of color, spatial frequency, orientation, location, and phase.

### 1.6.4 Pattern Adaptation

Pattern adaptation adjusts the sensitivity of the visual system in response to the prevalent stimulation patterns. For example, adaptation to patterns of a certain frequency can lead to a noticeable decrease of contrast sensitivity around this frequency [59], [60], [61]. Similar in effect to masking, adaptation is a significantly slower process.

An interesting study in this respect was reported in Reference [62]. Natural images of outdoor scenes (both distant views and close-ups) were used as adapting stimuli. It was found that exposure to such stimuli induces pronounced changes in contrast sensitivity. The effects can be characterized by selective losses in sensitivity at lower to medium spatial frequencies. This is consistent with the characteristic amplitude spectra of natural images, which decrease with frequency approximately as $1/f$.

Likewise, Reference [63] examined how color sensitivity and appearance might be influenced by adaptation to the color distributions of images. It was found that natural scenes exhibit a limited range of chromatic distributions, so that the range of adaptation states is normally limited as well. However, the variability is large enough so that different adaptation effects may occur for individual scenes or for different viewing conditions.

## 1.7 Color Perception

In its most general form, light can be described by its spectral power distribution. The human visual system (and most imaging devices for that matter) use a much more compact representation of color, however. This representation and its implications for digital imaging are discussed here.

### 1.7.1 Color Matching

Color perception can be studied by the color-matching experiment, which is the foundation of color science. In this experiment, the observer views a bipartite field, half of which

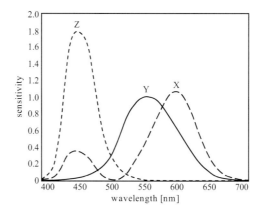

**FIGURE 1.20**

CIE *XYZ* color-matching functions.

is illuminated by a test light, the other half by an additive mixture of a certain number of primary lights. The observer is asked to adjust the intensities of the primary lights to match the appearance of the test light.

It is not *a priori* clear that it will be possible for the observer to make a match when the number of primaries is small. In general, however, observers are able to establish a match using only three primary lights. This is referred to as the trichromacy of human color vision.[2] Trichromacy implies that there exist lights with different spectral power distributions that cannot be distinguished by a human observer. Such physically different lights that produce identical color appearance are called metamers.

As was first established in Reference [64], photopic color matching satisfies homogeneity and superposition and can thus be analyzed using linear systems theory. Assume the test light is known by $N$ samples of its spectral distribution $x(\lambda)$, expressed as vector $\mathbf{x}$. The color-matching experiment can then be described in matrix form:

$$\mathbf{t} = \mathbf{Cx}, \tag{1.11}$$

where $\mathbf{t}$ is a three-dimensional vector whose coefficients are the intensities of the three primary lights found by the observer to visually match $\mathbf{x}$. They are also referred to as the tristimulus coordinates of the test light. The rows of matrix $\mathbf{C}$ are made up of $N$ samples of the so-called color-matching functions $C_k(\lambda)$ of the three primaries.

The Commission Internationale de l'Eclairage (CIE) carried out such color-matching experiments, on the basis of which a two-degree (foveal vision) "standard observer" was defined in 1931. The result was a set of three primaries $X$, $Y$, and $Z$ that can be used to match all observable colors by mixing these primaries with positive weights. $Y$ represents the luminance, $X$ and $Z$ are idealized chromaticity primaries. The CIE *XYZ* color-matching functions are shown in Figure 1.20.

---

[2]There are certain qualifications to this empirical generalization that three primaries are sufficient to match any test light. The primary lights must be chosen so that they are visually independent, that is, no additive mixture of any two of the primary lights should be a match to the third. Also, "negative" intensities of a primary must be allowed, which is just a mathematical convention of saying that a primary can be added to the test light instead of to the other primaries.

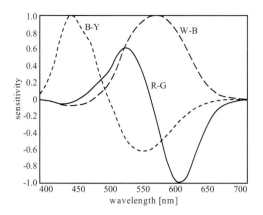

**FIGURE 1.21**

Normalized spectral sensitivities of white-black (W-B), red-green (R-G), and blue-yellow (B-Y) components of the opponent color space.

The mechanistic explanation of the color-matching experiment is that two lights match if they produce the same absorption rates in the L-, M-, and S-cones. To relate these cone absorption rates to the tristimulus coordinates of the test light, a color-matching experiment is performed with primaries $\mathbf{P}$, whose columns contain $N$ samples of the spectral power distributions of the three primaries $P_k(\lambda)$. It turns out that the cone absorption rates $\mathbf{r}$ are related to the tristimulus coordinates $\mathbf{t}$ of the test light by a linear transformation:

$$\mathbf{r} = \mathbf{Mt}, \tag{1.12}$$

where $\mathbf{M} = \mathbf{RP}$ is a $3 \times 3$ matrix. This also implies that the color-matching functions are determined by the cone sensitivities up to a linear transformation, which was first verified empirically in Reference [13]. The spectral sensitivities of the three cone types thus provide a satisfactory explanation of the color-matching experiment.

### 1.7.2   Opponent Colors

Reference [65] was the first to point out that some pairs of hues can coexist in a single color sensation (for example, a reddish yellow is perceived as orange), while others cannot (a reddish green is never perceived, for instance). This led to the conclusion that the sensations of red and green as well as blue and yellow are encoded as color difference signals in separate visual pathways, which is commonly referred to as the theory of opponent colors.

Empirical evidence in support of this theory came from a behavioral experiment designed to quantify opponent colors, the so-called hue cancellation experiment [66]. In this experiment, observers are able to cancel, for example, the reddish appearance of a test light by adding certain amounts of green light. Thus, the red-green or blue-yellow appearance of monochromatic lights can be measured.

Physiological experiments revealed the existence of opponent signals in the visual pathways [67]. It was demonstrated that cones may have an excitatory or an inhibitory effect on ganglion cells in the retina and on cells in the lateral geniculate nucleus. Depending on the cone types, certain excitation/inhibition pairings occur much more often than others:

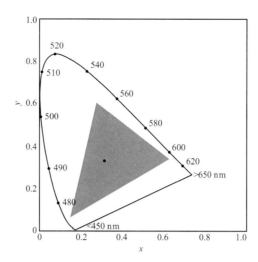

**FIGURE 1.22**

CIE *xy* chromaticity diagram. Monochromatic colors, indicated by their wavelengths, along the curve form the border. The dot in the center denotes the location of standard white. The gray triangle represents the gamut of a typical CRT monitor; the color gamut of printers is even smaller.

neurons excited by "red" L-cones are usually inhibited by "green" M-cones, and neurons excited by "blue" S-cones are often inhibited by a combination of L- and M-cones. Hence, the receptive fields of these neurons suggest a connection between neural signals and perceptual opponent colors.

The reason for an opponent-signal representation of color information in the human visual system may be that the resulting decorrelation of cone signals improves the coding efficiency of the visual pathways. In fact, this representation can also be traced back to the properties of natural image spectra [68]. The spectral sensitivities of an opponent color space derived in Reference [69] are shown in Figure 1.21 as an example. The principal components are white-black (W-B), red-green (R-G), and blue-yellow (B-Y) differences. The W-B channel, which encodes luminance information, is determined mainly by medium to long wavelengths. The R-G channel discriminates between medium and long wavelengths, while the B-Y channel discriminates between short and medium wavelengths.

### 1.7.3   Color Spaces and Conversions

CIE *XYZ* color space is very useful as the basis for color conversions because it is device independent. To visualize this color space, chromaticity coordinates are used. The chromaticity coordinates of a color are defined by the normalized weights of the three primaries:

$$x = \frac{X}{X+Y+Z}, \quad y = \frac{Y}{X+Y+Z}, \quad z = \frac{Z}{X+Y+Z} \tag{1.13}$$

Since by definition $x + y + z = 1$, the third chromaticity coordinate can now be derived from the first two. All possible sets of tristimulus values can thus be represented in a two-dimensional plot, the so-called chromaticity diagram (Figure 1.22). The monochromatic colors form its border, the spectral locus.

The chromaticity diagram is also used to define color gamuts, which are simply polygons on this diagram. All colors that are additive mixtures of the vertices of a gamut are located inside the gamut. As is evident from the shape of the chromaticity diagram, it is impossible to choose three real primaries such that all possible colors can be matched with additive mixtures of those primaries. The CIE *XYZ* primaries are actually imaginary. In any real additive color reproductive system such as television, only a limited gamut of colors can be displayed (see Figure 1.22).

An image is typically represented in nonlinear $R'G'B'$ space to compensate for the behavior of CRT displays ("gamma correction"). For conversion to linear *RGB* space, each of the resulting three components has to undergo a power-law nonlinearity of the form $x^\gamma$ with $\gamma \approx 2.5$.

*RGB* space is device dependent, because it is determined by the particular spectral power distribution of the light emitted from the display phosphors. Once the phosphor spectra of the monitor of interest have been determined, the device-independent CIE *XYZ* tristimulus values can be calculated. The primaries of contemporary monitors are closely approximated by the following transformation [70]:

$$\begin{bmatrix} X \\ Y \\ Z \end{bmatrix} = \begin{bmatrix} 0.412 & 0.358 & 0.180 \\ 0.213 & 0.715 & 0.072 \\ 0.019 & 0.119 & 0.950 \end{bmatrix} \begin{bmatrix} R \\ G \\ B \end{bmatrix}. \tag{1.14}$$

The responses of the L-, M-, and S-cones on the human retina (Section 1.3.1) can then be computed according to Equation 1.12 [71]:

$$\begin{bmatrix} L \\ M \\ S \end{bmatrix} = \begin{bmatrix} 0.240 & 0.854 & -0.044 \\ -0.389 & 1.160 & 0.085 \\ -0.001 & 0.002 & 0.573 \end{bmatrix} \begin{bmatrix} X \\ Y \\ Z \end{bmatrix}. \tag{1.15}$$

The *LMS* values can be converted to an opponent color space (Section 1.7.2). A variety of opponent color spaces have been proposed, which use different ways to combine the cone responses. The opponent color model from Reference [69] has been designed for maximum pattern-color separability, which has the advantage that color perception and pattern sensitivity can be decoupled. These components are computed from *LMS* values via the following transformation [69]:

$$\begin{bmatrix} W-B \\ R-G \\ B-Y \end{bmatrix} = \begin{bmatrix} 0.990 & -0.106 & -0.094 \\ -0.669 & 0.742 & -0.027 \\ -0.212 & -0.354 & 0.911 \end{bmatrix} \begin{bmatrix} L \\ M \\ S \end{bmatrix}. \tag{1.16}$$

One shortcoming of all the above color spaces is their perceptual nonuniformity in terms of color differences. To alleviate this problem, more uniform color spaces were proposed, for example, CIE $L^*a^*b^*$. It includes a component of perceived lightness (Section 1.5.5) and a color difference equation that is approximately uniform for large color patches. CIE lightness $L^*$ is computed according to a power law:[3]

$$L^* = 116 \, (Y/Y_0)^{1/3} - 16. \tag{1.17}$$

---

[3]Strictly speaking, the power law is only valid when the argument is greater than 0.008856. Below this value, a linear approximation is used due to the singularity of the power function at zero.

The two chromaticity coordinates $a^*$ and $b^*$ in CIE $L^*a^*b^*$ space are computed as follows:

$$a^* = 500 \left[ (X/X_0)^{1/3} - (Y/Y_0)^{1/3} \right],$$
$$b^* = 200 \left[ (Y/Y_0)^{1/3} - (Z/Z_0)^{1/3} \right], \qquad (1.18)$$

where the 0-subscript refers to the *XYZ* values of the reference white being used. The CIE $L^*a^*b^*$ color difference is given by

$$\Delta E_{ab}^* = \sqrt{(\Delta L^*)^2 + (\Delta a^*)^2 + (\Delta b^*)^2}. \qquad (1.19)$$

## 1.8  Depth Perception

There are a large number of depth cues that the human visual system uses when viewing a three-dimensional scene [72]. These can be classified into oculomotor cues coming from the eye muscles, and visual cues from the scene content itself. They can also be classified into monocular and binocular cues.

Oculomotor cues include accommodation and vergence. As discussed in Section 1.2.2, accommodation refers to the variation of lens shape and thickness (and thus focal length), which allows the eye to focus on an object at a certain distance. Vergence refers to the muscular rotation of the eyeballs, which serves to converge both eyes on the same object.

The visual depth cues also consist of monocular and binocular cues. There are many monocular depth cues, such as:

- relative size of objects (close objects usually appear bigger on the retina than the same objects further away),

- size of familiar objects,

- occlusion (close objects cover parts of objects farther away),

- perspective,

- texture gradients (the density and perspective of patterns change with distance),

- atmospheric blur (objects in the distance appear less sharp because of scatter and turbulence in the atmosphere),

- lighting, shading, and shadows, and

- motion parallax (when moving, nearby objects cover a larger distance in the field of view than faraway objects).

The most important binocular depth cue is the retinal disparity between points of the same object viewed from slightly different angles by the left and right eye. The resulting apparent displacement of an object viewed along two different lines of sight is called parallax.

Three-dimensional capture and display technology for the digital imaging of scenes with depth rely on (re)producing two or more separate views for the two eyes. An overview of the different technologies can be found in Reference [73]. The separation of the views for processing and display may result in mismatches of various depth cues, which are the main reason for the discomfort many people experience in stereoscopic viewing [74], [75].

## 1.9 Conclusion

Human vision has intrinsic limitations with respect to the visibility of stimuli, which should be taken into account in digital imaging applications. However, comprehensive modeling of vision processes is made difficult by the sheer complexity of the human visual system. While vision research has produced a large number of models, it has focused mainly on the perception of simple stimuli, such as sinusoids and Gabor patches at threshold; as a result, existing models and data cannot directly be extended to the perception of natural images. This is still very much an active topic of research.

The discussions on human vision in this chapter are necessarily limited in scope; for a more detailed overview, the reader is referred to the abundant literature, such as References [76], [77], and [78].

An exhaustive introduction to optics and vision can be found in Reference [79]. For more details about color science and color perception, the reader is referred to References [80], [81], and [82]. The more specific aspects of image and video quality are addressed in References [41] and [83].

## Acknowledgment

This chapter is adapted from Reference [41], with the permission of John Wiley & Sons. Figures 1.4 and 1.9 are adapted from Reference [84]. Figure 1.7 is reprinted from Reference [17], with the permission of John Wiley & Sons.

## References

[1] A.C. Guyton and J.E. Hall, *Textbook of Medical Physiology*. Philadelphia, PA, USA: Saunders, 12th edition, 2010.

[2] D.R. Williams, D.H. Brainard, M.J. McMahon, and R. Navarro, "Double-pass and interferometric measures of the optical quality of the eye," *Journal of the Optical Society of America A*, vol. 11, no. 12, pp. 3123–3135, December 1994.

[3] J.K. Ijspeert, T.J. van den Berg, and H. Spekreijse, "An improved mathematical description of

the foveal visual point spread function with parameters for age, pupil size and pigmentation," *Vision Research*, vol. 33, no. 1, pp. 15–20, January 1993.

[4] J. Rovamo, H. Kukkonen, and J. Mustonen, "Foveal optical modulation transfer function of the human eye at various pupil sizes," *Journal of the Optical Society of America A*, vol. 15, no. 9, pp. 2504–2513, September 1998.

[5] D.H. Marimont and B.A. Wandell, "Matching color images: The effects of axial chromatic aberration," *Journal of the Optical Society of America A*, vol. 11, no. 12, pp. 3113–3122, December 1994.

[6] R.H.S. Carpenter, *Movements of the Eyes*. London, UK: Pion, 1988.

[7] A.L. Yarbus, *Eye Movements and Vision*. New York, USA: Plenum Press, 1967.

[8] L.B. Stelmach and W.J. Tam, "Processing image sequences based on eye movements," *Proceedings of SPIE*, vol. 2179, pp. 90–98, February 1994.

[9] C. Endo, T. Asada, H. Haneishi, and Y. Miyake, "Analysis of the eye movements and its applications to image evaluation," in *Proceedings of the Color Imaging Conference*, Scottsdale, AZ, USA, November 1994, pp. 153–155.

[10] L. Itti and C. Koch, "A saliency-based search mechanism for overt and covert shifts of visual attention," *Vision Research*, vol. 40, no. 10–12, pp. 1489–1506, June 2000.

[11] J.M. Henderson, J.R. Brockmole, M.S. Castelhano, and M. Mack, "Visual saliency does not account for eye movements during visual search in real-world scenes," in *Eye Movements: A Window on Mind and Brain*, R. van Gompel, M. Fischer, W. Murray, and R. Hill (eds.), Amsterdam, The Netherlands: Elsevier, 2007, pp. 537–562.

[12] L.W. Renninger, P. Verghese, and J. Coughlan, "Where to look next? Eye movements reduce local uncertainty," *Journal of Vision*, vol. 7, no. 3, February 2007.

[13] D.A. Baylor, "Photoreceptor signals and vision," *Investigative Ophthalmology & Visual Science*, vol. 28, no. 2, pp. 34–49, January 1987.

[14] A. Stockman, D.I.A. MacLeod, and N.E. Johnson, "Spectral sensitivities of the human cones," *Journal of the Optical Society of America A*, vol. 10, no. 12. pp. 2491–2521, December 1993.

[15] A. Stockman, L.T. Sharpe, and C. Fach, "The spectral sensitivity of the human short-wavelength sensitive cones derived from thresholds and color matches," *Vision Research*, vol. 39, no. 17, pp. 2901–2927, August 1999.

[16] A. Stockman and L.T. Sharpe, "Spectral sensitivities of the middle- and long-wavelength sensitive cones derived from measurements in observers of known genotype," *Vision Research*, vol. 40, no. 13, pp. 1711–1737, June 2000.

[17] C.A. Curcio, K.R. Sloan, R.E. Kalina, and A.E. Hendrickson, "Human photoreceptor topography," *Journal of Comparative Neurology*, vol. 292, no. 4, pp. 497–523, February 1990.

[18] P.K. Ahnelt, "The photoreceptor mosaic," *Eye*, vol. 12, no. 3B, pp. 531–540, May 1998.

[19] C.A. Curcio, K.A. Allen, K.R. Sloan, C.L. Lerea, J.B. Hurley, I.B. Klock, and A.H. Milam, "Distribution and morphology of human cone photoreceptors stained with anti-blue opsin," *Journal of Comparative Neurology*, vol. 312, no. 4, pp. 610–624, October 1991.

[20] D.H. Hubel and T.N. Wiesel, "Receptive fields of single neurons in the cat's striate cortex," *Journal of Physiology*, vol. 148, no. 3, pp. 574–591, October 1959.

[21] D.H. Hubel and T.N. Wiesel, "Receptive fields and functional architecture of monkey striate cortex," *Journal of Physiology*, vol. 195, no. 1, pp. 215–243, March 1968.

[22] J.G. Daugman, "Two-dimensional spectral analysis of cortical receptive field profiles," *Vision Research*, vol. 20, no. 10, pp. 847–856, October 1980.

[23] G.C. Phillips and H.R. Wilson, "Orientation bandwidth of spatial mechanisms measured by masking," *Journal of the Optical Society of America A*, vol. 1, no. 2, pp. 226–232, February 1984.

[24] D.J. Field, "Relations between the statistics of natural images and the response properties of cortical cells," *Journal of the Optical Society of America A*, vol. 4, no. 12, pp. 2379–2394, December 1987.

[25] J.H. van Hateren and A. van der Schaaf, "Independent component filters of natural images compared with simple cells in primary visual cortex," *Proceedings of the Royal Society of London B*, vol. 265, no. 1394, pp. 1–8, 1998.

[26] M.A. Losada and K.T. Mullen, "The spatial tuning of chromatic mechanisms identified by simultaneous masking," *Vision Research*, vol. 34, no. 3, pp. 331–341, February 1994.

[27] M.A. Losada and K.T. Mullen, "Color and luminance spatial tuning estimated by noise masking in the absence of off-frequency looking," *Journal of the Optical Society of America A*, vol. 12, no. 2, pp. 250–260, February 1995.

[28] R.L. Pandey Vimal, "Orientation tuning of the spatial-frequency mechanisms of the red-green channel," *Journal of the Optical Society of America A*, vol. 14, no. 10, pp. 2622–2632, October 1997.

[29] A.B. Watson, "The cortex transform: Rapid computation of simulated neural images," *Computer Vision, Graphics, and Image Processing*, vol. 39, no. 3, pp. 311–327, September 1987.

[30] E.P. Simoncelli, W.T. Freeman, E.H. Adelson, and D.J. Heeger, "Shiftable multi-scale transforms," *IEEE Transactions on Information Theory*, vol. 38, no. 2, pp. 587–607, March 1992.

[31] K.T. Mullen, "The contrast sensitivity of human colour vision to red-green and blue-yellow chromatic gratings," *Journal of Physiology*, vol. 359, no. 1, pp. 381–400, February 1985.

[32] J. Yang and W. Makous, "Implicit masking constrained by spatial inhomogeneities," *Vision Research*, vol. 37, no. 14, pp. 1917–1927, July 1997.

[33] F.L. van Nes and M.A. Bouman, "Spatial modulation transfer in the human eye," *Journal of the Optical Society of America*, vol. 57, no. 3, pp. 401–406, March 1967.

[34] M.A. Berkley and F. Kitterle, "Grating visibility as a function of orientation and retinal eccentricity," *Vision Research*, vol. 15, no. 2, pp. 239–244, February 1975.

[35] J.L. Mannos and D.J. Sakrison, "The effects of a visual fidelity criterion on the encoding of images," *IEEE Transactions on Information Theory*, vol. 20, no. 4, pp. 525–536, July 1974.

[36] P.G.J. Barten, "Evaluation of subjective image quality with the square-root integral method," *Journal of the Optical Society of America A*, vol. 7, no. 10, pp. 2024–2031, October 1990.

[37] S. Daly, "The visible differences predictor: An algorithm for the assessment of image fidelity," in *Digital Images and Human Vision*, A.B. Watson (ed.), Cambridge, MA, USA: MIT Press, 1993, pp. 179–206.

[38] E. Peli, "In search of a contrast metric: Matching the perceived contrast of Gabor patches at different phases and bandwidths," *Vision Research*, vol. 37, no. 23, pp. 3217–3224, December 1997.

[39] E. Peli, "Contrast in complex images," *Journal of the Optical Society of America A*, vol. 7, no. 10, pp. 2032–2040, October 1990.

[40] J.G. Daugman, "Uncertainty relation for resolution in space, spatial frequency, and orientation optimized by two-dimensional visual cortical filters," *Journal of the Optical Society of America A*, vol. 2, no. 7, pp. 1160–1169, July 1985.

[41] S. Winkler, *Digital Video Quality – Vision Models and Metrics*. Chichester, UK: John Wiley & Sons, 2005.

[42] P. Vandergheynst, M. Kutter, and S. Winkler, "Wavelet-based contrast computation and application to digital image watermarking," *Proceedings of SPIE*, vol. 4119, pp. 82–92, August 2000.

[43] M. Kutter and S. Winkler, "A vision-based masking model for spread-spectrum image watermarking," *IEEE Transactions on Image Processing*, vol. 11, no. 1, pp. 16–25, January 2002.

[44] W.F. Schreiber, "Image processing for quality improvement," *Proceedings of the IEEE*, vol. 66, no. 12, pp. 1640–1651, December 1978.

[45] J.M. Foley, "Human luminance pattern-vision mechanisms: Masking experiments require a new model," *Journal of the Optical Society of America A*, vol. 11, no. 6, pp. 1710–1719, June 1994.

[46] J.M. Foley and Y. Yang, "Forward pattern masking: Effects of spatial frequency and contrast," *Journal of the Optical Society of America A*, vol. 8, no. 12, pp. 2026–2037, December 1991.

[47] E. Switkes, A. Bradley, and K.K. De Valois, "Contrast dependence and mechanisms of masking interactions among chromatic and luminance gratings," *Journal of the Optical Society of America A*, vol. 5, no. 7, pp. 1149–1162, July 1988.

[48] G.R. Cole, C.F. Stromeyer III, and R.E. Kronauer, "Visual interactions with luminance and chromatic stimuli," *Journal of the Optical Society of America A*, vol. 7, no. 1, pp. 128–140, January 1990.

[49] M.P. Eckstein and J.S. Whiting, "Visual signal detection in structured backgrounds. I. Effect of number of possible spatial locations and signal contrast," *Journal of the Optical Society of America A*, vol. 13, no. 9, pp. 1777–1787, September 1996.

[50] A.B. Watson, R. Borthwick, and M. Taylor, "Image quality and entropy masking," *Proceedings of SPIE*, vol. 3016, pp. 2–12, February 1997.

[51] M.J. Nadenau, J. Reichel, and M. Kunt, "Performance comparison of masking models based on a new psychovisual test method with natural scenery stimuli," *Signal Processing: Image Communication*, vol. 17, no. 10, pp. 807–823, November 2002.

[52] M. Masry, D. Chandler, and S.S. Hemami, "Digital watermarking using local contrast-based texture masking," in *Proceedings of the Asilomar Conference on Signals, Systems and Computers*, Pacific Grove, CA, USA, November 2003, pp. 1590–1594.

[53] S. Winkler and S. Süsstrunk, "Visibility of noise in natural images," *Proceedings of SPIE*, vol. 5292, pp. 121–129, January 2004.

[54] S.A. Klein, T. Carney, L. Barghout-Stein, and C.W. Tyler, "Seven models of masking," *Proceedings of SPIE*, vol. 3016, pp. 13–24, February 1997.

[55] D.J. Heeger, "Half-squaring in responses of cat striate cells," *Visual Neuroscience*, vol. 9, no. 5, pp. 427–443, November 1992.

[56] D.J. Heeger, "Normalization of cell responses in cat striate cortex," *Visual Neuroscience*, vol. 9, no. 2, pp. 181–197, August 1992.

[57] P.C. Teo and D.J. Heeger, "Perceptual image distortion," *Proceedings of SPIE*, vol. 2179, pp. 127–141, February 1994.

[58] A.B. Watson and J.A. Solomon, "Model of visual contrast gain control and pattern masking," *Journal of the Optical Society of America A*, vol. 14, no. 9, pp. 2379–2391, September 1997.

[59] M.W. Greenlee and J.P. Thomas, "Effect of pattern adaptation on spatial frequency discrimination," *Journal of the Optical Society of America A*, vol. 9, no. 6, pp. 857–862, June 1992.

[60] H.R. Wilson and R. Humanski, "Spatial frequency adaptation and contrast gain control," *Vision Research*, vol. 33, no. 8, pp. 1133–1149, May 1993.

[61] R.J. Snowden and S.T. Hammett, "Spatial frequency adaptation: Threshold elevation and perceived contrast," *Vision Research*, vol. 36, no. 12, pp. 1797–1809, June 1996.

[62] M.A. Webster and E. Miyahara, "Contrast adaptation and the spatial structure of natural images," *Journal of the Optical Society of America A*, vol. 14, no. 9, pp. 2355–2366, September 1997.

[63] M.A. Webster and J.D. Mollon, "Adaptation and the color statistics of natural images," *Vision Research*, vol. 37, no. 23, pp. 3283–3298, December 1997.

[64] H.G. Grassmann, "Zur theorie der farbenmischung," *Annalen der Physik und Chemie*, vol. 89, pp. 69–84, 1853.

[65] E. Hering, *Zur Lehre vom Lichtsinne*. Vienna, Austria: Carl Gerolds & Sohn, 1878.

[66] L.M. Hurvich and D. Jameson, "An opponent-process theory of color vision," *Psychological Review*, vol. 64, no. 6, pp. 384–404, November 1957.

[67] G. Svaetichin, "Spectral response curves from single cones," *Acta Physiologica Scandinavica*, vol. 39, no. 134, pp. 17–46, February 1956.

[68] T.W. Lee, T. Wachtler, and T.J. Sejnowski, "Color opponency is an efficient representation of spectral properties in natural scenes," *Vision Research*, vol. 42, no. 17, pp. 2095–2103, August 2002.

[69] A.B. Poirson and B.A. Wandell, "Appearance of colored patterns: Pattern-color separability," *Journal of the Optical Society of America A*, vol. 10, no. 12, pp. 2458–2470, December 1993.

[70] C.A. Poynton, *A Technical Introduction to Digital Video*. New York, USA: John Wiley & Sons, 1996.

[71] R.W.G. Hunt, *The Reproduction of Colour*. Kingston upon Thames, UK: Fountain Press, 5th edition, 1995.

[72] S. Reichelt, R. Häussler, G. Fütterer, and N. Leister, "Depth cues in human visual perception and their realization in 3D displays," *Proceedings of SPIE*, vol. 7690, pp. 850094:1–12, April 2010.

[73] L. Onural, T. Sikora, J. Ostermann, A. Smolic, M.R. Civanlar, and J. Watson, "An assessment of 3DTV technologies," in *Proceedings of the NAB Broadcast Engineering Conference*, Las Vegas, NV, USA, April 2006, pp. 456–467.

[74] P.A. Howarth, "Potential hazards of viewing 3-D stereoscopic television, cinema and computer games: A review," *Ophthalmic and Physiological Optics*, vol. 31, no. 2, pp. 111–122, March 2011.

[75] M. Lambooij and W. Ijsselsteijn, "Visual discomfort and visual fatigue of stereoscopic displays: A review," *Journal of Imaging Science and Technology*, vol. 53, no. 3, pp. 030201:1–14, May 2009.

[76] B.A. Wandell, *Foundations of Vision*. Sunderland, MA, USA: Sinauer Associates, 1995.

[77] S.E. Palmer, *Vision Science: Photons to Phenomenology*. Cambridge, MA, USA: MIT Press, 1999.

[78] R. Snowden, P. Thompson, and T. Troscianko, *Basic Vision: An Introduction to Visual Perception*. New York, USA: Oxford University Press, 2006.

[79] M. Bass, C. DeCusatis, J. Enoch, V. Lakshminarayanan, G. Li, C. MacDonald, V. Mahajan, and E. Van Stryland, *Handbook of Optics: Vision and Vision Optics*. New York, USA: McGraw-Hill, vol. III, 3rd edition, 2009.

[80] K.R. Gegenfurtner and L.T. Sharpe, *Color Vision: From Genes to Perception*. Cambridge, UK: Cambridge University Press, 2001.

[81] R.W.G. Hunt, *The Reproduction of Colour*. New York, USA: John Wiley & Sons Inc., 6th edition, 2004.

[82] H.C. Lee, *Introduction to Color Imaging Science*. Cambridge, UK: Cambridge University Press, 2005.

[83] B. Keelan, *Handbook of Image Quality: Characterization and Prediction*. Boca Raton, FL, USA: CRC Press, 2002.

[84] A.C. Guyton, *Textbook of Medical Physiology*. Philadelphia, PA, USA: Saunders, 8th edition, October 1990.

# 2

# An Analysis of Human Visual Perception Based on Real-Time Constraints of Ecological Vision

Haluk Öğmen

## 2.1   Introduction

Motion is ubiquitous in normal, ecological, viewing conditions. Many objects in the environment are in motion. The motion of objects is supplemented by the motion of the observer's body, head, and eyes. Therefore, ecological vision is highly dynamic and an understanding of biological vision requires an analysis of the implications of these afore-mentioned movements on the formation and processing of images. The dynamic nature of ecological vision was the cornerstone of several theories of visual perception, in particular the ecological perception theory [1]. This chapter will review some basic facts about eco-logical vision and discuss a neural model that offers a computational basis for the analysis of dynamic stimuli.

Image formation by the human eye can be characterized by using the laws of optics and projective geometry. Essentially, the optics of the eye works like a camera and a two-dimensional image of the environment is generated on each retina. The projections from retina to early visual cortex preserve neighborhood relations so that two neighboring points in space activate two neighboring photoreceptors in the retina. These neighboring recep-tors, in turn, send their signals through retino-cortical projections to two neighboring neu-rons in early visual areas. This is called a *retinotopic map* or a *retinotopic representation*.

Given the importance of motion in ecological viewing, it is essential to analyze the effects of various sources of movement on the retinotopic representation of the environment. Since the retinotopic image is the result of the *relative* position of the imaging system, the eye, with respect to the environment, sources of retinotopic motion can be broken down into two broad classes:

- retinotopic motion resulting from endogenous sources, that is, caused by the motion of the observer, and

- retinotopic motion resulting from exogenous sources, that is, caused by motion in the environment.

The next two sections analyze these two cases and outline their implications. Section 2.4 highlights two interrelated problems arising from these dynamic constraints, the problems of motion blur and moving ghosts. Section 2.5 explores further the motion blur problem in human vision and suggests mechanisms that can contribute to its control. Section 2.6 introduces an exposition of a model of retino-cortical dynamics (RECOD) that provides a mathematical framework for dealing with motion blur in human vision. Section 2.7 applies this model to critical data on motion deblurring to illustrate its operation. Finally, this chapter concludes with Section 2.8.

## 2.2 Endogenous Sources of Retinotopic Motion

First, each source of motion will be introduced and analyzed in isolation, assuming everything else being constant. Then, the combined operation of different types of motion will be discussed.

Consider the movement of the body, such as walking, turning, and leaning; the eyes shift their positions with respect to the environment. This is equivalent to moving a tripod on which a camera is positioned. These movements alter the relative position of the camera with respect to the environment and cause a *global* shift in the retinotopic image. In photography, it is well known that these types of camera movements result in a blurred image. This is because, as the camera moves, a continuously shifting image impinges on the film, and each point in space imprints its characteristics not on a single point in the image but over an extended surface, resulting in blur. The same situation occurs when the head moves.

In normal viewing conditions, human eyes also undergo complex movement patterns that can be decomposed into a combination of different types. A first general classification is based on the relative motion of the two eyes. Namely, when the two eyes move in opposite directions, the eye movements are called *disjunctive movements*. This occurs, for example, when shifting the gaze from a point far in the scene to a point near in the scene. In this case, to keep the point of interest in the central part of each retina (fovea), the two eyes need to move toward each other; this movement is known as a *convergence movement*. When the opposite occurs, the eyes move away from each other, creating a *divergence movement*. When the two eyes move in the same direction — for example, when the gaze is shifted from left to right — the eye movements are called *conjunctive movements*.

Conjunctive eye movements can be further categorized into two main subtypes: *smooth pursuit* and *saccadic* eye movements. Smooth pursuit is, as the name indicates, a smooth tracking movement to keep a target of interest in central vision. On the other hand, saccades are very fast ballistic movements to shift the gaze rapidly from one point in the scene to another point of interest. All these movements also generate *global* shifts in the retinotopic image. Depending on whether the eye movement is conjunctive or disjunctive, the global shifts in the retinotopic image can have same or opposite directions. Depending on the type of eye movement, the global shift can be relatively slow and smooth, or fast and discontinuous. Therefore, a negative effect of all these movements is to create a *global* blur on the retinotopic image.

The nervous system uses a multitude of mechanisms to compensate for the negative effects of relative movement between the retina and the environment. When this relative movement is caused by body or head movements, an obvious strategy is to move the eyes in the opposite direction so as to keep the retinotopic image steady; this behavior is known as *motor compensation*. In fact, when the body or the head moves, the *vestibular system* senses these movements and generates compensatory signals to move the eyes in the opposite direction so as to keep a steady gaze position. This is a reflexive system called the *vestibulo-ocular reflex* (VOR). As a reflexive movement, vestibulo-ocular reflex has a very short latency (approximately 10 ms) and an approximate unit gain resulting in a very effective compensation for these movements [2].

On the other hand, when the eyes themselves move, mechanisms other than motor compensation are needed. Here two sources of information are particularly relevant. The first is *proprioceptive signals* from eye muscles that carry information about eye position and movement. The second is the *efference copy signal* that provides a copy of the command signals that are sent to eye muscles for the intended eye movements. In fact, these two sources of information take part, in varying degrees, in neural mechanisms devoted to the analysis of visual signals under normal viewing conditions.

It should be pointed out that endogenous motion also produces useful information for visual processing. While endogenous motion generates a global shift in the retinotopic image, this global shift is not uniform but instead depends on the distance of a point in the scene with respect to the observer. The retinotopic motion of a near object will be larger than the retinotopic motion of a far object. This distance-dependent differential motion pattern, known as *motion parallax*, is a very important cue for the visual system to infer depth from two-dimensional retinotopic images.

## 2.3    Exogenous Sources of Retinotopic Motion

The second source of retinotopic motion is exogenous in that it is due to motion whose source is in the environment. Different objects in the scene can undergo different motion patterns. Three fundamental differences can be established between endogenous and exogenous motion:

- Endogenous motion generates global retinotopic motion, that is, covering the whole field, while exogenous motion generates *local* retinotopic motion, that is, confined to the area corresponding to the motion trajectory of the moving object.

- The retinotopic motion field generated by endogenous sources is correlated across the visual field in that one common motion vector is applied to the entire stimulus to generate depth-dependent retinotopic motion vectors. On the other hand, retinotopic motion generated by exogenous sources can be both correlated and uncorrelated. As an example, assume a jogger and a car passing nearby. The arms, legs, and the body of the jogger share both a common and a differentiated motion, while the motion of the car can be completely independent of that of the jogger.

- Because endogenous motion is generated by the observer, motor planning, execution, and feedback signals are available before and during the motion to the observer. As mentioned in the previous section, this information can be used before and during retinotopic motion. On the other hand, exogenous motion information is not directly available and needs to be estimated *during* retinotopic motion.

These important distinctions between exogenous and endogenous sources of motion suggest that different mechanisms are at play in dealing with each case. However, under normal viewing conditions, exogenous and endogenous motion occur often in conjunction, implying that these mechanisms need to work in concert. As an example, *optokinetic nystagmus* (OKN) is a reflexive movement of the eyes, which usually occurs in conjunction with vestibulo-ocular reflex. As mentioned above, when the body or the head moves, a large field motion in the retinotopic image is created. This can be compensated both by vestibulo-ocular reflex using internal sources of information, as well as by external sources of information by estimating large-field motion directly from the visual input. In fact, the latter generates optokinetic nystagmus. Similarly, when an object moves in the environment, the eye can track the object (smooth pursuit), thereby stabilizing the moving object in the fovea by motor compensation. However, because different objects may move in different directions and some objects are stationary (for example, background), motor compensation cannot stabilize everything in the scene. In fact, it involves trade-offs in the sense that smooth pursuit can stabilize a moving object of choice but will generate additional movement for all other objects that differ in their velocity.

## 2.4 Motion Blur and Moving Ghosts Problems

The camera analogy is helpful for understanding why blur occurs. Consider a camera with a photographic film. The development of the image on the film is not an instantaneous event but a *process*. Absorption of light by light sensitive chemicals on the film initiates a chemical reaction that takes time depending on the intensity of the light and the sensitivity of the chemicals to light. To capture a successful picture, exposure duration and the aperture of the camera need to be adjusted according to light intensity. For a given aperture, short and long exposure durations lead to underdeveloped and overdeveloped (saturated) pictures. When there is relative motion between the camera and the environment, the moving image pattern will expose different parts of the film, resulting in a smeared image. Furthermore, compared to a static object, moving objects will stay on a given region of the film only briefly, not allowing sufficient time for the chemical process to capture the form information of the moving object. The situation is similar for digital cameras and biological systems, which have, respectively, electronic and biological sensors with limited sensitivities. This implies that light patterns need to be exposed for sufficient duration in order to generate reliable signals from these sensors. Furthermore, the visual system does not just register a retinotopic image but processes it in order to give rise to percepts and to visually guided behaviors. This processing also takes time, putting additional temporal constraints on vision.

These observations can be summarized by highlighting two interrelated aspects of the problem:

- The problem of *motion blur*, which refers to the spatial aspect of the problem. In other words, the image of the moving object is spatially spread out.

- The problem of *moving ghosts*, which refers to the figural aspect of the problem. In other words, since the moving object does not stay long enough on a retinotopic locus, processes computing the form of the object do not receive sufficient input to synthesize the form of the moving object. The result is a spatially extended smear without a significant form information, known as a moving ghost.

The basic technique in photography for taking *static* pictures of moving objects is to use a fast shutter speed with a high-sensitivity film (depending on the lighting conditions). For dynamic imaging (videos), the approach is similar to biological systems; namely, tracking the moving object with the camera, much like the way humans track moving objects with smooth-pursuit eye movements. Examination of individual frames of videos of moderately fast-moving objects show extensive smear for objects that are not stabilized. However, in natural viewing, humans are not aware of this smear due to brain mechanisms that are developed to deal with this very problem. But how does the visual system deal with motion blur and moving ghosts problems?

## 2.5 Motion Blur in Human Vision

Several studies, going back to the 19th century, examined the motion blur problem by using a variety of stimuli and techniques. For example, in Reference [3], a bright line segment was moved in front of a dark field, and it was noticed that the line appeared extensively blurred. Several other studies reported similar results. One exception was a study in Reference [4], where observers were presented with an array of random dots in motion. It was found that perceived blur depended on the exposure duration of the stimulus, initially increasing as a function of exposure duration for up to approximately 30 ms, decreasing

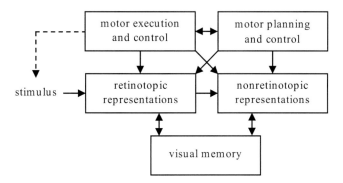

**FIGURE 2.1**

Schematic of the model. Visual inputs are first registered in retinotopic representations corresponding to early visual areas. A motion segmentation process transforms these into nonretinotopic representations. Visual memory is used to buffer information during the three-phase operation discussed in the next section. Motor systems include stages of motor planning and control that send command signals for motor execution and control, for example, for moving the eyes. Copies of efferent command and proprioceptive signals are available to modulate retinotopic and nonretinotopic representations. A motor action can either directly modify the stimulus, as in grasping, or shift the sensor (for example, the eye), thereby affecting the retinotopic stimulus indirectly. This relation between motor action and the stimulus is shown by a dashed arrow.

thereafter for longer exposure durations. The study called this *motion deblurring*. In order to reconcile early reports of extensive perceived blur for isolated moving targets and reduced blur for more complex stimuli (array of random dots), a study in Reference [5] varied systematically the density of the dots in the stimulus and showed that perceived blur depends both on exposure duration and the *density of dots*. As the dot density increased, perceived blur decreased. This finding was attributed to inhibitory interactions between activities generated by neighboring dots. Thus, the visual system uses stimulus-driven inhibitory interactions to reduce the spatial extent of motion blur caused by exogenous sources.

Using a similar stimulus, various studies investigated motion blur under conditions that include endogenous sources of blur [6], [7], [8], [9], [10], [11]. It was found that extra-retinal signals are used to reduce the spatial extent of motion blur.

Taken together, these results suggest that the human visual system uses stimulus-driven inhibitory interactions to limit the spatial extent of blur caused by exogenous sources. This strategy is augmented by use of extra-retinal signals to limit the spatial extent of blur caused by endogenous sources.

Figure 2.1 illustrates a schematic diagram of the model for how the visual system solves the problems of motion blur and moving ghosts. Visual inputs are first registered in retinotopic representations. Within retinotopic representations, moving targets generate extensively blurred (motion blur problem) and formless (moving ghosts problem) activities. The motion blur problem is addressed by retinotopic inhibitory interactions within retinotopic representations. In order to compute the form of moving objects, a motion grouping algorithm transforms retinotopic representations into nonretinotopic representations and stores them in visual short-term memory (VSTM). Endogenous sources of motion blur and mov-

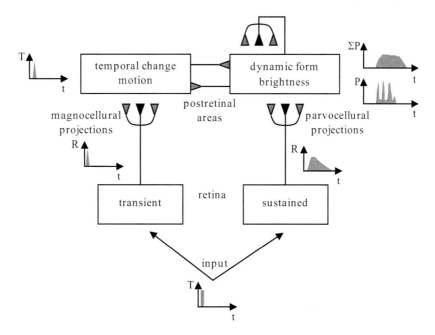

**FIGURE 2.2**

The RECOD model. The input is processed by retinal cells with transient and sustained response characteristics, giving rise to magnocellular and parvocellular pathways, respectively. The top layers show lumped representations of postretinal areas receiving their primary inputs from magnocellular and parvocellular pathways. Responses to a brief pulse input are shown in different parts of the model. Dark-gray and light-gray synaptic symbols indicate excitatory and inhibitory connections, respectively.

ing ghosts problems are shown at the top of the figure. Motor control systems of the brain send efferent command signals to motor systems (control of eye, head, and body muscles) and a copy of these command signals is available for internal compensation. In addition, the vestibular system provides sensory signals about self-movement to initiate both motor and neural compensation. The details of where these internal (retinotopic and nonretinotopic) signals act to contribute to the solution of the motion blur and moving ghosts problems are not fully understood. Therefore, in Figure 2.1, these signals are represented as potentially available to both retinotopic and nonretinotopic representations.

The rest of this chapter focuses on a computational model of how the spatial extent of motion blur can be controlled by stimulus-driven inhibitory interactions. For the nonretinotopic components of the general model shown in Figure 2.1, the reader is referred to recent reviews in References [12] and [13].

## 2.6 Retino-Cortical Dynamics Model

### 2.6.1 General Architecture

Figure 2.2 shows the schematic of the retino-cortical dynamics (RECOD) model [14]. The input is first processed by two populations of retinal ganglion cells, those with sus-

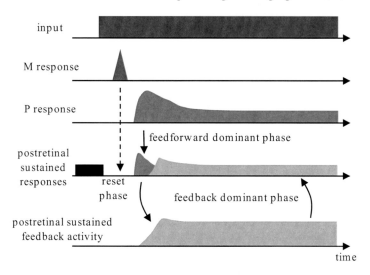

**FIGURE 2.3**

Schematic illustration of the three-phase operation of the RECOD model. The top trace shows a step input that generates a short latency transient (M) and a longer latency sustained (P) response. The sustained response initially peaks to a value and decays thereafter to a lower plateau. The transient activity inhibits via transient-on-sustained inhibition the ongoing activity (shown in black) in postretinal sustained areas. This is the reset phase. Following this reset, the initial rise to a peak activity in the P response transfers the new input information to postretinal sustained areas and also activates feedback loops. This is the feedforward dominant phase. As the feedforward sustained activity decays to a lower plateau, the feedback activities overtake postretinal sustained responses. This is the feedback-dominant phase.

tained and those with transient responses. The outputs of these two populations form two segregated pathways as they project from retina to visual cortex. These pathways are known as *parvocellular* and *magnocellular* pathways. In the visual cortex, one can distinguish neurons with sustained and those with transient responses. These populations are referred to as *sustained* and *transient channels* and correspond, mainly, to the cortical ventral and dorsal pathways. The ventral pathway is the main cortical system for processing figural properties, such as boundaries, surface features, color, etc. The dorsal pathway is the main cortical system for processing dynamic properties of stimuli, such as motion. Unlike pre-cortical areas, where parallel pathways remain segregated, in the cortex interactions exist between these two populations. These interactions can be both excitatory and inhibitory. The presented model emphasizes inhibitory interactions that play a critical role in motion deblurring. Thus, reciprocal inhibitory connections are shown between sustained and transient cortical channels. Since the problem of motion blur consists of smear in figural representations, the extent of smear in sustained pathways for moving targets is analyzed. It is also known that extensive feedback, including widespread *positive* feedback, exists in cortical areas. Because positive feedback can be a major source of hysteresis, and thus blur, these feedback connections are included in the analysis presented below.

The operation of the model dynamics is summarized in Algorithm 2.1. The following discusses the original motivations for proposing the three-phase operation of the model.

**ALGORITHM 2.1** The operation of the model dynamics.

1. Through the afferent retino-geniculate pathways, the onset of a stimulus evokes a fast transient magnocellular (M) activity and a slower sustained parvocellular (P) activity. These activities are shown schematically in Figure 2.3. The sustained response consists of an initial peak followed by a decay to a lower plateau value.

2. P and M activities flow predominantly to the dorsal and the ventral streams of cortical processing to initiate form/brightness and temporal-change/motion computations, respectively. Other P and M inputs to the dorsal and ventral pathways also exist to implement other functions and interactions between form and motion systems. Specifically, the model postulates reciprocal connections conveying M (P) input to the ventral (dorsal) pathway to implement reciprocal inhibition between these two pathways.

3. The temporal dynamics of the model in response to stimulus onset consists of three phases: reset phase, feedforward dominant phase, and feedback dominant phase.

4. At the stimulus onset, the fast transient activity inhibits via *transient-on-sustained inhibitory* connections any pre-existing activity at the retinotopic location of the new stimulus. This prevents the *temporal integration* and blurring of activities generated by different stimuli presented in temporal succession. This phase of operation is called the reset phase, because it is "resetting" the retinotopic locus of the new activity by removing any pre-existing activity.

5. As the sustained activity reaches its cortical locus, the initial peak of this activity transfers information from retina to cortex in a feedforward manner. This phase of operation is called the feedforward-dominant phase, because signals in the feedforward connections dominate signals in the feedback connections.

6. As the feedforward signals become registered in cortical areas, they activate feedback loops. The decay of the afferent sustained activity to a lower plateau level allows feedback signals to dominate in cortex, giving rise to feedback dominant phase of operation.

7. When a change occurs in the input signal, the whole cycle repeats itself.

Form synthesis under natural viewing conditions is an extremely difficult problem. The appearance of the stimuli and the context can change drastically within and across scenes. Feedforward architectures do not have the flexibility required to handle these complex problems. As a result, feedback, both positive and negative, is required for processing natural visual stimuli. In fact, the visual cortex contains massive feedback connections. However, a major problem with positive feedback is that it can drive the system toward unstable behavior. To address this problem, it has been suggested that the visual system is designed to operate in a succession of transient regimes to avoid uncontrolled explosion of activity that would result from a sustained positive feedback signaling. When an input changes, a fast transient activity is used to reset the ongoing activity. In the feedforward-dominant phase, input signals are conveyed to cortical areas with the strong initial overshoot of the activity in the sustained pathway. This allows fast registration of new inputs. After this initial registration, the feedforward signal decays to a lower plateau, where the signal is strong enough to maintain a cortical activity but not too strong to override figural synthesis by feedback signaling. Thus, the third phase consists of feedback-dominant operation, where feedback signals carry out figural synthesis. Overall, these phases last a few hundred milliseconds. The next reset can occur either exogenously by changes in the stimulus, which generate reset signals in the retinotopic map at the location of the change, or endogenously by self-generated motion, which in turn causes global changes and thus a global reset in the retinotopic map. As discussed in the next subsection, this three-phase operation provides a natural explanation to how the visual system deals with the motion blur problem.

### 2.6.2 Mathematical Description

The model is based on general types of differential equations used to describe the behavior of neurons, populations of neurons, and the behavior of biochemical dynamics. The first type of equation used in the model has the form of a generic Hodgkin-Huxley equation

$$\frac{dV_m}{dt} = -(E_p + V_m)g_p + (E_d - V_m)g_d - (E_h + V_m)g_h, \tag{2.1}$$

where $V_m$ represents the membrane potential, $g_p$, $g_d$, and $g_h$ are the conductances for passive, depolarizing, and hyperpolarizing channels, respectively, with $E_p$, $E_d$, and $E_h$ representing their Nernst potentials.

The above equation has been used extensively in neural modeling to characterize the dynamics of membrane patches, single cells, as well as networks of cells [15]. To put the model in an equivalent but simpler form, let $E_p = 0$ and use the symbols $B$, $D$, and $A$ for $E_d$, $E_h$, and $g_p$, respectively, to obtain the generic form for *multiplicative* or *shunting* equation [15]:

$$\frac{dV_m}{dt} = -AV_m + (B - V_m)g_d - (D + V_m)g_h. \tag{2.2}$$

The depolarizing and hyperpolarizing conductances are used to represent the excitatory and inhibitory inputs, respectively. The second type of equation, called the *additive, leaky-integrator* model,

$$\frac{dV_m}{dt} = -AV_m + I_d - I_h, \tag{2.3}$$

is a simplified version of Equation 2.2. The influence of external inputs on membrane potential occurs directly as depolarizing ($I_d$) and hyperpolarizing ($I_h$) currents, instead of conductance changes.

Shunting networks can automatically adjust their dynamic range, allowing them to process small and large inputs [15]. Accordingly, shunting equations can be used in the case of interactions among a large number of neurons so that a given neuron can maintain its sensitivity to a small subset of its inputs without running into saturation when a large number of inputs become active. The proposed model uses the simplified additive equations when the interactions involve few neurons.

Finally, a third type of equation is used to express biochemical reactions of the form

$$S + Z \rightarrow Y \rightarrow X \rightarrow S + Z, \tag{2.4}$$

(a)

(b)

(c)

(d)

**FIGURE 3.3**

The effect of the contrast masking property of human vision on visual quality using JPEG compression: (a) original *Building* image, (b) its distorted version with $MSE = 134.61$, (c) original *Caps* image, and (d) its distorted version with $MSE = 129.24$. The distorted images clearly differ in their visual quality, although they have similar MSE with respect to the original.

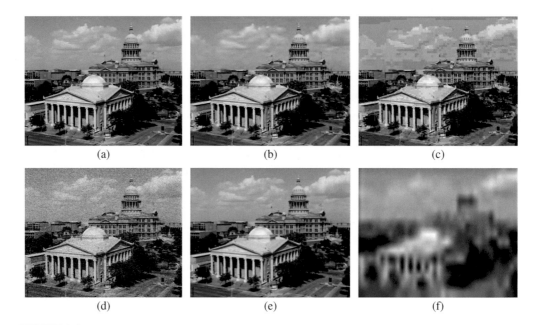

**FIGURE 3.4**

Example distortions from the LIVE IQA database: (a) original image, (b) JPEG2000 compression, (c) JPEG compression, (d) white noise, (e) blur, and (f) Rayleigh fading.

**FIGURE 3.5**

Sample frames from the LIVE VQA database.

**FIGURE 4.1**

Images with different aesthetic qualities. Images on the first row are taken by professional photographers while images on the second row are taken by amateur photographers.

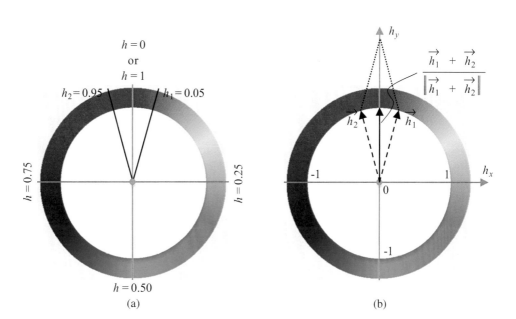

**FIGURE 4.4**

Different representations of the hue values: (a) scalar representation and (b) vector representation.

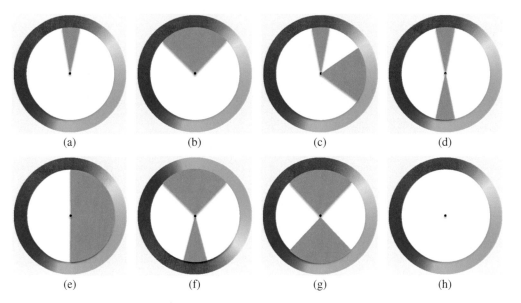

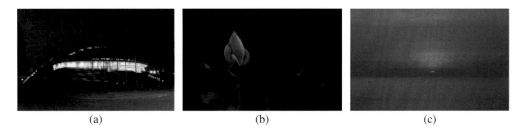

**FIGURE 4.5**

Harmony hue distribution models with gray indicating the efficient regions: (a) *i* type, (b) *V* type, (c) *L* type, (d) *I* type, (e) *T* type, (f) *Y* type, (g) *X* type, and (h) *N* type. All the models can be rotated by an arbitrary angle.

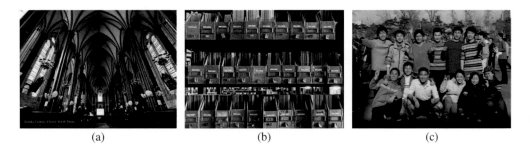

**FIGURE 4.6**

Image examples that fit with different types of hue harmony models in Figure 4.5: (a) fitting *I*-type harmony, (b) fitting *Y*-type harmony, and (c) fitting *i*-type harmony.

**FIGURE 4.9**

Texture and patterns. Patterns in these examples evoke aesthetic feelings.

**FIGURE 4.10**

Linear perspective in photographs.

**FIGURE 5.5**

An illustration of exposure: (a) portrait with a dark background and perceptual exposure, (b) portrait with a light background and perceptual exposure, (c) portrait with a dark background and naive mean value exposure, and (d) portrait with a light background and naive mean value exposure.

**FIGURE 5.14**

Example images with: (a) noise, (b) light RGB smoothing, (c) aggressive RGB smoothing, and (d) simple color opponent smoothing.

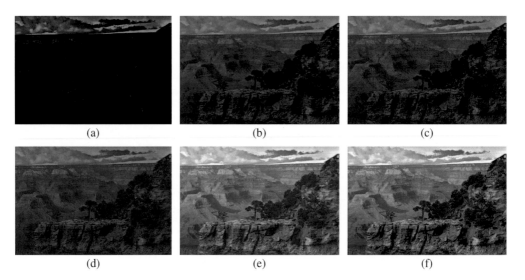

**FIGURE 5.15**

A series of images illustrating how an image approximately appears when displayed on a monitor at different steps in the color rendering, where (a–c) simulate images viewed within a flareless viewing environment and images (d–f) simulate images viewed with flare. (a) The original linear camera image, (b) after gamma correction but not color corrected, (c) after color correction and gamma correction, (d) after color correction and gamma correction (with flare), (e) with boosted RGB scales, and (f) after flare correction.

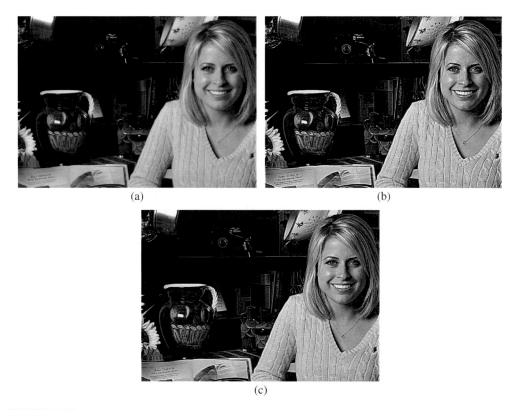

(a)

(b)

(c)

**FIGURE 5.18**

Example images with: (a) no edge enhancement, (b) linear edge enhancement, and (c) soft thresholding and edge-limited enhancement.

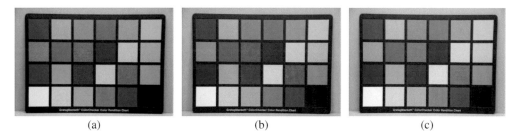

(a)

(b)

(c)

**FIGURE 6.1**

The Macbeth color checker taken under the (a) daylight, (b) tungsten, and (c) fluorescent illumination.

**FIGURE 6.3**

Chromatic adaptation. Stare for twenty seconds at the cross in the center of the top picture. Then focus on the cross in the bottom picture; the color casts should disappear and the image should look normal.

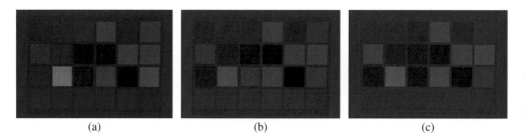

(a)                                    (b)                                    (c)

**FIGURE 6.8**

Chrominance representation of the images shown in Figure 6.1 for the (a) daylight, (b) tungsten, and (c) fluorescent illumination.

**FIGURE 6.9**

Examples of the images taken in different lighting conditions without applying white balancing.

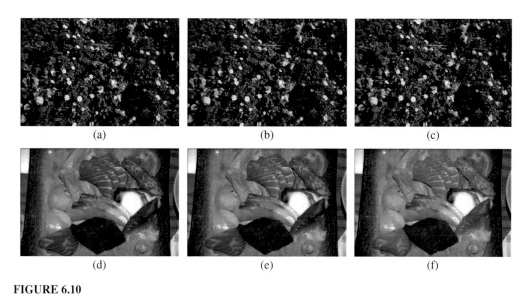

(a)          (b)          (c)

(d)          (e)          (f)

**FIGURE 6.10**

Joint white balancing and color enhancement using the spectral modeling approach: (a,d) $\rho = 0$, (b,e) $\rho = 0.25$, and (c,f) $\rho = 1.0$.

**FIGURE 6.11**

Joint white balancing and color enhancement using the spectral modeling approach: (a,d) $\rho = -0.5$, (b,e) $\rho = 0$, and (c,f) $\rho = 0.5$.

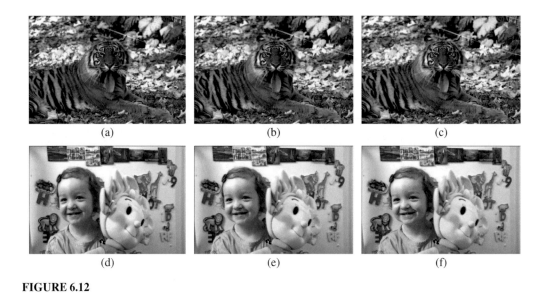

**FIGURE 6.12**

Joint white balancing and color enhancement using the combinatorial approach: (a,d) $w = 0.5$, (b,e) $w = 0$, and (c,f) $w = -0.5$.

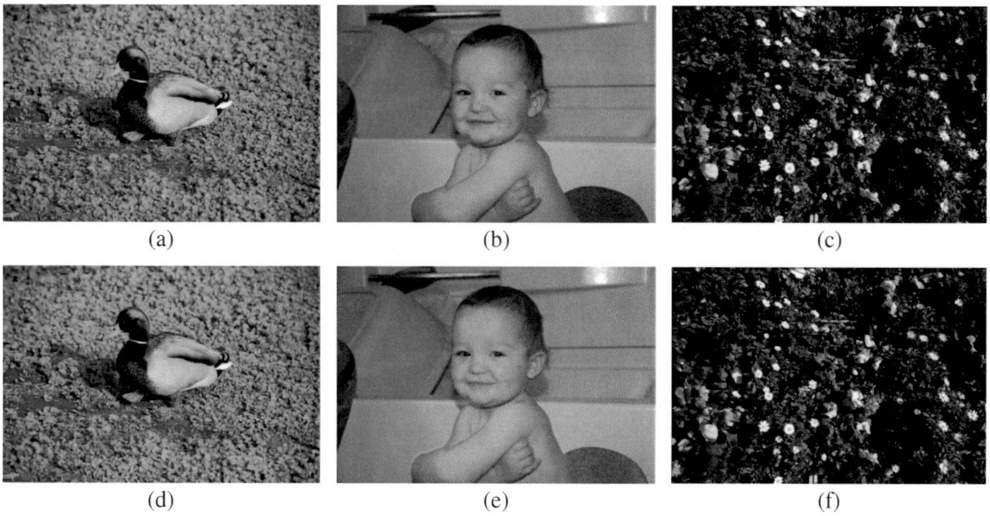

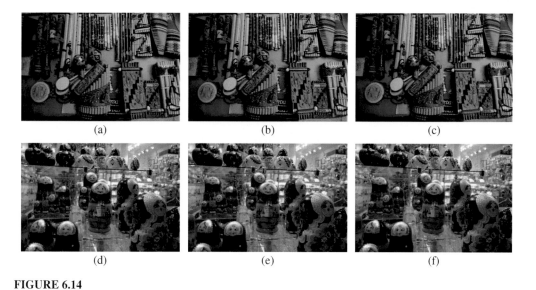

(a) (b) (c)

(d) (e) (f)

**FIGURE 6.13**

Color enhancement of images with inaccurate white balancing: (a–c) input images and (d–f) their enhanced versions.

(a) (b) (c)

(d) (e) (f)

**FIGURE 6.14**

Color adjustment of the final camera images: (a,d) suppressed coloration, (b,e) original coloration, and (c,f) enhanced coloration.

**FIGURE 6.15**

Color enhancement of the final camera images: (a–c) output camera images and (d–f) their enhanced versions.

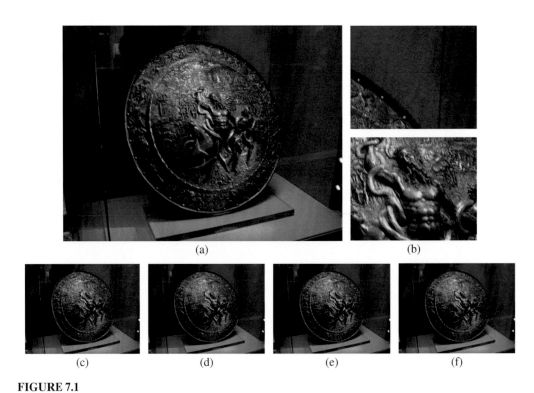

**FIGURE 7.1**

Signal-level thumbnail generation: (a) original image, (b) two magnified regions cropped from the original image, and (c–f) thumbnails. Thumbnails are downsampled using (c) decimation, (d) bilinear, (e) bicubic, and (f) Lanczos filtering.

(a)            (b)            (c)

**FIGURE 7.2**

ROI-based thumbnail generation: (a) high-resolution image, (b) automatic cropping, and (c) seam carving.

(a)            (b)

(c)            (d)

**FIGURE 7.3**

Quality-indicative visual cues: (a) blur, (b) noise, (c) bloom, and (d) red-eye defects.

**FIGURE 7.12**

Highlighting blur in the thumbnail when the original image is blurred: (a) conventional thumbnail at a resolution of 423 × 317 pixels, (b) two magnified regions from the original image with 2816 × 2112 pixels, (c) perceptual thumbnail obtained using gradient-based blur estimation, (d) corresponding blur map, (e) perceptual thumbnail obtained using the scale-space-based blur estimation, and (f) corresponding blur map.

**FIGURE 7.13**

Highlighting blur in the thumbnail when the original image is sharp: (a) conventional thumbnail with $341 \times 256$ pixels, (b) magnified region from the original image with $2048 \times 1536$ pixels, (c) perceptual thumbnail via gradient-based blur estimation, and (d) perceptual thumbnail via scale-space-based blur estimation.

**FIGURE 7.14**

Highlighting blur in the thumbnail when the original image with $1536 \times 2048$ pixels is sharp: (a) conventional thumbnail with $256 \times 341$ pixels, (b) perceptual thumbnail via gradient-based blur estimation, and (c) perceptual thumbnail via scale-space-based blur estimation.

**FIGURE 7.18**

Highlighting blur and noise in the thumbnail: (a) original image with both blur and noise, (b) two magnified regions from the original image, (c) conventional thumbnails obtained using bilinear filtering, (d) conventional thumbnails obtained using bicubic filtering, (e) conventional thumbnails obtained using Lanczos filtering, (f) gradient-based estimation blur map, (g) perceptual thumbnail obtained using gradient-based blur estimation, (h) perceptual thumbnail obtained using gradient-based blur estimation and region-based noise estimation, (i) scale-space-based estimation blur map, (i) perceptual thumbnail obtained using scale-space-based blur estimation, and (k) perceptual thumbnail obtained using scale-space-based blur estimation and multirate noise estimation.

(a)　　　　　　　　　　　　　(b)

**FIGURE 11.2**

Top six ranks of correlogram retrieval in two image databases with (a) 10000 images and (b) 20000 images. Top-left is the query image.

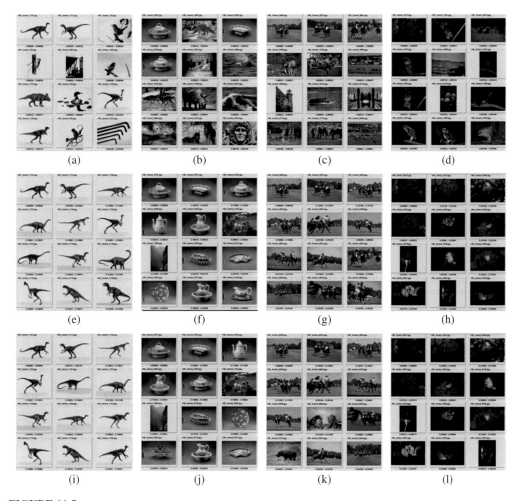

(a)　　　　(b)　　　　(c)　　　　(d)

(e)　　　　(f)　　　　(g)　　　　(h)

(i)　　　　(j)　　　　(k)　　　　(l)

**FIGURE 11.5**

Four queries in *Corel 10K* database: (a–d) correlogram for $d = 10$, (e–h) perceptual correlogram for $d = 10$, and (i–l) perceptual correlogram for $d = 40$. Top-left is the query image.

**FIGURE 11.6**

Retrieval result of the perceptual correlogram for the query in Figure 11.2.

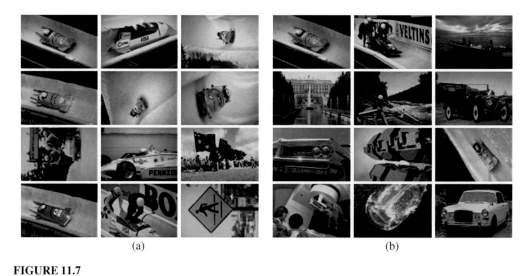

(a)                                                                    (b)

**FIGURE 11.7**

A special case where the correlogram works better than the perceptual correlogram descriptor: (a) correlogram and (b) perceptual correlogram.

**FIGURE 11.15**

Four typical queries using three descriptors in *Corel 10K* database: (a–d) dominant color, (e–h) correlogram, and (i–l) proposed method. Top-left is the query image.

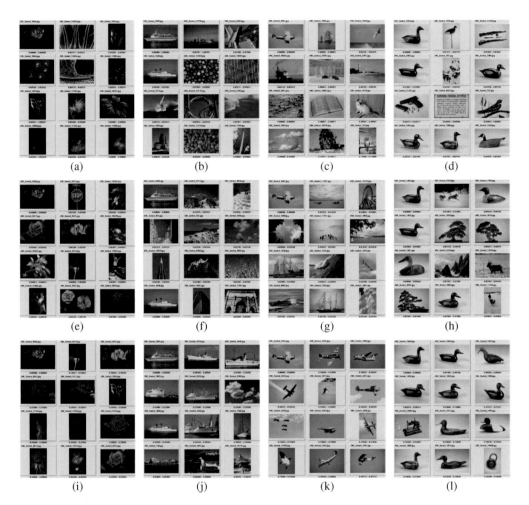

**FIGURE 11.16**

Four typical queries using three descriptors in *Corel 20K* database: (a–d) dominant color, (e–h) auto-correlogram, and (i–l) proposed method. Top-left is the query image.

**FIGURE 11.17**

Two queries in (a,b,e,f) *Corel 10K* and (c,d,g,h) *Corel 20K* databases where (auto-)correlogram performs better than SCD: (a,e) correlogram, (b,f) proposed method, (c,g) auto-correlogram, and (d,h) proposed method. Top-left is the query image.

**FIGURE 12.1**

An input image and its coarse segmentation.

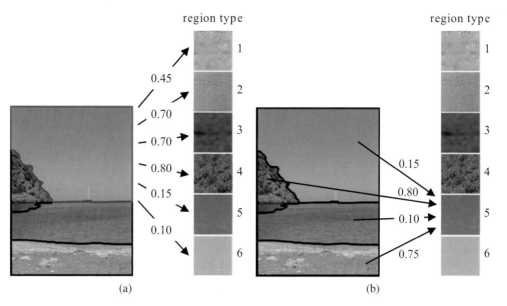

**FIGURE 12.4**

Distances between regions and region types: (a) distances between an image region and all region types; (b) distances between all regions and a specific region type.

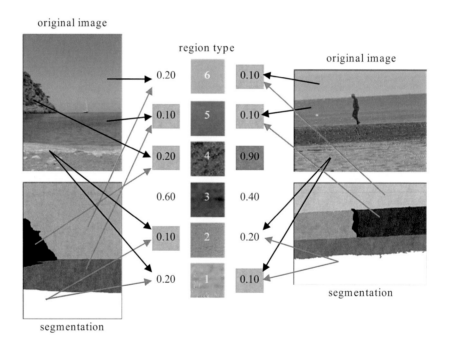

**FIGURE 12.5**

Construction of model vectors for two images and a visual thesaurus of six region types; lowest values of model vectors are highlighted (light gray) to note which region types of the thesaurus are most contained in each image, whereas a high value (dark gray) indicates a high distance between the corresponding region type and the image.

(a)            (b)

**FIGURE 12.10**

An example from the *Beach* domain, where the region types of an image are different than a typical *Beach* image.

(a)          (b)          (c)

(d)          (e)          (f)

**FIGURE 12.15**

Three examples from the *Beach* domain. Initial images and their segmentation maps.

**FIGURE 12.16**

Indicative Corel images.

**FIGURE 12.17**

Indicative TRECVID images.

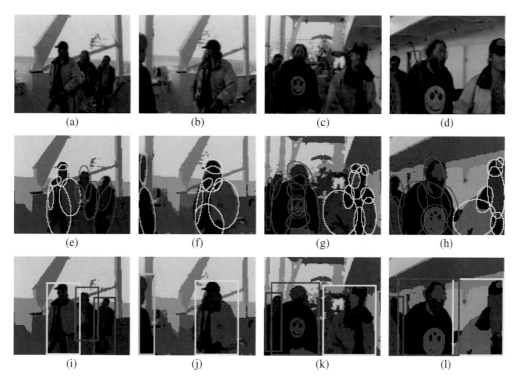

**FIGURE 13.4**

PGA performance demonstration on a sequence from *Titanic* movie: (a,e,i) frame 77, (b,f,j) frame 103, (c,g,k) frame 142, and (d,h,l) frame 175. (a–d) Original frames, (e–h) PGA-identified foreground blobs on the DSCT color-clustered frame, and (i–l) perceptual clusters marked with the same color as the blobs belonging to the distinct foreground clusters. Note that the person wearing the red jacket, initially seen on the right of the person wearing the yellow jacket, comes to the left starting from frame 110.

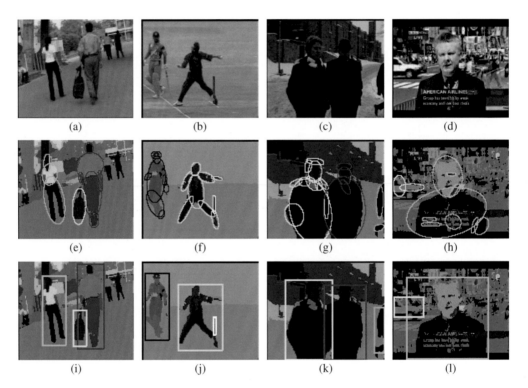

**FIGURE 13.5**

PGA performance demonstration on various scenes: (a–d) original frames in four different scenes, (e–h) identified foreground blobs, and (i–l) perceptual clusters.

**FIGURE 13.8**

Scene categorization: (a–d) subject-centric scenes, as they all have one or two prominent subjects, which exist throughout the duration of the shot, and (e–h) frame-centric scenes because the subjects in these scenes last only for a small duration in the shot. Prominence values for the subjects taken from left to right: (a) {0.96,0.1}, (b) {0.3, 0.8, 0.95}, (c) {0.8, 0.8, 0.18}, (d) {0.61, 0.66}, and (e–h) values smaller than 0.5 according to the proposed prominence model which includes life span as an attribute.

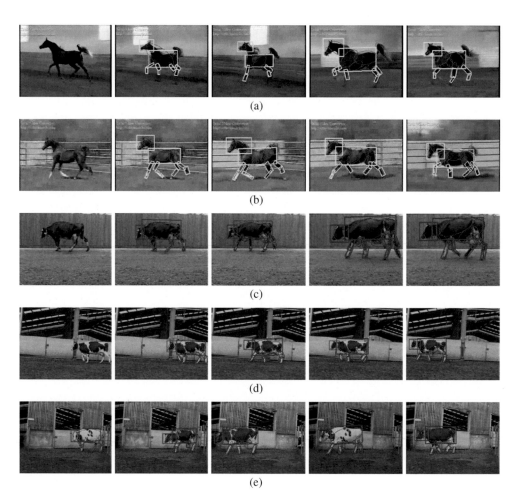

**FIGURE 13.11**

Results obtained for various horse and cow videos. Probability at node E: (a) 0.814, (b) 0.795, (c) 0.910, (d) 0.935, and (e) 0.771, 0.876, 0.876, 0.800, and 0.911 for different videos with the object model cow.

(a)  (b)  (c)

(d)  (e)  (f)

**FIGURE 15.3**

Two examples of how the Cornsweet illusion can enhance the contrast of tone-mapped images are presented on the left. (a–c) Example processed using a global tone mapping operator, where countershading restores contrast in the heavily compressed dynamic range of the sky and highlights. (d–f) Example processed using a local tone mapping operator, which emphasizes on local details at expense of losing global contrast between the landscape and sky. In this case countershading restores brightness relations in this starkly detailed tone mapping result. Notice that the contrast restoration preserves the particular style of each tone mapping algorithm.

(a)  (b)

**FIGURE 15.4**

The Cornsweet illusion used for color contrast enhancement: (a) original image and (b) its enhanced version using Cornsweet profiles in the chroma channel. In this example, the higher chroma contrast improves the sky and landscape separation and enhances the impression of scene depth.

**FIGURE 15.9**

The glare appearance example [22]. ©The Eurographics Association 2009

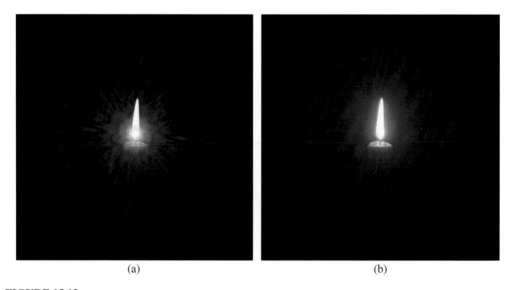

(a)                                                                                   (b)

**FIGURE 15.12**

The glare appearance for a luminous object of non-negligible spatial size [22]: (a) the light diffraction pattern (point-spread function) is just imposed on the background candle image, (b) the light diffraction pattern is convolved with the image. ©The Eurographics Association 2009.

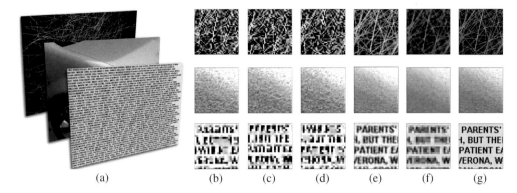

(a)  (b)  (c)  (d)  (e)  (f)  (g)

**FIGURE 15.17**

Resolution enhancement: (a) test images, (b–d) subimages obtained in the optimization process, (e) resolution enhancement via simulation of the images created on the retina once the three subimages are shown on a high frame rate display, (f) Lanczos downsampling as a standard method, and (g) original high-resolution images. Note, that even though the resolution was reduced three times, the presented method was able to recover fine details in contrast to the standard downsampling method.

(a)  (b)  (c)  (d)

**FIGURE 15.20**

Backward-compatible anaglyph stereo (c) offers good depth-quality reproduction with similar appearance to the standard stereo image. The small amount of disparity leads to relatively high-quality images when stereo glasses are not used. To achieve equivalent depth-quality reproduction in the traditional approach significantly more disparity is needed (b), in which case the appearance of anaglyph stereo image is significantly degraded once seen without special equipment.

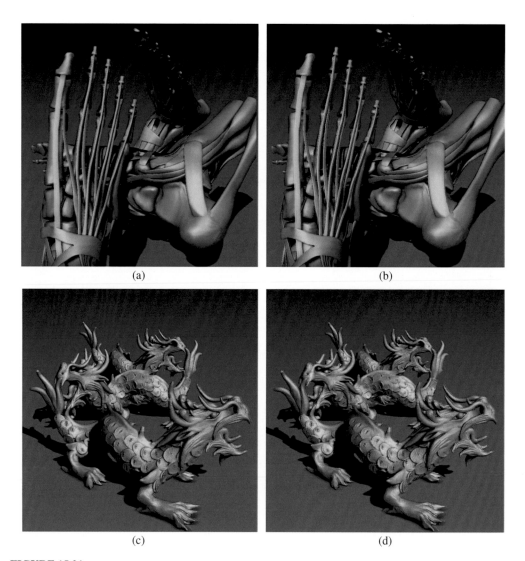

$$(a) \qquad\qquad (b)$$

$$(c) \qquad\qquad (d)$$

**FIGURE 15.21**

Depth enhancement using the Cornsweet illusion for two different scenes with significant depth range: (a,c) original images and (b,d) enhanced anaglyph images. A better separation between foreground and background objects and a more detailed surface structure depiction can be perceived in enhanced images.

where a biochemical agent $S$, activated by the input, interacts with a transducing agent $Z$ (e.g., a neurotransmitter) to produce an active complex $Y$ that carries the signal to the next processing stage. This active complex decays to an inactive state $X$, which in turn dissociates back into $S$ and $Z$. It can be shown (see Appendix in Reference [16]) that when the active state $X$ decays very fast, the dynamics of this system can be written as follows:

$$\frac{1}{\tau}\frac{dz}{dt} = \alpha(\beta - z) - \gamma Sz, \tag{2.5}$$

with the output given by $y(t) = \frac{\gamma}{\delta}S(t)z(t)$, where $S(t),z(t)$, and $y(t)$ represent the concentrations of $S$, $Z$, and $Y$, respectively. Parameters $\gamma$, $\delta$, and $\alpha$ denote rates of complex formation, decay to inactive state, and dissociation, respectively. The parameter $\tau$ dictates the overall speed of the reaction. This equation has been used in a variety of neural models, in particular to represent temporal adaptation, or gain control property, occurring for example through synaptic depression.

These fundamental equations can be now used to develop model equations. The retinotopic representation is a spatially two-dimensional representation. For simplicity, a one-dimensional model is built, and an index for each variable is used to denote its location in the retinotopic representation.

The first step in retinal processing is temporal adaptation realized by using Equation 2.5 as follows:

$$\frac{1}{\tau}\frac{dz_i}{dt} = \alpha(\beta - z_i) - \gamma(J + I_i)z_i, \tag{2.6}$$

where $z_i$ represents the concentration of a transducing agent, such as a neurotransmitter, for a neuron positioned at the $i$th retinotopic location, $J$ is a baseline signal determining baseline concentration, and $I_i$ is the external input (luminance) impinging on the $i$th retinotopic location. This transducing stage provides input to a retinal network that generates the output of the sustained, or parvocellular, cells. These cells have concentric center-surround receptive field organization and are described by the shunting equation (Equation 2.2) adapted as follows:

$$\frac{1}{\tau}\frac{dw_i}{dt} = -A_s w_i + (B_s - w_i)\sum_{j=i-n_s}^{i+n_s} G^{se}_{j-i}\Psi(J^s + I_j)z_j$$

$$-(D_s + w_i)\sum_{j=i-n_s}^{i+n_s} G^{si}_{j-i}\Psi(J^s + I_j)z_j. \tag{2.7}$$

The conductance values $g_d$ and $g_h$ controlling the depolarization and hyperpolarization of the membrane potential in Equation 2.2 have been replaced by depolarizing and hyperpolarizing terms that consist of the sum of baseline signal $J^s$ and external input $I_j$ at the $j$th retinotopic location multiplied by the concentration of the transducing agent $z_j$. These inputs are then weighted by Gaussian functions $G^{se}_{j-i}$ and $G^{se}_{j-i}$ of retinotopic distance $j-i$. The superscript $s$ denotes that the baseline input and the Gaussian function is for sustained cells. The superscripts $e$ and $i$ indicate that the Gaussian function relates to the excitatory and inhibitory components, respectively, of the difference of Gaussian receptive field profile of sustained cells. The parameter $n_s$ determines the spatial extent of the Gaussian in the retinotopic space. The output of each sustained cell is fed to an additive equation

$$\frac{dv_i}{dt} = \sigma(-v_i + \lambda([w_i - \Gamma_s]^+)^2),\qquad(2.8)$$

where the parameter $\sigma$ determines the decay rate of the activity, $\lambda$ is a constant representing the gain the the transformation of membrane potential to spike rate and $[.]^+$ is the half-wave rectification function with parameter $\Gamma_s$ representing the spiking threshold of the neuron. These equations together generate the sustained response profile shown in Figure 2.3, where a step input generates a peak that decays to a lower plateau due to the adaptation dynamics described by Equation 2.6.

Retinal cells with transient activities (magnocellular) obey a similar but simpler equation

$$\frac{dy_i}{dt} = -A_t y_i + (B_t - y_i) \sum_{j=i-n_t}^{i+n_t} G_{j-i}^{tse}\{I_j(t) - I_j(t - \delta)\}.\qquad(2.9)$$

Here the substraction of the delayed version of the input (delay is equal to $\delta$) implements a backward differentiation scheme and generates *transient* responses as shown schematically in Figures 2.2 and 2.3.

The postretinal network provides a lumped representation of visual areas beyond the retina (LGN and visual cortical areas). Here again, cells with sustained and transient activities are represented as those found in various visual cortical areas. However, unlike the retina where sustained and transient systems are isolated from each other, direct interactions between them are introduced here. In fact, it is known that parvo (sustained) and magno (transient) pathways remain separate until they reach visual cortex and constitute the main inputs to ventral and dorsal cortical streams. On the other hand, a variety of evidence shows that there exists a certain level intermixing between the pathways at the cortical level supporting the view of reciprocal interactions. While, both excitatory and inhibitory interactions can occur between these two systems, the focus here is on reciprocal inhibitory interactions, which are critical in explaining the motion deblurring phenomenon.

Accordingly, postretinal cells mainly driven by the parvocellular (sustained) pathway are described by the shunting equation

$$\frac{1}{\tau}\frac{dp_i}{dt} = -A_p p_i + (B_p - p_i)[\Phi(p_i) + 2v_i(t - \eta)] - p_i\left[\sum_{j=i-n_{pf};j!=i}^{i+n_{pf}} \Phi(p_j)\right.$$

$$\left. + \sum_{j=i-n_p}^{i+n_p} H_{j-i}^{pi} v_j(t - \eta - \kappa_p) + \sum_{j=i-n_p}^{i+n_p f} Q_{j-i}^{mp} m_j\right].\qquad(2.10)$$

The excitatory term contains positive feedback $\Phi(p_i)$ and the afferent signal from the sustained parvocellular pathway $v_i$, which is subject to delay $\eta$. The inhibitory term consists of inhibitory feedback $\Phi(p_j)$, inhibitory surround component of the afferent sustained signal $v_j$, and the inhibition from transient postretinal cells $m_j$. Terms $H_{j-i}^{pi}$ and $Q_{j-i}^{mp}$ are Gaussian weights of the inhibitory signals used to shape the receptive field.

Postretinal inhibitory interneurons carry inhibition from postretinal sustained cells to postretinal transient cells. They are described by the additive equation

$$\frac{dq_i}{dt} = -A_q q_i + B_q p_i,\qquad(2.11)$$

where $q_i$ is the activity of the $i$th postretinal interneuron.

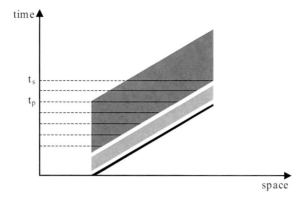

**FIGURE 2.4**

Space-time diagram depicting activities in the RECOD model for a single dot moving to the right with a constant speed. The black line represents the stimulus and the light-gray and dark-gray areas represent the transient and sustained responses, respectively. The length of horizontal lines across the sustained activity at different time instants indicate the spatial extent of motion blur for corresponding exposure durations. The spatial extent of motion blur increases linearly until the $t_p$ and remains constant until the time corresponding to the last visible point of the stimulus ($t_s$).

Finally, the equation for postretinal neurons mainly driven by magnocellular pathway is given by

$$\frac{dm_i}{dt} = -A_m m_i + (B_m - m_i) 2 [y_i(t)]^{++} -$$
$$m_i [ \sum_{j=i-n_p}^{i+n_p} H_{j-i}^{mi} [y_j(t - \kappa_m)]^{++} + \sum_{j=i-n_p}^{i+n_p} Q_{j-i}^{pm} q_j ], \qquad (2.12)$$

where $m_i$ is the activity of the $i$th postretinal transient cell and the function $[.]^{++}$ denotes full wave rectification generating on-off response characteristics of these transient cells. The parameter $\kappa_m$ is the relative delay of the intrachannel inhibition with respect to the excitatory signal.

## 2.7 Motion Deblurring

In order to understand why an isolated target generates motion blur but a dense array of targets do not, consider the following simplified scenario. Assume one-dimensional space and, as an isolated target, consider a single dot moving with a constant speed. The activities that would be generated by the model for such a stimulus are depicted in the space-time diagram of Figure 2.4. A dot moving at a constant speed corresponds to an oriented line in the space-time diagram. In response to this stimulus, a fast-transient and a slower-sustained activity are generated. The sustained activity decays slowly and, as a result, persists for a considerable time period (for example, 150 ms). As shown by the horizontal lines at

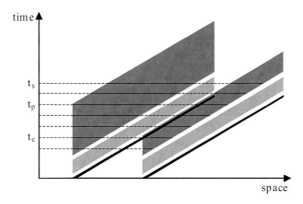

**FIGURE 2.5**

Space-time diagram depicting activities in the RECOD model for two dots moving to the right with a constant speed. The black lines represent the stimuli and the light-gray and dark-gray areas represent the transient and sustained responses, respectively. The length of horizontal lines across the sustained activity at different time instants indicate the spatial extent of motion blur for corresponding exposure durations. For the trailing dot, the spatial extent of motion blur increases linearly with exposure duration. The spatial extent of motion blur for the leading dot increases linearly like the trailing dot until $t_c$ when the transient activity of the trailing dot starts inhibiting the sustained activity of the leading dot.

various time instants in the figure, the spatial extent of motion blur increases as a function of exposure duration until it reaches a plateau at an exposure duration equal to the duration of visible persistence ($t_p$). The spatial extent of blur can be converted to an equivalent *duration of blur* by dividing it by the speed of the stimulus. If the duration of blur is plotted as a function of exposure duration, one would find a linear increase for the duration of blur with a slope of one until it reaches the critical value of the duration of visible persistence. After this value, it remains constant. This explains why an isolated target appears blurred and why the extent of spatial blur increases with exposure duration.

To understand why an array of dots does not appear blurred, consider the simple scenario where two dots travel at the same speed as shown in Figure 2.5. For the trailing dot, one obtains the same result as a single dot. However, as one can see from the figure, there will be a spatiotemporal overlap between the sustained activity of the leading dot and the transient activity of the trailing dot. According to the RECOD model, the transient-on-sustained inhibition (term $\sum_{j=i-n_p}^{i+n_p f} Q_{j-i}^{mp} m_j$ in Equation 2.10) will inhibit the sustained activity of the leading dot, thereby reducing the spatial extent of motion blur. If the duration of blur is plotted as a function of exposure duration, one can see that, initially, the duration of blur increases the same way for the leading and the trailing dots until the critical time $t_c$, where the transient activity of the trailing dot starts inhibiting the sustained activity of the leading dot. After $t_c$, the duration of motion blur for the leading dot remains constant, while the duration of blur for the trailing dot continues to increase.

These simple predictions of the RECOD model were tested by psychophysical experiments [5]. The results are shown in Figure 2.6. The left and right panels show data for the leading and trailing dots, respectively, for two spatial separation values between these two dots. For the trailing dot, the duration of blur increases linearly as a function of ex-

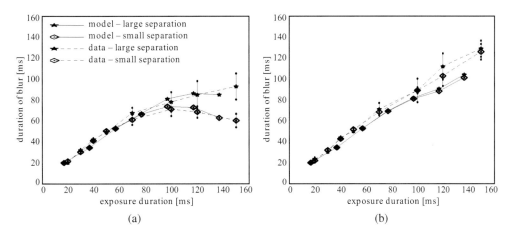

**FIGURE 2.6**

Motion deblurring data and model predictions for two dot-to-dot separations [17]: (a) the leading dot and (b) the trailing dot. ©1998 Elsevier

posure duration with little effect of dot-to-dot separation. In contrast, for the trailing dot, the linear increase is curtailed beyond exposure duration of approximately 100 ms. Moreover, this motion deblurring is more prominent when the dot-to-dot separation is small. As it can be seen in Equation 2.10, transient-on-sustained inhibition is weighted by a retinotopic distance-dependent Gaussian function $Q^{mp}_{j-i}$. Moreover, as the dynamics of differential equations in Section 2.6 suggests, the decay in the activity of the neurons is gradual. As a result, the effects depicted in Figures 2.4 and 2.5 occur in a more gradual way in the model compared to the schematic representations shown in these figures. In order to obtain quantitative predictions from the model and compare them to data, the model was simulated by using a numerical ordinary differential equation solver [17]. The results are plotted against the data in Figure 2.6. The good fit between model predictions and data provides support for the proposition that spatial extent of motion blur is controlled by inhibitory interactions between sustained and transient systems within retinotopic representations.

## 2.8   Conclusion

This chapter reviewed conditions that prevail under normal ecological vision by human observers. The sources of motion can be divided into exogenous and endogenous categories. The former refers to motion outside the observer; for example, when a car approaches the observer. The latter refers to motion resulting from the observer; for example, when the observer moves his/her head or eyes. This categorization is useful in that it allows to draw fundamental distinctions in resulting dynamic images. Namely, endogenous motion generates global retinotopic motion with a correlated retinotopic motion field, and it can be predicted and/or corrected by use of internal signals that control the motion of the observer. In contrast, exogenous motion is often localized in the visual field and lacks global correlation. Given these fundamental differences, it was suggested that the human nervous system uses a variety of strategies suited to each condition.

The second part of the chapter focused on the problem of motion blur and outlined a model of retino-cortical dynamics, called RECOD. It was shown that inhibitory interactions within retinotopic representations can reduce the spatial extent of motion blur. In this ongoing research, the complementary problem of moving ghosts is investigated. As shown in Figure 2.1, retinotopic representations are transformed into nonretinotopic representations — a fundamental factor in this transformation is the establishment of a new nonretinotopic reference frame. It was further suggested that motion grouping plays a critical role in establishing these reference frames. Gestalt psychologists noticed the tendency of stimuli sharing a common motion direction to group together and called this *the grouping law of common faith*. As mentioned above, for exogenous sources of motion, the resulting retinotopic patterns are local and necessitate a segmentation and a grouping process to identify and isolate each locus of common motion. Grouping by common faith can achieve this goal thereby leading to the transformation of retinotopic representations to nonretinotopic ones. Motion deblurring in retinotopic representations along with dynamic form computation in nonretinotopic representations can together explain how the visual system solves simultaneously the motion blur and moving ghosts problems.

## Acknowledgment

Figure 2.6 is reprinted from Reference [17] with the permission from Elsevier.

## References

[1] J. Gibson, "Ecological optics," *Vision Research*, vol. 1, no. 3–4, pp. 253–262, October 1961.

[2] H. Collewijn and J. Smeets, "Early components of the human vestibulo-ocular response to head rotation: Latency and gain," *Journal of Neurophysiology*, vol. 84, no. 1, pp. 376–389, July 2000.

[3] A. Charpentier, "Recherches sur la persistence des impressions retiniennes et sur les excitations lumineuses de courte duree," *Arch Ophthalmol*, vol. 10, pp. 108–135, 1890.

[4] D. Burr, "Motion smear," *Nature*, vol. 284, no. 5752, pp. 164–165, March 1980.

[5] S. Chen, H. Bedell, and H. Ogmen, "A target in real motion appears blurred in the absence of other proximal moving targets," *Vision Research*, vol. 35, no. 16, pp. 2315–2328, August 1995.

[6] H.E. Bedell, J. Tong, and M. Aydin, "The perception of motion smear during eye and head movements," *Vision Research*, vol. 50, no. 24, pp. 2692–2701, December 2010.

[7] J. Tong, S.B. Stevenson, and H.E. Bedell, "Signals of eye-muscle proprioception modulate perceived motion smear," *Journal of Vision*, vol. 8, no. 14, pp. 7.1–6, October 2008.

[8] J. Tong, M. Aydin, and H.E. Bedell, "Direction and extent of perceived motion smear during pursuit eye movement," *Vision Research*, vol. 47, no. 7, pp. 1011–1019, March 2007.

[9] J. Tong, S. Patel, and H. Bedell, "The attenuation of perceived motion smear during combined eye and head movements," *Vision Research*, vol. 46, no. 26, pp. 4387–4397, December 2006.

[10] H. Bedell and S. Patel, "Attenuation of perceived motion smear during the vestibulo-ocular reflex," *Vision Research*, vol. 45, no. 16, pp. 2191–2200, July 2005.

[11] H. Bedell, S. Chung, and S. Patel, "Attenuation of perceived motion smear during vergence and pursuit tracking," *Vision Research*, vol. 44, no. 9, pp. 895–902, April 2004.

[12] H. Ogmen, "A theory of moving form perception: Synergy between masking, perceptual grouping, and motion computation in retinotopic and non-retinotopic representations," *Advances in Cognitive Psychology*, vol. 3, no. 1–2, pp. 67–84, July 2007.

[13] H. Ogmen and M.H. Herzog, "The geometry of visual perception: Retinotopic and nonretinotopic representations in the human visual system," *Proceedings of the IEEE*, vol. 98, no. 3, pp. 479–492, March 2010.

[14] H. Ogmen, "A neural theory of retino-cortical dynamics," *Neural Networks*, vol. 6, no. 2, pp. 245–273, 1993.

[15] S. Grossberg, "Nonlinear neural networks – Principles, mechanisms, and architectures," *Neural Networks*, vol. 1, no. 1, pp. 17–61, 1988.

[16] M. Sarikaya, W. Wang, and H. Ogmen, "Neural network model of on-off units in the fly visual system: Simulations of dynamic behavior," *Biological Cybernetics*, vol. 78, no. 5, pp. 399–412, May 1998.

[17] G. Purushothaman, H. Ogmen, S. Chen, and H. Bedell, "Motion deblurring in a neural network model of retino-cortical dynamics," *Vision Research*, vol. 38, no. 12, pp. 1827–1842, June 1998.

# 3

# *Image and Video Quality Assessment: Perception, Psychophysical Models, and Algorithms*

**Anush K. Moorthy, Kalpana Seshadrinathan, and Alan C. Bovik**

## 3.1 Introduction

As of June 2010, Internet video accounted for over one-third of all consumer Internet traffic, and predictions indicate that Internet video will account for more than 57% of all consumer traffic by 2014 [1]. Predictions also indicate that advanced Internet video, which includes high-definition and stereoscopic video, will increase by a factor of 23 by 2014, compared to 2009. The sum of all forms of video, including television, video on demand, Internet, and peer-to-peer video streaming, will constitute more than 91% of global consumer traffic [1]. At the end of 2010, video accounted for almost 50% of mobile data traffic, and it is predicted that more than 75% of mobile traffic will be video in 2015 [2]. These numbers indicate that video – either on mobile devices or otherwise – is fast becoming an integral part of a daily life and will continue to do so for the next few years as well.

Today's average user is spoiled for choice, since visual entertainment is available at the touch of a button over a variety of devices that encompass a range of resolutions and which are capable of displaying visual stimuli with varying degrees of acceptability. The palatability of a visual signal is not a function of the display device alone however, since the stimulus that the user receives passes through a variety of stages that could possibly induce certain artifacts in the signal that make the stimulus unacceptable or unpalatable. Think of the blocky video or the sudden freezes that one experiences when viewing movies on the Internet. The former is due to the heavy quantization (and hence high compression) that was used in the creation and storage of the video, while the latter is due to a decrease in the allocated bandwidth leading to an empty buffer. In either case, the introduction of these compression or frame-freeze artifacts reduce the palatability of the visual stimulus. Visual quality assessment refers to gauging such palatability of a visual stimulus. Since humans are the ultimate viewers of such stimuli, human opinion on quality is representative of the palatability of visual signals. Such human opinion on visual quality is generally gauged using a large-scale study where human observers view and rate visual stimuli (images/videos) on a particular rating scale (say, 1 to 5, where 1 is bad and 5 is excellent) and such assessment of visual quality is referred to as *subjective quality assessment*. Subjective assessment of quality is not only time-consuming and cumbersome, but is also impractical and hence there is a need to develop algorithms that are capable of assessing the perceptual quality of a visual stimulus such that the scores generated by the algorithm correlate well with subjective opinion. Such automatic assessment of visual quality is referred to as *objective quality assessment*.

Automatic image quality assessment (IQA) and video quality assessment (VQA) algorithms are generally classified into three broad categories based on the amount of information that the algorithm has access to:

- Full-reference (FR) approaches, where the algorithm has access to not only the distorted stimulus whose quality is to be assessed but also the clean, pristine, reference stimulus with respect to which such quality assessment is performed.

- Reduced-reference (RR) approaches, where apart from the distorted stimulus, the algorithm has access to *some* information regarding the original, pristine stimulus (for example, from a side-channel), but not the original reference stimulus itself.

- No-reference (NR) approaches, which only have access to the distorted stimulus.

The three approaches are summarized in Figure 3.1. The reader will readily infer that as one moves from the FR to the NR realm, the difficulty of the problem increases, and it is no surprise that most research has revolved around the development of FR algorithms. Although NR algorithms are the holy grail of quality assessment (QA) research, development of FR algorithms lends one insight into various aspects of QA algorithm design and hence FR QA algorithms pave the way for the development of state-of-the-art NR QA algorithms. Finally, RR QA algorithms remain of interest, not only as a stepping-stone toward NR QA but also as feasible solutions to the problem.

Apart from predicting perceptual quality, QA algorithms can be deployed for a variety of purposes. For example, a QA algorithm can be used for the perceptual optimization of

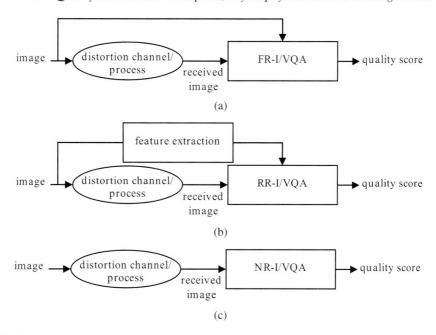

**FIGURE 3.1**

Taxonomy of quality assessment approaches. Full-reference approaches have access to both the original and the distorted signal, reduced reference approaches have access to some information regarding the original (through the supplementary channel here for example), and no-reference approaches have access only to the distorted stimulus whose quality is be assessed. (a) Full-reference quality assessment, (b) reduced-reference quality assessment, and (c) no-reference quality assessment.

a multimedia service; that is, transmission/display parameters of the signal may be varied in order to maximize perceptual quality, and hence the quality of experience (QoE). Other applications include benchmarking image processing algorithms, utilization of perceptual quality measures as optimization objectives for image enhancement algorithms, rate control using quality, rate-distortion optimization, visually lossless compression and so on.

Having described FR QA algorithms, the reader would presume that the simple mean-squared-error (MSE), which has been an engineer's tool of choice for comparison of two signals is an ideal candidate for FR visual quality assessment. The MSE between two $N$-dimensional vectors $x$ and $y$ is defined as

$$MSE = \frac{1}{N} \sum_{i=1}^{N} (x_i - y_i)^2. \tag{3.1}$$

A low MSE value indicates that the two vectors are similar. In order to follow a convention where a higher value indicates greater similarity, the peak signal-to-noise ratio (PSNR) is utilized. The PSNR is defined as follows:

$$PSNR = 10 log_{10} \left( \frac{L^2}{MSE} \right), \tag{3.2}$$

where $L$ is the dynamic range of the pixel values (for example, $L = 255$ for grayscale images). Throughout this chapter, MSE and PSNR are used interchangeably.

**FIGURE 3.2**

Example of why MSE is a poor measure of quality. The distortion is visible on the face of the woman, but not in the background on the chair or on the scarf where MSE still accounts for it.

The engineer's attachment to the MSE is not without reason. The MSE has several elegant properties, including but not limited to, simplicity, memoryless-ness, convexity, symmetry, and differentiability. With so much going for it, MSE should have been an ideal FR measure of quality, however, it is not [3], [4]. The major reason for this is that while MSE is a good indicator of differences between two signals, in the case of visual signals existence of such differences does not imply that these differences are perceived by the human observer, since not all differences are equally significant for the human. This is best illustrated by the image in Figure 3.2. The image has been distorted uniformly with additive white noise; however, the noise on the face of the woman is prominent, while one would be hard-pressed to quantify the distortion on the chair in the background or the scarf. MSE, however, would identify the distortion in the background and quantify it with equal weight as the distortion on the face, thereby making it a poor measure of perceptual quality. This phenomena, where the content of the image *masks* the presence of distortion, is referred to as contrast masking and will be discussed in the next section. It is hence that one needs to design algorithms that are capable of assessing the human perceived quality of a visual signal.

Quality assessment algorithms are of tremendous practical interest, and hence their design has attracted tremendous research interest. Since humans are the ultimate receivers of these visual signals, it is only logical that researchers first approached the QA problem by attempting to model the various features of the human visual system (HVS). Hence,

Section 3.2 presents various salient features of the HVS. Section 3.3 summarizes how one would evaluate a quality assessment algorithm and directs the reader to many publicly available quality assessment databases for images and videos. Then, Section 3.4 describes a number of image and video quality assessment algorithms, not only those based on the HVS but also those that are based on feature-extraction and natural scene statistics as well. Since FR QA has been an active area of research for some time, leading FR QA approaches are discussed, and since it is believed that NR approaches for QA are rapidly achieving competitive performance, approaches to NR QA are summarized as well. Section 3.5 outlines current and future areas of interest in the field of quality assessment research. Finally, this chapter concludes with Section 3.6.

## 3.2   Computational Modeling of Human Vision

The goal of perceptual quality assessment is to model and predict human behaviour in assessing the quality of multimedia signals. With this goal in mind, all IQA/VQA algorithms utilize models of the properties of human vision in some form. This section discusses certain basic properties of human vision and corresponding computational models that have found their way into IQA/VQA algorithms. As shown in this chapter, certain algorithms utilize statistical models of natural images/video in developing an IQA metric. However, since human vision has evolved in response to viewing these natural scenes, natural scene statistical models are also hypothesized to be dual of human vision models, and this duality has been studied in the literature [5].

Vision is one of the most important senses that humans use to perceive the outside world. The HVS is an extremely complex system that has evolved to perform a number of different functions using visual stimuli that are acquired using the optical components of the eyes. Vision in humans and various animals has been studied by neuroscientists for centuries and while tremendous progress has been achieved over the years, visual processing is still not completely understood.

Within the context of visual quality assessment, several properties of human vision are particularly relevant and research has focused on adoption of computational models of these in IQA/VQA. Two types of models of the HVS – behavioral models and neuronal models – have both been used extensively in IQA/VQA. Behavioral models attempt to use data derived from experiments performed with human users and examples include determination of threshold sensitivity models for luminance masking and contrast sensitivity. Neuronal models are derived from experimental recordings from neurons in animals and such data have also been used in IQA/VQA [6], [7]. Several models of the HVS that have been incorporated into IQA/VQA algorithms have focused on modeling the early stages of visual processing of which a fairly good understanding has been reached over the years and this includes processing that occurs in the retina and the so-called area V1 of the visual cortex. More recent algorithms attempt to focus on some of the higher-order processing that occurs farther downstream, including in the extra-striate cortex, to account for motion perception, saliency, and so on [7], [8], [9].

**FIGURE 3.3 (See color insert.)**

The effect of the contrast masking property of human vision on visual quality using JPEG compression: (a) original *Building* image, (b) its distorted version with $MSE = 134.61$, (c) original *Caps* image, and (d) its distorted version with $MSE = 129.24$. The distorted images clearly differ in their visual quality, although they have similar MSE with respect to the original.

Three properties of human vision – luminance sensitivity, contrast sensitivity, and contrast masking – have received a good deal of attention in the literature. Human perception of brightness is not linear, and a well-known law governing perception known as the Weber-Fechner law stipulates that over a large dynamic range, and for many parameters, the threshold of discrimination between two stimuli increases linearly with stimulus intensity. A commonly used model to account for Weber's law stipulates that the ratio of discrimination threshold to the input stimulus intensity is a constant over a specified range of stimulus intensity. The sensitivity of the human eye for sinusoidal illuminance changes as a function of spatial frequency was studied in Reference [10], and it was shown that human perception of brightness follows Weber's law over a broad range of stimulus strength.

Spatial contrast sensitivity of vision refers to differences in sensitivities to stimuli of varying spatial frequencies of equal strength. The spatial contrast sensitivity function of human vision shows a bandpass shape with reduced sensitivity to low and high spatial frequencies [11], [12]. In fact, the reduced sensitivity of human vision to higher spatial frequencies is exploited in lossy compression of images and video using Joint Photographic Experts Group (JPEG) or Moving Picture Experts Group (MPEG) coding to provide substantial gains in compression efficiency.

Contrast masking, also termed as texture masking, refers to the reduction in visibility of one signal component due to the presence of a similar signal component at adjacent spatial locations, spatial frequency, or orientation. Contrast masking, as relevant to image quality, is illustrated in Figure 3.3. The distorted images shown in Figures 3.3b and 3.3d have approximately the same MSE with respect to the original images shown in Figures 3.3a and 3.3c, respectively. However, the visibility of artifacts in these images are clearly very different. The reason for this difference is contrast masking in vision, where strong edges and textures in the *Buildings* image mask the visibility of distortions. However, artifacts are clearly visible in the predominantly smooth *Caps* image. This simple example clearly shows the importance of accounting for contrast masking in visual quality assessment.

Contrast masking has been studied extensively in the literature and psychophysical studies of this phenomenon that have influenced QA can be found, for example, in References [13], [14], [15], [16], and [17]. Masking models often take the form of a divisive normalization of the energy/intensity of the test or noise signal by the reference image and predict the level of distortion to which an image can be exposed before the alteration is apparent to a human observer. Recent work has also studied the relationship between masking models and IQA indices, such as SSIM and VIF, which are not derived using the HVS models, but based on hypotheses regarding human response to structural distortions and natural scene statistical models, respectively [18]. Most existing IQA/VQA methods utilize some modeling of the contrast masking property of human vision, which is a critical component in achieving the ultimate objective of matching human perception of image/video quality.

Several human vision-based VQA algorithms incorporate elaborate models of spatial properties of human vision. However, until recently, VQA algorithms have not successfully incorporated meaningful models of temporal vision that account for motion processing. Many algorithms utilized computational models of the first stage of temporal processing that occurs in area V1 of the visual cortex. These models typically have the form of one or two separable linear filters applied along the temporal dimension of the video. The two filters are usually lowpass and bandpass, to account for the sustained and transient mechanisms that have been identified in human vision [19], [20]. However, the HVS is extremely sensitive to motion and one of its critical functions is the computation of motion information (speed and direction of motion) from visual inputs. It is well known that a large number of neurons in visual area MT/V5, which is part of the extra-striate cortex, are directionally sensitive, and it is believed that area MT plays a significant role in motion perception [21]. This sensitivity to motion is clearly revealed in QA since humans are extremely sensitive to motion-related artifacts in video, such as ghosting, jerkiness, mosquito noise, and motion compensation mismatches. Recently, models of area MT/V5 have been utilized to develop a sophisticated VQA algorithm known as the motion-based video integrity evaluation (MOVIE) index [7], which is described later in this chapter.

Since human sensitivity to color is generally lower than the sensitivity to changes in the luminance information, most IQA/VQA models primarily focus on the luminance channel of images/video with only rudimentary inclusion of color information (for example, a similar or simplified version of the model is applied to the chrominance channels) [22], [23]. Several HVS based algorithms use different color spaces, such as the CIELAB and CIEYUV color spaces, that attempt to approximate the spectral response properties of human vision. In addition to these, the effect of viewing distance on perceived quality has also been studied [22], [24]. The spatial resolution of the HVS is nonuniform with increased resolution at the fovea and with resolution dropping off with eccentricity in the peripheral regions of vision. Many VQA algorithms incorporate models of the viewing distance and the resolution of the display to convert pixel units into units of cycles per degree of viewing angle, which are then used in the rest of the algorithm. A conservative approach taken by several models is to assume the same resolution for the entire image at the specified viewing distance since human subjects can achieve this resolution using eye movements.

## 3.3 Subjective Quality and Performance Evaluation

Before discussing various visual quality assessment algorithms, the following describes how the performance of a QA algorithm is gauged. This section assumes an algorithm that is capable of producing a measure of quality when given as input the image or video whose quality is to be assessed (and the original for FR algorithms, or additional information for RR algorithms). Given such an algorithm, one seeks to describe its performance in terms of its correlation with human visual perception.

Human perception of quality is generally gauged by large-scale subjective studies in which a number of human observers view and rate distorted visual stimuli on a particular rating scale. Subject rejection procedures are applied in order to reject those subjects whose opinions are markedly different from the average observer, and the opinion scores are averaged across the remaining subjects in order to produce a mean opinion score (MOS), which is representative of the perceived quality of the stimulus. Such subjective studies are generally carried out in controlled environments, where the stimuli are viewed at a fixed viewing distance on a calibrated monitor. Most of the human studies listed below follow the recommendations of the International Telecommunications Union (ITU) [25].

Subjective studies can be single stimulus or double stimulus in nature. Single-stimulus studies are those in which only one stimulus is seen by the subject at a time, and the subject is required to rate that stimulus on its own. In double-stimulus studies, the subject is required to rate the (possibly) distorted stimulus with respect to the pristine reference stimulus, which is displayed as well (usually in a side-by-side arrangement) [25]. It should be obvious that double-stimulus studies generally take twice as much time as single-stimulus studies, which is one reason why single-stimulus studies are common in the literature. Another reason is that single-stimulus studies better simulate the real-world experience of a consumer, who ordinarily does not compare the watched videos side-by-side unless purchasing a display device, such as a television.

A popular methodology that has been adopted by many researchers is the single-stimulus with hidden reference methodology [26], [27]. Here, a single-stimulus study is carried out, and the pristine original reference stimulus is embedded among the stimuli that the subject sees; the score that the subject gives this reference is representative of his or her bias, and in these studies, the score given to the distorted stimulus is subtracted from that given to the reference to form the differential mean opinion score (DMOS), which is representative of the perceived quality of the stimulus – a smaller DMOS indicates better quality.

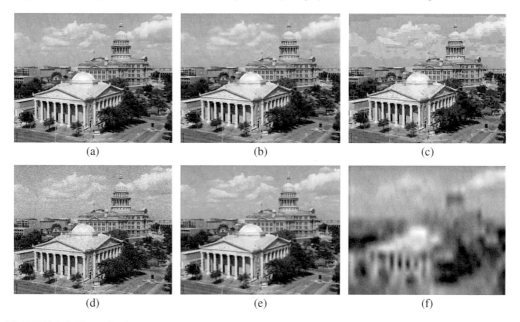

(a)          (b)          (c)

(d)          (e)          (f)

**FIGURE 3.4  (See color insert.)**

Example distortions from the LIVE IQA database: (a) original image, (b) JPEG2000 compression, (c) JPEG compression, (d) white noise, (e) blur, and (f) Rayleigh fading.

Given a large database of (distorted) visual stimuli with associated DMOS/MOS scores, the scores produced by the algorithm under test for these stimuli are correlated with the MOS/DMOS. Generally, researchers use the following measures of correlation between MOS and algorithm scores: Spearman's rank ordered correlation coefficient (SROCC) and linear (Pearson's) correlation coefficient (LCC) [28], [29]. Algorithm scores could correlate nonlinearly or linearly with MOS/DMOS and since SROCC is a nonlinear measure of correlation, SROCC can be directly applied. However, LCC is a linear measure of quality and hence algorithm scores need to be passed through a nonlinearity (referred to as a logistic function) prior to LCC computation. A similar procedure is applied when computing the root mean squared error (RMSE) between the (linearized) algorithm scores and MOS/DMOS. A value close to one for SROCC and LCC and a value close to zero for RMSE indicate good performance in terms of correlation with human perception. The outlier ratio has been sporadically used as a measure of performance as well [29]. Apart from correlations and errors, in order to gauge whether the performance of two algorithms are statistically different, that is, one algorithm is provably better than the other, researchers have used the F-statistic between the residuals obtained from the predicted MOS/DMOS and actual MOS/DMOS [30]. It is here believed that a statistical analysis of performance is a must when evaluating algorithm performance, since higher correlation numbers have no merit, unless backed by the approval of statistical superiority.

Although human studies are time-consuming and cumbersome, their need as a tool for benchmarking QA algorithms emboldens researchers to undertake such studies. In the past, the video quality experts group (VQEG) performed such a large-scale human study in order to benchmark the performance of VQA algorithms [29]. Unfortunately, owing to its poor design [27], [26], none of the algorithms demonstrated better performance than the

**TABLE 3.1**
Summary of various publicly available IQA databases.

| Database | Reference Images | Distorted Images | Distortion Categories |
|---|---|---|---|
| LIVE [30] | 29 | 779 | 5 |
| Cornell-A57 [32] | 3 | 54 | 6 |
| IVC [34] | 10 | 235 | 4 |
| Toyama-MICT [33] | 14 | 168 | 2 |
| TID2008 [35] | 25 | 1700 | 17 |
| CSIQ [36] | 30 | 866 | 6 |

simple PSNR. The VQEG conducted many other studies, and although some of the data have recently been released for research purposes, copyright issues with regard to content abound [31]. In general, a database has value only if it is publicly available for future benchmarking. A closed dataset, with restricted submissions being evaluated, does not help in algorithm development. By recognizing this fact, various databases have been proposed for image quality assessment in the recent past [30], [32], [33], [34], [35], [36]. One of the first IQA databases was proposed by researchers at the Laboratory for Image and Video Engineering (LIVE) [30], and the LIVE IQA database remains one of the most popular databases for benchmarking, with more than 1500 download requests as of April 2011. The LIVE IQA database was created using 29 reference images, which were distorted using five different distortions – JPEG2000 compression, JPEG compression, white noise, blur, and a Rayleigh fading channel model – leading to a total of 779 distorted images. In order to illustrate how each of these distortions affects visual perception, Figure 3.4 shows a reference image and its various distorted versions.

Although the LIVE IQA database is a de facto standard to measure the performance of an IQA algorithm, many other databases have followed. Table 3.1 lists the various IQA databases that are publicly available and details some specifics about these databases, including the number of reference images used, the number of distorted images, and the various distortion categories that were used to simulate the distorted images.

**FIGURE 3.5 (See color insert.)**

Sample frames from the LIVE VQA database.

Unfortunately, despite the existence of various public databases for benchmarking IQA algorithms, this abundance is not seen in VQA. As mentioned above, the VQEG was primarily responsible for creating one of the first publicly available databases for VQA, and the VQEG phase-I dataset, as it is called, was until recently the de facto standard on which VQA algorithms were tested. However, owing to the poor design of the database, algorithms and humans had difficulty in making perceptual judgments on the quality of the distorted videos. Recently, researchers at LIVE released a database for VQA [27] that complements the popular LIVE IQA database. The LIVE VQA database consists of 10 reference, uncompressed, videos in YUV420p format and 150 distorted videos that were distorted using a variety of distortion techniques, such as MPEG4 compression, H.264 compression, IP packet-losses, and wireless packet-losses [27]. Figure 3.5 shows a subset of the sample frames from the LIVE VQA database in order to give the reader a visual example of the content that the database incorporates.

The LIVE VQA database, which is publicly available for download at no cost to researchers (see Reference [27] for details), also provides the associated DMOS for each of these distorted videos, and it is fast becoming the database on which VQA algorithms are benchmarked. The LIVE VQA database is extremely challenging and clearly demonstrates that PSNR is a poor measure of perceptual quality (see next section). The diverse distortions, along with the carefully conducted human study and the lack of good publicly available VQA databases, makes the LIVE VQA an invaluable asset to researchers in the field of video quality assessment.

## 3.4    Algorithmic Measurement of Perceptual Quality

Having described the human visual system mechanisms that are relevant to visual quality assessment techniques and having detailed how one assesses the performance of a QA algorithm, this section summarizes some QA algorithms. The summary is broadly divided into image and video quality assessment algorithms.

### 3.4.1    Image Quality Assessment

A taxonomy of image quality assessment (IQA) algorithms includes those algorithms that seek to model HVS mechanism, those that are based on feature extraction procedures, as well as those that seek to model natural-scene statistics (NSS), which is seen as a dual to the problem of HVS modeling. Figure 3.6 provides a classification of the various IQA algorithms that are summarized below.

#### 3.4.1.1    FR IQA: Human Visual System-Based Approaches

FR QA algorithms based on the HVS generally proceed in the following manner. The image is decomposed using a scale-space orientation decomposition, which is then followed by a masking model that accounts for contrast masking in each of the frequency channels. The masked coefficients of the reference and distorted image are then compared

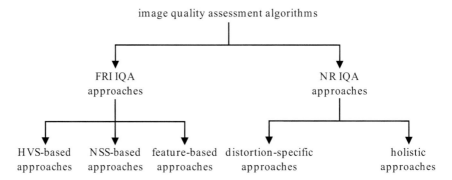

**FIGURE 3.6**

Taxonomy of image quality assessment algorithms summarized in this chapter.

using an error measure, which is generally pooled using a Minkowski summation to produce an error (quality) score. The major difference between the approaches is the choice of the decomposition basis, the parameters of the masking models, and the pooling approach used for collapsing spatial error scores.

One of the first attempts at modeling human visual system mechanisms for quality assessment was presented in Reference [37]. In this pioneering approach, the authors utilize luminance adaptation and a contrast sensitivity function. However, the model incorporates only a single visual channel, as opposed to a set of channels over the frequency space, and fails to model contrast masking. As noted in Reference [38], the method is equivalent to a weighted MSE, and hence does not perform very well.

The so-called visual difference predictor (VDP) [39] applies a pointwise nonlinearity to the images to model the inherent nonlinearity of visual sensitivity and perception of lightness to luminance, which is followed by a contrast sensitivity function. The image is then decomposed using a modified version of the Cortex transform [40], which models the initial stage of the human detection mechanisms. A masking procedure ensues. The dependence of probability of detection to stimulus contrast is modeled using a psychometric function, and a probability summation is undertaken to produce a quality score.

Reference [41] proposed DCTune as a measure of perceptual quality so that one may design image-dependent perceptual quantization values for DCT-based compression algorithms such as JPEG compression. Luminance masking and contrast masking are modeled in the DCT domain and the obtained coefficients are converted into units of just noticeable differences (JNDs), which are then pooled using probability summations across space and frequencies. As discussed in the original work, the model may be used to design quantization parameters for different images.

The other popular model for HVS-based IQA is presented in Reference [6], where the reference and test images are decomposed using a hex-quadrature mirror filter. The responses are squared and contrast normalization is performed. The vector distance between the reference and test image forms a measure of error and hence of quality.

Reference [42] proposed an FR IQA algorithm that first applies a set of lowpass filters and fan filters to the input images; this leads to thirty filtered versions with five spatial frequency bands and six orientations. In the next step, a modified Peli's contrast is computed.

A threshold elevation computation ensues, followed by a pooling stage that produces the perceptual error measure (PEM). The authors evaluate their approach using human opinion scores from a small-scale study and demonstrate that this measure does better than PSNR in terms of correlation with human perception.

Visual signal-to-noise ratio (VSNR) [43] aims at evaluating the effect of *supra-threshold* distortions using parameters for the HVS model derived from experiments where the stimulus was an actual image as against sinusoidal gratings or Gabor patches. First, a difference image from the reference and distorted images is computed and subjected to a discrete wavelet transform. Within each subband, the visibility of distortions is evaluated by comparing the contrast of the distortion to the detection threshold and then computing the root mean squared contrast of the error signal. Finally, using a strategy inspired from what is termed as *global precedence* in the HVS, a global precedence preserving contrast is computed. The final index is a linear combination of the error signal and contrast.

### 3.4.1.2    FR IQA: NSS-Based Approaches

Natural images (that is, any image that is not produced synthetically on a computer) follow certain statistical properties that do not vary with varying content. For example, the image of a mountain range and the close-up of a dog, both may exhibit the property that their power spectral density falls of as $1/f^\gamma$. Such content-independent statistical properties of natural scenes are referred to as natural scene statistics (NSS), and researchers have been trying to understand the relationship between the statistical properties of natural scenes and the properties of the human visual system [44]. The hypothesis is that since humans evolved in natural environments, the HVS is optimized to extract the statistical properties of natural scenes, and hence modeling scene statistics would give one insight into human visual processing. NSS-based QA algorithms utilize such NSS properties for quality assessment.

- Single-scale structural similarity index (SS-SSIM)

For two image patches drawn from the same location of the reference and distorted images, that is, $\mathbf{x} = \{x_i | i = 1, 2, \ldots, N\}$ and $\mathbf{y} = \{y_i | i = 1, 2, \ldots, N\}$, respectively, SS-SSIM evaluates luminance, contrast, and structure as follows [45]:

$$l(\mathbf{x}, \mathbf{y}) = \frac{2\mu_x\mu_y + C_1}{\mu_x^2 + \mu_y^2 + C_1}, \tag{3.3}$$

$$c(\mathbf{x}, \mathbf{y}) = \frac{2\sigma_x\sigma_y + C_2}{\sigma_x^2 + \sigma_y^2 + C_2}, \tag{3.4}$$

$$s(\mathbf{x}, \mathbf{y}) = \frac{\sigma_{xy} + C_3}{\sigma_x\sigma_y + C_3}, \tag{3.5}$$

where $C_1$, $C_2$, and $C_3$ ($C_3 = C_2/2$) are small constants included to prevent instabilities from arising when the denominator tends to zero. The terms $\mu_x$, $\mu_y$, $\sigma_x^2$, $\sigma_y^2$, and $\sigma_{xy}$ are the means of $\mathbf{x}$ and $\mathbf{y}$, the variances of $\mathbf{x}$ and $\mathbf{y}$, and the covariance between $\mathbf{x}$ and $\mathbf{y}$, respectively, computed using a sliding window approach. The window that is commonly used is a $11 \times 11$ circular-symmetric Gaussian weighting function $w = \{w_i | i = 1, 2, \ldots, N\}$, with standard deviation of 1.5 samples, normalized to sum to unity ($\sum_{i=1}^{N} w_i = 1$).

Finally, the SSIM index between signal $\mathbf{x}$ and $\mathbf{y}$ is defined as:

$$SSIM(\mathbf{x},\mathbf{y}) = \frac{(2\mu_x\mu_y + C_1)(2\sigma_{xy} + C_2)}{(\mu_x^2 + \mu_y^2 + C_1)(\sigma_x^2 + \sigma_y^2 + C_2)}. \tag{3.6}$$

The above computation produces a quality map of scores, which is spatially collapsed using the simple mean in order to produce a final quality measure for the image, although many other pooling strategies have been explored [8], [46], [47].

- Multi-scale structural similarity index (MS-SSIM)

Since images are multi-scale and the human visual system processes visual information at multiple scales, Reference [48] proposed to evaluate image quality at multiple resolutions using the so-called multi-scale SSIM (MS-SSIM) index. This method performs quality assessment over multiple scales of the reference and distorted image patches (the signals defined as $\mathbf{x}$ and $\mathbf{y}$ in the previous discussion on SS-SSIM) by iteratively lowpass filtering and downsampling the signals by a factor of two. The original image scale is indexed as 1, the first down-sampled version is indexed as 2, and so on. The highest scale $M$ is obtained after $M - 1$ iterations.

At each scale $j$, the contrast (Equation 3.4) and structure (Equation 3.5) comparison terms are calculated and denoted as $c_j(\mathbf{x},\mathbf{y})$ and $s_j(\mathbf{x},\mathbf{y})$, respectively. The luminance comparison (Equation 3.3) term is computed only at scale $M$ and is denoted as $l_M(\mathbf{x},\mathbf{y})$. The overall SSIM evaluation is obtained by combining the measurement over scales:

$$SSIM(\mathbf{x},\mathbf{y}) = [l_M(\mathbf{x},\mathbf{y})]^{\alpha_M} \cdot \prod_{j=1}^{M} [c_j(\mathbf{x},\mathbf{y})]^{\beta_j} \cdot [s_j(\mathbf{x},\mathbf{y})]^{\gamma_j}. \tag{3.7}$$

The highest scale used here is $M = 5$. The exponents $\alpha_j$, $\beta_j$, and $\gamma_j$ are selected such that $\alpha_j = \beta_j = \gamma_j$ and $\sum_{j=1}^{M} \gamma_j = 1$. Variants of SSIM include SSIM computation in the complex-wavelet domain [49], [50], a gradient-based approach [51], and techniques based on pooling three perceptually important parameters [52], [53].

- Visual information fidelity (VIF)

Reference [54] models wavelet coefficients, obtained from a subband decomposition of the reference and distorted images, using the Gaussian scale mixture (GSM) model [55] and utilizes the properties of natural scenes [56] in order to produce a quality measure. Each subband in the reference is modeled as $C = S \cdot U$, where $S$ is a random field (RF) of scalars and $U$ is a Gaussian vector RF. The distortion model is $D = GC + v$, where $G$ is a scalar gain field and $v$ is additive Gaussian noise RF. It is assumed that the distorted and source images pass through the human visual system, and the HVS uncertainty is modeled as *visual noise*, that is, $N$ and $N'$ for the source and distorted image, respectively, where $N$ and $N'$ are zero-mean uncorrelated multivariate Gaussians. The method then computes $E = C + N$ and $F = D + N'$ and evaluates the following criterion:

$$VIF = \frac{\sum_{j \in all\,subbands} I(C^j; F^j | s^j)}{\sum_{j \in all\,subbands} I(C^j; E^j | s^j)}, \tag{3.8}$$

where $I(X; Y | Z)$ is the conditional mutual information between $X$ and $Y$, conditioned on $Z$. The term $s^j$ is a realization of $S^j$ for a particular image, where $j$ is the index that runs through all the subbands in the decomposed image.

Although VIF is directly based on NSS properties, it may not be clear to the reader why the SSIM indices are classified as NSS-based approaches. While an explanation is beyond the scope of this chapter, the reader is directed to Reference [57] for an analysis of the relationship between SSIM and VIF.

### 3.4.1.3  FR IQA: Feature-Based Approaches

Feature-based approaches for QA are generally based on extracting information relevant to the task of QA. Such feature extraction procedures seek inspiration from HVS properties and use certain heuristics while avoiding an explicit modeling of the HVS.

In Reference [58], the authors first compute the difference between the reference and distorted image, and then filter this difference using directional filters to form oriented edge images. A mask is applied to each of these oriented responses independently based on the activity measure and brightness. The resulting output is passed through a nonlinearity to obtain directional error, which is then pooled to produce a final error (quality) score.

In another work [59], the reference and distorted images are decomposed using the singular value decomposition (SVD), which is applied to $8 \times 8$ blocks. Distortion per block is computed as $D_i = \sqrt{\sum_{i=1}^{n}(s_i - \widehat{s_i})}$, where $s_i$ and $\widehat{s_i}$ are the singular values for block $i$ from the reference and test images. The final quality score is computed as $M - SVD = \sum_{i \in all\_blocks} |D_i - D_{mid}|$, where $D_{mid}$ represents the median of the block distortions. Even though the authors claim that the algorithm performs well, its relation to the HVS is unclear as is the significance of the SVD for IQA.

The IQA index proposed in Reference [60] uses the discrete wavelet transform to decompose the reference and test images. In each subband, mean surface curvature maps are obtained as follows:

$$H = \frac{I_{uu} + I_{vv} + I_{uu}I_v^2 + I_{vv}I_u^2 - 2I_u I_v I_{uv}}{2(1 + I_u^2 + I_v^2)^{3/2}}, \tag{3.9}$$

where $I_{uu}$, $I_{vv}$, $I_u$, and $I_v$ are the partial derivatives of the image $I$. The correlation coefficient between the curvatures of the original and distorted images is then evaluated. These correlation coefficients are then collapsed across the subbands to produce a quality score.

### 3.4.1.4  NR IQA: Distortion-Specific Algorithms

Having discussed FR algorithms, the focus is now shifted to no-reference (NR) QA algorithms. NR QA algorithms are broadly classified into those that are distortion specific and those that are distortion agnostic or holistic in nature. Distortion specific means that the algorithms can assess the quality of an image under the assumption that the image is affected by distortion $X$, where $X$ could be JPEG compression, blur, and so on. Unlike this approach, which has been extensively studied, there exist just a few holistic QA algorithms, as discussed below.

- JPEG IQA

Reference [61] proposed a blockiness measure for JPEG coded images under the assumption that although compression leads to multiple artifacts, blockiness is the most important attribute. To measure blockiness, a hermite transform-based approach is taken in which the one-dimensional (horizontal/vertical) blurred edges are modeled as a combination of the mean signal value, the step amplitude, blur parameters, and distance from the center of the window being used for computation. Once these properties are measured, simple statistics, such as the peak, mean, and standard deviation, are computed and evaluated for their performance. The authors also evaluate a Minkowski sum with various exponents to tune its performance relative to correlation with human perception.

Another work [62] proposed a JPEG QA algorithm that measures blocking and blurring due to JPEG compression and relates these measures to quality. The authors compute a horizontal first-order difference across the image and then estimate blockiness as the average difference across block boundaries. Further, activity is computed using two factors: the average absolute difference between in-block image samples and the zero crossing rate. These measures are combined using a nonlinear pooling strategy where the parameters are estimated using a training set. The authors report good performance on a relatively large database.

In Reference [63], the images are first converted from the RGB space to the perceptually relevant LMS color-space and analysis is performed on the achromatic component only. Neighboring differences in luminance along horizontal and vertical directions are computed and combined to produce a local blocking measure, which is pooled using a Minkowski summation to produce a final blocking measure. The authors also compute a blurring and ringing measure, using an importance map weighting spatial activity. A nonlinear pooling follows to produce an objective score.

In another work [64], edges are first classified as either block-boundaries or true edges using a threshold on the gradient value. The so-called plainness measure is then computed, which is a measure of the activity; this accounts for masking. Block artifact quality is then estimated by a combination of the gradient measure and the plainness measure.

Reference [65] proposed a visually significant blocking artifact metric (VSBAM), which decomposes a JPEG compressed image into a combination of primary edges, undistorted image edges, and blocking artifacts. In order to evaluate quality, the image is transformed into the Fourier domain along with the DC-only image, which is computed using local DCT. Vertical and horizontal blocking artifacts are then defined in the frequency domain using differences, and these are then transformed back to the spatial domain to produce a measure of quality.

- JPEG2000 IQA

In Reference [66], the authors compute the gradient image and then obtain the direction of the gradient, followed by a Canny edge detector. For each edge pixel, a search is performed along the gradient direction as well as in the perpendicular direction to estimate the edge-spread at local extrema locations, which is a measure of blur. The average edge spread is then computed and pooled spatially to produce a measure of quality.

In Reference [67], the authors utilize a Canny edge detector to detect edges in an image and analyze the edge points using a window-based approach. In the training phase, these windowed edge points are computed for original images ($S^+$) and distorted images ($S^-$). A covariance matrix $\Sigma = (N^- \cdot \Sigma^- + N^+ \cdot \Sigma^+)/(N^+ + N^-)$ is then computed, where $N^+$ and $N^-$ denote the number of windowed vectors in $S^+$ and $S^-$, respectively, whereas $\Sigma^+$ and $\Sigma^-$ denote the covariance matrices for vectors in $S^+$ and $S^-$, respectively. Principal component analysis (PCA) is then performed on $\Sigma$, and the second and third principal components are used as axes for projection. During the testing phase, windowed vectors are projected onto these principal axes and the mean and variance of the distributions along these axes are computed. Finally, the distortion metric is pooled to produce a quality score, with parameters obtained from a training set.

A small subjective study for JPEG2000 IQA to produce mean opinion scores (MOS) was conducted in Reference [68]. The objective measure computes several features, namely, the standard deviation of luminance in a region around each pixel, the absolute difference of a central pixel from the second closest neighborhood pixel, and zero crossing in horizontal and vertical directions. These features are then combined using a nonlinear approach. The parameters of the model are trained and performance is demonstrated on various datasets, including the JPEG2000 portion of the LIVE IQA database [30].

A natural-scene statistics (NSS)-based approach to JPEG2000 quality assessment was proposed in Reference [69], where the measure of quality is evaluated in the wavelet domain. The authors exploited the dependency between a wavelet coefficient and its neighbors, and the fact that the presence of distortion will alter these dependencies. The dependencies are captured using a threshold and offset approach, where the parameters are estimated using a training set. Good performance of the presented approach was demonstrated on the LIVE IQA dataset.

Reference [70] presented an edge-based approach to blur quality assessment; and applied this approach to JPEG2000 compressed images. The measure is based on the number of edges found along a scanline as well as the blur width, which is measured using the local extrema around the detected edge point. A simple combination of these measures leads to a blur estimate, which can also be used as a measure of quality for JPEG2000 compressed images.

- Sharpness/blur IQA

The authors of Reference [71] developed a sharpness measure based on the two-dimensional kurtosis of block DCT AC-coefficients. First, edge detection is performed, followed by thresholding, then the two-dimensional kurtosis of the DCT coefficients are computed over $8 \times 8$ blocks. The average kurtosis over the image forms a measure of sharpness.

In Reference [72], a Sobel filter is applied to compute the number of edge pixels and the gradient threshold. Then the gradient threshold is modified and the edge detection process is reapplied. This loop continues until the number of edge pixels becomes stable. Once the loop ends, the number of edge pixels found is compared to a parameter to check if the block is smooth or an edge block. For each edge block, contrast and edge width are estimated and a probability summation is used to collapse the contrast scores, which are finally subjected to a Minkowski summation to produce a sharpness metric.

In another work [73], the above algorithm is utilized with the addition of a saliency model. Specifically, once the amount of perceived blur is computed as in Reference [72], a saliency mask/weighting similar to that from Reference [74] is applied to the computed values to collapse the scores using a Minkowski summation.

Reference [75] proposed a measure that estimates the quality of image blur in the presence of noise using a gradient-based approach. Within a local window, the gradient covariance matrix is computed and is subjected to a singular value decomposition to estimate the local dominant orientation; the authors contend that the first two singular values are intimately related to sharpness. The final model for quality involves a contrast-gain-control mechanism using a combination of predetermined and estimated parameters.

A no-reference IQA algorithm, which measures image sharpness based on a cumulative probability of blur detection, was proposed in Reference [76]. Again, the algorithm follows an edge detection-based approach where blocks are categorized as smooth or non-smooth based on a predetermined threshold on the number of edge pixels. Once edge blocks are discovered, contrast computation follows. This is accompanied by a measure of edge width and a cumulative probability of edge detection, which is simply a summation over probabilities of detecting blur. The major difference here with respect to previous approaches [72], [73] is the absence of a Minkowski pooling stage at the end of the chain.

Reference [77] proposed a just-noticeable blur model that is an extension of the work in Reference [73], where the authors account for viewing distance as well, though the saliency model is dropped. Again, a variety of predetermined and computed thresholds are used to produce a quality score for the image, which is a Minkowski sum.

### 3.4.1.5  NR IQA: Holistic Algorithms

Reference [78] utilized a series of heuristic measures to quantify visual quality, which include edge sharpness, random noise level (impulse/additive white Gaussian noise), and structural noise. Edge detection is used to measure the sharpness of an edge, while impulse noise is measured using a local smoothness-based approach. A partial differential equation (PDE) model is used to measure Gaussian noise, whereas blocking and ringing artifacts from compression techniques such as JPEG and JPEG2000 are indicators of structural noise. While this work describes these measures to encompass a variety of distortions, it fails to propose a method to combine these measures to produce a quality assessment algorithm and fails to analyze the performance of the algorithm.

An interesting approach to the NR IQA problem was proposed in Reference [79], where a Renyi entropy measure along different orientations was used to measure anisotropy (that is, rotational asymmetry) and related the measured anisotropy to visual quality. The approach uses measure of mean, standard deviation, and range of the Renyi entropy along four orientations in the spatial domain. Although the authors demonstrate that their measure correlates with perceptual quality, a thorough performance evaluation is missing.

A different approach to blind IQA, which utilizes a two-stage framework that first identifies the proportion of each distortion present in the image and then performs distortion-specific quality assessment, was proposed in References [80] and [81]. This so-called distortion identification-based image verity and integrity evaluation (DIIVINE) index was shown to perform extremely well in terms of correlation with human perception on the LIVE IQA database [30], [81]. The DIIVINE index first decomposes an image using the wavelet transform and then performs the perceptually relevant step of divisive normalization. A series of statistical features, such as spatial correlation structure and across orientation correlation, are computed, and these features are then used to predict quality. The authors demonstrated that the DIIVINE approach is statistically equivalent in performance to the FR SSIM index and is statistically superior to the often-used PSNR, across all the distortions on the LIVE IQA database.

**TABLE 3.2**

Median Spearman's rank ordered correlation coefficient across 1000 train-test trials on the LIVE image quality assessment database. Symbol * indicates NR IQA algorithms, others are FR IQA algorithms.

| Algorithm | Distortion | | | | | |
| --- | --- | --- | --- | --- | --- | --- |
| | JP2K | JPEG | WN | Gblur | FF | All |
| PSNR | 0.87 | 0.89 | 0.94 | 0.76 | 0.88 | 0.87 |
| SSIM (SS) [45] | 0.94 | 0.95 | 0.96 | 0.91 | 0.94 | 0.91 |
| SSIM (MS) [48] | 0.97 | 0.96 | 0.98 | 0.95 | 0.94 | 0.95 |
| VIF [54] | 0.97 | 0.96 | 0.98 | 0.97 | 0.97 | 0.96 |
| BIQI-4D [26]* | 0.81 | 0.87 | 0.96 | 0.82 | 0.73 | 0.82 |
| Anisotropic IQA [79]* | 0.17 | 0.09 | 0.69 | 0.60 | 0.54 | 0.32 |
| BLIINDS [83]* | 0.95 | 0.94 | 0.98 | 0.94 | 0.93 | 0.91 |
| DIIVINE [81]* | 0.91 | 0.91 | 0.98 | 0.92 | 0.86 | 0.92 |

An alternate distortion-agnostic approach to blind IQA was proposed to compute statistical features in the DCT domain [82], [83]. This pragmatic approach, titled blind image integrity notator using DCT statistics (BLIINDS–II), extracts shape and variance from generalized Gaussian fits to DCT coefficients as well as energy ratios between frequency bands in the DCT. A small number of features are extracted at two scales in order to account for the multi-scale nature of both images and human visual perception; these statistical features are then regressed on to DMOS. As demonstrated on the LIVE IQA database, BLIINDS performs competitively with the FR SSIM index.

In order to provide the reader with a taste for how well some of the algorithms described here perform, Table 3.2 lists the median SROCC values across 1000 train-test trials on the LIVE IQA database [30]. Since NR algorithms considered here require training, partitioning the dataset is necessary to use the training images in order to set the parameters of the NR QA algorithm. In order to ensure performance across content, such an evaluation is repeated across multiple train-test iterations and the median value is reported. It is pertinent to note that the BLIINDS and DIIVINE NR models are competitive with the FR SSIM and beat the FR PSNR in terms of correlation with human perception.

### 3.4.2 Video Quality Assessment

Video quality assessment algorithms summarized in this section follow a similar taxonomy to IQA algorithms, as seen in Figure 3.7.

#### 3.4.2.1 FR VQA: Models Based on the Human Visual System

Human visual system (HVS) based models for VQA generally follow a series of operations akin to that for IQA. Examples of such operations include linear decomposition, contrast sensitivity modeling, and so on, although the models take into account the temporal dimension of the video as well. Generally, the spatial contrast sensitivity function (CSF) is modeled using a lowpass filter (since the HVS is not as sensitive to higher frequencies), and the temporal CSF is modeled using bandpass filters. It is generally easier to model the spatial and temporal CSFs separately instead of modeling a spatiotemporal CSF [84]. A good overview of HVS-based models for IQA/VQA can be found in Reference [85].

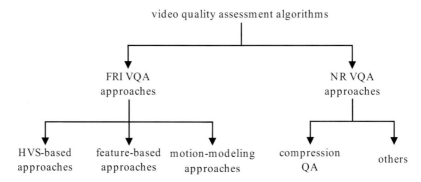

**FIGURE 3.7**

Taxonomy of video quality assessment algorithms summarized in this chapter.

The Sarnoff visual discrimination model (VDM) [86] was extended to the Sarnoff JND metric for video in Reference [87]. This approach applies a point-spread function (PSF) to the image, followed by a model of retinal cone-sampling, and then the image is decomposed using a Laplacian pyramid from which a contrast energy measure is computed. A masking procedure is then applied, and a JND distance measure is computed to produce a quality index.

Reference [88] proposed the digital video quality metric (DVQ), where quality is evaluated in the DCT-domain. The DVQ metric first transforms the image into the rarely used YOZ opponent color space [89], which is then decomposed using an $8 \times 8$ block DCT. The ratio of the AC amplitude to the DC amplitude is computed and labeled as local contrast, which is converted into JND units using thresholds derived from a human study, followed by contrast masking. The error scores between the reference and distorted frames are then computed using a Minkowski sum.

In Reference [90], a video is decomposed using Gabor filters for spatial decomposition and two temporal channels, which is followed by a spatiotemporal CSF. A simple intra-channel contrast masking is modeled as well. The moving picture quality metric (MPQM), as it is called, also segments the video into various regions, such as uniform areas, contours, and textures. The error scores from the decompositions between the reference and distorted videos for each of these regions are pooled separately to form a final error (quality) score.

Reference [91] applies a lowpass filter to a video, and this filtered version is subjected to a wavelet transform. A subset of the resulting coefficients are selected for distortion evaluation, which is computed using an error measure between the masked coefficients in the reference and distorted videos. The scalable wavelet-based distortion metric differs from other HVS-based approaches in that the parameters used for contrast masking and the contrast sensitivity function are derived from human responses to natural videos, as opposed to those derived from human responses to sinusoidal gratings.

### 3.4.2.2   FR VQA: Feature-Based Approaches

Feature-based approaches for VQA are similar to those for IQA, and generally extract features on a frame-by-frame basis. Some of these models incorporate motion information to model the temporal dimension of the video as well.

The perceptual video quality metric (PVQM) [92] extracts three features: edginess, a temporal indicator, and a chrominance indicator. Edginess is measured using the difference between the dilated edges of the reference and distorted frames computed using a local gradient measure, and reflects the loss in spatial detail. Temporal indicator is simply a measure of the correlation coefficient between adjacent frames, and the chrominance indicator is a function of the color saturation. Each of these three measures is pooled separately across the video and then combined to produce a quality measure.

The video quality metric (VQM) proposed in Reference [93] was one of the top performers in the video quality experts group (VQEG) phase-II studies [31] and has also been standardized by the American National Standards Institute; the International Telecommunications Union (ITU) has included VQM as a normative measure for digital cable television systems [94]. VQM extracts a series of features from the videos, including those that describe the loss of spatial detail, a shift in the orientation of the edges, the spread of chrominance components, and so on. Each of these features is thresholded and passed through nonlinearities, and the final quality score is a weighted sum of these values. VQM also includes elementary motion information in the form of the difference between frames and a quality improvement feature that accounts for improvements arising, for example, from sharpness operations.

Reference [95] proposed a measure of video quality that models the temporal variations in spatial distortions and relates such variations to visual quality. Temporal distortions, such as mosquito noise, flickering, and jerkiness, are modeled as an evolution of spatial distortions over time. A spatiotemporal tube, consisting of a spatiotemporal chunk of the video computed from motion vector information, is created, which is then evaluated for its spatial distortion. The spatial distortion is computed using the WQA quality index [96]. A temporal filtering of the spatial distortion is then undertaken, followed by a measurement of the temporal variation of the distortion. The quality scores are then pooled across the video to produce the final quality index.

The tetra video quality metric [97] is one of the few measure of video quality that utilize motion information. In this approach, motion estimates are determined using a block-based motion estimation algorithm, and such motion information from each heuristically determined shot [98] is utilized for temporal trajectory evaluation of the distorted video. This information is logged in a temporal information buffer, which is followed by a temporal visibility mask. Spatial distortion is evaluated by MSE. A spatiotemporal distortion map is then created, which is masked using a foveation model and temporal summation is then performed. The approach also accounts for the frame rate, frame pauses, and frame skips. All of the indicators are pooled to create a measure of quality.

### 3.4.2.3 FR VQA: Motion Modeling-Based Approaches

It should be clear from the above sections that while there exist many models for quality assessment of videos, most of these models are simply extensions of IQA algorithms that are applied on a frame-by-frame basis. Some of these models indeed utilize motion information, however, none of these models systematically investigate the effect that motion has on visual perception and how motion modeling may result in a better VQA algorithm. The reader will appreciate that simple spatial analysis of a video is insufficient to gauge the perceived spatiotemporal quality. In this section, two recent motion-modeling based approaches that seek to predict such spatiotemporal quality, are described.

- Speed-weighted structural similarity index (SW-SSIM)

SW-SSIM is based on the notion that motion-based weighting of spatial scores obtained from the structural similarity index [45] is sufficient to model motion effects on visual quality [99]. In order to obtain this weighting scheme, the authors draw inspiration from human visual speed perception studies conducted by visual psychophysicists [100]. A histogram-based approach is used to identify the global motion vector for each frame, where the motion vector estimates are computed using the optical flow computation algorithm from Reference [101]. Relative motion is then extracted as the difference between the absolute motion vectors (computed from optical flow) and the global motion vectors. The weight for each pixel is a function of the relative motion, the global motion, and the stimulus contrast. Once such a weight map is obtained at each pixel location, the SSIM map applied on that frame is weighted by the computed speed-weight map. The weighted scores are then pooled across the video and normalized to produce a quality score for the video. The authors demonstrated that the approach well correlates with human visual perception.

- Motion-based video integrity evaluation (MOVIE)

Although SW-SSIM uses motion information to weight SSIM scores, this kind of motion-based weighting of spatial quality may not suffice in successfully modeling the effect of spatiotemporal distortions on visual perception. To address this concern, Reference [102] introduced the MOVIE index, which is loosely based on models of motion perception in visual area MT of the primary visual cortex.

In the MOVIE approach, the reference and distorted videos are first decomposed using a multi-scale and multi-orientation spatiotemporal Gabor filter set, akin to what is hypothesized to occur in visual area V1 [103]. The MOVIE index is computed using two separate channels, one for spatial quality and the other for temporal quality, where the segregation of computation is inspired from the hypothesized spatiotemporal separability of the human visual system (HVS). Spatial quality is computed in much the same way as MS-SSIM. A modified version of the algorithm for optical flow [104] is used to compute motion estimates. Such optical flow computation is efficiently performed using the same set of Gabor filters that were used to decompose the videos. Temporal MOVIE is based on the principle that translational motion in the spatiotemporal domain manifests as a plane in the frequency domain [105] and that deviations from the reference motion will cause a shearing of this plane. Quantifying the shear leads to a measure of temporal quality. In order to do this, positive excitatory weights are assigned to the response of those filters which lie close to the spectral plane defined by the computed motion vectors and negative inhibitory weights to those filter responses which lie farther away from the spectral plane. Such weighting results in a strong response if the motion in the test and reference video coincide and a weak response is produced if the test video has motion deviant from that in the reference video. The mean-squared error between the responses from the test and reference filter banks then provides the temporal quality estimate. The final MOVIE index is the product of the spatial and temporal quality indices. The authors demonstrated that the MOVIE index performs extremely well in terms of correlation with human visual perception.

**TABLE 3.3**
FR algorithm performance on the LIVE VQA database [27] using Spearman's rank ordered correlation coefficient.

| Algorithm | Distortion | | | | |
|---|---|---|---|---|---|
| | Wireless | IP | H.264 | MPEG2 | All |
| PSNR | 0.66 | 0.42 | 0.46 | 0.39 | 0.54 |
| SSIM [45] | 0.52 | 0.46 | 0.65 | 0.56 | 0.53 |
| MS-SSIM [48] | 0.73 | 0.65 | 0.71 | 0.66 | 0.74 |
| Speed SSIM [99] | 0.56 | 0.47 | 0.71 | 0.62 | 0.59 |
| VSNR [43] | 0.70 | 0.69 | 0.65 | 0.59 | 0.68 |
| VQM [93] | 0.72 | 0.64 | 0.65 | 0.78 | 0.70 |
| V-VIF [106] | 0.55 | 0.47 | 0.68 | 0.61 | 0.57 |
| MOVIE [102] | 0.81 | 0.72 | 0.77 | 0.77 | 0.79 |

Having described a handful of FR VQA algorithms, Table 3.3 lists the value of Spearman's rank ordered correlation coefficient (SROCC) of some of these approaches on the LIVE VQA database [27], to provide the reader with a sense of how well the various approaches predict visual quality. It should be noted that MOVIE is currently the best performer on the LIVE VQA database, as well as the VQEG phase-I dataset [29].

### 3.4.2.4 NR VQA: Compression

The majority of NR VQA algorithms in literature model quality due to compression artifacts and hence measure blockiness and bluriness. Some of these models also attempt to model temporal quality and chrominance artifacts. There exist no general-purpose NR VQA algorithms yet and hence this section describes algorithms that have specifically been designed for compression while those that serve various other purposes are described in the next section.

- Blockiness-based techniques

Reference [107] uses a Sobel operator to compute the gradient of each frame which is then subjected to a Fourier transform. A harmonic analysis follows, in which the ratio of the sum of all harmonics to all AC components within a block is computed in both the horizontal and vertical directions. The variance of the empirical pdf obtained from the phase of the harmonics across frames is an indicator of blockiness. The authors demonstrated that their technique works well for both I- and P-frames in compressed video.

The method presented in Reference [108] creates a set of subsampled images from each frame, so that each subimage consists of a particular pixel from one of the $8 \times 8$ blocks used for MPEG compression, and correlates them to produce a blockiness measure for MPEG compressed SD videos. Reference [109] compared this measure with two other measures and concluded that this measure is a poor indicator of quality.

The perceptually significant block-edge impairment metric (PS-BIM) [110] computes horizontal and vertical block strengths as weighted sums of luminance differences in a frame, where the weights are derived from luminance masking effects. Although no subjective evaluation is conducted, the authors use I-frames from an encoded video to demonstrate performance.

Another gradient-based approach was proposed in Reference [111] where a normalized horizontal gradient is computed at each pixel as the ratio of the absolute gradient and the average gradient over a neighboring region in a frame. This normalized gradient is then summed across rows and the ratio of this sum at block boundaries (assuming a fixed block-size) to the sum at intermediate positions leads to a quality measure. The authors demonstrated in a small human study that this metric has good performance.

- Multiple artifact measurement-based algorithms

The method presented in Reference [112] computes a series of measures, including measures of blocking, blur, and sharpness. Blocking is computed as weighted pixel differences between neighboring blocks, ringing artifacts are computed using an edge detection-based approach, and a kurtosis-based approach is used for evaluation of sharpness. The parameters for each of these measures are trained using a small set of training videos with subjective MOS. The authors demonstrated a high degree of correlation with human perception.

Reference [113] evaluates two measures of quality; one for blocking and the other for packet loss. In this approach, blockiness is a function of the variance of the block, while packet loss is quantified using a binary-edge image and row-differences for each of the encoded blocks. A comparison of the proposed metrics is performed with others known from the literature; however, this comparison does not involve any subjective correlations.

Reference [114] evaluates blockiness, blur, and noise in a similar way as Reference [112]. The method from Reference [108] is modified to measure blockiness. Edge-widths are used to measure blur, and variances of subblocks within an $8 \times 8$ block to measure noise. A weighted Minkowski pooling ensues and weights are estimated from human data. The algorithm is shown to perform well on a small database of human observations.

Another similar approach was proposed in Reference [115], which evaluated noise, blocking, moving artifacts, and color impairments in videos. The individual color channels are inspected and thresholded to form a color measure, while spatiotemporal regions are evaluated to study the effect of noise on quality. Although the authors demonstrated how these measures can be incorporated into a system, no such pooling was actually undertaken and no subjective evaluation was performed either.

Reference [116] presents a model for NR VQA consisting of frame quality measure and correction, asymmetric tracking, and mean value filtering. The frame quality measure is simply neighboring pixel differences with and without edge preservation filtering. This is subject to histogram calculations, pooling, and corrections to obtain the final measure. Asymmetric tracking accounts for the fact that humans tend to perceive poorer regions with greater severity than good ones [8]. Finally, mean value filtering removes high-frequency ingredients from the measure to produce the quality index. Evaluation on a small dataset shows good correlation with perception.

### 3.4.2.5 NR VQA: Other Techniques

An NR VQA that uses motion information was proposed in Reference [117]. This approach uses block-based motion estimates from a lowpass version of the distorted video along with luminance thresholds to identify translation regions of high spatial complexity. The sum of squared error between the block under consideration and its motion-compensated block in the previous frame is lowpass filtered to produce a spatial distortion measure. A temporal distortion measure is computed using a function of the mean of the motion vectors. Parameters and thresholds are trained using a part of the VQEG dataset [29] and when tested on the remaining part of the database, the approach demonstrates high correlation with human perception.

Reference [118] evaluated the blur caused due to video compression and imaging by first reducing the blocking artifacts by lowpass filtering. A subset of pixels from a frame are selected to evaluate blur due to compression (as opposed to intentional blur effects, such as those related to focus settings). Blur is then computed by quantifying edge-strength, which is shown to correlate well with PSNR for three videos.

Reference [119] proposed a measure for videos that have been subjected to frame-drops. The dropping severity is modeled as a function of the number frame drops using timestamp information from the video. The video is then segmented into cuts [98], and the motion activity of such a segment is determined using motion vector information and used to map the dropping severity to perceptual quality. A series of pooling steps, which include empirically determined thresholds and constants, follow, leading to a final quality score. The measure is shown to correlate well with human perception on a small dataset.

Channel-induced distortion for H.264/AVC compressed [120] videos is modeled in Reference [121], which attempts to estimate the MSE between the received and transmitted video using information from the macroblocks in the distorted video. The authors classify the macroblocks as either correctly received or lost. The lost macroblocks are modeled as a sum of distortions arising from motion vectors, prediction residuals, and distortion propagation. When the macroblocks are correctly received, the distortion is simply due to its propagation from the previous frame. Correlation between estimated channel distortion and measured channel distortion is used as a measure of performance.

Finally, Reference [122] modeled jerkiness between frames using absolute difference between adjacent frames and a threshold. Picture loss is similarly detected with another threshold. Blockiness is detected using a technique similar to those proposed previously. The test methodology is nonstandard and requires users to identify number of picture freezes, blocks, and picture losses in the videos; and perceptual quality is not evaluated.

## 3.5 Current and Future Research in Quality Assessment

Currently, FR IQA algorithms, such as the SSIM index, perform extremely well in terms of correlation with human perception, and although researchers have made attempts at bettering performance [8], [51], [123], the gains achieved have been minimal and statistically insignificant. Hence, it seems as if the field of two-dimensional IQA is currently at a plateau; that is, any additional gains that may be achieved will be at the cost of significant increase in processing that might not be justified by the end result. However, efficient implementation of IQA algorithms [124], [125] and use of quality indices as a criterion in the algorithm development [126], [127] remain exciting fields of research. Investigation of eye movements for the task of QA is another field that seemingly has great potential [8], [128], [129], [130], [131], [132], [133].

The field of NR IQA remains one of interest, since only recently distortion-agnostic approaches that perform competitively have been proposed [80], [82]. Development of efficient distortion-agnostic NR IQA algorithms that perform at least as well as leading FR IQA algorithms, such as MS-SSIM or VIF, are interesting avenues of future research. Further, evaluation of algorithm performance (of both FR and NR algorithms) on multiply-distorted images (for example, blur and compression) remains another interesting avenue of research [134], [135].

The field of FR VQA still has great potential for research, since the best algorithms do not correlate as well as the best IQA algorithms with human perception. As with FR IQA, efficient computation of FR VQA algorithms is an interesting direction of future research, especially when it comes to practical application of VQA algorithms [125]. Predicting visual attention in videos remains of interest, more so in VQA than IQA, since one would imagine that extracting low-level features from the reference video would make possible improved prediction of fixations and hence of quality. Further, the effect of distortions on visual attention is another ill-explored area of research [133].

As already discussed, existing NR VQA algorithms are generally geared toward compressed videos [110], [119], and a truly blind NR VQA algorithm that is distortion agnostic does not exist. Research on spatiotemporal perception and models for motion masking, such as the one used in Reference [27] and explored in Reference [136], may provide researchers with clues on how to model distortion-agnostic NR VQA algorithms. Temporal pooling of quality scores for VQA remains of interest and would complement existing literature on spatial pooling strategies for IQA [8], [46].

Apart from two-dimensional QA, stereoscopic or three-dimensional QA is an emerging research area [137], [138], [139], [140]. Device-specific QA remains of interest, as does utilization of QA algorithms for perceptual optimization for video delivery, video conferencing, and so on [141], [142], [143]. Studying visual aesthetics and context, as well as content-aware QA, is another interesting direction to pursue [144], [145], [146], [147], [148].

## 3.6  Conclusion

This chapter focused on perceptual quality evaluation of digital images and videos. Perceptual quality assessment is critical in the face of the explosion of media applications that are bringing digital images and video to people via the Internet, television, and cellular phones. The discussion started by focusing on perception of images and videos by humans and computational models of visual processing of these signals within the context of quality. Then, this chapter surveyed quality assessment algorithms that automatically predict the quality of a signal with or without the presence of a pristine reference based on perceptual models. It also discussed subjective methods to obtain ground truth data from human subjects on perceptual quality and presented comparative evaluations of different algorithms in matching visual perception. This was followed by discussing recent advances in the field of quality assessment and research directions in this field that remain open.

Although there exists tremendous scope for research in the area of QA, it is believed that a better understanding of human visual perception as well as better models of visual processing will tremendously benefit QA research. Researchers would do well to incorporate lessons learned from visual psychophysics in the development of visual quality assessment algorithms – for, after all, the ultimate receiver of the visual signal is the human observer!

## References

[1] Cisco Corp., "Cisco visual networking index: Forecast and methodology, 2009-2014." Available online, http://www.cisco.com/en/US/solutions/collateral/ns341/ns525/ns537/ns705/ns827/white_paper_c11-481360_ns827_Networking_Solutions_White_Paper.html.

[2] Cisco Corp., "Cisco visual networking index: Global mobile data traffic forecast update, 2010–2015." Available online, http://www.cisco.com/en/US/solutions/collateral/ns341/ns525/ns537/ns705/ns827/white_paper_c11-520862.html.

[3] B. Girod, "What's wrong with mean-squared error?," in *Digital Images and Human Vision*, A.B. Watson (ed.), Cambridge, MA, USA: MIT Press, 1993, pp. 207–220.

[4] Z. Wang and A.C. Bovik, "Mean squared error: Love it or leave it? – A new look at signal fidelity measures," *IEEE Signal Proceessing Magazine*, vol. 26, no. 1, pp. 98–117, January 2009.

[5] O. Schwartz and E.P. Simoncelli, "Natural signal statistics and sensory gain control," *Nature Neuroscience*, vol. 4, no. 8, pp. 819–825, August 2001.

[6] P.C. Teo and D.J. Heeger, "Perceptual image distortion," in *Proceedings of the IEEE International Conference on Image Processing*, Austin, Texas, USA, November 1994, pp. 982–986.

[7] K. Seshadrinathan and A.C. Bovik, "Motion tuned spatio-temporal quality assessment of natural videos," *IEEE Transactions on Image Processing*, vol. 19, no. 2, pp. 335–350, February 2010.

[8] A.K. Moorthy and A.C. Bovik, "Visual importance pooling for image quality assessment," *IEEE Journal on Selected Topics in Signal Processing, Special Issue on Visual Media Quality*, vol. 3, no. 2, pp. 193–201, April 2009.

[9] O. Le Meur, P. Le Callet, D. Barba, and D. Thoreau, "A coherent computational approach to model bottom-up visual attention," *IEEE Transactions on Pattern Analysis and Machine Intelligence*, vol. 28, no. 5, pp. 802–817, May 2006.

[10] F.L. Van Nes and M.A. Bouman, "Spatial modulation transfer in the human eye," *Journal of Optical Society of America*, vol. 57, no. 3, pp. 401–406, March 1967.

[11] H.A.W. Schober and R. Hilz, "Contrast sensitivity of the human eye for square-wave gratings," *Journal of Optical Society of America*, vol. 55, no. 9, pp. 1086–1090, September 1965.

[12] J.G. Robson, "Spatial and temporal contrast-sensitivity functions of the visual system," *Journal of Optical Society of America*, vol. 56, no. 8, pp. 1141–1142, August 1966.

[13] J. Nachmias and R.V. Sansbury, "Grating contrast: Discrimination may be better than detection," *Vision Research*, vol. 14, no. 10, pp. 1039–1042, October 1974.

[14] G.E. Legge and J.M. Foley, "Contrast masking in human vision," *Journal of Optical Society of America*, vol. 70, no. 12, pp. 1458–1471, December 1980.

[15] J. Ross and H.D. Speed, "Contrast adaptation and contrast masking in human vision," *Proceedings of the Biological Society*, vol. 245, no. 1315, pp. 61–70, October 1991.

[16] A.B. Watson *Digital Images and Human Vision*. Cambridge, MA, USA: MIT Press, 1993.

[17] J. Foley, "Human luminance pattern-vision mechanisms: Masking experiments require a new model," *Journal of Optical Society of America A*, vol. 11, no. 6, pp. 1710–1719, June 1994.

[18] K. Seshadrinathan and A.C. Bovik, "Unifying analysis of full reference image quality assessment," in *Proceedings of the IEEE International Conference on Image Processing*, San Diego, California, USA, October 2008, pp. 1200–1293.

[19] G.E. Legge, "Sustained and transient mechanisms in human vision: Temporal and spatial properties," *Vision Research*, vol. 18, no. 1, pp. 69–81, January 1978.

[20] J.J. Kulikowski and D.J. Tolhurst, "Psychophysical evidence for sustained and transient detectors in human vision," *Journal of Physiology*, vol. 232, no. 1, pp. 149–162, July 1973.

[21] B.A. Wandell, *Foundations of Vision*. Sunderland, MA, USA: Sinauer Associates Inc., 1995.

[22] J. Lubin, "The use of psychophysical data and models in the analysis of display system performance," in *Digital Images and Human Vision*, A.B. Watson (ed.), Cambridge, MA, USA: MIT Press, 1993, pp. 163–178.

[23] Z. Wang, A.C. Bovik, H.R. Sheikh, and E.P. Simoncelli, "Image quality assessment: From error visibility to structural similarity," *IEEE Transactions on Image Processing*, vol. 13, no. 4, pp. 600–612, April 2004.

[24] S. Daly, "The visible difference predictor: An algorithm for the assessment of image fidelity," in *Digital Images and Human Vision*, A.B. Watson (ed.), Cambridge, MA, USA: MIT Press, 1993, pp. 176–206.

[25] International Telecommunication Union, "Methodology for the subjective assessment of the quality of television pictures," Recommendation ITU-R BT.500-11, 2002.

[26] A.K. Moorthy, K. Seshadrinathan, R. Soundararajan, and A.C. Bovik, "Wireless video quality assessment: A study of subjective scores and objective algorithms," *IEEE Transactions on Circuits and Systems for Video Technology*, vol. 20, no. 4, pp. 513–516, April 2010.

[27] K. Seshadrinathan, R. Soundararajan, A.C. Bovik, and L.K. Cormack, "Study of subjective and objective quality assessment of video," *IEEE Transactions on Image Processing*, vol. 19, no. 6, pp. 1427–1441, June 2010.

[28] D. Sheshkin, *Handbook of Parametric and Nonparametric Statistical Procedures*. Boca Raton, FL, USA: CRC Press, 2004.

[29] Video Quality Experts Group (VQEG), "Final report from the video quality experts group on the validation of objective quality metrics for video quality assessment phase I," 2000. Available online, http://www.its.bldrdoc.gov/vqeg/projects/frtv-phase-i/frtv-phase-i.aspx.

[30] H.R. Sheikh, M.F. Sabir, and A.C. Bovik, "A statistical evaluation of recent full reference image quality assessment algorithms," *IEEE Transactions on Image Processing*, vol. 15, no. 11, pp. 3440–3451, November 2006.

[31] Video Quality Experts Group (VQEG), "Final report from the video quality experts group on the validation of objective quality metrics for video quality assessment phase II," 2003. Available online, http://www.its.bldrdoc.gov/vqeg/projects/frtv-phase-ii/frtv-phase-ii.aspx.

[32] D.M. Chandler and S.S. Hemami, "A57 database," 2007. Available online, http://foulard.ece.cornell.edu/dmc27/vsnr/vsnr.html.

[33] "Toyama image database." Available online, http://mict.eng.u-toyama.ac.jp/mict/index2.html.

[34] P. Le Callet and F. Autrusseau, "Subjective quality assessment irccyn/ivc database," 2005. Available online, http://www.irccyn.ec-nantes.fr/ivcdb/.

[35] N. Ponomarenko, V. Lukin, A. Zelensky, K. Egiazarian, M. Carli, and F. Battisti, "TID2008 – a database for evaluation of full reference visual quality assessment metrics," *Advances of Modern Radioelectronics*, vol. 10, pp. 30–45, 2009.

[36] E.C. Larson and D.M. Chandler, "Most apparent distortion: Full-reference image quality assessment and the role of strategy," *Journal of Electronic Imaging*, vol. 19, no. 1, p. 0011006, January 2010.

[37] J.L. Mannos and D. Sakrison, "The effects of a visual fidelity criterion of the encoding of images," *IEEE Transactions on Information Theory*, vol. 20, no. 4, pp. 525–536, April 1974.

[38] M. Eckert and A. Bradley, "Perceptual quality metrics applied to still image compression," *Signal Processing*, vol. 70, no. 3, pp. 177–200, March 1998.

[39] S. Daly, "Visible differences predictor: An algorithm for the assessment of image fidelity," *Proceedings of SPIE*, vol. 1666, p. 179-206, February 1992.

[40] A.B. Watson, "The cortex transform: Rapid computation of simulated neural images," *Computer Vision, Graphics, and Image Processing*, vol. 39, no. 3, pp. 311–327, March 1987.

[41] A. Watson, "DCTune: A technique for visual optimization of DCT quantization matrices for individual images," *Society for Information Display International Symposium Digest of Technical Papers*, vol. 24, pp. 946–946, 1993.

[42] S.J.P. Westen, R.L. Lagendijk, and J. Biemond, "Perceptual image quality based on a multiple channel HVS model," in *Proceedings of the IEEE International Conference on Acoustics, Speech and Signal Processing*, Detroit, Michigan, USA, May 1995, pp. 2351–2354.

[43] D.M. Chandler and S.S. Hemami, "VSNR: A wavelet-based visual signal-to-noise ratio for natural images," *IEEE Transactions on Image Processing*, vol. 16, no. 9, pp. 2284–2298, September 2007.

[44] A. Srivastava, A.B. Lee, E.P. Simoncelli, and S.C. Zhu, "On advances in statistical modeling of natural images," *Journal of Mathematical Imaging and Vision*, vol. 18, no. 1, pp. 17–33, January 2003.

[45] Z. Wang, A.C. Bovik, H.R. Sheikh, and E.P. Simoncelli, "Image quality assessment: From error measurement to structural similarity," *IEEE Transactions on Image Processing*, vol. 13, no. 5, pp. 600–612, April 2004.

[46] Z. Wang and X. Shang, "Spatial pooling strategies for perceptual image quality assessment," in *Proceedings of the IEEE International Conference on Image Processing*, Atlanta, GA, USA, October 2006, pp. 2945–2948.

[47] A.K. Moorthy and A.C. Bovik, "Perceptually significant spatial pooling techniques for image quality assessment," *Proceedings of SPIE*, vol. 7240, pp. 724012:1–11, January 2009.

[48] Z. Wang, E.P. Simoncelli, and A.C. Bovik, "Multi-scale structural similarity for image quality assessment," in *Proceedings of the IEEE Asilomar Conference on Signals, Systems, and Computers*, Pacific Grove, CA, USA, November 2003, vol. 2, pp. 1398–1402.

[49] Z. Wang and E.P. Simoncelli, "Translation insensitive image similarity in complex wavelet domain," in *Proceedings of the IEEE International Conference on Acoustics, Speech, and Signal Processing*, Philadelphia, PA, USA, March 2005, pp. 573–576.

[50] M.P. Sampat, Z. Wang, S. Gupta, A.C. Bovik, and M.K. Markey, "Complex wavelet structural similarity: A new image similarity index," *IEEE Transactions on Image Processing*, vol. 18, no. 10, pp. 2385–2401, October 2009.

[51] G.H. Chen, C.L. Yang, and S.L. Xie, "Gradient-based structural similarity for image quality assessment," in *Proceedings of the IEEE International Conference on Image Processing*, Atlanta, GA, USA, October 2006, pp. 2929–2932.

[52] Q. Li and Z. Wang, "Reduced-reference image quality assessment using divisive normalization-based image representation," *IEEE Journal of Selected Topics in Signal Processing, Special Issue on Visual Media Quality Assessment*, vol. 3, no. 2, pp. 202–211, April 2009.

[53] X. Gao, T. Wang, and J. Li, "A content-based image quality metric," *Lecture Notes in Computer Science*, vol. 3642, pp. 231–240, September 2005.

[54] H.R. Sheikh and A.C. Bovik, "Image information and visual quality," *IEEE Transactions on Image Processing*, vol. 15, no. 2, pp. 430–444, February 2006.

[55] M.J. Wainwright and E.P. Simoncelli, "Scale mixtures of Gaussians and the statistics of natural images," *Advances in Neural Information Processing Systems*, vol. 12, pp. 855–861, January 2000.

[56] E.P. Simoncelli and B.A. Olshausen, "Natural image statistics and neural representation," *Annual Review of Neuroscience*, vol. 24, no. 1, pp. 1193–1216, 2001.

[57] K. Seshadrinathan and A.C. Bovik, "Unifying analysis of full reference image quality assessment," in *Proceedings of the IEEE International Conference on Image Processing*, San Diego, CA, USA, October 2008, pp. 1200–1203.

[58] S. Karunasekera and N. Kingsbury, "A distortion measure for blocking artifacts in images based on human visual sensitivity," *IEEE Transactions on Image Processing*, vol. 4, no. 6, pp. 713–724, June 1995.

[59] A. Shnayderman, A. Gusev, and A.M. Eskicioglu, "A multidimensional image quality measure using singular value decomposition," *Proceedings of SPIE*, vol. 5294, pp. 82–92, January 2003.

[60] S. Yao, W. Lin, Z. Lu, E. Ong, M. Locke, and S. Wu, "Image quality measure using curvature similarity," in *Proceedings of the IEEE International Conference on Image Processing*, San Antonio, TX, USA, September 2007, vol. 3, pp. 437–440.

[61] L. Meesters and J.B. Martens, "A single-ended blockiness measure for JPEG-coded images," *Signal Processing*, vol. 82, no. 3, pp. 369–387, March 2002.

[62] Z. Wang, H.R. Sheikh, and A.C. Bovik, "No-reference perceptual quality assessment of JPEG compressed images," in *Proceedings of the IEEE International Conference on Image Processing*, Rochester, NY, USA, September 2002, vol. 1, pp. 477–480.

[63] R. Barland and A. Saadane, "A reference free quality metric for compressed images," in *Proceedings of the International Workshop on Video Processing and Quality Metrics for Consumer Electronics*, Sydney, Australia, August 2005, pp. 351–354.

[64] J. Chen, Y. Zhang, L. Liang, S. Ma, R. Wang, and W. Gao, "A no-reference blocking artifacts metric using selective gradient and plainness measures," in *Proceedings of the 9th Pacific Rim Conference on Multimedia: Advances in Multimedia Information Processing*, Tainan, Taiwan, December 2008, pp. 894–897.

[65] S. Suthaharan, "No-reference visually significant blocking artifact metric for natural scene images," *Signal Processing*, vol. 89, no. 8, pp. 1647–1652, August 2009.

[66] E. Ong, W. Lin, Z. Lu, S. Yao, X. Yang, and L. Jiang, "No-reference JPEG-2000 image quality metric," in *Proceedings of the IEEE International Conference on Multimedia and Expo*, Baltimore, MD, USA, July 2003, pp. 6–9.

[67] H. Tong, M. Li, H.J. Zhang, and C. Zhang, "No-reference quality assessment for JPEG2000 compressed images," in *Proceedings of the IEEE International Conference of Image Processing*, Singapore, October 2004, pp. 24–27.

[68] Z.M.P. Sazzad, Y. Kawayoke, and Y. Horita, "No reference image quality assessment for JPEG2000 based on spatial features," *Signal Processing: Image Communication*, vol. 23, no. 4, pp. 257–268, April 2008.

[69] H.R. Sheikh, A.C. Bovik, and L.K. Cormack, "No-reference quality assessment using natural scene statistics: JPEG 2000," *IEEE Transactions on Image Processing*, vol. 14, no. 11, pp. 1918–1927, November 2005.

[70] P. Marziliano, F. Dufaux, S. Winkler, and T. Ebrahimi, "Perceptual blur and ringing metrics: Application to JPEG2000," *Signal Processing: Image Communication*, vol. 19, no. 2, pp. 163–172, February 2004.

[71] J. Caviedes and S. Gurbuz, "No-reference sharpness metric based on local edge kurtosis," in *Proceedings of the IEEE International Conference on Image Processing*, Rochester, NY, USA, September 2002, pp. 53–56.

[72] S. Varadarajan and L. Karam, "An improved perception-based no-reference objective image sharpness metric using iterative edge refinement," in *Proceedings of the IEEE International Conference on Image Processing*, San Diego, CA, USA, October 2008, pp. 401–404.

[73] N.G. Sadaka, L.J. Karam, R. Ferzli, and G.P. Abousleman, "A no-reference perceptual image sharpness metric based on saliency-weighted foveal pooling," in *Proceedings of the IEEE International Conference on Image Processing*, San Diego, CA, USA, October 2008, pp. 369–372.

[74] L. Itti, C. Koch, and E. Niebur, "A model of saliency-based visual attention for rapid scene analysis," *IEEE Transactions on Pattern Analysis and Machine Intelligence*, vol. 20, no. 11, pp. 1254–1259, November 2002.

[75] X. Zhu and P. Milanfar, "A no-reference sharpness metric sensitive to blur and noise," in *Proceedings of the International Workshop on Quality of Multimedia Experience*, San Diego, CA, USA, July 2009, pp. 64–69.

[76] N. Narvekar and L. Karam, "A no-reference perceptual image sharpness metric based on a cumulative probability of blur detection," in *Proceedings of the International Workshop on Quality of Multimedia Experience*, San Diego, CA, USA, July 2009, pp. 87–91.

[77] R. Ferzli and L. Karam, "A no-reference objective image sharpness metric based on the notion of just noticeable blur (JNB)," *IEEE Transactions on Image Processing*, vol. 18, no. 4, pp. 717–728, April 2009.

[78] X. Li, "Blind image quality assessment," in *Proceedings of the IEEE International Conference on Image Processing*, Rochester, NY, USA, September 2002, pp. 449–452.

[79] S. Gabarda and G. Cristóbal, "Blind image quality assessment through anisotropy," *Journal of the Optical Society of America A*, vol. 24, no. 12, pp. 42–51, December 2007.

[80] A.K. Moorthy and A.C. Bovik, "A two-step framework for constructing blind image quality indices," *IEEE Signal Processing Letters*, vol. 17, no. 5, pp. 587–599, May 2010.

[81] A.K. Moorthy and A.C. Bovik, "Blind image quality assessment: From natural scene statistics to perceptual quality," *IEEE Transactions on Image Processing*, to appear.

[82] M.A. Saad, A.C. Bovik, and C. Charrier, "A perceptual DCT statistics based blind image quality metric," *IEEE Signal Processing Letters*, vol. 17, no. 6, pp. 583–586, June 2010.

[83] M.A. Saad, A.C. Bovik, and C. Charrier, "Model-based blind image quality assessment using natural DCT statistics," *IEEE Transactions on Image Processing*, submitted.

[84] D.H. Kelly, "Spatiotemporal variation of chromatic and achromatic contrast thresholds," *Journal of the Optical Society of America*, vol. 73, no. 6, pp. 742–750, June 1983.

[85] M. Nadenau, S. Winkler, D. Alleysson, and M. Kunt, "Human vision models for perceptually optimized image processing – a review," Available online, http://citeseerx.ist.psu.edu/viewdoc/summary?doi=10.1.1.5.2376.

[86] J. Lubin, "A visual discrimination model for imaging system design and evaluation," in *Vision Models for Target Detection and Recognition: In Memory of Arthur Menendez*, E. Peli (ed.), River Edge, NJ, USA: World Scientific Publishing, 1995, p. 245.

[87] J. Lubin and D. Fibush, "Sarnoff JND vision model," *T1A1*, Technical Report T1A1, vol. 5, 1997, pp. 97–612.

[88] A.B. Watson, J. Hu, and J.F. McGowan III, "Digital video quality metric based on human vision," *Journal of Electronic Imaging*, vol. 10, no. 1, pp. 20–29, January 2001.

[89] H.A. Peterson, A.J. Ahumada Jr., and A.B. Watson, "An improved detection model for DCT coefficient quantization," *Proceedings of SPIE*, vol. 1913, pp. 191–201, February 1993.

[90] C. Van den Branden Lambrecht and O. Verscheure, "Perceptual quality measure using a spatiotemporal model of the human visual system," *Proceedings of SPIE*, vol. 2668, pp. 450–461, January 1996.

[91] M. Masry, S.S. Hemami, and Y. Sermadevi, "A scalable wavelet-based video distortion metric and applications," *IEEE Transactions on Circuits and Systems for Video Technology*, vol. 16, no. 2, pp. 260–273, February 2006.

[92] A. Hekstra, J. Beerends, D. Ledermann, F. De Caluwe, S. Kohler, R. Koenen, S. Rihs, M. Ehrsam, and D. Schlauss, "PVQM – a perceptual video quality measure," *Signal Processing: Image Communication*, vol. 17, no. 10, pp. 781–798, October 2002.

[93] M.H. Pinson and S. Wolf, "A new standardized method for objectively measuring video quality," *IEEE Transactions on Broadcasting*, vol. 50, no. 3, pp. 312–313, September 2004.

[94] International Telecommunications Union, "Objective perceptual video quality measurement techniques for digital cable television in the presence of a full reference," ITU-T Recommendation 144, 2004.

[95] A. Ninassi, O.L. Meur, P.L. Callet, and D. Barba, "Considering temporal variations of spatial visual distortions in video quality assessment," *IEEE Journal on Selected Topics in Signal Processing*, vol. 3, no. 2, pp. 253–265, April 2009.

[96] A. Ninassi, O. Le Meur, P. Le Callet, and D. Barba, "On the performance of human visual system based image quality assessment metric using wavelet domain," *Proceedings of SPIE*, vol. 6806, pp. 680610:1–12, January 2008.

[97] M. Barkowsky, B.E.J. Bialkowski, R. Bitto, and A. Kaup, "Temporal trajectory aware video quality measure," *IEEE Journal on Selected Topics in Signal Processing*, vol. 3, no. 2, pp. 266–279, April 2009.

[98] C. Cotsaces, N. Nikolaidis, and I. Pitas, "Video shot detection and condensed representation. A review," *IEEE Signal Processing Magazine*, vol. 23, no. 2, pp. 28–37, March 2006.

[99] Z. Wang and Q. Li, "Video quality assessment using a statistical model of human visual speed perception," *Journal of Optical Society of America*, vol. 24, no. 12, pp. B61–B69, December 2007.

[100] A.A. Stocker and E.P. Simoncelli, "Noise characteristics and prior expectations in human visual speed perception," *Nature Neuroscience*, vol. 9, no. 4, pp. 578–585, April 2006.

[101] M.J. Black and P. Anandan, "The robust estimation of multiple motions: Parametric and piecewise-smooth flow fields," *Computer Vision and Image Understanding*, vol. 63, no. 1, pp. 75–104, January 1996.

[102] K. Seshadrinathan, *Video Quality Assessment Based on Motion Models*. PhD thesis, The University of Texas at Austin, TX, USA, 2008.

[103] M. Carandini, J. Demb, V. Mante, D. Tolhurst, Y. Dan, B. Olshausen, J. Gallant, and N. Rust, "Do we know what the early visual system does?," *Journal of Neuroscience*, vol. 25, no. 46, pp. 10577–10597, November 2005.

[104] D.J. Fleet and A.D. Jepson, "Computation of component image velocity from local phase information," *International Journal of Computer Vision*, vol. 5, no. 1, pp. 77–104, January 1990.

[105] A.B. Watson and A.J. Ahumada, "Model of human visual-motion sensing," *Journal of the Optical Society of America A*, vol. 2, no. 2, pp. 322–342, February 1985.

[106] H.R. Sheikh and A.C. Bovik, "A visual information fidelity approach to video quality assessment," in *Proceedings of the International Workshop on Video Processing and Quality Metrics for Conusmer Electronics*, Scottsdale, AZ, USA, January 2005, pp. 23–25.

[107] K.T. Tan and M. Ghanbari, "Blockiness detection for MPEG2-coded video," *IEEE Signal Processing Letters*, vol. 7, no. 8, pp. 213–215, August 2000.

[108] T. Vlachos, "Detection of blocking artifacts in compressed video," *Electronics Letters*, vol. 36, no. 13, pp. 1106–1108, June 2000.

[109] S. Winkler, A. Sharma, and D. McNally, "Perceptual video quality and blockiness metrics for multimedia streaming applications," in *Proceedings of the International Symposium on Wireless Personal Multimedia Communications*, Aalborg, Denmark, September 2001, pp. 553–556.

[110] S. Suthaharan, "Perceptual quality metric for digital video coding," *Electronics Letters*, vol. 39, no. 5, pp. 431–433, March 2003.

[111] R. Muijs and I. Kirenko, "A no-reference blocking artifact measure for adaptive video processing," in *Proceedings of the European Signal Processing Conference*, Antalya, Turkey, September 2005.

[112] J. Caviedes and F. Oberti, "No-reference quality metric for degraded and enhanced video," *Proceedings of SPIE*, vol. 5150, pp. 621–632, July 2003.

[113] R. Babu, A. Bopardikar, A. Perkis, and O. Hillestad, "No-reference metrics for video streaming applications," in *Proceedings of the International Packet Video Workshop*, Irvine, CA, USA, December 2004.

[114] M.C.Q. Farias and S.K. Mitra, "No-reference video quality metric based on artifact measurements," in *Proceedings of the IEEE International Conference on Image Processing*, Genoa, Italy, September 2005, pp. 141–144.

[115] R. Dosselmann and X. Yang, "A prototype no-reference video quality system," in *Proceedings of the Canadian Conference on Computer and Robot Vision*, Montreal, QC, Canada, May 2007, pp. 411–417.

[116] Y. Kawayoke and Y. Horita, "NR objective continuous video quality assessment model based on frame quality measure," in *Proceedings of the IEEE International Conference on Image Processing*, San Diego, CA, USA, October 2008, pp. 385–388.

[117] F. Yang, S. Wan, Y. Chang, and H. Wu, "A novel objective no-reference metric for digital video quality assessment," *IEEE Signal Processing Letters*, vol. 12, no. 10, pp. 685–688, October 2005.

[118] J. Lu, "Image analysis for video artifact estimation and measurement," *Proceedings of SPIE*, vol. 4301, pp. 166–174, January 2001.

[119] K. Yang, C. Guest, K. El-Maleh, and P. Das, "Perceptual temporal quality metric for compressed video," *IEEE Transactions on Multimedia*, vol. 9, no. 7, pp. 1528–1535, November 2007.

[120] I.E.G. Richardson, *H.264 and MPEG-4 Video Compression: Video Coding for Next-generation Multimedia*. Chichester, EK: John Wiley & Sons Inc., 2003.

[121] M. Naccari, M. Tagliasacchi, F. Pereira, and S. Tubaro, "No-reference modeling of the channel induced distortion at the decoder for H. 264/AVC video coding," in *Proceedings of the IEEE International Conference on Image Processing*, San Diego, CA, USA, October 2008, pp. 2324–2327.

[122] E. Ong, S. Wu, M. Loke, S. Rahardja, J. Tay, C. Tan, and L. Huang, "Video quality monitoring of streamed videos," in *Proceedings of the IEEE International Conference on Acoustics, Speech and Signal Processing*, Taipei, Taiwan, April 2009, pp. 1153–1156.

[123] C. Li and A.C. Bovik, "Three-component weighted structural similarity index," *Proceedings of SPIE*, vol. 7242, pp. 72420Q, January 2009.

[124] D.M. Rouse and S.S. Hemami, "Understanding and simplifying the structural similarity metric," in *Proceedings of the IEEE International Conference on Image Processing*, San Diego, CA, USA, October 2008, pp. 1188–1191.

[125] M.J. Chen and A.C. Bovik, "Fast structural similarity index algorithm," in *Proceedings of the IEEE International Conference on Acoustics Speech and Signal Processing*, Dallas, TX, USA, March 2010, pp. 994–997.

[126] S.S. Channappayya, A.C. Bovik, and R.W. Heath, "A linear estimator optimized for the structural similarity index and its application to image denoising," in *Proceedings of the IEEE International Conference on Image Processing*, Atlanta, GA, USA, October 2006, pp. 2637–2640.

[127] S. Channappayya, A.C. Bovik, C. Caramanis, and R.W. Heath, "Design of linear equalizers optimized for the structural similarity index," *IEEE Transactions on Image Processing*, vol. 17, no. 6, pp. 857–872, June 2008.

[128] K. Miyata, M. Saito, N. Tsumura, H. Haneishi, and Y. Miyake, "Eye movement analysis and its application to evaluation of image quality," in *Proceedings of the IS&TSID Color Imaging Conference: Color Science, Systems and Applications*, Scottsdale, AZ, USA, November 1997, pp. 116–119.

[129] P.L. Callet, S. Perchard, S. Tourancheau, A. Ninassi, and D. Barba, "Towards the next generation of video and image quality metrics: Impact of display, resolution, content and visual attention in subjective assessment," in *Proceedings of the Second International Workshop on Image Media Quality and its Applications*, Chiba, Japan, March 2007, pp. 10–51.

[130] T. Vuori and M. Olkkonen, "The effect of image sharpness on quantitative eye movement data and on image quality evaluation while viewing natural images," *Proceedings of SPIE*, vol. 6059, pp. 605903, January 2006.

[131] C.T. Vu, E.C. Larson, and D.M. Chandler, "Visual fixation patterns when judging image quality: Effects of distortion type, amount, and subject experience," in *Proceedings of the IEEE Southwest Symposium on Image Analysis and Interpretation*, Santa Fe, NM, USA, March 2008, pp. 73–76.

[132] A. Ninassi, O. Le Meur, P. Le Callet, D. Barba, and A. Tirel, "Task impact on the visual attention in subjective image quality assessment," in *Proceedings of the European Signal Processing Conference*, Florence, Italy, September 2006.

[133] A.K. Moorthy, W.S. Geisler, and A.C. Bovik, "Evaluating the task dependence on eye movements for compressed videos," in *Proceedings of the International Workshop on Video Processing and Quality Metrics for Consumer Electronics*, Scottsdale, AZ, USA, January 2010.

[134] V. Kayargadde and J. Martens, "Perceptual characterization of images degraded by blur and noise: Model," *Journal of the Optical Society of America A*, vol. 13, no. 6, pp. 1178–1188, June 1996.

[135] M. Farias, M. Moore, J. Foley, and S. Mitra, "Perceptual contributions of blocky, blurry, and fuzzy impairments to overall annoyance," *Proceedings of SPIE*, vol. 5292, pp. 109–120, June 2004.

[136] M.A. Saad and A.C. Bovik, "Natural motion statistics for no-reference video quality assessment," in *Proceedings of the International Workshop on Quality of Multimedia Experience*, San Diego, CA, USA, July 2009, pp. 163–167.

[137] Z.M.P. Sazzad, Y. Kawayoke, and Y. Horita, "Spatial features based no reference image quality assessment for JPEG2000," in *Proceedings of the IEEE International Conference on Image Processing*, San Antonio, TX, USA, September 2007, pp. 517–520.

[138] P. Gorley and N. Holliman, "Stereoscopic image quality metrics and compression," *Proceedings of SPIE*, vol. 6803, pp. 680305:1–12, January 2008.

[139] P. Campisi, P. Le Callet, and E. Marini, "Stereoscopic images quality assessment," in *Proceedings of the European Signal Processing Conference*, Poznan, Poland, September 2007.

[140] P. Seuntiens, L.M.J. Meesters, and W. Ijsselsteijn, "Perceived quality of compressed stereoscopic images: Effects of symmetric and asymmetric JPEG coding and camera separation," *ACM Transactions on Applied Perception*, vol. 3, no. 2, pp. 109–125, April 2006.

[141] H. Hoffmann, T. Itagaki, D. Wood, and A. Bock, "Studies on the bit rate requirements for a HDTV format with $1920 \times 1080$ pixel resolution, progressive scanning at 50 Hz frame rate targeting large flat panel displays," *IEEE Transactions on Broadcasting*, vol. 52, no. 4, pp. 420–434, December 2006.

[142] J. Okamoto, K. Watanabe, A. Honda, M. Uchida, and S. Hangai, "HDTV objective video quality assessment method applying fuzzy measure," in *Proceedings of the International Workshop on Quality of Multimedia Experience*, San Diego, CA, USA, July 2009, pp. 168–173.

[143] M.H. Pinson, S. Wolf, and G. Cermak, "HDTV subjective quality of H.264 vs. MPEG-2, with and without packet loss," *IEEE Transactions on Broadcasting*, vol. 56, no. 1, pp. 86–91, March 2010.

[144] R. Datta, D. Joshi, J. Li, and J. Wang, "Studying aesthetics in photographic images using a computational approach," *Lecture Notes in Computer Science*, vol. 3953, pp. 288–301, May 2006.

[145] Y. Ke, X. Tang, and F. Jing, "The design of high-level features for photo quality assessment," in *Proceedings of the IEEE Conference on Computer Vision and Pattern Recognition*, New York, USA, June 2006, pp. 419–426.

[146] C. Li and T. Chen, "Aesthetic visual quality assessment of paintings," *IEEE Journal of Selected Topics in Signal Processing, Special Issue on Visual Media Quality*, vol. 3, no. 2, pp. 236–252, April 2009.

[147] Y. Luo and X. Tang, "Photo and video quality evaluation: Focusing on the subject," in *Proceedings of the European Conference on Computer Vision*, Marseille, France, October 2008, pp. 386–399.

[148] A.K. Moorthy and A.C. Bovik, "Statistics of natural image distortions," in *Proceedings of the IEEE International Conference on Acoustics Speech and Signal Processing*, Dallas, TX, USA, March 2010, pp. 962–965.

# 4

# *Visual Aesthetic Quality Assessment of Digital Images*

**Congcong Li and Tsuhan Chen**

## 4.1 Introduction

Visual aesthetic quality is a measure of visually perceived beauty. Judgment of the visual aesthetic quality of images is highly subjective, involving sentiments and personal taste [1]. However, some images are often believed, by consensus, to be visually more appealing than others. This serves as one of the principles in the emerging research area of computational

**FIGURE 4.1** (See color insert.)

Images with different aesthetic qualities. Images on the first row are taken by professional photographers while images on the second row are taken by amateur photographers.

aesthetics. Computational aesthetics is concerned with exploring computational techniques to predict emotional response to a visual stimulus, and developing methods to create and enhance pleasing impressions [2], [3]. This chapter focuses on exploring computational solutions to automatically infer the aesthetic quality of images. The greatest challenge in this research lies in the gap between low-level image properties and the high-level human perception of aesthetics.

This chapter starts by defining the problem of interest and giving a brief review of related works. After the introduction, Section 4.2 discusses some key aspects of the problem, including database and human judgments, and introduces the general flow of a computational solution for learning the image aesthetics. Section 4.3 introduces the principles and approaches being employed to extract computational features for representing the aesthetic quality. After analyzing the feature extraction, Section 4.4 explores different learning models that can be utilized for mapping from the computational features to the perceived aesthetics. Section 4.5 provides some examples among the large amount of applications for which the aesthetic quality assessment technique is highly desirable. Finally, this chapter concludes with Section 4.6.

### 4.1.1 Visual Aesthetics

The *Oxford Advanced Learners Dictionary* defines aesthetics as "concerned with beauty and art and the understanding of beautiful things, and made in an artistic way and beautiful to look at" [4]. Generally speaking, visual aesthetics is a measure of the perceived beauty of a visual stimulus. The visual aesthetic quality of an image measures how visually appealing the image is in people's eyes. As shown in Figure 4.1, different images arouse different emotional responses in people. Some images give the viewer more pleasing impressions than other images.

Unlike measuring the visual quality, that is, evaluating the level of image degradation, the aesthetic quality of an image is more related to its semantic content [5]. Given the abstract and subjective properties of aesthetics, it is unclear even for the professional photographers which characteristics are more correlated with the aesthetic quality of an image. However, professional photographers have summarized some general rules and techniques to guide the adjustment of color, composition, depth, and so on.

Psychologists and photography researchers have tried to explore how humans judge visual aesthetics. From studies in neuroscience, Reference [6] suggests that there is not one aesthetic sense only but many senses, each tied to a different specialized processing system. Different artistic attributes excite different groups of cells in the brain, so there is also a functional specialization in aesthetics. Reference [2] summarizes the basic dimensions of visual aesthetics, called aesthetic primitives, as the different modularities of the human visual system, and points out the most important modularities are color, form, spatial organization, motion, depth, and the human body. There are also studies that focus on how some of these characteristics affects image aesthetics. For example, the role of color preferences has been studied [7], [8], [9], and the role of the spatial composition has been analyzed [10], [11]. In previous works on computational aesthetics, psychology and photography have been important sources that inspire the design of computational features. More analysis on the aesthetic characteristics will be given in Section 4.3.

What is visual aesthetic quality assessment of images? The goal of image aesthetic quality assessment is to have an automatic computational algorithm provide images aesthetics scores that are consistent with human consensus. As will be discussed, to simplify the task, in some cases the algorithm only needs to predict the aesthetics labels for the images, indicating whether they belong to the high-quality class or low-quality class. The aesthetic quality assessment technique is highly useful in many applications, such as image search, photo enhancement, photo management, photography, and so on.

## 4.1.2   Related Works

Recent years have shown rapidly increasing interest in how to computationally predict the aesthetic quality of images. Assessing the aesthetic quality of images is a highly subjective task. Although it has long been studied in photography and visual psychology [6], [7], [12], [13], [14], [15], and many guidelines have been provided, there are no fixed rules that can directly connect the image content with the perceived aesthetics. Recent works seeking for computational solutions address the challenge of automatically predicting image aesthetic quality using their visual content as a machine learning problem. Generally, these techniques first extract computational features from the images and then learn the mappings from the features to the human-rated aesthetic quality through learning techniques.

One of the earliest attempts in this direction was made in Reference [16], which aimed at classifying photos as professional or amateur. Without analyzing how the semantic content in the image relates to the two quality classes, a large group of low-level features was collected and used to train standard learning algorithms for the classification. However, this black-box approach did not provide insight on how to design better features to represent the image appeals. Recent research studies in this area have focused on designing features to represent some high-level concepts that are highly correlated to the image aesthetics [4], [17]. The intuition of feature extraction often comes from concepts in psychology, photography, or human studies. For example, Reference [17] extracted low-level features, such as

average hue and edge distribution, to represent the high-level properties, like color preferences or simplicity. Similarly, Reference [4] extracted another group of visual features to
represent different characteristics related to the image's composition, object shape, color,
and so on. These ideas were further extended in Reference [18] which proposed features
related to the subject region of the image that is assumed to receive more attentions from
the viewer and is thus more relevant to the aesthetic quality. Addressing digital images of
paintings, Reference [19] introduced a set of features motivated from psychology, art, and
controlled human study, to predict common people's aesthetics judgment toward different
impressionistic paintings. Many efforts have also been made to study features specifically
to assess the aesthetic quality of consumer photographs [20], [21], [22], [23], [24].

In Reference [25], the authors proposed adding the high-level describable image attributes instead of using only the low-level features to represent the image appeal. Existing
algorithms were utilized to extract high-level attributes, such as the presence of salient objects, the presence of faces, and the scene category to which the respective image belongs.
The significant improvement in performance demonstrates the power of utilizing high-level
semantic attributes. The more the content and structure of the image is understood, the better performance can be achieved in the quality prediction. Therefore, the present work also
encourages closer collaboration between aesthetic quality assessment with other areas that
are dedicated to the holistic scene understanding.

Many applications based on the assessment have been developed, such as aesthetics-
assisted image search [17], [24], automatic photo enhancement [26], [27], [28], [29], [30],
and album management [20], [28], [31]. There have been also works that extend the aesthetic quality assessment techniques from images to videos [18], [32]. However, the automatic aesthetics assessment itself is still far from being solved. Section 4.2 discusses the
challenges in data collection and human study, and introduces the general techniques used
for aesthetics assessment.

## 4.2 Database, Human Study, and System Design

### 4.2.1 Database and Human Study

The subjectivity in predicting aesthetics raises a critical issue: what data should be used
to study image aesthetics? In order to train and evaluate any computational algorithms, a
database of images with aesthetic ratings based on human's judgments is needed. Unlike
other research areas in computer vision, such as scene recognition and object detection,
where standard datasets are available for evaluation, no benchmark database has been explicitly created for aesthetic quality assessment. Constructing such a database with manual
aesthetic labels is challenging. Ideally, the database needs to be a large-scale collection of
photos coming from a diverse group of photographers, including amateurs and professionals, and representing different contents and styles, from portrait to scenery, from abstract
to realistic, and so on. Moreover, a large-scale human study is needed in order to reach any
consensus in the aesthetic scores/labels given to an image. Participants in the study should
be chosen based on the definition of aesthetics. For example, if the goal is to understand

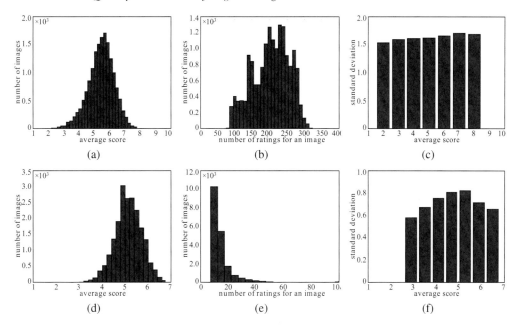

**FIGURE 4.2**

Statistic analysis on two large subsets [34] from two photo sharing websites: (a–c) DPChallenge.com and (d–f) Photo.net.

aesthetics in the eyes of photographers, a study should be conducted among a large group of photographers. However, if the goal is to study aesthetics in the eyes of common people who are non-experts in photography, the diversity in the demography of the participants has to be increased. The human study also needs to be well controlled in terms of participant constitution, viewing conditions, the viewing time, monitor screen resolutions, and other factors. Although crowdsourcing Internet marketplaces, such as Amazon Mechanical Turk,[1] have been widely utilized to quickly collect labels for other computational vision tasks [33], using such services to help build a dataset for aesthetics would pose further challenges in the design of a human study due to the uncontrolled noisy feedbacks. Although there have not been standard datasets in image aesthetics, researchers have made alternative efforts to push forward the research in this area.

Given the fact that a larger-scale dataset is more difficult for a controlled human study, prior works have started by creating datasets that contain a certain group of images. For example, Reference [35] presented a dataset with consumer photos (that is, non-professional photos) and aesthetic labels collected through a well-controlled human study. Reference [22] introduced a dataset of consumer photos containing human faces and the corresponding aesthetic labels collected through a human study via Amazon mechanical turk, with delicate designs to ensure that the participants used consistent standards to assess the images across the whole survey. Studying photos of paintings, an offline rating survey was conducted to collect the participants' general ideas about factors affecting the painting aesthetics and the aesthetic scores for all images [19]. However, these existing datasets are either not large enough or are not yet publicly available.

---

[1] https://www.mturk.com/mturk/

Another option is to use images from popular photo-sharing websites, such as DPChallenge[2] and Photo.net[3] [4], [17], [18]. The former one allows users to participate in theme-based photography contests, in which winners are determined based on the peer ratings of overall quality on a one to ten scale. The latter one serves as a platform for photography enthusiasts to post their shots, which are peer rated on a one to seven scale in terms of the aesthetic quality. Some other websites also provide measures related to the photo's aesthetic quality. For example, Flickr[4] provides a measure called "interestingness," which is computed through analysis of social interactions (such as viewing patterns, popularity of the content owner, favoring behavior, and so on) with that photo. The originality scores for images are highly correlated with the aesthetics scores given the photos on the Photo.net website [4]. All of these datasets have the advantage of containing large-scale data. However, there is not much control in the rating conditions, which results in biased ratings under certain circumstances. Moreover, the diversity of the rating participants is limited; many of the ratings are given by professional photographers instead of the general community. To mitigate this problem, Reference [34] used large subsets from three web-based sources (Photo.net, DPChallenge, and Terragalleria[5]) and reported the nature of user ratings on each dataset.

Here, the samples from Reference [34] are utilized and some statistical results are generated for the subsets from DPChallenge and Photo.net. Namely, Figures 4.2a and 4.2d show the distributions of the average scores in the respective dataset, Figures 4.2b and 4.2e show the distributions of number of ratings received by an image in the respective dataset, and Figures 4.2c and 4.2f show the standard deviation of the user scores corresponding to different average scores. Note that for Photo.net, scores are concentrated in the high end, perhaps because people are more likely to rate high-quality photos without specific instructions. Moreover, most images in the Photo.net subset received fewer than 20 ratings, while images in DPChallenge receive a larger amount of ratings. In Figures 4.2c and 4.2f, the variance provides some idea about the consensus of the ratings. The ratings in DPChallenge are less consistent than those in Photo.net. Another interesting result in the variance trend for Photo.net is that the photos with average scores at the two extreme ends (extremely low/high) received better consensus. This result is consistent with what has been found in controlled user studies in References [21] and [22].

### 4.2.2  System Design

This section introduces the general flow for designing a visual aesthetic quality assessment system, which is adopted by most of the current aesthetics assessment algorithms. As shown in Figure 4.3, building such a system comprises two parts: conducting a human study to collect human ratings and developing the computational algorithms. Note that the voting and rating activities on those photo-sharing websites mentioned earlier are generally considered as human studies.

---

[2]http://www.dpchallenge.com/

[3]http://photo.net/

[4]http://www.flickr.com/

[5]http://www.terragalleria.com/

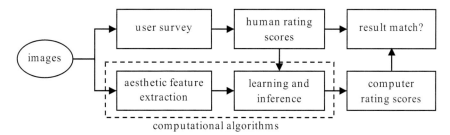

**FIGURE 4.3**

The general flow used for design an aesthetic visual quality assessment system.

The human ratings gained from the study are used to generate ground-truth labels, which are important inputs to the computational algorithms. In general, previous works have used the average score of the human ratings given to an image as the ground-truth aesthetic score for the image. For the computational part, a series of features representing various characteristics of the image appeal are extracted, followed by a regression or classification algorithm to map the extracted features to the ground-truth aesthetic scores. The system first extracts features from the test image, and then uses the corresponding inference algorithm to predict the aesthetic score. The performance of such a system is evaluated by measuring how well the predicted scores match with the average human-rated scores. In this chapter, Section 4.3 details how to extract effective computational features to represent the image aesthetic appeal, whereas Section 4.4 describes how to make use of a learning algorithm to predict the image aesthetics.

## 4.3 Computational Features for Visual Aesthetics

Extracting computational features to represent the image appeal is a crucial part of the aesthetic quality assessment task. Previous works have designed such features mainly by using human intuition, conducting human studies, and gaining inspirations from photography or psychology literatures. Reference [2] analyzed the modularities of the human visual system and derived the six basic dimensions of visual aesthetics called aesthetic primitives: color, form, spatial organization, motion, depth, and human body. Reference [17] found three distinguishing factors that affect the image appeals: simplicity, realism, and basic photographic techniques. Similarly, using questionnaires, Reference [19] found that color, composition, meaning, texture, and shape are the most frequently mentioned factors that affect judgment of aesthetic quality. Besides considering the global image characteristics, Reference [18] emphasized the importance of the main subject in an image on the aesthetic quality and summarized findings presented in References [13], [14], and [36]. According to those findings, "high quality photos generally satisfy these three principles: a clear topic, gathering most attention on the subject, and removing objects that distract attention from the subject." Thus, Reference [18] considered four criteria related to the different treatment of the subject and background: composition, lighting, focus controlling, and color.

**TABLE 4.1**
Summary of the low-level features introduced in Section 4.3.1, which are divided into five groups according to the characteristics which the respective features represent.

| Color Statistics | Golden-Section Rule (Rule of Thirds) |
|---|---|
| color distribution | visual balance |
| color simplicity | shapes |
| color harmony | texture and pattern |

| Illumination | Depth | Main Subject Region |
|---|---|---|
| average lightness | linear perspective | subject colorfulness |
| lightness contrast | sharpness-blurring distributions | background simplicity |
| | | subject spatial composition |
| | | subject-background contrasts |
| | | face-related characteristics |

Based on the analysis of criteria used for assessing the image aesthetics, various algorithms, such as those presented in References [4], [17], [18], and [19], have been proposed to extract low-level image features related to the criteria. Some recent works [22], [25] also proposed higher-level semantic attributes to describe the image appeal. The following sections will analyze the low-level features and high-level attributes.

### 4.3.1 Low-Level Features

This section introduces features that have been popularly used in previous works on aesthetic quality assessment. Depending on their representing characteristics, the features can be mainly divided into the following groups: color-related features, illumination-related features, spatial composition-related features, depth-related features, and main subject-related features, as summarized in Table 4.1. However, features in different groups are not completely independent. Instead, they can be related and contain overlapping information.

#### 4.3.1.1 Color-Related Features

Color is probably the most direct information that can be catched from an image, even before its close inspection. Creating more appealing color composition is an important artifice for a professional photographer.

- Color statistics

A rough measure of an image's color characteristics is to calculate the mean value of the pixel colors across the image. From the artistic aspect, the average color more or less represents the color keynote of the image. Assuming the RGB representation, the mean value can be computed for each of the three channels. In the case of the HSV representation, similar operations can be used to get the mean for the S and V channels, but not for the H channel. Consider the two hue values $h_1$ and $h_2$ shown in Figure 4.4a. Though their perceived colors are close to each other, the average of these two hue values is $h^* = (h_1 + h_2)/2 = 0.5$, which is not even close to either of these two hue values. To avoid this problem, the vector repre-

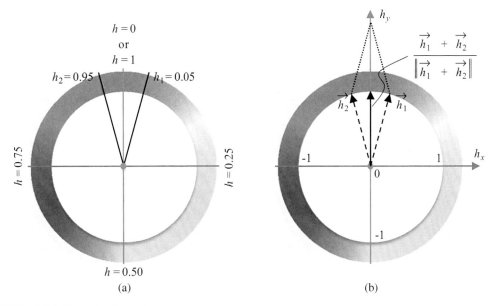

**FIGURE 4.4 (See color insert.)**

Different representations of the hue values: (a) scalar representation and (b) vector representation.

sentation for the hue value should be used, as shown in Figure 4.4b. The average hue vector can be defined as $\vec{h}^* = \sum_i \vec{h}_i / \|\sum_i \vec{h}_i\|$, where the normalization is operated because only the direction of the resulted vector matters. The scalar $h^*$ can be obtained by converting $\vec{h}^*$ back to the scalar representation.

- Color distribution

Prior works used a three-dimensional histogram with each axis corresponding to a color channel, or considered the histogram for the different channels individually. The histograms can be generated based on all the pixels in the image to represent the global color composition of the image, or based on the pixels within a presegmented region to represent local color composition.

- Hue count

The hue count of an image is a measure of the image's simplicity [17]. Most professional photos are often so simple that the main focusing objects can easily stand out of the background, while amateurish photos often seem cluttered due to the improper mixture of the different colors from different objects. The hue count of an image is calculated as follows. First, color images are converted to the HSV space. Second, only pixels with brightness values in the range $[0.15, 0.95]$ and saturation larger than $0.2$ are considered. This is because a pixel not satisfying these requirements looks close to gray color no matter what hue value it has. Third, a $K$-bin histogram is computed on the hue values of the requirement-satisfied pixels. The hue count is the number of bins with values larger than a certain threshold $t$, that is,

$$N_{hue} = \|\{i | H_{hue}(i) > t, 1 \le i \le K\}\|, \tag{4.1}$$

where $\| \cdot \|$ indicates the number of elements in a set.

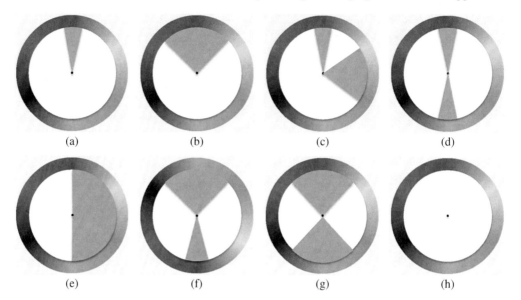

| (a) | (b) | (c) | (d) |
|-----|-----|-----|-----|

| (e) | (f) | (g) | (h) |
|-----|-----|-----|-----|

**FIGURE 4.5 (See color insert.)**

Harmony hue distribution models with gray indicating the efficient regions: (a) *i* type, (b) *V* type, (c) *L* type, (d) *I* type, (e) *T* type, (f) *Y* type, (g) *X* type, and (h) *N* type. All the models can be rotated by an arbitrary angle.

- Color harmony

Harmony in color is an important factor that affects the image appeal. Some patterns, such as combining complementary colors or using only neighboring colors, may bring better harmony than other randomly spread color patterns. Previous works [19], [26], [37] utilized the models from Reference [38] to evaluate the extent of color harmony in an image. This prior work studied various color schemes that were adopted as print clothes and dresses for girl students, and classified them into some groups, respectively, for the hue distribution and the tone-lightness distribution. As an example, Figure 4.5 shows the eight hue distribution models. In this figure, the type-N model corresponds to gray-scale images while the other seven models, each of which consists of one or two sectors, are related to color images. Gray regions indicate that the respective model prefers the pixel hues to agglomerate within the specific regions. Since all the models can be rotated by an arbitrary angle, what they are measuring is the relative relationship of the hues in the image rather than the specific color distribution modeled by the hue histogram itself. Figure 4.6 gives three example images that fit well with different types of harmony models.

As discussed in Reference [26], the set of hue distribution models can be used to evaluate the color harmony. Given an image, each of these models can be fit to the hue histogram of the image. Let $T_k(\alpha)$ be defined as the $k$th hue model rotated by an angle $\alpha$ and $E_{T_k(\alpha)}(p)$ as the hue in the model $T_k(\alpha)$ that is closest to the hue of the $p$th pixel $h(p)$; that is,

$$E_{T_k(\alpha)}(p) = \begin{cases} h(p) & \text{if } h(p) \in G_k, \\ B_{nearest} & \text{if } h(p) \notin G_k, \end{cases} \tag{4.2}$$

where $G_k$ is the gray region of model $T_k(\alpha)$ and $B_{nearest}$ is the hue of the sector border in model $T_k(\alpha)$ that is closest to the hue of pixel $p$. Further, let the distance between the hue histogram and the $k$th model rotated by $\alpha$ be defined as follows:

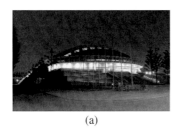  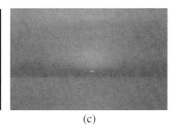

(a)                                   (b)                                   (c)

**FIGURE 4.6 (See color insert.)**

Image examples that fit with different types of hue harmony models in Figure 4.5: (a) fitting *I*-type harmony, (b) fitting *Y*-type harmony, and (c) fitting *i*-type harmony.

$$F_{k,\alpha} = \sum_{p} \|h(p) - E_{T_k(\alpha)}(p)\| \cdot s(p), \tag{4.3}$$

where $\| \cdot \|$ refers to the arc-length distance on the hue wheel and $s(p)$ is the saturation of pixel $p$, which appears here as a weight since the difference between colors with low saturation is perceptually less noticeable. Now, in order to fit the $k$th model with the current image, it is necessary to look for the best rotation angle $\alpha_k^*$ that minimizes $F_{k,\alpha}$ as

$$\alpha_k^* = \underset{\alpha}{\operatorname{argmin}}(F_{k,\alpha}). \tag{4.4}$$

The above operations is repeated for all harmony models to find out how each of them fits with the image; the procedure forms a set of harmony scores $\{F_{k,\alpha_k^*}\}$, for $k = 1, 2, \ldots, K$. These harmony scores can be directly used as the hue harmony features for the image, or summarized to other features, such as choosing the minimal score to indicate how harmonious the image is (as in Reference [26]) or forming a $K$-dimension binary feature to indicate which harmony model fits the image best (as in Reference [19]). Besides using the models from Reference [38], some other methods also proposed to measure color harmony in an image [18].

#### 4.3.1.2 Illumination-Related Features

Choosing proper lighting conditions for capturing photo and controlling the exposure are basic techniques for professional photographers. Professional photographers also make good use of the lighting contrast to emphasize the main objects on which they expect the audience to focus. Therefore, the lighting effects on an image are important measures for its appeal.

- Average lightness
Considering the HSV color representation, the arithmetic average lightness value of an image can be computed. Another way to measure the average lightness of an image is to compute the logarithmic average lightness, that is, $\frac{1}{N_p}\sum_p \log(\varepsilon + v(p))$, where $v(p)$ is the lightness value of the pixel $p$, the term $N_p$ denotes the number of pixels in the image, and $\varepsilon$ is a small number to prevent from computing $\log(0)$. The logarithmic average can provide some complementary information about the dynamic range of the lighting in the image. For example, two images with the same arithmetic average lightness can have very different logarithmic average lightness, due to their different dynamic ranges.

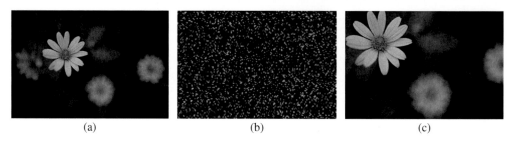

**FIGURE 4.7**

The effects of different spatial compositions: (a) original professional photograph, (b) image created by shuffling the pixels in (a), and (c) image created by changing the spatial composition in (a).

- Lightness contrast

Human color vision relies on both hue and lightness [39]. Without hue contrast, it would be difficult for human eyes to recognize different objects; without lightness contrast, it would be difficult for human eyes to decide the exact place for something. Looking at an image with flat lightness over it, human eyes cannot easily figure out a proper point to focus on. However, this does not mean that low lightness contrast definitely decreases the aesthetics of an image. An overexposed or underexposed photograph under certain scenarios may yield very original and beautiful shots. In artistic creation, low luminance contrast may create an illusion of vision [39]. This suggests that the lightness contrast is an important feature related to the image appeals although the preference of high contrast or low contrast is data dependent. One way to measure the lightness contrast is as follows. Consider the HSV color representation of the image, with $H_v$ being the histogram of the lightness values of the pixels. The histogram is used to search for the minimal region $[b_l, b_r]$ that centralizes 98% energy of the lightness histogram. The lightness contrast to be measured can be defined as $|b_r - b_l|$.

### 4.3.1.3 Spatial Composition-Related Features

Not only the color and lighting compositions are important for the image appeals, but also the locations of objects and their mutual spatial relations play an important role in the aesthetic quality of the image. Professional photographers follow a wealth of composition knowledge and techniques [1]. By shuffling the pixels of an aesthetically appealing professional photograph (Figure 4.7a), it is possible to end up getting a noisy and no-longer appealing image, although the previous color and lighting features are not changed (Figure 4.7b). Even when keeping the complete shapes of the semantic objects but just changing their spatial configuration, the aesthetic quality of the resulted image (Figure 4.7c) would be quite different from the original one. The following discussion introduces some popular features used to measure the spatial composition of an image. Many features are inspired by the guidelines and principles in photography, such as golden-section rules, shapes and lines, visual balance, and so on [12].

- Golden-section rule

Golden section refers to dividing a line at a particular ratio $(\sqrt{5} - 1)/2$, so that the ratio of the smaller part to the larger part is the same as the ratio of the larger part to the whole. The rule can be commonly found in nature, and has been widely used in photography and

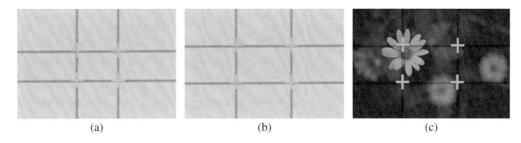

| (a) | (b) | (c) |

**FIGURE 4.8**

Demonstrations of the golden-section rule and its approximation, rule of thirds, in photography: (a) golden-section rule, (b) rule of thirds, and (c) subject at one of the intersections.

art as a tool in order to achieve balance and beauty. The golden-section rule specifies that the focus (main object of the interest) should be located at one of the four intersections as shown in Figure 4.8. In photography, this rule is often approximated by the so-called rule of thirds, which approximates the golden ratio as $1/3$. So, the first spatial composition-related feature is to measure how well the composition of an image fits with this rule. The procedure starts with deciding the main object of interest in the image. Some methods of extracting the main subject of the image were proposed in Reference [18]. Another work [22] considered face as the main subject whenever there is face within the image and performed face detection to extract it. Various object detection and region segmentation techniques can also be used to identify the main object of the image. Once the main object has been found, the golden rule-based feature can be computed as the minimum among the distances between the object center and all four golden-section line intersections:

$$f_{\text{golden}} = \min_{i=1,2,3,4} \sqrt{((O_x - P_x^i)^2 + (O_y - P_y^i)^2)}, \qquad (4.5)$$

where $(O_x, O_y)$ denotes the coordinates for the center of the object of interest and $\{(P_x^i, P_y^i)\}$ is the set of coordinates for the four intersection points. All the coordinates are normalized by the width and height of the image, respectively. If considering multiple objects/regions of interest, the above feature can be computed for each salient object/region and then combined using some appropriate techniques. An optional method for combining the set of features $\{f_{\text{golden}}^j\}$, where $j = 1, 2, \ldots, J$ for the $J$ salient objects, is given by

$$f_{\text{golden}}^c = \frac{1}{\sum_j w_j} \sum_j w_j G(f_{\text{golden}}^j). \qquad (4.6)$$

The term $G(\cdot)$ denotes a function performing a certain transform on the input features, and $w_j$ is the weight given to the $j$th salient object. Both $G(\cdot)$ and $w_j$ can be chosen based on the applications and the specific algorithms. For example, some previous works [27], [28] use $G(x) = e^{-x^2}$. In Reference [27], $w_j$ is set to be the mass of the $j$th salient region, while in Reference [28], $w_j$ indicates an importance calculated by some algorithm.

- Visual balance

Balance is an important concept related to visual harmony and aesthetics. Reference [12] discusses that salient objects are preferred to be distributed near-evenly around the center of the image for the sake of visual balance. The statement on the "right and left" balance in visual psychology says: an object with a fixed size seems to have more weight when put on the right, thus its size needs increasing when put on the left for balance [15]. This statement suggests larger objects on the left can balance with smaller objects on the right. A simple way to measure the visual balance is as follows. Consider the top $J$ most salient regions (objects) in the image. The procedure first computes the center $(C_x^j, C_y^j)$ and the respective mass $M^j$ for the $j$th region. This is followed by computing the offset between the weighted average center for these $J$ regions and the image center $(C_{im}^x, C_{im}^y)$, as follows:

$$f_{bal}^x = \frac{\sum_j M^j C_x^j}{\sum_j M^j} - C_{im}^x, \quad f_{bal}^y = \frac{\sum_j M^j C_y^j}{\sum_j M^j} - C_{im}^y, \tag{4.7}$$

with the coordinates normalized, respectively, by the width or height of the image.

- Shapes

Shapes are crucial components in an image, so they can significantly affect the aesthetic quality of the image. Reference [19] computed the high-order central moments on the co-ordinates within a segmented region to represent the shape of the region. Reference [4] computed the convex hull of a segmented region, and used the ratio of pixels within the hull to measure the degree of convexity for the region. However, both metrics can only roughly measure some characteristics of the shapes. Moreover, there is still not enough understanding of which kind of shape enhances the image appeals. To overcome this draw-back, it is possible to measure the shape in more detail and leave the learning algorithm to figure out the useful properties. Borrowing the technique "bag of features" [40], [41] popularly used for image matching, each region can be represented as a collection of local features, such as scale-invariant feature transform (SIFT) [42] and maximally stable ex-tremal region (MSER) [43] features. Another problem is that the segmentation process is not always reliable and can break one semantic region into multiple segments or merging semantically non-related regions into one. Potential solutions to these problems can replace the segmentation technique with the spatial pyramid matching technique [44] to compose the shape-related features within spatially divided regions at multiple scales. Alternatively, other techniques, such as object detection and saliency detection, can also be used to better divide the image into meaningful elements.

- Texture and patterns

Texture describes the structural characteristics of a surface, which can be classified as graininess, smoothness, line-likeness, regularity, and others. As discussed in Reference [4], whether an image looks grainy or smooth as a whole is related to the camera settings and techniques used by the photographer. A grainy image due to high ISO settings or a smooth image due to out-of-focus capturing is generally not visually preferred. Moreover, texture is not only related to the photo-shooting conditions, but also an important artistic element of the image composition in photography. Patterns, which are regular texture with more or less repetitive entity, can often evoke aesthetic feelings in humans. Figure 4.9 shows some examples with patterns. Although Figure 4.9c is not even a professional photograph, it is still pleasing to the eyes of the audience because its regular organization of smiling faces into rows. To extract texture- and pattern-related features, Reference [4] used Daubechies

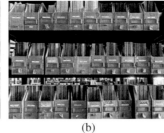

(a)                                    (b)                                    (c)

**FIGURE 4.9** (See color insert.)

Texture and patterns. Patterns in these examples evoke aesthetic feelings.

wavelet transform [45] to measure the spatial smoothness in the image. The reader should also refer to the literature on texture analysis [46], [47], [48], [49], [50], [51], [52]. However, prior works in texture analysis mainly focus on basic pattern, such as dots, lines, and lattice. Only very few studies have explored the detection of patterns formed by complex entities, such as the faces in Figure 4.9c. An example can be found in Reference [53], which proposed a graph-based algorithm for finding rows of people in group images; this concept can be extended to other object entities.

### 4.3.1.4 Depth-Related Features

Image is represented as a two-dimensional plane. However, the real world is composed of three dimensions. Therefore, a number of techniques have been used in photography to create the illusion of depth in an image. For example, professional photographers often compose images in linear perspective, as shown in Figure 4.10. Linear perspective is useful in extending human's imagining space out of the limited image space. Another commonly used technique to create depth illusion is to compose the sharpness and blurring properly. The contrast between sharp regions and blurring regions can help the viewer to catch easily the subject emphasized in the image. Besides these two techniques, professional photographers also use light, shadow, and color to enhance the three-dimensional impression of an

**FIGURE 4.10** (See color insert.)

Linear perspective in photographs.

image [54]. The following discussion focuses on two features related to the depth characteristics of an image: linear perspective and sharpness-blurring distribution.

- Linear perspective

Linear perspective is characterized by converging lines. In reality, the edges of the road, the edges of the skywalk, and the vertical lines of the building sides in Figure 4.10 are actually parallel. However, they appear to converge to a vanishing point in human eyes, though they are still perceived as straight and parallel. Since the image recorded by the camera is also optical, the same vanishing effect appears in the image, and it can be strengthened or weaken by various skills used by the photographers. The linear perspective effect of an image can be represented by estimating the vanishing point or the vanishing line of the image, and use their locations as features. The problem of vanishing point and vanishing line detection and estimation have been addressed in many previous studies [55], [56], [57], [58], [59], and [60].

- Sharpness-blurring distribution

To represent the sharpness, the presence of high-frequency edges is of interest. Professional photographers often reduce the depth of field while focusing on only a single object or region in the scene, which results in a clarified composition with sharp focusing region and blurred background and avoids producing an unpleasant cluttered scene. In Reference [4], the authors measure the ratio of high-frequency elements within the center regions over the whole image, assuming that the object of interest is usually near the center of the image. However, the location of the object of interest may vary from image to image. Another way to measure the spatial distribution of edges is to measure how compact the sharp region is. The following introduces a method similar to that proposed in Reference [17]. First, the Laplacian filter is applied to the grayscale version of the image, and take the absolute values of the results to generate an edge map $E(x,y)$ for the image. The next step aims at finding the smallest bounding box that encloses a high ratio $r_e$ of the edge energy. To do this, the edge map $E(x,y)$ can be projected onto the vertical and horizontal directions, respectively, as follows:

$$P_X(x) = \sum_y E(x,y), \quad P_Y(y) = \sum_x E(x,y). \qquad (4.8)$$

Figure 4.11 illustrates the implementation on an example image. For the projected vector $P_X (or P_Y)$, the procedure search for the minimal region $[b_X^l, b_X^r]$ (or $[b_Y^l, b_Y^r]$) that centralizes the ratio $\sqrt{r_e}$ of energy of the vector as follows:

$$\left[b_X^l, b_X^r\right] = \underset{[b_X^1, b_X^2]}{\arg\min} |b_X^2 - b_X^1|, \text{ subject to } \sum_{b_X^1 \le x \le b_X^2} P_X(x) \ge \sqrt{r_e} \sum_x P_X(x), \qquad (4.9)$$

$$\left[b_Y^l, b_Y^r\right] = \underset{[b_Y^1, b_Y^2]}{\arg\min} |b_Y^2 - b_Y^1|, \text{ subject to } \sum_{b_Y^1 \le y \le b_Y^2} P_Y(y) \ge \sqrt{r_e} \sum_y P_Y(y). \qquad (4.10)$$

Note that $r_e$ is set to 96.04% in Reference [17]. Then, the area of the bounding box $(b_X^2 - b_X^1)(b_Y^2 - b_Y^1)$ is computed to measure the compactness of the bounding box. Note that the image height and width are both normalized to one in the whole process. With this metric,

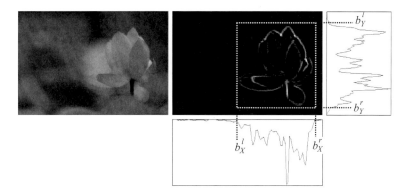

**FIGURE 4.11**

Illustration for the computation of the sharpness-blurring distribution feature.

for a cluttered image the area of the bounding box would be large while for a well-focused image the area of the bounding box would be smaller.

### 4.3.1.5  Main Subject-Related Features

Professional photographers make use of various techniques to highlight the main subject of the photo. Therefore, not every part in an image contributes equally to the aesthetic quality of the image. The subject region, which may receive more attention from the viewer, can have a more significant impact on the image appeals than the background regions. Before discussing any further about the characteristics of the subject region, various ways of separating the subject region from the background should be considered.

- Subject region detection

Reference [18] presents an efficient subject detection algorithm based on the assumption that the subject regions appear as a clearer area in the photo while the background looks blurrier. In the literature on saliency detection, many computational algorithms [61], [62], [63], [64], [65] have been proposed to extract the subject region in the image based on the characteristics of color, edge, contrast, and frequency distribution. Furthermore, if some prior information about the content of the images or the object categories of interest is available, object detection algorithms can be used to identify objects in certain categories. Among the thousands of object categories present in a daily life, research in visual psychology [54] presents that the most common high-attractant subjects are the key parts of the human face. Research in the nervous systems also shows that there are specific brain modules for recognizing faces. Thus, when faces are present in an image, some works [21], [22], [23] consider faces as salient subjects, and extract features within the detected facial regions for predicting the image aesthetics. In these cases, face detection algorithms are used to detect the subject regions of the image. Besides the faces, many other objects can be present as the main subject of the image, depending on the scenario where the image is taken. For example, if an image is already known to be taken in the zoo, then it is more likely that animals are the subjects; similarly, if it is taken in a garden, then it is more likely for the flowers to be the subjects. All these contextual information have not been thoroughly exploited to help aesthetics prediction, which will be among the future directions.

The following discussion presents certain features proposed in prior works [18], [22], [66] based on the separation of the foreground subject region and the background. While the features discussed earlier in this chapter are mostly extracted from the whole image, many of them can also be extracted within the subject region correspondingly to represent the color, illumination, texture, and depth characteristics of the subject region. Besides, it is possible to measure the contrast between the subject regions and the background in sense of lighting, blurring, and so on.

- Colorfulness of the subject region

The hue count feature introduced in Equation 4.1 is computed for the subject region to indicate how colorful this region is.

- Background simplicity

In general, simple background can help the viewer better focus on the subject region. The simplicity of the background can be measured using the color distribution of the background region. Each of the RGB channels can be quantized into $K$ bins, creating a histogram of $K^3$ bins. Then, the procedure evaluates the number of bins that have counts larger than a threshold, which is set to be a small ratio (for example, 0.1) of the maximum count in the histogram.

- Spatial location of the subject region

As discussed earlier, the location of the subject region has a high impact on the spatial composition of the image in the sense of visual balance and the fitness to rule of thirds. Here the minimal distance between the subject center and one of the four intersection points in Figure 4.8 is calculated, as described in Equation 4.5. The offset from the subject center to the image center is also measured as features related to visual balance.

- Subject-background contrast

In this case, the contrast between the subject region and the background is evaluated in terms of color, lightness, and sharpness. For color contrast, the color histogram of the subject region $Hist_s$ and that of the background region $Hist_b$ can be extracted to compute the feature $1 - <Hist_s, Hist_b> / (\|Hist_s\| \|Hist_b\|)$, where $< \cdot, \cdot >$ is the operation of inner product. For lightness contrast, the average brightness of the subject region $L_s$ and $L_b$ is computed to measure the contrast as $\log(L_s/L_b)$. For the sharpness contrast, the procedure first computes $Hf_s$, which is the ratio of the area of high-frequency components in the subject region over the subject area and $Hf_I$, which is the ratio of the area of high-frequency components across the whole image over the image area. The high-frequency components can be detected through Fourier transform or wavelet techniques. Then, the sharpness contrast feature can be defined as $Hf_s/Hf_I$.

- Face-related features

Faces are among the most effective objects that evoke the viewer's emotion. The number of faces, the sizes of the faces, the spatial relationship between multiple faces, and the facial expressions and poses, all have impacts on the image aesthetics. Among these features, the spatial relationship between multiple faces more or less indicate how close the people in the photo are in their social relationship. An intuition is that if the people within an image

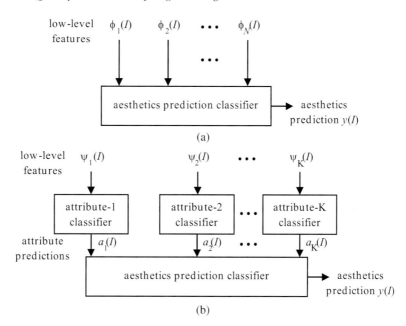

**FIGURE 4.12**

Aesthetics prediction schemes based on different inputs: (a) aesthetic prediction using low-level features and (b) aesthetic prediction using high-level attributes. The term $y(I)$ denotes the aesthetics prediction for the input image $I$. Low-level features $\phi_n(I)$, for $n = 1, 2, ..., N$, are extracted based on human intuition to represent the image appeal. Low-level features $\psi_k(I)$, for $k = 1, 2, ..., K$, are used as inputs for the $k$th attribute classifier, whose outputs are the attribute predictions $a_k(I)$.

look close to each other, it may arouse a pleasant feeling in the viewer. In Reference [22], the authors first extracted a tree graph connecting neighbor faces together, and then measured the average distance across all pairs of connected faces. Furthermore, to measure facial characteristics, such as expressions and poses, one can take advantage of the existing face expression classification and face pose estimation techniques and generate high-level attributes based on the outputs from these modules, as will be discussed below.

In summary, this section has presented five groups of low-level features to represent the image appeal in terms of color, illumination, spatial composition, depth, and the subject region. It has focused on explaining the intuition and knowledge that inspire the design of the features, rather than introducing all features described in previous works, to encourage the reader to come up with more interesting and effective features.

### 4.3.2 High-Level Attributes

The previously discussed features are designed to represent high-level attributes in aesthetics, such as image composition, subject-background contrast, and so on. However, these features themselves may not explicitly describe the aesthetic properties. Inspired by the recent progress of using describable attributes to improve the performance of object recognition [67], [68], Reference [25] proposed three groups of describable image attributes for aesthetics: i) compositional attributes related to the image layout or configuration, ii) content attributes related to the objects or scene types depicted, and iii) sky-illumination

**TABLE 4.2**
The three groups of attributes proposed in Reference [25].

| Composition Attributes | Content Attributes | Sky-Illumination Attributes |
|---|---|---|
| presence of a salient object | presence of people | clear skies |
| rule of thirds | portrait depiction | cloudy skies |
| low depth of field | presence of animals | sunset skies |
| opposing colors | indoor-outdoor classification | |
| | scene type (15 categories) | |

attributes related to the natural lighting conditions. Instead of designing low-level features related to these characteristics (Figure 4.12a) as discussed in Section 4.3.1, the proposed method in Reference [25] explicitly trained classifiers to recognize various high-level describable attributes (Figure 4.12b).

Table 4.2 summarizes the three groups of attributes used in Reference [25]. For each attribute, a classifier or predictor is trained on data that contain manual labels for the respective attribute. For example, in order to build a classifier for the "low depth of field" attribute, the authors first collected a dataset of thousands of images with manual labels indicating whether an image has low depth of field. Then, the authors trained a support vector machine (SVM) classifier taking Daubechies wavelet-based features as inputs and outputting a score to indicate how confident the image has low depth of field. Similarly, to build a classifier for the "clear skies" attribute, which describes whether a photo is taken in sunny clear weather conditions, three-dimensional color histograms are computed on roughly segmented sky regions, followed by a SVM classifier trained using the manually labeled images.

Experiments in Reference [25] demonstrated that using the high-level attributes produced a significantly more accurate ranking than using the low-level features proposed in Reference [17]. Moreover, the combination of both the low-level features and the high-level attributes resulted in an even stronger classifier for predicting aesthetics. This suggests that the low-level features and the high-level attributes can provide useful complementary information for each other.

High-level attributes to be used in the aesthetic quality assessment are not limited to those proposed in Reference [25]. Consider color characteristics of images as an example. Although the only color-related attribute in Table 4.2 is the presence of opposing colors, there exist works, such as References [19] and [26], that use all the color harmony models proposed in Reference [38] to represent "the presence of color harmony" attribute within the image. As another example one can consider attributes related to the presence of different objects. The presence of people and animals can influence the aesthetic impression of an image, and other objects, such as plants and decorations, can also be correlated with image aesthetics. Given that the second-stage aesthetic classifier can automatically learn the importance of different attributes, the more information about the visual content of the image is derived, the better the aesthetic quality can be predicted. This makes the task of aesthetic quality assessment no longer an isolated task in computer vision, but can be benefited from many other tasks that have been more thoroughly explored, such as scene categorization, object detection, depth estimation, saliency detection, geometric layout prediction, and

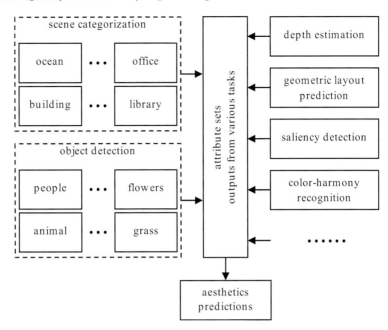

**FIGURE 4.13**

Seeking help from various tasks in computer vision toward the aesthetic assessment.

perspective estimation. One can simply take advantage of the existing state-of-the-art algorithms for all these different tasks and use their outputs as attribute inputs for the aesthetics classifier, as shown in Figure 4.13.

## 4.4   Learning Models for Aesthetic Quality Assessment

The previous section has identified a set of features, including the low-level features and high-level attributes, to represent the image appeal. Although the design of some features are inspired by rules in photography, art, and psychology, no rules are manually set here for the mapping from the computational features to the aesthetics scores. This section will introduce the learning modules applied to the extracted features.

### 4.4.1   Feature Selection

In the earlier stage, various aspects potentially related to image aesthetics can be taken into account, thus likely resulting in a large-size feature set. However, some features may cover overlapping information with others, and some may be less correlated to the human ratings than others. Especially when considering outputs from some relevant tasks, such as high-level attributes (for example, the predicted depths across the image from the depth estimation task), the dimension of the features becomes very high and can suffer from overfitting in building learning models. Dimension reduction algorithms and feature selection algorithms are both well-studied topics in machine learning. Compared to popular dimension reduction algorithms like principle component analysis (PCA), feature selection approaches, which try to find a subset of the original features, provide greater power in interpreting the resulted low-dimension features. There are two common strategies in feature selection [69]: filter-based approaches and wrapper-based approaches. Filter-based approaches work as a preprocessing step to remove redundant features based on metrics, such as correlation and mutual information, or remove irrelevant features based on the class separability of individual features. Wrapper-based approaches choose features by searching through the space of possible features and evaluating the subsets by running the classifier on the subset. For example, among previous works in aesthetics assessment, Reference [4] used the forward selection algorithm, References [16] and [19] used boosting algorithms [70], [71], and Reference [28] used L1-sparsity based linear model [72], in order to select features.

### 4.4.2 Classification

Many works on assessing visual aesthetics have formulated the quality assessment problem as a classification problem. Formulating the task into a classification problem not only provides an easier starting point for this new research area but also reduces the noise due to the inconsistency of the human ratings. The ratings are highly subjective, and different people may have different standards for their ratings. Even the ratings given by the same person toward different images are likely to be made based on not the same standards. Therefore, the slight differences in the average scores within a certain range, are not necessarily meaningful. Instead, considering images whose average scores are within a certain range as a group, can help to eliminate the effect from inconsistent ratings. In this case, each group of images represent the aesthetic quality at a certain level, corresponding to a certain quality category. Hence, by dividing the average scores in to $M$ regions, one ends up with an $M$-class aesthetic quality classification problem, defined below.

Consider a set of images with aesthetic quality labels $(I_1, y_1), (I_2, y_2), \dots, (I_N, y_N)$, where $I_i$ indicates the features extracted from the respective image and $y_i$ is the quality class label for the respective image. In the training stage, a classification algorithm is learned to discriminate the data in different classes. In the testing stage, given a testing image with extracted features $I_{test}$, the learned algorithm predicts its quality label $\hat{y}_{test}$.

Many previous works treat the problem as a binary classification problem, that is, $M = 2$. The two classes are defined as "high-aesthetic quality" and "low-aesthetic quality." As discussed in Section 4.2.2, the human ratings from some controlled studies [21], [22] have shown that the rating variance values for images with extremely low or extremely high average scores are lower than those for images with average scores in the middle. Therefore, to reduce ambiguity, previous works often choose images with top average scores as the high-quality class and images with bottom average scores as the low-quality class. Classifiers used in previous works include decision trees [4], Bayesian classifiers [16], [17], [18], [19], [73], SVM [4], [16], [18], [22], [25], boosting [16], [18], [19], and so on.

To evaluate the performance of an aesthetic quality assessment algorithm whose learning module performs classification, one can use the general metrics in classification. For the two-class aesthetics classification, the receiver-operating characteristic curve or precision-recall curve are generally used. Take the precision-recall curve as an example. At any threshold set to separate the testing outputs, the precision and recall are computed as

$$recall = \frac{\text{\# high-quality photos above threshold}}{\text{total \# high-quality photos}},$$ (4.11)

$$precision = \frac{\text{\# high-quality photos above threshold}}{\text{\# photos above threshold}}.$$ (4.12)

To evaluate a multi-class aesthetics classification algorithm, some previous works [22] used the metric cross-category error (CCE). Consider $M$ categories and the category label increasing from 1 to $M$ to indicate that the respective quality varies from lowest to highest, that is, Category 1 indicates the lowest aesthetic-quality category while Category $M$ indicates the highest aesthetic-quality category. The CCE is then defined as follows:

$$CCE(k) = \frac{1}{N_{test}} \sum_{i}^{N_{test}} \mathscr{I}(\hat{y}_i - y_i == k),$$ (4.13)

where $N_{test}$ denotes the number of photos for testing, $y_i$ is the ground-truth category label for the $i$th test image, and $\hat{y}_i$ is the estimated category label for the $i$th test image. The function $\mathscr{I}(\cdot)$ equals to one when its content stands, otherwise it equals to zero. The term $CCE(k)$ indicates the ratio of images whose estimated labels has a $k$-category offset to their respective ground-truth labels.

### 4.4.3 Regression

Although the classification solution can help to eliminate the learning model from overfitting to the noisy scores, the ideal target is to have the computer robustly predict aesthetic scores as human do. Especially for some tasks like ranking photos based on aesthetics, one may want more accurate score outputs other than just rough quality levels since the learning in classification problems do not ensure to catch the relative ranking between images in the same category. Therefore, this section describes regression models for connecting the extracted features and the human ratings on the image aesthetics.

*Regression for aesthetic quality prediction.* Consider a set of images with aesthetic quality labels $(I_1, s_1), (I_2, s_2), \ldots, (I_N, s_N)$, where $I_i$ indicates the features extracted from the respective image and $s_i$ is the average rating score on aesthetics for the respective image. In the training stage, a regression model is learned to approximate the mapping between the image features and the desired scores. In the testing stage, given a testing image with extracted features $I_{test}$, the learned model predicts its aesthetic quality score $\hat{s}_{test}$.

Previous attempts on using regression algorithms for aesthetics prediction relied on linear regression [4], [22], [28], [73], SVM regression [22], and so on. To evaluate the performance of the predicted aesthetics, the residual sum-of-squares error defined as

$$R_{res} = \frac{1}{N_{test} - 1} \sum_{i=1}^{N_{test}} (\hat{s}_i - s_i)^2,$$ (4.14)

is commonly used. Here, $N_{test}$ is the number of photos for testing, $s_i$ is the ground-truth aesthetic score for the $i$th test image, and $\hat{s}_i$ is the respective estimated score. To verify whether the extracted features have contribution to the prediction, Reference [4] proposed the processing flow depicted in Algorithm 4.1.

**ALGORITHM 4.1** Verification of the extracted features contributing to the aesthetic quality prediction.

1. Shuffle the ground-truth scores for the training images, so that the correlation between the scores and the features are broken.

2. Train the regression model based on the shuffled scores.

3. Test the resulted model on the testing data and compute the $R_{res}^{shfl}$.

4. Repeat the above operations for several rounds and calculate the average error as $\bar{R}_{res}^{shfl}$.

5. Compare $\bar{R}_{res}^{shfl}$ with the performance $R_{res}$ from the normal algorithm. If $R_{res}$ is obviously smaller, it indicates that the extracted features contribute to the aesthetic quality prediction.

## 4.5 Applications

With the proliferation of digital cameras, the number of photos that can be accessed online and that are stored in personal photo albums is growing explosively. In 2010, an average of 6.5 million photographs were uploaded daily by the users of Flickr. People instinctually seek for visual aesthetics in daily life. Automatic aesthetic quality assessment can help people better search, browse, manage, and create visually appealing photos. This section discusses several real-life applications of aesthetic quality assessment.

### 4.5.1 Aesthetics Assisted Web Search

Many of the previous works on visual aesthetics assessment [17], [18], [25] have demonstrated the effectiveness of their algorithms in refining the ranks for images retrieved from popular web search applications. A common situation of using current image search engines is as follows. The search engine retrieves a large amount of related results based on the user's query; however, the top results do not look visually appealing. In that case one need to browse more results to find the one that is both relevant and visually satisfying. In this situation, automatic aesthetic quality assessment can reduce this effort. Acting as a postprocessing step, the aesthetic quality assessment algorithm re-ranks the top retrieved images based on their aesthetics. As a result, the top ranked images are of higher quality. In the future, the ranking based on aesthetics is expected to be combined with other ranking criteria in order to provide a more enjoyable experience in image search experience.

### 4.5.2 Automatic Photo Enhancement

Photo editing tools are commonly used by photographers and designers to modify some characteristics of a photo according to the user's intent. Commercial software like Adobe Photoshop provides such tools but often requires the user to have good knowledge of design concepts and photography theory. For general users, it is unclear which elements need to be edited and how to edit them in order to make their images more appealing. In this case, an automatic photo-editing tool for increasing the image aesthetics would be helpful. As discussed in Reference [27], developing such a tool is a twofold problem: how to edit the image and how to measure the aesthetics for different editing proposals. The answer to the latter is to employ the aesthetic quality assessment technique. Some success has been achieved using the aesthetic quality assessment technique to help photo editing [26], [27], [28], [29], [30]. Reference [26] focused on editing the color aesthetics in terms of harmony; the best editing scheme is chosen by measuring the different aesthetic qualities resulting from fitting different color harmony schemes. Other works [27], [28], [29], [30] utilized the assessment of composition-based aesthetics to aid in the automatic editing of photo composition. A common implementation is to iterate between two operations: selecting a candidate scheme and evaluating the aesthetic quality of the candidate scheme. All of these attempts demonstrate the promising potential of the aesthetics assessment technique, which allows a computer not only to tell the user how good a photo looks but also to help the user to enhance the photo appeal automatically.

### 4.5.3   Photo Selection and Album Management

As discussed in Section 4.5, the proliferation of personal photos raises an issue: manually managing a large amount of photos would be time consuming. Thus, it becomes necessary and important to develop automated tools for efficient photo selection and management. Automated tools, such as those in References [20] and [31], have been proposed to better organize and retrieve photos based on various photo characteristics. Consider the manual photo selection process for general people. Aesthetic visual quality actually plays an important role in this process. Reference [28] presented a photo selection application to select quintessential photos from large personal photo collections. The input of the application, a personal album, can be a collection of photos taken during a trip with friends, a collection of photos from a series of family events, and so on. The photos might contain different individuals and can be taken in different places and by different cameras. The core algorithm in this application measures the quintessence in terms of appeal and representation. Regarding photo appeal, the selected photos are expected to have high scores predicted by the aesthetic quality assessment algorithm. Regarding representation, the selected photos are supposed to cover as many different scenes and as many different people that have appeared in the collection as possible. As a result, the top selected photos are not only aesthetically appealing but also diverse in sense of scene ad people inside the image. This application demonstrates the potential of using aesthetic quality as an optional metric for photo indexing in album management.

### 4.5.4   Personalized Aesthetics

So far, this chapter has considered a learning setting trained on the average ratings and the consensus in a general community. As a result, a model for predicting the consensus-based aesthetics has been learned. However, given sufficient information and feedback from a single user or a small group of users that share similar opinions in aesthetics, personalized models of aesthetics can be learned. Since the aesthetic quality assessment is a highly subjective task, individual models can help adapt to different tastes. Personalization was discussed in Reference [34], but no computational implementation was provided there. A personalized photograph ranking and selection system was proposed in Reference [24], where information given by the users were employed to train and adjust the parameters inside the score prediction module. The system provided two personalized ranking user interfaces: one is feature based, and the other is example based. Users can provide either their preferences on some specific aspects of the image (feature based), or select their preferred and non-preferred examples (example based). With the extra user input, the system outperformed the previous algorithms [17], [18] that studied only the general aesthetics. Also, a human study showed that the example-based interface was preferred by twice as many people than the feature-based one. This result is understandable, since the feature-based input may be confusing for the general users who are not familiar with the underlying characteristics of images. Recently, a preliminary attempt on learning the personalized aesthetics of videos was conducted [32]. Although only two persons were investigated in this initial trial, the preliminary results showed that the personalized aesthetics learners can produce more accurate predicted aesthetic scores than the generalized aesthetics learner.

### 4.5.5 Video Aesthetics

With the increasing attention on image aesthetics and the proliferation of videos on the Internet, research has been extended from image aesthetics to video aesthetics [18], [32]. Two databases used for video aesthetics have been constructed. The dataset used in Reference [18] contains 4000 professional movie clips and 4000 amateurish clips, labeled as high aesthetic quality and low aesthetic quality videos, respectively. However, as pointed out in Reference [32], there is not enough information on how the dataset was obtained, what the video contents are, or how well the assumption of equating professional videos to aesthetically high quality videos stands. To address these problems, a smaller but better-controlled consumer video dataset was proposed in Reference [32]. This database contains 160 videos from YouTube with aesthetic scores collected via a controlled human study. Although the existing datasets are not well controlled or relatively small, they provide a starting point to explore video aesthetics.

The main difference between video content and image content lies in the additional temporal and motion information across the frames in a video sequence. Therefore, besides using previously mentioned image-based features, the recent works [18], [32] proposed a series of features related to the temporal and motion characteristics, including the length of the subject region motion in the video, the motion stability, the actual frame-rate in the video, the blockiness effect from video compression, and so on. While Reference [18] combined the frame-based features by taking the average, Reference [32] proposed a two-stage pooling strategy: first, generating shot-based features by combining the frame-based features and second, generating the video feature from the shot-based features. The promising results reported in these works help to envision many potential applications in the near future. For example, Reference [32] presented some plans to develop aesthetics-assisted hierarchical user interfaces to allow end-users to efficiently navigate their personal video collections. Furthermore, as the image aesthetics assessment technique can help to re-rank the results from the image search engines, video aesthetics assessment techniques can be also utilized to refine the results returned by the online video search engines.

Applications that can benefit from aesthetic quality assessment are not limited to the above examples. For instance, a real-time aesthetic quality assessment module could be integrated into the cameras to feature new functions, such as providing users with real-time composition or lighting suggestions. However, continuous efforts are needed to advance the development of the aesthetic quality assessment techniques.

## 4.6 Conclusion

This chapter discussed the assessment of the visual aesthetic quality. Because of the subjectivity of aesthetics, the problem of aesthetic quality assessment is far more difficult compared to objective quality assessment, which measures the image quality or its lack due to various degradations, for example, introduced during image acquisition, or caused by faulty transmission or aggressive processing / compression. Recent efforts [4], [17], [18], [19], [22], [24], [32] in this area have achieved some progress, by addressing the automatic evaluation of aesthetics as a machine learning problem. Following the general flow used by recent algorithms, this chapter discussed the basic modules in developing algorithms for aesthetics assessment: data collection and human study, computational feature extraction, and learning techniques. The focus was on the computational feature extraction, since the features designed to represent the aesthetic characteristics have immediate influence on the amount of useful information passed to the following learning modules. Efforts toward feature extraction span from designing effective image features according to the knowledge and experience in visual psychology, photography, and art, to creating high-level attributes by employing state-of-the-art techniques from other computational vision areas.

Although promising results have been demonstrated by recent works in predicting the visual aesthetics, there are many avenues that still need to be explored. Computational aesthetics is a challenging multidisciplinary research topic. Closer communication and collaboration between researchers in the areas of psychology, photography, and computational vision will be of great help for further progress. On the visual computing side, researchers have been making continuous efforts to exploit, through computational intelligent algorithms, the underlying structures, and characteristics within the images, which relate to the perceived quality by humans. Furthermore, it is also worthwhile to develop more real-life applications based on the assessment of visual aesthetics in order to attract more attention from the general community, from whom stronger support in terms of data collection, human study, and algorithm design can be achieved.

## Acknowledgment

Professional photographs presented in this chapter have been taken by Shun Zeng. The remaining photographs have been taken by the authors.

## References

[1] B. Martinez and J. Block, *Visual Forces, an Introduction to Design*. New York, USA: Prentice-Hall, 1998.

[2] G. Peters, "Aesthetic primitives of images for visualization," in *Proceedings of the International Conference on Information Visualization*, Zurich, Switzerland, July 2007, pp. 316–325.

[3] V. Rivotti, J. Proença, J.A. Jorge, and M.C. Sousa, "Composition principles for quality depiction and aesthetics," in *Proceedings of the Eurographics Workshop on Computational Aesthetics in Graphics, Visualization and Imaging*, Banff, AB, Canada, June 2007, pp. 37–44.

[4] R. Datta, D. Joshi, J. Li, and J.Z. Wang, "Studying aesthetics in photographic images using a computational approach," in *Proceedings of the European Conference on Computer Vision*, Graz, Austria, May 2006, pp. 21–26.

[5] A.K. Moorthy and A.C. Bovik, "Visual quality assessment algorithms: What does the future hold?," *Multimedia Tools Applications*, vol. 51, no. 2, pp. 675–696, January 2011.

[6] S. Zeki, *Inner Vision: An Exploration of Art and the Brain*. New York, USA: Oxford University Press, 1999.

[7] K.B. Schloss and S.E. Palmer, "Aesthetic preference for color combinations," in *Proceedings of the 7th Annual Meeting of the Vision Science Society*, Sarasota, FL, USA, May 2007, pp. 123–128.

[8] C. Nothelfer, K.B. Schloss, and S.E. Palmer, "The role of spatial composition in preference for color pairs," in *Proceddings of the 9th Annual Meeting of the Vision Science Society*, Naples, FL, USA, May 2009, pp. 623–627.

[9] R.M. Poggesi, K.B. Schloss, and S.E. Palmer, "Preference for three-color combinations in varying proportions," in *Proceedings of the 9th Annual Meeting of the Vision Science Society*, Naples, FL, USA, May 2009, pp. 613–617.

[10] O. Axelsson, "Towards a psychology of photography: Dimensions underlying aesthetic appeal of photographs," *Perceptual and Motor Skills*, vol. 105, no. 2, pp. 411–434, January 2007.

[11] J. Gardner, C. Nothelfer, and S.E. Palmer, "Exploring aesthetic principles of spatial composition through stock photography," in *Proceedings of the 8th Annual Meeting of the Vision Science Society*, Naples, FL, USA, May 2008, pp. 337–338.

[12] B. Krages, *Photography: The Art of Composition*. New York, USA: Allworth Press, 2005.

[13] M. Freeman, *The Photographer's Eye: Composition and Design for Better Digital Photos*. Lewes, UK: The Ilex Press, 2007.

[14] B. London, J. Upton, J. Stone, K. Kobre, and B. Brill, *Photography*. London, UK: Pearson Prentice Hall, 2005.

[15] R. Arnheim, *Art and Visual Perception: A Psychology of the Creative Eye*. Berkeley, CA, USA: University of California Press, 1974.

[16] H. Tong, M. Li, H. Jiang Zhang, J. He, and C. Zhang, "Classification of digital photos taken by photographers or home users," in *Proceedings of Pacific Rim Conference on Multimedia*, Tokyo, Japan, November 2004, pp. 198–205.

[17] Y. Ke, X. Tang, and F. Jing, "The design of high-level features for photo quality assessment," in *Proceedings of the IEEE Conference on Computer Vision and Pattern Recognition*, New York, USA, June 2006, pp. 419–426.

[18] Y. Luo and X. Tang, "Photo and video quality evaluation: Focusing on the subject," in *Proceedings of the European Conference on Computer Vision*, Marseille, France, October 2008, pp. 97–100.

[19] C. Li and T. Chen, "Aesthetic visual quality assessment of paintings," *IEEE Journal of Selected Topics in Signal Processing*, vol. 3, no. 2, pp. 236–252, April 2009.

[20] A. Loui, M. Wood, A. Scalise, and J. Birkelund, "Multidimensional image value assessment and rating for automated albuming and retrieval," in *Proceedings of the International Conference on Image Processing*, San Diego, CA, USA, October 2008, pp. 97–100.

[21] C. Cerosaletti and A. Loui, "Measuring the perceived aesthetic quality of photographic images," in *Proceedings of the First International Workshop on Quality of Multimedia Experience*, San Diego, CA, USA, July 2009, pp. 282–289.

[22] C. Li, A. Gallagher, A. Loui, and T. Chen, "Aesthetic quality assessment of consumer photos with faces," in *Proceedings of the IEEE International Conference on Image Processing*, Hong Kong, September 2010, pp. 3221–3224.

[23] W. Jiang, A. Loui, and C. Cerosaletti, "Automatic aesthetic value assessment in photographic images," in *Proceedings of the IEEE International Conference on Multimedia and Expo*, Singapore, July 2010, pp. 920–925.

[24] C.H. Yeh, Y.C. Ho, B.A. Barsky, and M. Ouhyoung, "Personalized photograph ranking and selection system," in *Proceedings of the International Conference on Multimedia*, Firenze, Italy, October 2010, pp. 211–220.

[25] S. Dhar, V. Ordonez, and T.L. Berg, "High level describable attributes for predicting aesthetics and interestingness," in *Proceedings of the IEEE Conference on Computer Vision and Pattern Recognition*, Colorado Springs, CO, USA, June 2011, pp. 122–133.

[26] D. Cohen-Or, O. Sorkine, R. Gal, T. Leyvand, and Y.Q. Xu, "Color harmonization," *ACM Transactions on Graphics*, vol. 25, no. 3, pp. 624–630, July 2006.

[27] L. Liu, R. Chen, L. Wolf, and D. Cohen-Or, "Optimizing photo composition," *Computer Graphic Forum*, vol. 29, no. 2, pp. 469–478, January 2010.

[28] C. Li, A.C. Loui, and T. Chen, "Towards aesthetics: A photo quality assessment and photo selection system," in *Proceedings of the International Conference on Multimedia*, Firenze, Italy, October 2010, pp. 827–830.

[29] S. Bhattacahrya, R. Sukthankar, and M. Shah, "A framework for photo-quality assessment and enhancement based on visual aesthetics," in *Proceedings of the International Conference on Multimedia*, Firenze, Italy, October 2010, pp. 271–280.

[30] M. Nishiyama, T. Okabe, Y. Sato, and I. Sato, "Sensation-based photo cropping," in *Proceedings of the 17th ACM International Conference on Multimedia*, Beijing, China, October 2009, pp. 669–672.

[31] W.T. Chu, C.H. Lin, and J.Y. Yu, "Feature classification for representative photo selection," in *Proceedings of the 17th ACM International Conference on Multimedia*, Beijing, China, October 2009, pp. 509–512.

[32] A.K. Moorthy, P. Obrador, N. Oliver, and A.C. Bovik, "Towards computational models of the visual aesthetic appeal of consumer videos," in *Proceedings of the 11th European conference on Computer vision*, Heraklion, Crcte, Greece, September 2010, pp. 1–14.

[33] A. Sorokin and D. Forsyth, "Utility data annotation with amazon mechanical turk," in *Proceedings of the Computer Vision and Pattern Recognition Workshop*, Anchorage, Alaska, USA, June 2008, pp. 392–398.

[34] R. Datta, J. Li, and J. Wang, "Algorithmic inferencing of aesthetics and emotion in natural images: An exposition," in *Proceedings of the IEEE International Conference on Image Processing*, San Diego, CA, USA, October 2008, pp. 105–108.

[35] A.E. Savakis, S.P. Etz, and A.C. Loui, "Evaluation of image appeal in consumer photography," *Proceedings of SPIE*, vol. 3959, January 2000, pp. 111–120.

[36] M. Freeman, *The Complete Guild to Light and Lighting in Digital Photography*. Lewes, UK: The Ilex Press, 2007.

[37] M. Tokumaru, N. Muranaka, and S. Imanishi, "Color design support system considering color harmony," in *Proceedings of the IEEE International Conference on Fuzzy System*, Honolulu, Hawaii, USA, May 2002, pp. 378–383.

[38] Y. Matsuda, *Color Design*. Tokyo, Japan: Asakura Shoten, 1995.

[39] M. Livingstone, *Vision and Art: The Biology of Seeing*. New York, USA: Harry N. Abrams, 2002.

[40] G. Csurka, C.R. Dance, L. Fan, J. Willamowski, and C. Bray, "Visual categorization with bags of keypoints," in *Proceedings of the European Conference on Computer Vision – Workshop on Statistical Learning in Computer Vision*, Prague, Czech Republic, May 2004, pp. 1–22.

[41] J. Winn, A. Criminisi, and T. Minka, "Object categorization by learned universal visual dictionary," in *Proceedings of the IEEE International Conference on Computer Vision*, Beijing, China, October 2005, pp. 1800–1807.

[42] D.G. Lowe, "Distinctive image features from scale-invariant keypoints," *International Journal of Computer Vision*, vol. 60, no. 2, pp. 91–110, November 2004.

[43] J. Matas, O. Chum, M. Urban, and T. Pajdla, "Robust wide baseline stereo from maximally stable extremal regions," in *Proceedings of British Machine Vision Conference*, Cardiff, UK, September 2002, pp. 384–393.

[44] S. Lazebnik, C. Schmid, and J. Ponce, "Beyond bags of features: Spatial pyramid matching for recognizing natural scene categories," in *Proceedings of the IEEE Conference on Computer Vision and Pattern Recognition*, New York, USA, June 2006, pp. 2169–2178.

[45] I. Daubechies, *Ten Lectures on Wavelets*. Philadelphia, PA, USA: Society for Industrial and Applied Mathematics, 1992.

[46] T. Leung and J. Malik, "Representing and recognizing the visual appearance of materials using three-dimensional textons," *International Journal of Computer Vision*, vol. 43, no. 1, pp. 29–44, June 2001.

[47] C. Schmid, "Constructing models for content-based image retrieval," in *Proceedings of the IEEE Conference on Computer Vision and Pattern Recognition*, Kauai, HI, USA, December 2001, pp. 39–45.

[48] S. Lazebnik, C. Schmid, and J. Ponce, "Affine-invariant local descriptors and neighborhood statistics for texture recognition," in *Proceedings of the IEEE International Conference on Computer Vision*, Nice, France, October 2003, pp. 649–655.

[49] O.G. Cula and K.J. Dana, "3D texture recognition using bidirectional feature histograms," *International Journal of Computer Vision*, vol. 59, no. 1, pp. 33–60, August 2004.

[50] M. Varma and A. Zisserman, "A statistical approach to texture classification from single images," *International Journal of Computer Vision*, vol. 62, no. 1–2, pp. 61–81, April 2005.

[51] J. Hays, M. Leordeanu, A.A. Efros, and Y. Liu, "Discovering texture regularity as a higher-order correspondence problem," in *Proceedings of European Conference on Computer Vision*, Graz, Austria, May 2006, pp. 522–535.

[52] M. Park, K. Brocklehurst, R. Collins, and Y. Liu, "Deformed lattice detection in real-world images using mean-shift belief propagation," *IEEE Transactions on Pattern Analysis and Machine Intelligence*, vol. 31, no. 10, pp. 1804–1816, October 2009.

[53] A. Gallagher and T. Chen, "Finding rows of people in group images," in *Proceedings of the IEEE International Conference on Multimedia and Expo*, New York, USA, June 2009, pp. 602–605.

[54] M. Freeman, *The Photographer's Eye: Composition and Design for Better Digital Photos.* Lewes, UK: Ilex Press, 2007.

[55] R. Collins and R. Weiss, "Vanishing point calculation as a statistical inference on the unit sphere," in *Proceedings of International Conference on Computer Vision*, Osaka, Japan, December 1990, pp. 400–403.

[56] B. Brillault-O'Mahony, "New method for vanishing point detection," *CVGIP Image Understanding*, vol. 54, no. 2, pp. 289–300, January 1991.

[57] M. Antone and S. Teller, "Automatic recovery of relative camera rotations for urban scenes," in *Proceedings of IEEE Conference on Computer Vision and Pattern Recognition*, Hilton Head, SC, USA, June 2000, pp. 282–289.

[58] C. Rother, "A new approach for vanishing point detection in architectural environments," in *Proceedings of the 11th British Machine Vision Conference*, Bristol, UK, September 2000, pp. 382–391.

[59] J. Kosecka and W. Zhang, "Video compass," in *Proceedings of the European Conference on Computer Vision*, Copenhagen, Denmark, May 2002, pp. 476–490.

[60] D. Hoiem, A. Efros, and M. Hebert, "Automatic photo pop-up," in *International Conference on Computer Graphics and Interactive Techniques*, Los Angeles, CA, USA, July 2005, pp. 522–535.

[61] L. Itti, C. Koch, and E. Niebur, "A model of saliency-based visual attention for rapid scene analysis," *IEEE Transactions on Pattern Analysis and Machine Intelligence*, vol. 20, no. 11, pp. 1254–1259, November 1998.

[62] Y.F. Ma and H.J. Zhang, "Contrast-based image attention analysis by using fuzzy growing," in *Proceedings of the 11th ACM International Conference on Multimedia*, Berkeley, CA, USA, November 2003, pp. 374–381.

[63] X. Hou and L. Zhang, "Saliency detection: A spectral residual approach," in *Proceedings of the IEEE Conference on Computer Vision and Pattern Recognition*, Minneapolis, MN, USA, June 2007, pp. 1–8.

[64] R. Achanta, S. Hemami, F. Estrada, and S. Susstrunk, "Frequency-tuned salient region detection," in *In Proceedings of IEEE Conference on Computer Vision and Pattern Recognition*, Miami, FL, USA, June 2009, pp. 524–531.

[65] T. Liu, Z. Yuan, J. Sun, J. Wang, N. Zheng, X. Tang, and H.Y. Shum, "Learning to detect a salient object," *IEEE Transactions on Pattern Analysis and Machine Intelligence*, vol. 33, no. 2, pp. 353–367, February 2011.

[66] X. Sun, H. Yao, R. Ji, and S. Liu, "Photo assessment based on computational visual attention model," in *Proceedings of the 17th ACM International Conference on Multimedia*, Beijing, China, October 2009, pp. 541–544.

[67] A. Farhadi, I. Endres, D. Hoiem, and D. Forsyth, "Describing objects by their attributes," in *Proceedings of the IEEE Conference on Computer Vision and Pattern Recognition*, Miami, FL, USA, June 2009, pp. 1778–1785.

[68] C.H. Lampert, H. Nickisch, and S. Harmeling, "Learning to detect unseen object classes by betweenclass attribute transfer," in *Proceedings of the IEEE Conference on Computer Vision and Pattern Recognition*, Miami, FL, USA, June 2009, pp. 951–958.

[69] I. Guyon and A. Elisseeff, "An introduction to variable and feature selection," *Journal of Machine Learning Research*, vol. 3, pp. 1157–1182, March 2003.

[70] J. Friedman, T. Hastie, and R. Tibshirani, "Additive logistic regression: A statistical view of boosting," *Annals of Statistics*, vol. 28, no. 2, pp. 337–407, April 2000.

[71]  Y. Freund and R.E. Schapire, "A short introduction to boosting," in *Proceedings of the 16th International Joint Conference on Artificial Intelligence*, Stockholm, Sweden, July 1999, pp. 1401–1406.

[72]  J. Mairal, M. Leordeanu, F. Bach, M. Hebert, and J. Ponce, "Discriminative sparse image models for class-specific edge detection and image interpretation," in *Proceedings of the European Conference on Computer Vision*, Marseille, France, October 2008, pp. 297–302.

[73]  R. Datta, J. Li, and J.Z. Wang, "Learning the consensus on visual quality for next-generation image management," in *Proceedings of the 15th International Conference on Multimedia*, Augsburg, Germany, September 2007, pp. 533–536.

# 5

*Perceptually Based Image Processing Algorithm Design*

**James E. Adams, Jr., Aaron T. Deever, Efraín O. Morales, and Bruce H. Pillman**

## 5.1  Introduction

The human visual system plays a pivotal role in the creation of images in a digital camera. Perceptually based image processing helps ensure that an image generated by a digital camera "looks right" from a human visual system perspective. In achieving this goal, perceptual considerations are present in all aspects of a digital image processing chain.

In this chapter, the components of a digital image processing chain are described, with particular emphasis on how perceptual considerations influence algorithm design at each stage in the processing chain. There are a variety of ways in which the human visual system impacts processing decisions. Even before the capture button is pressed, perceptual considerations affect the decisions made by the automatic camera control algorithms that adjust the exposure, focus, and white balance settings of the camera. The design of the color filter array on the image sensor, as well as the associated demosaicking algorithm, are influenced by the human visual system. Computational efficiency is also important in a digital image processing chain, and algorithms are designed to execute quickly without sacrificing perceptual quality. Even image storage is influenced by perceptual considerations, as image compression algorithms exploit the characteristics of the human visual system to discard perceptually insignificant data and reduce the size of the image file.

The remainder of this chapter is organized as follows. In Section 5.2, a canonical digital camera image processing chain is presented. This section introduces the processing steps that are described in greater detail throughout the remainder of the chapter. Namely, Section 5.3 discusses automatic camera control, Section 5.4 focuses on demosaicking, Section 5.5 discusses noise reduction, and Section 5.6 presents color rendering. Edge enhancement is detailed in Section 5.7 and compression is discussed in Section 5.8. Concluding remarks are provided in Section 5.9.

## 5.2  Digital Camera Image Processing Chain

A digital camera image processing chain is constructed to simulate the most important functions of the *human visual system* (HVS). What follows is an overview of a representative chain [1], with detailed discussions of the components occurring in subsequent sections.

Figure 5.1 is a diagram of an example chain. The input image of this chain is a "raw" color filter array (CFA) image produced by the camera sensor. From an HVS perspective, much may have already occurred prior to the CFA image being produced. If the camera is in an automatic setting, autofocus, autoexposure, and automatic white balancing algorithms (collectively referred to as "A*") will have sensed the scene edge content, brightness level, and illuminant color and have adjusted the corresponding camera capture parameters before the shutter button has been pressed. This can be thought of as simulating the automatic adjustment of HVS parameters, such as the eye's pupil diameter and lens thickness, as well as the gains within the retina.

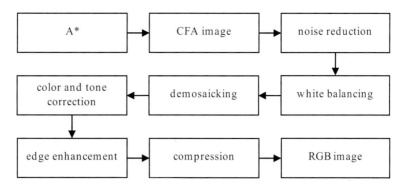

**FIGURE 5.1**

Digital camera image processing chain.

Once the CFA image has been read from the camera sensor, the first of possibly many *noise reduction* operations is performed. While the name of this operation carries an obvious meaning and intent, noise reduction is a far more subtle and sophisticated task than one might presume. From the HVS perspective, not all kinds of *noise* are equally important, and married with the additional HVS perspective that not all kinds of *scene detail* are equally important, the noise reduction task now becomes one of modifying the aggressiveness of the cleaning process to respect the integrity of the underlying scene detail that is visually most important. At the same time, the effectiveness of the noise reduction operation must be ensured.

Following noise reduction, white-point correction or *white balancing*, is performed. This simulates the *chromatic adaptation* of the HVS. Regardless of the color of the scene illuminant, be it a bluish daylight or a yellowish incandescent lightbulb glow, colors must appear as expected. This means that a piece of white paper needs to appear white regardless of the color of the illuminating light. While the HVS performs this task automatically, it is the role of the white balancing operation to mimic this capability.

CFA interpolation, or *demosaicking*, is perhaps the image processing operation most unique to digital cameras. To reduce cost and size, most cameras employ a single sensor covered with a CFA. The CFA restricts each pixel in the sensor to a single color from a small set of colors, typically red, green, and blue. Since each pixel records only one color channel value, and three color channel values are needed to describe a color, it is the job of the demosaicking operation to restore the missing two color channel values at each pixel. This is accomplished by taking advantage of the inherent *color channel correlation* between neighboring pixels and the luminance-chrominance signal processing of the HVS.

The color values produced by the demosaicking operation are generally in a device-specific, nonstandard color space. The role of *color correction* is to transform the non-standard color data into a standard color space appropriate for the final use or uses of the image. These uses could include display on a computer monitor, hardcopy printing, or storage and archiving. When the image is rendered to a device, the limitations of the destination equipment usually mean not all colors can be produced faithfully. Depending on the sophistication of the color correction process, colors that are *out of gamut* of the destination device can end up producing rather significant artifacts. Additionally, as with most signal amplifying operations, precautions must be taken against undue noise amplification. Therefore, the color correction operation usually compromises among color accuracy, color gamut limitations, and noise amplification. The color sensitivity of the HVS, both objectively and subjectively, becomes the primary tool for guiding such color correction compromises.

*Tone scaling* is the means in which the image processing chain simulates the *brightness adaptation* mechanism of the HVS. The HVS is capable of sensing scene brightnesses over a range of ten orders of magnitude! However, it cannot sense that entire range simultaneously. Instead, it uses brightness adaptation to properly sense the portion of the brightness range that is currently being observed, be it a brightly lit beach or a candle-lit birthday cake. In each case, the contrast of the scene is adjusted to appear similar. Tone scaling generally addresses the situation of rendering a scene captured under outdoor lighting so that it looks correct when shown on a display that is dimmer by a number of magnitudes. This is achieved by applying an appropriate nonlinear transformation (a *tone scale*) to the image data. The design of the tone scale naturally follows from the characteristics of the HVS.

There are a number of elements of the digital camera imaging chain, both hardware and software, that act as lowpass (blurring) filters. The genuine high-frequency spatial components of the image can be significantly attenuated, if not outright eliminated, by factors such as lens aberrations and aggressive noise reduction. *Edge enhancement* (sharpening) attempts to restore at least some of the fidelity of this high-frequency spatial image content. Since this sharpening operation is, by its nature, a signal amplifying step, it is important that some effort is made to amplify the real image detail and not the image noise. Again, the HVS can be referenced so as to note which spatial frequencies in an image are important to sharpen, and which can be left unaltered in the name of avoiding noise amplification and image processing artifacts.

From the perspective of the HVS, there is a lot of redundancy in the image produced by a digital camera and its image processing chain. This redundancy takes the form of spatial frequency detail that is visually unimportant, if not outright undetectable. Additionally, what is important spatial frequency detail in the luminance portion of the image is not necessarily as important in the chrominance portions of the image. *Lossless compression* of a digital camera image looks for mathematical redundancies in the image data and removes them in order to produce a smaller image file for reduced storage requirements. Since lossless compression is exactly reversible, no information is lost, although this also means the amount of compression achieved is generally modest. *Lossy compression* uses knowledge of the HVS to discard data that are perceptually unimportant in order to achieve a much higher compression ratio. While the result is no longer *mathematically* reversible, an image with little to no *visual* degradation can be reconstructed from a much smaller image file.

Taken together, the image processing chain of a digital camera simulates the major operations of the HVS. The exact composition of the chain is not fixed, and Figure 5.1 must be taken as just a starting place. Some operations such as noise reduction can easily occur at different locations and more than once within the chain. Additional operations can be added, such as motion deblurring. Still, the image processing chain and its components discussed here provide a foundation from which to discuss subsequent variations.

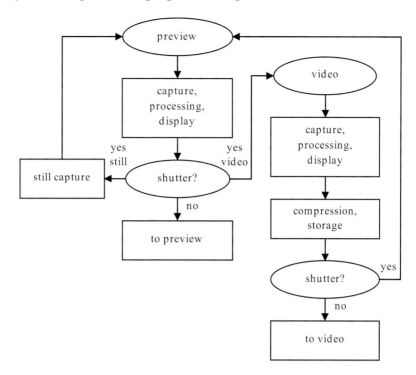

**FIGURE 5.2**

High-level flow for processing in a digital camera.

## 5.3 Automatic Camera Control

The HVS adapts to an extremely wide range of circumstances, placing similar demands on a camera for general consumer use. This section discusses automatic camera control for a general purpose compact digital camera. The primary algorithms are autoexposure, autofocus, and automatic white balancing, sometimes collectively referred to as "A*." Because the goal is reproduction of a scene for human viewing, perceptual considerations are key in developing the algorithms. This has been the case for decades in photography. For example, an early approach to standardization of film "speed" (measurement of film sensitivity) focused precisely on the photographic exposure required to make an excellent print [2]. On the other hand, the development of the zone system was based on exposing, processing, and printing a negative to achieve specific aims for the reproduced image [3].

While the goal is framed around the perceived quality of the reproduction, achievement of the goal begins with capturing (at least) the scene information a human sees. Maximizing the information captured is important, but a related goal is to minimize perceptually significant degradations in reproduced images. There are limits to the scene information that can be captured with conventional camera technology, and trade-offs must be made. For example, most exposure control algorithms maximize the exposure (light captured) to improve the signal-to-noise ratio in the reproduced image. However, still cameras also limit the exposure time to avoid motion blur, because the quality degradation caused by significant motion blur usually outweighs the quality improvement from increased exposure. These trade-offs vary with the scene and with viewing conditions, such as the size of the reproduced image, the image processing, and the viewing distance. For example, because perception of a stream of images is different from perception of a still image, video cameras usually control exposure time primarily to fit the frame rate of the capture and limit motion artifacts such as judder, rather than simply minimize motion blur. These trade-offs and others will be discussed in more detail later.

To discuss these decisions in context, a brief tutorial on the operation of compact digital cameras is presented. Figure 5.2 shows a simplified diagram of processing in a digital still camera. The central column of the figure illustrates the *preview* flow—the capture and presentation of images to display on a panel or in an electronic view finder. The right-hand column of the figure illustrates the *video* flow—the capture, display, and storage of a sequence of images in a video. The video and preview flows are largely the same; the primary difference is whether images are compressed and stored, or simply displayed. The block at the left of the figure represents all of the operations for capture and processing of a *still* image once a still capture is initiated.

In this diagram, camera operation begins in the *preview* oval at the top of the diagram. Normal operation continues through capture, processing, and display of a preview image, a test of the shutter trigger, and without shutter activation a return to the top of the *preview* column for another iteration. The diamond decision blocks in the figure show responses to shutter operation. If the camera is set in still capture mode, the preview flow is active until the shutter is actuated. When the shutter is triggered, the still capture process is executed, then processing returns to the preview flow. If the camera is set in video capture mode, triggering the shutter moves control to the *video* flow, which includes capture, processing, display, compression, and storage. Once in the video flow, the camera stays in video operation until the shutter is triggered again, finishing storage of the video sequence and returning the camera to preview mode.

From a camera control perspective, the preview and video flows are essentially identical. In both cases, there is a stream of images being captured, processed, and displayed. Both flows have the same constraints: operation at video rates, such as 30 frames per second, with images of low to moderate resolution (for example, $640 \times 480$ on up to $1920 \times 1080$). Still capture mode entails capture of a high-resolution image, however, it does not need to fit the same constraints. The exposure time for a still image may be different from the exposure time for a video frame, and flash or other active illumination may be used. However, exposure and focus decisions about capture of the still image are based primarily on analysis of the preview stream. The following discussion will describe the preview stream and the reader should understand the comments generally apply as well to a video capture.

Figure 5.3 shows a slightly more detailed diagram of processing in a typical compact digital camera while a scene is being composed for capture. The first step, in the upper left corner of Figure 5.3, is capture of a raw preview image, stored in a buffer denoted by the oval at the top of the figure. This image is fed to three analysis processes (exposure, focus, white balance) as well as to an image processing chain to render the preview image for display. The three analysis processes may interact with each other, but are shown this way because the parameters for controlling exposure, focus, and white balance are largely

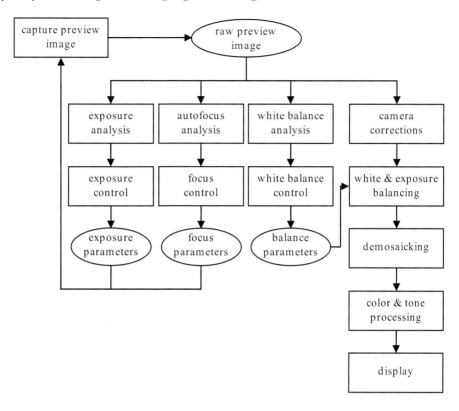

**FIGURE 5.3**

Nominal flow for preview analysis and display.

independent and will be discussed in turn. This largely independent processing for exposure and focus is also practiced in digital single-lens reflex (DSLR) cameras, which often have separate, dedicated subsystems for focus and exposure control. The perceptual issues with DSLRs are similar to those with compact digital cameras, although there is greater use of dedicated hardware in DSLRs. The current discussion will use the compact digital camera as an example system, to highlight algorithm development rather than hardware design.

After a preview image is analyzed, the capture parameters updated, and the image is displayed, the process is repeated. The processing chain to display the preview image normally uses parameters already determined from previous preview images, rather than hold up the display process waiting for analysis of the current image to be completed.

Other processing steps, such as face detection, face recognition, and automatic scene detection, not shown in the figure, are often used in digital cameras. These processes can provide important information for camera control, but are not tied to one specific camera control decision. This discussion will focus on the simpler algorithms tied more directly to control of a camera.

### 5.3.1 Automatic Exposure

As shown in Figure 5.3, exposure control is essentially a feedback control loop, repeatedly analyzing a captured image and updating the exposure parameters. The degree to which the preview image is underexposed or overexposed is considered the exposure error, the feedback signal for this control loop. The discussion here separately addresses the analysis of exposure error in a preview image and the use of different parameters to control exposure.

As with any control loop, stability is important, but for this loop, the criteria for closed loop behavior are based on human perception, since the goal is to produce a stream of rendered images that are all exposed well. Rapid adjustment to a new exposure when the scene changes significantly must be balanced against the desire for stable exposure in the face of minor scene changes.

When capturing video, this exposure control loop is all that is needed, since the video sequence is the captured product. In a digital still camera, the analysis performed on the preview stream can be a precursor to further exposure analysis for a still capture. Because the processing path for the still capture is different from the video path, the optimal exposure can be different, so exposure is often re-analyzed just before a still capture. In addition, still capture normally has different limitations on exposure parameters and active illumination than video capture. This will be discussed further under exposure control.

### 5.3.1.1  Exposure Analysis

In simplified form, the exposure analysis module considers whether the latest preview image is well exposed, overexposed, or underexposed. This can be quantified as a relative exposure error, and exposure parameters (discussed later) are adjusted to minimize the exposure error. To begin, the normal definition of proper exposure is simply that exposure which provides a suitable image when viewed in print or on display. This means the proper exposure depends on the image processing chain, especially the color and tone rendering steps.

Some exposure analysis, such as detection of faces, other specific content, or a pyramid decomposition, may be too computationally demanding to run at preview frame rates. In this case, the preview exposure analysis is often simplified to fit the available resources. Sometimes, the complex processes are run less often, with the results fed into the exposure control loop when available.

For decades, with relatively static rendering systems (film and paper tone scales), automatic exposure analysis focused on aligning the exposure of a midtone gray with the proper location on a fixed tone curve. Development of digital adaptive tone scale technology has made the exposure problem more complex as the rendering process becomes more complex, but most camera autoexposure algorithms still operate as extensions of the traditional process.

Human perception is not linear with exposure; a midtone gray is not 50% reflectance. This is illustrated in Figure 5.4, showing a plot of the conversion from linear relative exposure to CIE L*, a perceptually even measure of lightness [4]. As shown in Figure 5.4, a midtone gray near 50 L* corresponds to a relative exposure of 18%, highlighted by the small circle on the curve at L* 50. Most automatic exposure algorithms are based on some extension of the observation that images of many different scenes will average out to this 18% gray, often known as the gray world model. Many scenes taken together do average near 18% gray, but individual scenes vary widely from the average.

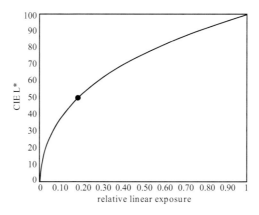

**FIGURE 5.4**

CIE L* vs. relative linear exposure.

Because the gray world model is so weak when applied to an individual scene, greater sophistication is needed. The approaches involve various schemes for weighting regions of the image differently, trying to improve the correlation between an estimated gray and a perceptual gray. In effect, the goal is to roughly discount what is unimportant and emphasize what is most important. Fully automatic metering systems use a combination of methods to improve metering estimation:

- Area weighting uses the spatial location of pixels within the image to determine significance. Spot metering and center-weighted metering are the simplest examples of area weighting.

- Tone weighting uses the tone value to determine significance. The simpler methods weight the central portion of the tone scale more heavily, although usually with less emphasis on shadows than lighter tones.

- Modulation weighting uses the local modulation (texture) in the scene to control weighting, to avoid the average being overly affected by large, flat areas in the scene.

These weighting methods are combined in practice, and even early metering systems, such as one presented in Reference [5], describe combinations of area weighting and modulation weighting, especially considering scene types with substantial differences between center areas and surround areas. Because the different weighting methods are motivated by different perceptual scenarios, they are discussed separately below.

Area weighting is motivated by the realization that the subject of a scene is more likely to be in the central portion of the image, the upper portion of the scene is more likely to contain sky, and so on. Other models of the probability of subject locations within images can also be used. An early example of area weighting can be found in Reference [5], with more recent implementations presented in References [6] and [7]. The use of multiple metering zones continues to develop, especially combined with modulation weighting and recognition of classified scene types.

Tonal weighting is motivated by the fact that different portions of the tone scale in a rendered image have different importance. Exposure analysis from a preview image is complicated by the fact that many scenes have a greater observed dynamic range than can be captured in a single exposure, and also greater dynamic range than can be displayed in the reproduced image. When scene content has an exposure beyond the captured or rendered range of an image, it is clipped, such as a white wedding dress being rendered without highlight detail. In practice, the problem of clipping tends to be one-sided, since only highlights are clipped at capture, while shadows recede into the dark noise. On the other hand, the large dynamic range of many scenes means a simple approach, such as calculating exposure to capture the brightest portion of the scene without clipping, is likely to fail. Images containing candles, the sun, and other light sources would be exposed "for the highlights," resulting in a much lower signal-to-noise ratio than desired for the rest of the scene.

As with all averaging processes, a small variation in the distribution of extreme scene exposure values affects the mean much more than similar variations closer to the mean. For example, exposure of an evening street scene would vary with the number of street lights included in the field of view. The effect is partially addressed by transforming data to a more visually even metric, such as CIE L*, before computing an average. Considering the shape of Figure 5.4, the compression of highlight information (decreasing slope with increased L*) mitigates the impact of highlight data on an average. The limited exposure range for many consumer cameras also tends to limit the problem, since these highlight values are often clipped in capture of a preview image. A clear example emphasizing the use of tone weighting is presented in Reference [8].

Modulation weighting is driven by the realization that people pay more attention to scene regions with modulation than to flat areas. Averaging image areas near edges also tends to make the results more robust, since scene tones on both sides of an edge are included. The challenge is to avoid confusing modulation caused by noise with modulation caused by the scene. This can be done by using a bandpass analysis that is sensitive to relatively low frequencies. More sophisticated use of modulation is described in Reference [9].

Modeling of the human visual response to scenes with high dynamic range has expanded with the development in high dynamic range capture and rendering technology. One example is a multi-scale pattern, luminance, and color processing model for the HVS [10]. A somewhat different approach is the development of a tone reproduction operator to generate suitable reproduction images from a high dynamic range scene [11]. Because of their complexity, these analyses are mostly used in development of adaptive renderings, rather than in precise exposure control of the initial captured image(s).

Many refinements of automatic exposure analysis use classification approaches, such as identifying scenes with faces, blue skies, or snow, and handling them differently. These approaches tend to operate heuristically, but they are driven more by perception than optimization of a single objective metric. Development of theory-based objective metrics that correlate with subjective decisions is an open area of research.

Figure 5.5 shows an example of the difference between naive gray world exposure and perceptually driven exposure. Two different portrait scenes are shown, one with a dark background and one with a light background. Figures 5.5a and 5.5b were captured with perceptually based exposure, resulting in fairly consistent exposure for the main subject. Figures 5.5c and 5.5d illustrate naive exposure control. Both of these images average to a mid-tone gray, but the overall captures are not as good as with perceptually based exposure.

(a)　　　　　　　　　　　　　　　　　(b)

(c)　　　　　　　　　　　　　　　　　(d)

**FIGURE 5.5 (See color insert.)**

An illustration of exposure: (a) portrait with a dark background and perceptual exposure, (b) portrait with a light background and perceptual exposure, (c) portrait with a dark background and naive mean value exposure, and (d) portrait with a light background and naive mean value exposure.

The result of exposure analysis can be quantified as an estimate of how far the preview image is from optimal exposure, providing a natural error signal for the preview exposure control loop.

### 5.3.1.2 Exposure Control

The parameters used for capture of the preview image and the estimated exposure error can be combined to compute the scene brightness. Once the scene brightness is estimated, the next challenge is to determine how to best utilize the camera exposure controls to capture the given scene. Physical camera exposure is controlled through aperture (f/number), gain, exposure time, and flash, or other active illumination. The parameters that control exposure have simple physical models, but they have multiple, often conflicting, perceptual effects in the final image, depending on the scene captured. For example, increasing gain allows rendering of images captured with less exposure, at the expense of increasing the visibility of noise in the image. Increasing the exposure time used to capture an image increases the amount of light captured, but also increases the potential for motion blur in the captured image. Active illumination, such as flash or light-emitting diodes, adds light to the scene being captured, increasing the signal and improving the signal-to-noise ratio, and also filling in some shadows cast by harsh ambient illumination. The exposure added falls off with the square of the distance to the scene, so active illumination has a very different impact on foreground and background scene content.

Exposure calculations for cameras are commonly based on the Additive Photographic Exposure (APEX) system [12]. In computing exposure for digital cameras, it is common to refer to changing the camera's ISO setting, but this is more accurately changing an exposure index (EI) setting. The notions of EI and ISO value have become confused in common usage associated with digital photography, so a brief tutorial is presented here.

ISO speed ratings were developed to quantify photographic film speed or sensitivity. The ISO value quantifies the exposure required at the film surface to obtain standard image quality. A film ISO rating can also be entered into a light meter and used to control exposure metering to obtain the desired exposure on the film. When the ISO value is used in metering, the precise descriptor for this value is exposure index (EI). While ISO represents the exposure a particular film needs for standard exposure, EI is a way to specify how much exposure is actually used for a single capture of a scene.

The standard for digital camera sensitivity [13] defines at least two pertinent ISO values. One is the saturation limited ISO, typically between 40 and 100 for consumer cameras, quantifying the lowest EI value at which the camera can operate. The other is the noise limited ISO, usually ranging up to 1600 or so for compact cameras, quantifying the maximum EI value at which the camera can operate and still achieve the signal-to-noise ratio stated in the sensitivity standard [13]. These two ISO values quantify the range of exposures within which the camera can produce normally acceptable images, although it is common for a camera to allow operation over a different range of EI values. The user interface for most digital cameras uses the term ISO, although as explained previously, EI is a more precise term.

Adjustment of EI in a digital camera has a dual role in exposure control. In the first role, selection of a specific EI for capture of a scene determines how much light is accumulated on the sensor for that capture, although several other parameters, such as exposure time and f/number, are used to control the exposure. In the other role, image processing parameters, such as gain, noise reduction, and sharpening, are usually adjusted in a camera as a function of EI. This allows a camera to provide its best image rendering for a given EI.

**ALGORITHM 5.1**  A simplified automatic exposure control algorithm.

1. Set f/number to the lowest available value and set EI to the camera's saturation-limited ISO (lowest EI value).

2. Estimate scene brightness from the preview stream.

3. Calculate exposure time $t$ from scene brightness, f/number, and EI.

4. If $t > t_{max}$, set $t = t_{max}$ and obtain EI from scene brightness, f/number, and $t$.

5. If $t < t_{min}$, set $t = t_{min}$ and obtain f/number from scene brightness, EI, and $t$.

Returning to exposure control, the optimum trade-offs between various capture parameters (for instance, f/number, exposure time, EI, flash) vary depending on the scene conditions, the design limitations of the camera, and the viewing conditions. Comprehensive understanding of the scene is an open research question, so most algorithms for choosing a set of exposure control parameters have been based on very simple heuristic algorithms, such as one depicted in Algorithm 5.1. In this example, $t_{max}$ is a threshold for maximum exposure time, normally calculated based on the focal length of the camera to limit the risk of motion blur. The term $t_{min}$ denotes a minimum exposure time, based on the shutter limitations of the camera. This very simple model works fairly well for many scenes in the context of a compact digital camera.

The relatively large depth of field for a compact digital camera means there is usually not a significant advantage from stopping down the lens. With the small pixels often used in compact digital cameras, decreasing the lens aperture (running at higher f/number) usually degrades resolution and sharpness, as well as decreasing the light available for capture with the sensor. The optical reasons for this are discussed in more detail in the Appendix.

These considerations lead to an exposure control algorithm that keeps the f/number at the lowest value possible. This leaves exposure time, EI, and active illumination available for exposure control. Of these parameters, EI is usually simply adjusted to provide an effective change in sensitivity. In the end, there are practically two variables to change: active illumination and exposure time. Before the development of practical motion estimation technology, the process of balancing the risk of motion blur associated with longer exposure time against signal-to-noise ratio was done heuristically based on the focal length of the camera. Some recent work explicitly models the quality impact of motion blur versus the quality impact of operation with higher exposure index [14], providing a framework for a more precise trade-off. This framework has been extended to include use of electronic flash along with ambient illumination [15].

As mentioned previously, video and still capture have different exposure constraints. While exposure time for a still capture can easily range from 1/1000 second to 1/15 second or longer, exposure time for video is usually limited. The exposure time for a video frame can approach the frame time, but it cannot be longer than the frame time (for instance, 1/30 second for video at 30 frames per second). There are also two perceptual artifacts that further constrain exposure time in video. When motion is captured in a video, longer exposure times increase the motion blur captured in each frame, which can lead to a soft, blurry video of low quality. Conversely, exposure times much shorter than the frame time can lead to frames with very little motion blur, but large motion displacement between frames. This presents the visual system with an artificial stop-action kind of motion. In practice, the exposure time for video capture is usually selected to minimize or balance both of these motion artifacts.

### 5.3.2 Automatic Focus

As with exposure, the goal for automatic focus is to adjust the camera to capture the scene at hand, with the most important portion of the scene in best focus. The depth of field depends on the focal length of the lens, the aperture (f/number), and the focus setting. Some cameras, particularly inexpensive ones with fixed wide angle lenses, are fixed focus, obviating automatic focus completely. Focus control is more critical at longer focal lengths, because depth of field is decreased. As a result, cameras with zoom lenses have focus requirements that vary with focal length. Finally, focus for a high-resolution capture is more critical than for a low-resolution capture. This is important for cameras that provide both still and video functions, since stills are usually much higher resolution than videos.

### 5.3.2.1  Focus Analysis

The HVS has greater sensitivity to luminance detail than chrominance detail, and the green channel has a greater impact on luminance than the red or blue channels. Focus analysis is simplified by choosing a single color band for which to optimize focus, normally the green channel.

Focus can be analyzed in a number of ways, but the most common approach in compact digital cameras is contrast maximization. A series of images are captured and analyzed with the lens at different focus settings. Local contrast values at the different focus settings are compared, and the focus setting that provides maximum local neighborhood contrast, or best sharpness, is chosen as being in best focus. A key step in the contrast maximization process is filtering to determine local contrast. The usual approach is to process a preview image with one or more bandpass filters, designed to maximize the ability to determine changes in sharpness. Focus changes affect higher more than lower spatial frequencies, but scene modulation is also reduced at higher frequencies, especially considering the modulation transfer function of the camera lens, reducing the signal available for focus detection. In addition, human perception of sharpness tends to be weighted toward lower frequencies, as discussed later in Section 5.7.

Human perception also influences focus analysis in the selection of a focus region within the image. When a scene has significant depth, it is crucial that the focus analysis emphasizes the scene content most likely to be the subject. Preferably, focus analysis is also performed on image data in a metric with a visually even tone scale, so that edges at different tone values produce similar contrast values. Calculating local contrast in a non-visual metric, such as relative linear exposure, will make brighter areas of the scene appear to have much greater contrast than midtones.

Selection of focus areas also considers analysis results from multiple focus settings. Regions that show a sharper peak in focus values are usually weighted more heavily than regions that show less of a focus peak, which usually have less edge content. Regions with peak contrast at a closer focus setting are also more likely to be the main subject.

Viewing a video raises temporal concerns not present with a still capture. In particular, stability of the focus setting is important. As described earlier, contrast maximization requires images collected at different focus settings, some of which must be measurably out of focus in order to ensure the best focus has been found. The challenge is to measure local sharpness changes to confirm focus, while minimizing the visibility of focus changes in the stored video stream. Analysis of frame or scene differences can be used to help the camera determine when the scene has shifted enough to warrant refocusing.

Most other interesting features of focus analysis algorithms are driven by optical and engineering concerns rather than perceptual ones. Reference [16] describes an autofocus system in a digital camera, including discussion of system hardware, focus evaluation, and logic for focus region weighting. Reference [17] describes the use of noise reduction and the use of filter banks for improved autofocus analysis. The temporal problems of maintaining focus in video capture are discussed in Reference [18].

#### 5.3.2.2 Focus Control

Focus control is essentially the movement of one or more elements in the camera or lens to bring the image into best focus. The perceptual concerns here are minor; the major concerns are mechanical and optical. The only impact human perception has on focus control is in tolerance analysis.

### 5.3.3 Automatic White Balancing

Automatic white balancing has been an active area of research since the development of color photography. It is the technology that allows images to be captured under a wide variety of illuminants and then reproduced in a pleasing way. Because the HVS has great flexibility in adaptation to different illuminants, cameras must provide roughly similar behavior. As with automatic exposure and focus, white balance correction must have the desired speed of response to scene and illuminant changes, while offering stability in the face of insignificant changes.

#### 5.3.3.1 White Balance Analysis

Much of the early work on white balance analysis focused on automatic creation of prints from color negatives. The gray world hypothesis (the average of many scenes will integrate to 18% gray) was first mentioned in Reference [19]. Models of human adaptation to different illuminants are considered in color science, although the models tend to be more complex than those commonly used in cameras. The problem usually addressed in a camera is more precisely illuminant estimation, rather than complete modeling of human adaptation to it.

White balance analysis during preview or video capture must be computationally lightweight, usually using dedicated hardware and very simple processing based on extensions of the original gray world model. The goal of the analysis is the estimation of a gain triplet, with an element for each of three color channels, which scales the native sensor response so neutral scene content will have equal mean values in all channels. Having approximately equal channel responses for neutral scene content improves robustness in the later color correction operation, allowing colors to move *around* the neutral axis. Tone correction can then render the tone scale for proper reproduction. Some camera sensors have more than three channels [20], but the automatic white balance analysis is quite similar.

Most white balance analysis is done in a luminance-chrominance color space in order to better separate exposure effects from illuminant balance effects, somewhat akin to the human perception of color [21]. Because this analysis is done before full color correction, it is not in a calibrated color space, but even an approximate separation of luminance from chrominance helps decorrelate the channels for analysis. A common color space used is the ratio of red to blue and green to magenta (combined red and blue). If the linear red, green, and blue values are converted to a log space, this is easily accomplished with a matrix transform similar to one proposed in Reference [22], shown below:

$$\begin{bmatrix} Y \\ C_1 \\ C_2 \end{bmatrix} = \frac{1}{4} \begin{bmatrix} 1 & 2 & 1 \\ -1 & 2 & -1 \\ -2 & 0 & 2 \end{bmatrix} \begin{bmatrix} R \\ G \\ B \end{bmatrix}. \tag{5.1}$$

Algorithms for white balance analysis usually employ averaging schemes with variable weighting of different content in the scene. The variable weighting schemes generally emphasize regions based on five perceptual phenomena:

- local modulation (texture),

- distance from the neutral axis,

- location in the tone scale, particularly highlights,

- recognizable scene content, such as faces or memory colors, and

- human sensitivity to different white balance errors

The motivation for emphasizing areas with modulation or texture is partly statistical. It avoids having the balance driven one way or another by large areas of a single color, such as red buildings, blue skies, or green foliage. Further, as with exposure analysis, there is greater likelihood that the main subject is in a region with modulation.

Colors closer to the neutral axis are weighted more heavily than saturated colors, because people are more sensitive to white balance errors in neutrals. It also helps stabilize the balance estimate a bit, as the extreme colors are weighted less. This concept complicates the algorithm, since the notion of neutral is not well defined before performing illuminant estimation. This can be approached by applying a calibrated daylight balance, then analyzing colors in the resulting image [23]. Another option is to estimate the probability of each of several illuminants being the actual scene illuminant [24], [25].

Highlight colors can be given special consideration, based on the theory that highlights are specular reflections that are the color of the illuminant [26], [27], [28]. This theory is not applicable for scenes that have no truly specular highlights. However, human sensitivity to balance errors tends to diminish in shadows, justifying a tonally based weighting even in the absence of specular highlights.

Some white balance approaches also consider the color of particular scene content. The most common of these is using face detection to guide balance adjustments, providing a reasonable balance for the face(s). This has other challenges, because faces themselves vary in color. Memory colors, such as blue skies and green foliage, may also help, but the complexity of reliably detecting specific scene content, especially before proper balance and color correction, makes this challenging.

Finally, because human sensitivity to balance errors is not symmetric, white balance algorithms are adjusted to avoid the worst quality losses. For example, balance errors in the green-magenta direction are more perceptually objectionable than balance errors in the red-blue direction. Preference can also enter into the tuning. For example, many people prefer a slightly warm (yellowish) reproduction of faces to a slightly cool (bluish) reproduction.

When a camera is used in a still capture mode, the white balance analysis for the full resolution still image can be more complex than the analysis run at video rates, since the regular preview stream is interrupted. If the camera uses a flash or other active illumination, the white balance analysis will yield very different results from analysis of preview images, captured without active illumination. Further, since flash is often a somewhat different color balance than the ambient illumination, still captures with flash are usually mixed illuminant captures, making the estimation problem more complex, although use of distance or range information can help.

#### 5.3.3.2 White Balance Control

Control of white balance is almost always the application of a global adjustment to each color channel. If the balance is applied to linear data, the white balance correction is a triplet of gain factors. Because the white balance operation is usually approached independently of the exposure operation, the triplet of channel gains has two degrees of freedom. Multiplying all channels by a constant gain is in effect an exposure adjustment; therefore, the gain triplet is usually normalized so the minimum gain in the triplet is one, leaving only two values greater than one.

Some algorithms deal with flash as a special case of mixed illuminant capture by using a spatially varying balance correction. This is perceptually helpful, since a human observer rarely ever "sees" the scene with the flash illumination, so it is especially desirable to minimize the appearance of mixed illuminant white balance problems in the reproduced image.

## 5.4 Color Filter Array Interpolation (Demosaicking)

The process of designing CFAs and the related task of CFA interpolation (or *demosaicking*) can derive significant inspiration from the HVS. Due to differences in how the HVS perceives edges based on their orientation and colors based on their luminance-chrominance composition, significant economies can be made in the CFA process without sacrificing high image quality in the final results. The details of HVS-driven array and algorithm designs are discussed below.

### 5.4.1 Color Filter Array

As stated in Section 5.2, CFA interpolation attempts to reconstruct the missing color values at each pixel that were not directly captured [29]. By far the most ubiquitous CFA is the Bayer pattern [30]. Referring to Fig. 5.6a, this pattern was initially introduced as a luminance-chrominance sampling pattern, rather than a red-green-blue system. It used the

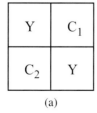

(a)

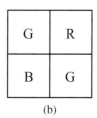

(b)

**FIGURE 5.6**

Bayer CFA pattern: (a) luminance-chrominance version, (b) red-green-blue version.

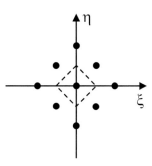

**FIGURE 5.7**

Checkerboard sampling Nyquist diagram.

HVS as the motivation: "In providing for a dominance of luminance sampling, recognition is taken of the human visual system's relatively greater ability to discern luminance detail (than chrominance detail)" [30]. This reasoning has set the tone for most subsequent efforts in CFA design. When designing a new CFA pattern, the first step is to establish a strong luminance sampling backbone. This frequently takes the form of a checkerboard of luminance pixels as in Figure 5.6a. This is attractive from an HVS perspective due to the *oblique effect* [31], [32]. The oblique effect is the observed reduction of the visibility of diagonal edges compared to horizontal and vertical edges. The relationship between the oblique effect and CFA sampling with a checkerboard array can be developed from the frequency response of this sampling pattern as follows [33]:

$$S(\xi, \eta) = \frac{1}{2}\text{comb}\left(\frac{\xi - \eta}{2}, \frac{\xi + \eta}{2}\right), \tag{5.2}$$

with the comb function defined as

$$\text{comb}\left(\frac{x}{b}, \frac{y}{d}\right) = |bd| \sum_{m=-\infty}^{\infty} \sum_{n=-\infty}^{\infty} \delta(x - mb, y - nd), \tag{5.3}$$

where $\delta(x, y)$ is the two-dimensional delta function [34]. The corresponding Nyquist diagram of Equation 5.2 is shown in Figure 5.7. The dots mark the centers of the fundamental frequency component (within the dashed diamond) and some of the sideband frequency components of the frequency response. The dashed lines indicate the *Nyquist sampling frequency* of the fundamental component, that is, the maximum detectable spatial frequency of the sampling array. It can be seen that the Nyquist frequency reaches its peak along the horizontal and vertical axes and is lowest in the diagonal directions. Therefore, in accordance with the oblique effect, a checkerboard array has its maximum frequency response when sampling horizontal and vertical edges, and its minimum frequency response when sampling diagonal edges.

Once the positions of the luminance pixels are fixed in the CFA pattern, the remaining locations are usually evenly divided among the chrominance channels. In the case of a $2 \times 2$ pattern such as Figure 5.6a, the locations of the chrominance pixels are essentially forced. However, with larger CFA patterns, additional degrees of freedom become available and multiple chrominance patterns are then possible for a given luminance pattern.

| B | $G_1$ | B |
|---|---|---|
| $G_2$ | $R_0$ | $G_3$ |
| B | $G_4$ | B |

**FIGURE 5.8**

Bayer green interpolation neighborhood.

### 5.4.2 Luminance Interpolation

It is natural to approach the task of interpolating the CFA luminance values by using neighboring luminance values within the CFA pattern, such as in Figure 5.6a. In Figure 5.6b, this would be equated to interpolating missed green values from neighboring green values. This can be posed as a linear or nonlinear interpolation problem. Figure 5.8 depicts a typical green pixel interpolation neighborhood. Linearly, one would simply compute $G_0' = (G_1 + G_2 + G_3 + G_4)/4$. Nonlinearly, the approach would be to perform directional edge detection and interpolate in the direction of least edge activity [35]. For example, if $|G_2 - G_3| \le |G_1 - G_4|$, then $G_0' = (G_2 + G_3)/2$. Otherwise, $G_0' = (G_1 + G_4)/2$. Any number of more sophisticated decision-making processes can be described, but the essence would be the same.

All of the preceding misses a point, however, from the HVS perspective. The red and blue pixels do *not* produce true chrominance values. Instead, they produce *color* values that can be thought of as consisting of a luminance component and a chrominance component. For example, in Figure 5.9, a full-color image (Figure 5.9a) is shown with its corresponding green (Figure 5.9b) and red (Figure 5.9c) color channels. Both color channels have strong luminance components. Therefore, it is possible for the red and blue pixel values to contribute to the interpolation of green pixel values. A simple, yet very useful, model is given below:

$$R = Y + C_R, \quad G = Y, \quad B = Y + C_B. \tag{5.4}$$

This model reflects the *color channel correlation* present in the HVS [1]. Over small distances in the image, $C_R$ and $C_B$ can be considered to be approximately constant, as shown in Figure 5.9d. Under this model, spatial variation is largely confined to the luminance channel with chrominance information changing slowly across the image. Using Equation 5.4 and Figure 5.10, an improved luminance interpolation calculation based on HVS considerations can be developed [36]. Assuming a symmetric interpolation kernel, the horizontally interpolated green value $G_5'$ can be computed as follows:

$$G_5' = a_1 (R_3 - C_R) + a_2 G_4 + a_3 (R_5 - C_R) + a_2 G_6 + a_1 (R_7 - C_R). \tag{5.5}$$

In this equation it is assumed that the chrominance term $C_R$ is constant over the depicted pixel neighborhood. Three constraints are required to determine the coefficients $a_1$ through $a_3$. Since $C_R$ is generally unknown, the first constraint is to eliminate it from the equation.

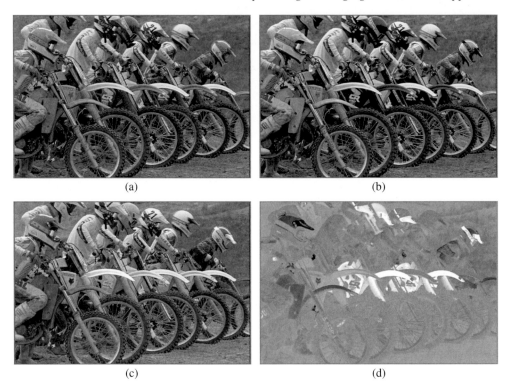

**FIGURE 5.9**

Full-color image and its components: (a) original image, (b) green channel, (c) red channel, and (d) $C_R$ channel.

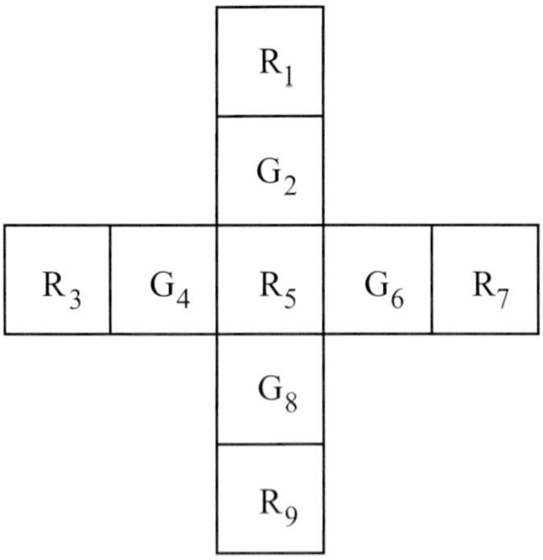

$$\begin{array}{c}
\boxed{R_1} \\
\boxed{G_2} \\
\boxed{R_3}\ \boxed{G_4}\ \boxed{R_5}\ \boxed{G_6}\ \boxed{R_7} \\
\boxed{G_8} \\
\boxed{R_9}
\end{array}$$

**FIGURE 5.10**

Bayer green-red interpolation neighborhood.

This is done by computing its coefficient and setting it to zero:

$$G_5' = a_1 R_3 + a_2 G_4 + a_3 R_5 + a_2 G_6 + a_1 R_7 - (2a_1 + a_3)C_R \quad \Rightarrow \quad a_3 = -2a_1, \qquad (5.6)$$

resulting in

$$G_5' = a_2 \left( G_4 + G_6 \right) + a_1 \left( R_3 - 2R_5 + R_7 \right). \qquad (5.7)$$

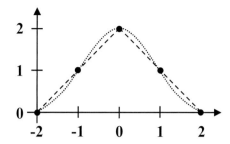

**FIGURE 5.11**

Triangular edge profile with collocated sinusoid.

The next constraint to add is that a neutral flat field input produces a neutral flat field output. In other words, if $G_4 = G_6 = R_3 = R_5 = R_7 = k$, then $G_5' = k$. This results in $a_2 = 1/2$. Finally, there is no one ideal constraint for determining $a_1$, which is the last coefficient. One reasonable constraint is the exact reconstruction of a neutral edge with a triangular profile, as shown in Figure 5.11. As indicated, this edge profile could also be a sampled version of $1 + \cos(2\pi x/4)$. Setting $G_4 = G_6 = 1$, $R_3 = R_7 = 0$, and $R_5 = G_5' = 2$ produces $a_1 = -1/4$. The final expression is given below [37]:

$$G_5' = \frac{G_4 + G_6}{2} - \frac{R_3 - 2R_5 + R_7}{4}. \tag{5.8}$$

The set of constraints used could also be interpreted as imposing exact reconstruction at spatial frequencies of 0 and 1/4 cycles/sample. While the constraint at 0 cycles/sample is probably best left intact, the second constraint could easily be changed to a different spatial frequency with a differing corresponding value for $a_1$. A vertical interpolation expression would be generated in the same way. For a linear interpolation algorithm, both horizontal and vertical expressions can be averaged to produce a two-dimensional interpolation kernel as follows:

$$G_5' = \frac{G_2 + G_4 + G_6 + G_8}{4} - \frac{R_1 + R_3 - 4R_5 + R_7 + R_9}{8}. \tag{5.9}$$

The color channel correlation of the HVS can also be used to improve the edge detectors used in a nonlinear, adaptive interpolation algorithm [38]. Using reasoning similar to that used above, the horizontal edge detector can be augmented from $|G_4 - G_6|$ to $|G_4 - G_6| + |R_3 - 2R_5 + R_7|$ and the vertical edge detector from $|G_2 - G_8|$ to $|G_2 - G_8| + |R_1 - 2R_5 + R_9|$. Again, the direction of least edge activity would be interpolated with the corresponding horizontal or vertical interpolation expression.

### 5.4.3  Chrominance Interpolation

Since the HVS is most sensitive to changes in chrominance at the lower spatial frequencies, linear interpolation of chrominance values should be sufficient under most circumstances. As stated in Section 5.4.2, it is important to interpolate *chrominances* and not simply *colors* [39]. This can be accomplished by returning to Equation 5.4 and rewriting the model in terms of chrominances as follows:

$$C_R = R - Y, \quad Y = G, \quad C_B = B - Y. \tag{5.10}$$

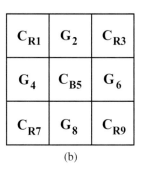

| $R_1$ | $G_2$ | $R_3$ |
|-------|-------|-------|
| $G_4$ | $B_5$ | $G_6$ |
| $R_7$ | $G_8$ | $R_9$ |

(a)

| $C_{R1}$ | $G_2$ | $C_{R3}$ |
|----------|-------|----------|
| $G_4$ | $C_{B5}$ | $G_6$ |
| $C_{R7}$ | $G_8$ | $C_{R9}$ |

(b)

**FIGURE 5.12**

Chrominance interpolation neighborhood: (a) colors, (b) chrominances.

Assuming luminance interpolation has been previously performed, each pixel will have an associated luminance (green) value. Therefore, at each red pixel location a chrominance value $C_R = R - G$ can be computed. Similarly, at each blue pixel location $C_B = B - G$ can be computed. The resulting pixel neighborhood is given in Figure 5.12b. Missing $C_R$ chrominance values are computed using averages of nearest neighbors, for example, $C'_{R2} = (C_{R1} + C_{R3})/2$, $C'_{R4} = (C_{R1} + C_{R7})/2$, and $C'_{R5} = (C_{R1} + C_{R3} + C_{R7} + C_{R9})/4$. Missing values for $C_B$ are computed in a similar manner. After chrominance interpolation, the chrominance values can be converted back to colors using Equation 5.4. It is possible to combine the color to chrominance and back to color transformations with the chrominance interpolation expressions to produce equivalent chrominance interpolation expressions using just color values. These resulting expressions are given below:

$$R'_2 = \frac{R_1 + R_3}{2} - \frac{G_1 - 2G_2 + G_3}{2}, \tag{5.11}$$

$$R'_4 = \frac{R_1 + R_7}{2} - \frac{G_1 - 2G_4 + G_7}{2}, \tag{5.12}$$

$$R'_5 = \frac{R_1 + R_3 + R_7 + R_9}{4} - \frac{G_1 + G_3 - 4G_5 + G_7 + G_9}{4}. \tag{5.13}$$

These expressions are written in a way to illustrate their differences from simply averaging color values of the same color, for example, averaging only neighboring red values to produce interpolated red values.

## 5.5 Noise Reduction

Noise reduction lessens random signal variation in an image while preserving scene content. The random variation, commonly called noise, diminishes the quality of a reproduced image viewed by a human observer [40]. Since noise reduction usually introduces undesirable side effects, the challenge in designing the noise reduction algorithm is to ensure that the perceptual quality improvement exceeds the quality degradation from side effects.

The primary side effects from noise reduction are perceived as loss of sharpness or loss of texture. Adaptive noise reduction normally preserves high-contrast edges but smooths lower-contrast edges, leading to texture loss. The quality degradation caused by loss of texture is a topic of current research, especially for development of ways to objectively measure image quality from digital cameras [41].

Other side effects can be caused by abrupt transitions in smoothing behavior. Usually, small modulations (small in spatial extent, contrast, or both) are smoothed, while larger modulations are preserved. When a single textured region (such as a textured wall, grass, or foliage) contains both small and large modulations, noise reduction will change the texture greatly if the transition from smoothing to preservation is too abrupt, thus causing switching artifacts. Another potential side effect is contouring, when a smooth gradation is broken into multiple flat regions with sharp transitions, that is, an image is heavily quantized. This can occur with noise reduction based on averaging values near a neighborhood mean, such as a sigma [42] or bilateral [43] filter, especially with iterative approaches.

In addition to the perceptual complexity of side effects, noise reduction must operate over a large range of conditions in a compact digital camera. When a camera operates at its lowest EI, the signal-to-noise ratio in the raw capture is often four to eight times the signal-to-noise ratio when operating at the maximum EI. Further, noise in a raw image and the *visibility* of noise in the rendered image vary with signal level. The wide range of signal-to-noise ratios and noise visibility requires noise reduction to adapt over a broad set of conditions to provide a pleasing reproduced image. For example, if a camera operates at a very low EI and the raw image has a good signal-to-noise ratio to begin with, the noise may degrade image quality only slightly. In this case, noise reduction should be tuned to minimize its side effects and ensure retention of maximum scene content.

Overall image quality has been modeled by considering multiple distinct degradations to image quality as perceptually independent visual attributes, such as sharpness, noise, contouring, and others [40]. Because noise reduction produces multiple effects that are not necessarily perceptually independent, this modeling approach will vary in effectiveness, depending on the precise nature of the noise reduction effects. However, the multivariate quality model of [40] has been applied with some success to modeling the overall quality of images from a camera operating over a range of EI values [14].

### 5.5.1   Contrast Sensitivity Function and Noise Reduction

Since human contrast sensitivity diminishes at higher spatial frequencies, noise reduction primarily reduces higher-frequency modulation and preserves lower-frequency modulation as much as possible. However, the relationship between a contrast sensitivity function (CSF) and optimal noise reduction is indirect for two reasons. First, the human CSF is defined in terms of a contrast sensitivity threshold as a function of angular frequency at the eye. The relationship between spatial frequency in a camera image (cycles/pixel) and angular frequency at the eye depends on the full image processing chain, including resizing, output or display, and viewing conditions. Second, contrast sensitivity functions are usually based on measurement of minimum detectable contrast in a uniform field. The modulation that is due to noise in a compact digital camera is often above the threshold of visibility, and the quality degradation that arises from different levels of supra-threshold noise is more complex than expressed in a simple CSF.

## 5.5.2   Color Opponent Image Processing

The human visual system sensitivity to luminance modulation is different from its sensitivity to chrominance modulation. The luminance CSF has a bandpass character with peak sensitivity between one and ten cycles/degree. The chrominance contrast sensitivity is also reduced as frequency increases, but the reduction begins at a lower frequency, roughly one third to one half the spatial frequency of the peak of the luminance CSF. Because of this, noise reduction is most effective when operating in a luminance-chrominance color space, treating the luminance and two chrominance channels differently. While a colorimetric uniform color space such as CIELAB is attractive for separating luminance from chrominance, the transformation is somewhat complex. Because noise reduction makes relatively small changes to mean signal values, an approximate transformation is usually adequate. A common approach is a matrix transform similar to one proposed in Reference [22], shown earlier in Equation 5.1. Another very common space is the $YC_BC_R$ space used for luma-chroma JPEG compression. The primary advantage of using the JPEG $YC_BC_R$ space for noise reduction is that the data are usually converted to that space for compression. Converting to that color space earlier in the chain and applying noise reduction there saves one or two color conversions in the processing chain.

After converting to a luma-chroma color space, it is normal to apply noise reduction to each of the resulting channels with different techniques. Noise reduction for the luminance channel seeks to attenuate high-frequency noise, rather than eliminate it completely, and seeks to preserve as much sharpness as possible within the cleaning requirements. Noise reduction for the chrominance channels is generally much more aggressive, and the associated degradation of sharpness and spatial resolution is quite acceptable. In practice, since chroma is often subsampled for compression, noise reduction is often applied to the subsampled chroma channels. This saves processing and makes control of lower spatial frequencies easier because the chroma channels are sampled at lower frequency.

To illustrate the effect of noise reduction in a digital camera, Figure 5.13 shows example hypothetical noise power spectra (NPS) for rendered images of flat patches with noise reduction applied, along with an example CSF. Noise in a raw digital camera image generally has a relatively flat (white) frequency response, so a non-flat shape for the final NPS is the result of noise reduction (and other operations) in the processing chain. The first curve in Figure 5.13 is a luminance CSF using the model from Reference [44]. For this figure, the CSF has been computed for a viewing distance of 400 mm and a pixel pitch of 0.254 mm, corresponding to viewing an image at full resolution on a typical LCD display, and normalized so the CSF response fills the range of the plot to overlay the NPS curves. The peak CSF response is at 0.11 cycles/pixel for this situation. Increasing the viewing distance would move the peak toward lower frequencies, as would using a higher resolution display.

The application of noise reduction in a luma-chroma color space causes the resulting image's NPS for the luma channel to be very different from those for the chroma channels. Figure 5.13 shows four sample NPS curves: a sample luma NPS at a low EI (*Luma 100*), a luma NPS at a high EI (*Luma 1600*), and corresponding chroma curves (*Chroma 100* and *Chroma 1600*). The $C_B$ and $C_R$ channels normally have similar NPS characteristics, so this

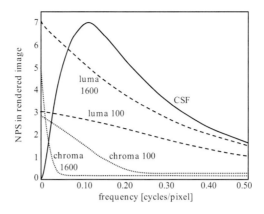

**FIGURE 5.13**

Example noise power spectrum for a still image.

example shows a single chroma NPS curve. All of the NPS plots show a lowpass characteristic, although of variable shape. At low EI, the luma NPS has relatively little reduction and a fairly wide passband. Since small amounts of noise have little impact on overall quality, there is relatively little need to reduce the luma noise lower than shown for EI 100. The chroma NPS shows no significant response past 0.25 cycles/pixel, consistent with subsampled chroma channels. Because of the perceptual degradation from low-frequency chroma noise, the preference is to make the chroma NPS similar to or lower than the luma NPS.

At high EI, the luma NPS is much more aggressively filtered. The noise in the raw image at EI 1600 is usually roughly four times the noise at EI 100, yet after noise reduction, it is less than three times the noise at EI 100. Further, the relatively flat NPS from white noise in the raw image has been modified much more than for EI 100. While attenuation is still fairly linear with frequency, there is more attenuation of higher frequencies and a tendency to be concave upward. The chroma NPS is much more aggressively filtered than for EI 100, deliberately sacrificing chrominance detail in order to reduce chrominance noise. At the higher EI, the quality degradation that is due to noise is substantial, so trading off sharpness and texture to limit the noise provides better overall quality. Modeling of the relationship between NPS and human perception is a topic of current research [45].

NPS analysis for a frame of a video shows some similar characteristics; noise reduction is more aggressive at higher frequencies and at a higher EI. In compact digital cameras, there is typically a greater emphasis on limiting the noise in the video than on preserving sharpness, partially driven by the greater compression applied to videos. As noise is difficult to compress and tends to trigger compression artifacts, there is a greater emphasis on avoiding these noise-induced artifacts than on preserving scene detail.

Another important feature illustrated in Figure 5.13 is the use of masking. Allowing luma noise power to be greater than chroma noise power tends to mask the chroma noise with the luminance modulation, making the chroma noise less visible. In addition, the development of noise reduction algorithms also depends on use of masking from scene content. Noise reduction algorithms tend to smooth more in areas with little scene modulation and preserve more modulation in regions with greater local contrast. The challenge is to discriminate as well as possible between noise and local contrast due to scene content.

**FIGURE 5.14 (See color insert.)**

Example images with: (a) noise, (b) light RGB smoothing, (c) aggressive RGB smoothing, and (d) simple color opponent smoothing.

An example illustrating some of the power of color opponent processing for noise reduction is shown in Figure 5.14. Namely, Figure 5.14a is an example noisy image, such as might come from a capture at a high EI, processed through demosaicking and white balance. The raw captured image was very noisy with a fairly flat NPS, and the early processing steps have reduced some noise, especially at higher frequencies, while leaving a fair bit of lower-frequency noise. Figure 5.14b illustrates a light smoothing (Gaussian blur) applied in RGB color space. There is modest sharpness loss, but also modest reduction in noise, especially in the lower frequencies. Low-frequency variations in color remain in areas of low scene texture, such as the blue wall, the face, and green background. Figure 5.14c illustrates a more aggressive smoothing (Gaussian blur) applied in RGB. The noise is significantly reduced, but the loss in sharpness degrades the quality substantially. Figure 5.14d illustrates light smoothing applied to a luma channel and more aggressive smoothing applied to both chroma channels. The luma channel smoothing matches that for Figure 5.14b and the chroma channel smoothing matches that for Figure 5.14c. The overall quality is significantly better than the other three figures, because the reduction in chroma noise is achieved without a similar reduction in sharpness.

### 5.5.2.1 Noise Probability Distributions

The probability distribution for noise has a great impact on noise reduction. The dominant noise in a digital camera raw image is usually shot noise with a Poisson distribution, plus an additive read noise with a Gaussian distribution. The read noise is signal independent, but the shot noise is dependent on signal level. Further, a small number of pixels in the sensor are abnormal, casually referred to as defective pixels. Defective pixels act like impulse noise, but have highly variable responses. Rather than two or three clear peaks (normal, dark defects, bright defects), a histogram of a low-light, flat-field image capture is more like a mixture of Gaussians, with a main peak and an extremely wide tail.

One treatment for defective pixels and other contributors that spread out the tails of the noise distribution is to apply an impulse filter before other operations, preventing spread of impulse noise into other pixels during smoothing. The challenge in performing impulse filtering is to avoid filtering scene detail unnecessarily. For example, small light sources, and their reflections in eyes, are easily misidentified as impulses. Despite their similarity to impulse noise, they have exceptional perceptual importance.

## 5.6   Color Rendering

Nearly all practical imaging systems have a small number of band-limited light detectors and emitters. The shape of the detectors' spectral efficiencies and the emitters' radiance characteristics determine the intrinsic color reproduction of the imaging system, and this reproduction is almost never accurate or visually pleasing without any signal processing. Additionally, there may be factors present when viewing a reproduction that alter its appearance and that may not have been present when capturing the corresponding original scene. The signal processing that modifies an image to account for the differences between the HVS and the imaging system and to account for the viewing environment is called color rendering. Knowledge of HVS color perception and a mathematical specification of color stimuli help to define the signal processing needed for rendering. These two topics are described in the following section, prior to a discussion of the specific steps to render colors of an image captured with a camera and displayed on a monitor.

### 5.6.1   Color Perception

In the seventeenth century, it was discovered that a color is a percept induced by light composed of one or more single-wavelength components [46]. In the early nineteenth century, the hypothesis was that there are three retinal light sensors in the human eye [47], even though the sensing mechanism was unknown. Later in the nineteenth century, it was determined that wavelengths that are usually associated with red, green, and blue compose a minimal set of primary wavelengths that can be mixed to induce the colors associated with the rest of the wavelength spectrum [48], [49]. It is now known that there are two types of light-sensitive cells in the human eye, called rods and cones because of their shapes. Rods and cones derive their light sensitivity from the bleaching of chromoproteins by light, and the rods have higher sensitivity than the cones [50], [51], [52]. Signals from the rods significantly contribute to human vision only under low illumination levels and do not contribute to color vision. The signals that contribute to color vision are produced by three types of cones, which are called rho, gamma, and beta cones. Each type of cone has a different chromoprotein, giving the rho, gamma, and beta cones spectral passbands with maximum sensitivities near 580 nm, 550 nm, and 445 nm, respectively. The three passbands correspond to light wavelength bands that are usually associated with the colors red, green, and blue, and therefore the light-sensitive cone cells are the biological cause of the early findings presented in References [48] and [49].

The signals created by the rods and cones during photo-stimulation are electrical nerve impulse trains, and the four impulse trains (one from the rods and three from the cones) are transformed into one achromatic signal and two chromatic signals. In the late nineteenth century, it was proposed that the chromatic signals represent a color-opponent space [21]. In the modern view, the achromatic signal is also included in the color-opponent space, making it three dimensional, and the most basic model specifies that each axis of the space represents a measure of two opponent, mutually exclusive sensations: redness and greenness, yellowness and blueness, and lightness and darkness. The basic model explains why humans do not perceive any color as being, for example, simultaneously greenish and reddish but may perceive a color as being, say, simultaneously greenish and bluish. The visual signals are transmitted through the optic nerve to the brain, and several effects originate during the perception of color. The effects are psychological and psychophysical in nature and three such effects that are relevant to color rendering are general brightness adaptation, lateral brightness adaptation, and chromatic adaptation.

General brightness adaptation of the visual system is an adjustment that occurs in response to the overall luminance of the observed scene. The overall sensitivity of the visual system increases or decreases as the overall luminance of the observed scene decreases or increases, respectively. An emergent effect of this type of adaptation is that a scene is perceived to have lower contrast and to be less colorful when viewed under a lower overall luminance than when viewed under a higher overall luminance. Another effect that arises is related to changes in instantaneous dynamic range. Even though the total dynamic range of the visual system is extremely high (on the order of $10^{10}$), at any given general brightness adaptation level the instantaneous dynamic range is much smaller (and varies, on the order of 10 for a very dark level and up to $10^3$ under full daylight) [53].

Lateral brightness adaptation is a change in the sensitivity of a retinal area as influenced by the signals from neighboring retinal areas. This type of adaptation is very complex and yields several visual effects, including Mach bands [54] and the Bezold-Brücke effect [55]. Another lateral brightness adaptation effect is simultaneous contrast, whereby the lightness, hue, and colorfulness of a region [56] depend on the color of its surround. Simultaneous contrast is the origin of the dark-surround effect, which occurs when an image is perceived as having lower contrast if viewed with a dark surround than if viewed with a light surround [57].

Chromatic adaptation is a reaction of the visual response to the overall chromaticity of the observed scene and the chromaticity of a stimulus that is perceived as neutral is called the chromatic adaptation point. This type of adaptation allows the visual system to recognize an object under a range of illumination conditions. For example, snow is perceived as white under daylight and incandescent lamps even though the latter source lacks power in the shorter wavelengths and therefore yields a radically different visual spectral stimulus than daylight.

For a more detailed description of the physiology, properties, and modeling of human vision, see References [32], [56], and [58].

### 5.6.2 Colorimetry

Human vision is an astonishingly complex mechanism of perception and a complete theoretical description has not yet been developed. However, the science of colorimetry can be used to predict whether two color stimuli will perceptually match [59], [60], [61]. Colorimetry is based on color-matching experiments, where an observer views test and reference stimuli and the reference stimulus is adjusted until the observer perceptually matches the two stimuli. The reference stimulus is composed of three independent light sources, which are called the color primaries. The observers adjust the intensity of each color primary to achieve the perceptual match. When a perceptual match to the test stimulus is achieved, the primary color intensities are recorded and are called the tristimulus values for the test color. If a color-matching experiment is carried out for all test colors composed of a single wavelength, the set of tristimulus values forms a set of three curves that are called color-matching functions (CMFs). Each set of color primaries will have an associated CMF set, and any CMF set may be converted with a linear transform into another CMF set having different color primaries [62].

Besides depending on the color primaries, CMFs can also vary because of differences in the eye responses among individuals, so the Commission Internationale de l'Eclairage (CIE) defined standard observers for unambiguous tristimulus specification [63]. The CIE XYZ tristimulus values for any arbitrary stimulus can be calculated as

$$X = k \int_{\lambda_1}^{\lambda_2} S(\lambda) \bar{x}(\lambda) d\lambda, \qquad (5.14)$$

$$Y = k \int_{\lambda_1}^{\lambda_2} S(\lambda) \bar{y}(\lambda) d\lambda, \qquad (5.15)$$

$$Z = k \int_{\lambda_1}^{\lambda_2} S(\lambda) \bar{z}(\lambda) d\lambda, \qquad (5.16)$$

where $\lambda$ is the wavelength variable, the interval $[\lambda_1, \lambda_2]$ defines the visible wavelength range, $S(\lambda)$ is the spectral power distribution of the stimulus in units of watts per nanometer, and $\bar{x}$, $\bar{y}$, and $\bar{z}$ are the CIE standard observer CMFs. The normalizing constant $k$ is sometimes chosen such that $Y$ is equal to 1 when $S(\lambda)$ is equal to the spectral power distribution of the illuminant, yielding luminance factor values. When this type of normalization is used, the tristimulus values specify *relative colorimetry*. The normalizing constant $k$ can also be chosen such that $Y$ yields luminance values in units of candelas per square meter $(cd/m^2)$, in which case the tristimulus values specify *absolute colorimetry* [58].

The CIE Y value, or luminance, correlates well with the perception of brightness. The magnitudes of the CIE X and Z values do not correlate well with any perceptual attribute, but the relative magnitudes of the CIE XYZ tristimulus values are correlated with hue and colorfulness [56]. For these reasons, the CIE defined relative tristimulus values, called *xyz* chromaticities, as

$$x = \frac{X}{X+Y+Z}, \quad y = \frac{Y}{X+Y+Z}, \quad z = \frac{Z}{X+Y+Z}. \qquad (5.17)$$

### 5.6.3 Color Rendering Using Relative Colorimetric Matching

One common imaging task is to display on a monitor an image that was captured with a digital camera. It is usually required to display a realistic, visually pleasing reproduction of the original scene. A reasonable strategy, but one that is not preferred as will be discussed later, is to apply a color rendering that achieves a relative colorimetric match between the displayed reproduction and the original scene. Relative rather than absolute colorimetry is used because the dynamic range and the gamut of many original scenes are greater than the monitor's dynamic range and gamut, and in those cases an absolute match cannot be achieved.

A monitor display is an additive color reproduction device that typically has three color primaries. Three control code values are input to the device to produce a color stimulus composed of the mixture of the three color primary intensities. The dynamic range of a cathode-ray tube (CRT) monitor in a darkened viewing environment is typically on the order of $10^3$ and that of a liquid-crystal display (LCD) is on the order of $10^2$, although when viewed with environmental flare an LCD monitor may actually have a larger dynamic range than a CRT monitor [64]. CRTs were the most popular monitor devices until recently but now LCDs dominate the monitor market. The neutral tone scale characteristic of many CRTs can be approximated by a simple power-law relationship:

$$Y_m = k_m \left( \frac{V}{V_{max}} \right)^{\gamma} + Y_{m0}, \tag{5.18}$$

where $Y_m$ is the CIE luminance factor produced by the monitor, $V$ is the input control code value, $V_{max}$ is the maximum input code value, $Y_{m0}$ is the CIE luminance factor when $V = 0$ and $k_m$ is a normalizing constant chosen such that $Y_m = 1$ when $V = V_{max}$. The relation between control code value and CIE luminance factor for LCD monitors is more complicated than for CRTs, but it can be made to follow the same power-law relationship as shown above with some signal processing between the input and the output [64].

A digital color camera is equipped with a light sensor and, most often, with an array of three types of light receptors (RGB or CMY) whose responses are linear with respect to irradiance. The dynamic range of a sensor depends on the receptors' electron-gathering capacity [65], but it is common for a sensor to have a dynamic range of at least $10^3$, which is sufficient to cover the instantaneous dynamic range of the human eye. Sensor spectral responsivities typically form approximations to a set of color-matching functions. If the approximation is good, a sensor's responses can be multiplied by a $3 \times 3$ matrix to obtain good approximations to CIE XYZ tristimulus values.

Suppose that an image is captured with a digital color camera that has spectral responsivities that are good approximations to color-matching functions and that the camera responses have been transformed to CIE XYZ tristimulus values. For neutral stimuli (that is, all stimuli that have the chromaticity of the chromatic adaptation point), the chromaticity ratios $x/y$ and $z/y$ are constant and therefore the tristimulus ratios $X/Y$ and $Z/Y$ are also constant. Thus, it is possible to gain the camera channels such that neutral stimuli produce the same response in all three channels as described in Section 5.3.3. Such channel balancing is assumed, as well as that the display monitor is calibrated such that the chromaticity of its output is the same for any triplet of equal input code values. Further suppose that the goal is to display the image such that a colorimetric, rather than a perceptual, match is achieved between the original scene and the monitor reproduction. For simplicity, it is assumed that the camera signals do not saturate, that all of the camera responses are within the monitor's gamut, that the observer is chromatically adapted to CIE illuminant D65 when viewing the original scene and the monitor reproduction, and that the monitor white point has the chromaticity of CIE illuminant D65.

The camera's output CIE luminance factor can be transformed with the inverse of the monitor neutral tone scale characteristic:

$$V = V_{max} \left( \frac{Y_c - Y_{m0}}{k_m} \right)^{\frac{1}{\gamma}}, \tag{5.19}$$

where $Y_c$ is the CIE luminance factor measured by the camera. This type of transformation, that accounts for the display nonlinearity, is usually called *gamma correction*. The code value $V$ is used as input to the monitor's three channels so that the output luminance factor of the monitor is $Y_m = Y_c$ (and, by monitor calibration, the output chromaticity is the same as that of CIE illuminant D65). In short, the relative colorimetric matching for the neutral tone scale is achieved with the transformation of camera responses to CIE XYZ approximations, camera channel balancing, monitor calibration, and gamma correction. Assuming a colorimetrically matched neutral tone scale, the relative colorimetric matching of non-neutral stimuli is described next.

A set of CMFs for the monitor may be derived by performing a color-matching experiment using the monitor to produce the reference stimuli. Once the display CMFs are obtained, relative colorimetric matching of non-neutral stimuli would be achieved if the camera responsivities are proportional to the display CMFs because the camera would directly measure how much of each monitor primary is present in a given color stimulus of an original scene. However, the proportionality is impossible for physically realizable cameras and monitors, but the matching can be achieved with additional signal processing [66]. The balanced camera responses may be transformed with a $3 \times 3$ matrix to tristimulus values associated with the monitor color primaries [62]. This transformation is usually called *color correction*. Once color correction is accomplished, the new tristimulus values are gamma-corrected and applied to the monitor to achieve relative colorimetric matching between all of the stimuli in the original scene and the monitor reproduction.

Figure 5.15a shows how an image as captured with a camera with no signal processing approximately appears when displayed on a monitor. The camera signals are linear with sensor irradiance, no gamma correction is applied, and there is a discrepancy between the camera responsivities and the monitor CMFs. Under these circumstances most colors are reproduced too dark with incorrect chromaticities. When that camera image is gamma-corrected before being displayed, the colors will have approximately the correct luminance factors but the chromaticities will remain incorrect as illustrated in Figure 5.15b. If color correction is performed before gamma correction, then the chromaticities will also be correct. Figure 5.15c shows an illustration of an image that has been processed with color and gamma corrections and thus simulates a perfect relative colorimetric match to the original scene. Yet, this image still looks like it could be improved – its luminance contrast and colorfulness seem low. These two attributes are perceived to be low because the observer's general brightness adaptation level corresponds to the display's overall luminance level,

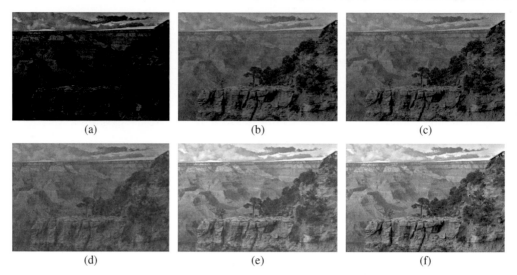

**FIGURE 5.15** (See color insert.)

A series of images illustrating how an image approximately appears when displayed on a monitor at different steps in the color rendering, where (a–c) simulate images viewed within a flareless viewing environment and images (d–f) simulate images viewed with flare. (a) The original linear camera image, (b) after gamma correction but not color corrected, (c) after color correction and gamma correction, (d) after color correction and gamma correction (with flare), (e) with boosted RGB scales, and (f) after flare correction.

which is much lower than the level of the original scene. Colorimetric matching is evidently not sufficient to achieve perceptual matching. While the system has accounted for chromatic adaptation (by balancing the camera channels for neutral stimuli and setting the monitor white point to have the chromaticity of CIE illuminant D65), no other adaptation effects have been considered in the system setup or the signal processing. Moreover, up to this point the viewing environment has been assumed to be flareless. If the colorimetrically matched image of Figure 5.15c is displayed in a viewing environment with flare, then the darker colors will look less dark as illustrated in Figure 5.15d. Viewing flare can have many forms, but here it is assumed that it is light with the same chromaticity as the monitor white point and that it has a uniformly distributed intensity over the monitor face.

### 5.6.4 Color Rendering with Adaptation and Viewing Flare Considerations

A way to simultaneously account for the low luminance contrast and low colorfulness that is perceived as a result of an observer's low general brightness adaptation level is to boost all three image channels. The boost can be achieved by gaining the color-corrected tristimulus values just before sending the signal to the display. The same gain is applied to all three channels to maintain the chromaticities of all colors. Figure 5.15e shows how the image in Figure 5.15d would approximately appear if the three channels are gained. The overall luminance contrast and the colorfulness have been increased, but the darker colors appear to not be dark enough because of the viewing flare. Since it is assumed that the viewing flare is uniform across the monitor, to correct for the flare it is sufficient to subtract the boosted tristimulus values of the flare from the boosted tristimulus values of the scene. The channels are gained again such that the white point remains the same before and after the flare tristimulus subtractions. Figure 5.15f illustrates the results of applying the flare correction to the image from Figure 5.15e, and it can be seen that while lighter colors appear about the same in both figures, the darker colors appear even darker after the correction. It is this last image that appears to be a perceptual match to the original scene.

## 5.7 Edge Enhancement

During image capture, edge detail is lost through optical and other effects that reduce the transfer of modulation from the scene to the reproduced image. Edge enhancement, also known as sharpening, is used to boost edge contrast, making the image appear sharper and partially compensating for the losses in the modulation transfer function (MTF). There are several main perceptual effects from edge enhancement: increased sharpness, increased noise, and spatial artifacts such as over-sharpening or ringing, visible as halos at high-contrast edges. Just as noise reduction is intended to reduce noise while preserving sharpness, edge enhancement is intended to increase sharpness while avoiding noise amplification and other artifacts. When applying edge enhancement, it is important that the quality improvement from increased sharpness is greater than the quality loss from increased noise or other artifacts.

### 5.7.1 Human Visual Sharpness Response

The human visual sharpness response depends more strongly on low and intermediate spatial frequencies in a viewed image. An early publication investigated sharpness in photographic images and concluded that both acutance and resolving power contributed to a prediction of sharpness [67]. Since then, further work relating system MTF and sharpness-related image quality included the development of an optical merit function based on camera or system MTF and the contrast sensitivity of the eye [68]. A more recent publication describing models of visual response for analysis of image processing algorithms included a conveniently parameterized model for a two-dimensional CSF [44].

Perception of sharpness in a reproduced image depends on viewing distance, the resolution of the reproduced image, and the MTF of the entire reproduction chain from camera lens through print or display. Figure 5.16 shows an example camera MTF (including lens, sensor, and processing effects) along with a CSF from Reference [44].

The frequency axis for the CSF is scaled for viewing the image at 100 pixels per inch on a display at a viewing distance of 20 inches, a common viewing distance for a computer display. The peak in the human CSF aligns fairly well with the peak of the camera MTF, which is desirable for a well-tuned image chain. The perceived improvement from edge enhancement is maximized if the edge enhancement has the greatest effect on frequencies with the greatest response in the human CSF.

During capture, viewing conditions for the reproduced image are generally unknown and variable, since users view digital images in a variety of ways on a variety of devices.

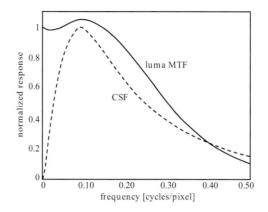

**FIGURE 5.16**

Example spatial frequency response with CSF overlay.

Because the viewing conditions are variable, digital camera edge enhancement is usually optimized for a selected, somewhat conservative, viewing condition. One common condition is viewing of the image on a high-resolution display at one camera pixel per display pixel and a viewing distance of twice the image height, cropping the image to fit the display. Another viewing condition, with images resized to fill a large display and viewing at a distance of 34 inches, was used in development of a quality ruler for subjective image evaluation [69], [70], while other options include viewing prints of various sizes.

### 5.7.2    Linear Edge Enhancement

Routine edge enhancement is based on a convolution operation to obtain an edge map, which is scaled and added to the original image, as follows:

$$\mathbf{A}' = \mathbf{A} + k\mathbf{E}, \tag{5.20}$$

where $\mathbf{A}'$ is the enhanced image, $\mathbf{A}$ is the image from the previous processing stage, $k$ is a scalar gain, and $\mathbf{E}$ is the edge map. There are several possible ways to create $\mathbf{E}$, one of which is an unsharp mask:

$$\mathbf{E} = \mathbf{A} - \mathbf{A} * \mathbf{b}, \tag{5.21}$$

where $\mathbf{b}$ is a lowpass convolution kernel and $*$ refers to a convolution. A second method is direct convolution:

$$\mathbf{E} = \mathbf{A} * \mathbf{h}, \tag{5.22}$$

where $\mathbf{h}$ is a highpass or bandpass convolution kernel. The design of the convolution kernel, either $\mathbf{b}$ or $\mathbf{h}$, and the choice of $k$, are the main tuning parameters in linear edge enhancement. The kernel controls which spatial frequencies to enhance, while $k$ controls the magnitude of the enhancement. Often, the kernel is designed to produce a bandpass edge map, providing limited gain or even zero gain at the highest spatial frequencies.

Consistency of edge response in the enhanced image is an important consideration in kernel design. If the camera has an anisotropic optical transfer function (OTF), such as caused by an anti-aliasing filter that blurs only in one direction, the kernel may be designed

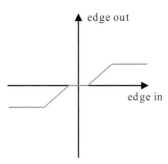

**FIGURE 5.17**

Soft thresholding and edge limiting edge enhancement nonlinearity.

to partially compensate for the anisotropy. If the camera OTF is isotropic, a rotationally symmetric kernel provides equal enhancement for edges of all orientations, preserving that consistency.

Equations 5.21 and 5.22 do not specify the color metric for **A**, such as linear camera RGB, rendered image RGB, or another color space. If the loss in system MTF is primarily in the camera optical system, then performing enhancement in a metric linear with relative exposure will allow for the most precise correction. However, MTF losses occur in multiple steps in the image processing chain, several of them not linear with the original exposure. Experience has shown that edge enhancement in a nonlinear space (for example, after tone rendering and gamma correction) provides a more perceptually even enhancement than applying it in a linear space [1].

### 5.7.3 Nonlinear Edge Enhancement

Even with a well-designed bandpass kernel, the formulation in Section 5.7.2 amplifies noise and can produce halo artifacts at high-contrast edges. While limiting the edge enhancement is possible, better perceptual results can be achieved by applying a nonlinearity to the edge map before scaling and adding it to the original image.

Figure 5.17 shows an example of nonlinearity to be applied to values in the edge map **E** with a simple lookup table operation, $\mathbf{E}' = f(\mathbf{E})$. Small edge map values are most likely to be the result of noise, while larger edge map values are likely to come from scene edges. The soft thresholding function shown in Figure 5.17 reduces noise amplification by reducing the magnitude of all edge values by a constant, and is widely used for noise reduction, such as in Reference [71]. Soft thresholding eliminates edge enhancement for small modulations, while continuing to enhance larger modulations. The nonlinearity shown in Figure 5.17 also reduces halo artifacts by limiting the largest edge values, since high-contrast edges are the most likely to exhibit halo artifacts after edge enhancement.

Figure 5.18 illustrates some of the effects discussed here. Namely, Figure 5.18a shows an image without edge enhancement applied. There is measurable noise in the image, but it does not detract substantially from the overall quality of the image under the intended viewing conditions. Figure 5.18b shows the same image with a simple linear convolution edge enhancement, as described in Section 5.7.2. The image is much sharper, with higher

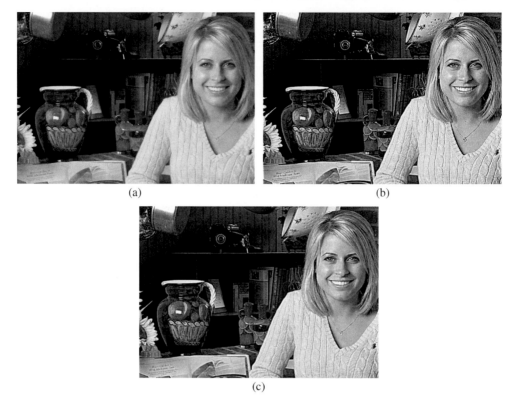

(a)                                                    (b)

(c)

**FIGURE 5.18** (See color insert.)

Example images with: (a) no edge enhancement, (b) linear edge enhancement, and (c) soft thresholding and edge-limited enhancement.

edge contrast, but also much higher noise and halo artifacts, particularly obvious in the text in the cookbook. Figure 5.18c shows the same image with an edge-limiting soft thresholding step used in edge enhancement. The sharpness of the image is only slightly less than the sharpness of Figure 5.18b, but the noise and halo artifacts are substantially reduced.

Application of edge enhancement in an RGB color space tends to amplify colored edges caused by chromatic aberration or other artifacts earlier in the capture chain, as well as colored noise. Because of this, edge enhancement is often applied to the luma channel of the image after converting to a luma-chroma color space. The specific luma-chroma color space is often the JPEG $YC_BC_R$ space, chosen for the same reasons as for noise reduction.

## 5.8   Compression

A digital image that has been captured and processed as described in the previous sections of this chapter contains both mathematically redundant as well as perceptually insignificant data. An image compression step can be included in the overall image processing chain to produce a more efficient representation of the image, in which the redundant and perceptually insignificant data have been removed. This step of image compression reduces the amount of physical storage space required to represent the image. This also increases the number of images that can be stored in a fixed-size storage space as well as to allow faster transmission of individual images from a digital camera to other devices.

Image compression can be divided into two categories: lossy and lossless. Lossless image compression is reversible, meaning that the exact original image data can be recovered from the compressed image data. This characteristic limits the amount of compression that is possible. Lossless image compression focuses on removing redundancy in an image representation. Images commonly have both spatial redundancy and spectral redundancy.

In addition to removing redundancy, lossy image compression uses knowledge of the HVS to discard data that are perceptually insignificant. This typically allows a much greater compression ratio to be achieved than with lossless image compression, while it still allows an image with little or no visual degradation to be reconstructed from the compressed image data. When no visual degradation occurs, the compression is referred to as visually lossless.

During (or even before) compression, images are often converted from an RGB color space to a luminance-chrominance color space, such as $YC_BC_R$ [72]. This conversion serves multiple purposes. First, it reduces the spectral redundancy among the image channels, allowing them to be independently encoded more efficiently [73]. Second, it transforms the image data into a space in which the characteristics of the HVS can be more easily exploited. The HVS is generally more sensitive to luminance information than chrominance information, in particular at higher spatial frequencies [74]. The chrominance channels can be compressed more aggressively than can the luminance channel without visual degradation. One way this is achieved is by spatially subsampling the chrominance channels prior to encoding.

Many lossy compression schemes transform the pixel data into the frequency domain. Digital cameras predominantly use the JPEG lossy image compression standard [72]. JPEG employs a two-dimensional discrete cosine transform (DCT) to convert pixel values into spatial frequency coefficients. For natural imagery containing mostly low-frequency information, this transformation compacts most of the signal energy into a small fraction of the transform coefficients, which is advantageous for compression. Representing images in the frequency domain is also beneficial for optimizing compression with respect to the HVS.

It is known that the HVS has varying sensitivity to different spatial frequencies [74]. In particular, the HVS has decreasing sensitivity at higher spatial frequencies. The HVS also has varying sensitivity according to the orientation of the spatial frequency information, being less sensitive at a diagonal orientation than at horizontal or vertical orientations.

In JPEG image compression, the DCT is performed on $8 \times 8$ blocks of image data. This produces a collection of 64 frequency coefficients, varying in horizontal and vertical spatial frequency. Each of these coefficients is quantized. This is the lossy step within JPEG compression, in which information is discarded. Quantization involves dividing each frequency coefficient by an integer quantization value, and rounding the result to the nearest integer. Reducing the accuracy by which the frequency coefficients are represented allows them to be more efficiently compressed. Increased quantization generally results in increased compression. On the other hand, the greater the quantization value applied to a given frequency coefficient, the more uncertainty and hence greater expected error there is when recovering the frequency coefficient from its quantized representation during decoding.

**TABLE 5.1**
Default JPEG luminance quantization table.

| 16 | 11 | 10 | 16 | 24 | 40 | 51 | 61 |
|----|----|----|----|----|----|----|----|
| 12 | 12 | 14 | 19 | 26 | 58 | 60 | 55 |
| 14 | 13 | 16 | 24 | 40 | 57 | 69 | 56 |
| 14 | 17 | 22 | 29 | 51 | 87 | 80 | 62 |
| 18 | 22 | 37 | 56 | 68 | 109 | 103 | 77 |
| 24 | 35 | 55 | 64 | 81 | 104 | 113 | 92 |
| 49 | 64 | 78 | 87 | 103 | 121 | 120 | 101 |
| 72 | 92 | 95 | 98 | 112 | 100 | 103 | 99 |

The characteristics of the HVS can be applied to this quantization process, exploiting the varying sensitivity of the HVS to different spatial frequencies and orientations, as well as to luminance and chrominance information. The contrast sensitivity function is used to indicate the base detection threshold at which a spatial frequency just becomes visible under certain viewing conditions [74]. These detection thresholds can be used to design a quantization table (also referred to as a quantization matrix) that allows maximum quantization of frequency coefficients while retaining a visually lossless representation of the information. Table 5.1 lists the default luminance quantization table associated with the JPEG image compression standard, designed to be at the threshold of visibility under specific viewing conditions [75]. The quantization values generally increase as spatial frequency increases (from left to right, and from top to bottom), corresponding to the decreased sensitivity of the HVS at high spatial frequencies. Also, there are relatively larger quantization values for spatial frequencies at 45° orientations (on the diagonal of the matrix), indicative of the decreased sensitivity of the HVS to diagonal frequency content.

In the JPEG compression standard, the quantization process is content independent. For a given color channel, a single quantization table is used throughout the entire image. Because of this, JPEG-based compression is unable to directly exploit the HVS characteristic of contrast masking. Contrast masking refers to the change in visibility of a target signal due to the presence of a background signal [74]. As the contrast of the masking background signal increases, it can be more difficult to detect the target signal. Contrast masking can be observed, for example, when random noise is added to an image. The noise is more visible in smooth regions with low contrast than in regions of high contrast and texture.

Contrast masking applies to compression. Studies have suggested three perceptually significant activity regions in an image: smooth, edge, and texture [76]. Because of the masking properties of the HVS, high-contrast textured image regions can tolerate more error without visual degradation than smooth regions of low contrast or isolated sharp edges in an image. In JPEG, high-contrast textured image regions can tolerate greater quantization of frequency coefficients than other regions. However, since the quantization tables used in JPEG are constant throughout the image, a compromise must be made. A conservative level of quantization can be applied throughout the image to avoid visual degradation in the most sensitive areas. This comes at a cost of poor compression efficiency, particularly in regions of texture where extra bits are wasted encoding perceptually insignificant information.

One method to exploit the masking properties of the HVS to increase compression efficiency while remaining compliant with the JPEG image compression standard is discussed in Reference [77]. In this approach, local texture measurements are used to compute visibility thresholds for the spatial frequency coefficients in each block. Coefficients that are smaller than their corresponding visibility threshold can be prequantized to zero without introducing any visual degradation. This prequantization allows some coefficients to be quantized to zero that otherwise would not be when the standard quantization matrix is applied. Zero values after quantization are more efficiently encoded than nonzero values, resulting in improved compression efficiency.

An alternative method to exploit the masking properties of the HVS within the framework of JPEG is discussed in Reference [78]. In this approach, the quantization matrix value applied to a particular frequency coefficient can be normalized based on previously encoded quantized frequency coefficients. The previously encoded information is used to estimate the amount of contrast masking present, and correspondingly the normalization factors that can be applied to subsequent quantization matrix values. In particular, when significant contrast masking is present, normalization values greater than one are used to increase the quantization applied to an individual frequency coefficient. The increased quantization results in increased compression efficiency, without introducing any visual degradation. Although this can be done using a single quantization matrix (for a given color channel) throughout the image as required by the JPEG standard, it does require a smart decoder to properly interpret the compressed image data. The smart decoder must duplicate the calculation of the normalization terms used during quantization in order to allow correct recovery of the frequency coefficients from their quantized representations.

## 5.9 Conclusion

The human visual system influences the design of image processing algorithms in a digital camera. This chapter illustrates the perceptual considerations involved in the many different steps of a digital camera image processing chain. Most tangibly, perceptual considerations affect processing algorithms to help ensure that the final image "looks right." Noise reduction, color rendering, and edge enhancement are all performed with the human visual system in mind. Other aspects of the processing chain are affected by the human visual system as well. The design of the image sensor color filter array and the corresponding demosaicking algorithm exploit human visual system characteristics. Even computational and compression efficiency are influenced by perceptual considerations. Throughout the processing chain, perceptual considerations affect the design of image processing algorithms in a digital camera.

## Appendix

Compact digital cameras usually have a pixel pitch between 1 and 2 $\mu$m, so the sharpness of the captured image is often limited by the lens even when the scene is at best focus. The

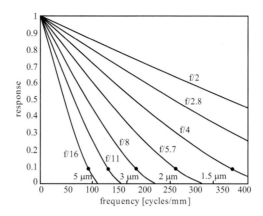

**FIGURE 5.19**

Diffraction-limited lens MTF.

MTF for a diffraction-limited lens with a circular aperture for a perfectly focused image can be expressed as follows [79]:

$$
\text{MTF}(v) = \begin{cases} \frac{2}{\pi}\left[\arccos(\frac{v}{v_C}) - \frac{v}{v_C}\sqrt{1 - (\frac{v}{v_C})^2}\right] & \text{for } v < v_C, \\ 0 & \text{otherwise}, \end{cases} \tag{5.23}
$$

where $v$ is spatial frequency and $v_C = 1/(\lambda N)$ denotes the cutoff frequency, which depends on wavelength $\lambda$ and the f/number $N$. This function is plotted for green light (550 nm) and a wide range of f/numbers in Figure 5.19.

Each MTF curve in the figure is labeled with the f/number used to generate it. Inexpensive lenses rarely have an f/number below 2.5, even for fairly short focal lengths. Inexpensive lenses with a longer focal length have higher minimum f/numbers, often at least 5.6. The half-sampling frequency is also shown for several different pixel pitches, each labeled with the appropriate pixel pitch. Finally, small circles are shown for f/numbers equal to four and above where MTF goes through the frequency corresponding to the Rayleigh resolution limit ($0.82v_C$). This figure shows that as the f/number is increased, the resolution limit is decreased and the MTF drops at all frequencies. While lenses are usually not diffraction-limited at their widest apertures, the MTF still tends to drop further as the f/number is increased.

# References

[1] J.E. Adams and J.F. Hamilton, "Digital camera image processing chain design," in *Single-Sensor Imaging: Methods and Applications for Digital Cameras*, R. Lukac (ed.), Boca Raton, FL, USA: CRC Press / Taylor & Francis, September 2008, pp. 67–103.

[2] L. Jones and C. Nelson, "Study of various sensitometric criteria of negative film speeds," *Journal of the Optical Society of America*, vol. 30, no. 3, pp. 93–109, March 1940.

[3] A. Adams, *The Negative*. New York, USA: Little, Brown and Company, 1995.

[4] Commission Internationale de l'Eclairage, "CIE publication no. 15.2, Colorimetry," Technical report, CIE, 1986.

[5] A. Stimson, "Photographic exposure measuring device," U.S. Patent 3 232 192, February 1966.

[6] B. Johnson, "Photographic exposure control system and method," U.S. Patent 4 423 936, January 1984.

[7] J.S. Lee, Y.Y. Jung, B.S. Kim, and S.J. Ko, "An advanced video camera system with robust AF, AE, and AWB control," *IEEE Transactions on Consumer Electronics*, vol. 47, no. 3, pp. 694–699, August 2001.

[8] F. Arai, "Video camera exposure control method and apparatus for preventing improper exposure due to changing object size or displacement and luminance difference between the object and background," U.S. Patent 5 049 997, September 1991.

[9] E. Gindele, "Digital image processing method and apparatus for brightness adjustment of digital images," U.S. Patent 7 289 154, October 2007.

[10] S. Pattanaik, J. Ferwerda, M. Fairchild, and D. Greenberg, "A multiscale model of adaptation and spatial vision for realistic image display," in *Proceedings of the 25th Annual Conference on Computer Graphics and Interactive Techniques*, Orlando, FL, USA, July 1998, pp. 287–298.

[11] E. Reinhard, M. Stark, P. Shirley, and J. Ferwerda, "Photographic tone reproduction for digital images," *ACM Transactions on Graphics*, vol. 21, no. 3, pp. 267–276, July 2002.

[12] D. Kerr, "APEX – the additive system of photographic exposure." Available online, http://dougkerr.net/Pumpkin/articles/APEX.pdf, 2007.

[13] International Organization for Standardization, "Photography – digital still cameras – determination of exposure index, ISO speed ratings, standard output sensitivity, and recommended exposure index," Technical report, ISO 12232, ISO TC42/WG 18, 2006.

[14] B. Pillman, "Camera exposure determination based on a psychometric quality model," in *Proceedings of IEEE Workshop on Signal Processing Systems*, San Francisco, CA, USA, October 2010, pp. 339–344.

[15] B. Pillman and D. Jasinski, "Camera exposure determination based on a psychometric quality model," *Journal of Signal Processing Systems*, submitted, 2011.

[16] K. Omata, T. Miyano, and M. Kiri, "Focusing device," U.S. Patent 6 441 855, August 2002.

[17] M. Gamadia and N. Kehtarnavaz, "Enhanced low-light auto-focus system model in digital still and cell-phone cameras," in *Proceedings of IEEE International Conference on Image Processing*, Cairo, Egypt, November 2009, pp. 2677–2680.

[18] M. Gamadia and N. Kehtarnavaz, "A real-time continuous automatic focus algorithm for digital cameras," in *Proceedings of IEEE Southwest Symposium on Image Analysis and Interpretation*, Denver, CO, USA, March 2006, pp. 163–167.

[19] R. Evans, "Method for correcting photographic color prints," U.S. Patent 2 571 697, October 1951.

[20] M. Kumar, E. Morales, J. Adams, and W. Hao, "New digital camera sensor architecture for low light imaging," in *Proceedings of IEEE International Conference on Image Processing*, Cairo, Egypt, November 2009, pp. 2681–2684.

[21] L.M. Hurvich and D. Jameson, "An opponent-process theory of color vision," *Psychological Review*, vol. 64, no. 6, pp. 384–404, November 1957.

[22] Y.I. Ohta, T. Kanade, and T. Sakai, "Color information for region segmentation," *Computer Graphics and Image Processing*, vol. 13, no. 3, pp. 222–241, July 1980.

[23] E. Gindele, J. Adams, Jr., J. Hamilton, Jr., and B. Pillman, "Method for automatic white balance of digital images," U.S. Patent 7 158 174, January 2007.

[24] T. Miyano and E. Shimizu, "Automatic white balance adjusting device," U.S. Patent 5 644 358, July 1997.

[25] G. Finlayson, S. Hordley, and P. Hubel, "Color by correlation: A simple, unifying framework for color constancy," *IEEE Transactions on Pattern Analysis and Machine Intelligence*, vol. 23, no. 11, pp. 1209–1221, November 2001.

[26] H. Lee, "Method for computing the scene-illuminant chromaticity from specular highlights," *Journal of the Optical Society of America A*, vol. 3, no. 10, pp. 1694–1699, October 1986.

[27] J.J. McCann, S.P. McKee, and T.H. Taylor, "Quantitative studies in retinex theory," *Vision Research*, vol. 16, no. 5, pp. 445–458, May 1976.

[28] T. Miyano, "Auto white adjusting device," U.S. Patent 5 659 357, August 1997.

[29] J.E. Adams, "Interaction between color plane interpolation and other image processing functions in electronic photography," *Proceedings of SPIE*, vol. 2416, pp. 144–151, February 1995.

[30] B.E. Bayer, "Color imaging array," U.S. Patent 3 971 065, July 1976.

[31] M.J. McMahon and D.I.A. MacLeod, "The origin of the oblique effect examined with pattern adaptation and masking," *Journal of Vision*, vol. 3, no. 3, pp. 230–239, April 2003.

[32] K.R. Boff and J.E. Lincoln, *Engineering Data Compendium: Human Perception and Performance*. Wright-Patterson AFB, Armstrong Aerospace Medical Research Laboratory, OH, USA, 1988.

[33] J.E. Adams, J.F. Hamilton, M. Kumar, E.O. Morales, R. Palum, and B.H. Pillman, "Single capture image fusion," in *Computational Photography: Methods and Applications*, R. Lukac (ed.), Boca Raton, FL, USA: CRC Press / Taylor & Francis, October 2010, pp. 1–62.

[34] J.D. Gaskill, *Linear Systems, Fourier Transforms, and Optics*. New York, USA: John Wiley & Sons, 1978.

[35] R.H. Hibbard, "Apparatus and method for adaptively interpolating a full color image utilizing luminance gradients," U.S. Patent 5 382 976, January 1995.

[36] J.E. Adams, "Design of practical color filter array interpolation algorithms for digital cameras," *Proceedings of SPIE*, vol. 3028, pp. 117–125, February 1997.

[37] J.E. Adams and J.F. Hamilton, "Adaptive color plan interpolation in single sensor color electronic camera," U.S. Patent 5 506 619, April 1996.

[38] J.F. Hamilton and J.E. Adams, "Adaptive color plan interpolation in single sensor color electronic camera," U.S. Patent 5 629 734, May 1997.

[39] D.R. Cok, "Signal processing method and apparatus for producing interpolated chrominance values in a sampled color image signal," U.S. Patent 4 642 678, February 1987.

[40] B.W. Keelan, *Handbook of Image Quality: Characterization and Prediction*. New York, USA: Marcel Dekker, 2002.

[41] J. McElvain, S. Campbell, J. Miller, and E. Jin, "Texture-based measurement of spatial frequency response using the dead leaves target: Extensions, and application to real camera systems," *Proceedings of SPIE*, vol. 7537, p. 75370D, January 2010.

[42] J.S. Lee, "Digital image smoothing and the sigma filter," *Computer Vision, Graphics, and Image Processing*, vol. 24, no. 2, pp. 255–269, November 1983.

[43] C. Tomasi and R. Manduchi, "Bilateral filtering for gray and color images," in *Proceedings of Sixth International Conference on Computer Vision*, Bombay, India, January 1998, pp. 839–846.

[44] S. Daly, "A visual model for optimizing the design of image processing algorithms," in *Proceedings of IEEE International Conference on Image Processing*, Austin, TX, USA, vol. 2, pp. 16–20, November 1994.

[45] R. Jenkin and B. Keelan, "Perceptually relevant evaluation of noise power spectra in adaptive pictorial systems," *Proceedings of SPIE*, vol. 7867, pp. 786708:1–12, January 2011.

[46] I. Newton, *Opticks*. London, UK: Smith & Walford, 1704.

[47] T. Young, "On the theory of light and colours," *Philosophical Transactions of the Royal Society of London*, vol. 92, pp. 12–48, 1802.

[48] H. von Helmholtz, *Helmholtz's Treatise on Physiological Optics*, vol. 2. Rochester, NY, USA: The Optical Society of America, 1924.

[49] J.C. Maxwell, "On the theory of compound colours, and the relations of the colours of the spectrum," *Philosophical Transactions of the Royal Society of London*, vol. 150, pp. 57–84, 1860.

[50] D. Baylor, T. Lamb, and K.W. Yau, "Responses of retinal rods to single photons," *Journal of Physiology*, vol. 288, pp. 613–634, March 1979.

[51] J. Nathans, "The evolution and physiology of human color vision: Insights from molecular genetic studies of visual pigments," *Neuron*, vol. 24, no. 2, pp. 299–312, October 1999.

[52] H. Okawa and A.P. Sampath, "Optimization of single-photon response transmission at the rod-to-rod bipolar synapse," *Physiology*, vol. 22, no. 4, pp. 279–286, August 2007.

[53] J.J. Sheppard, *Human Color Perception*. New York, USA: Elsevier, 1968.

[54] R.B. Lotto, S.M. Williams, and D. Purves, "Mach bands as empirically derived associations," *Proceedings of the National Academy of Sciences of the United States of America*, vol. 96, no. 9, pp. 5245–5250, April 1999.

[55] R.W. Pridmore, "Bezold-Brücke hue-shift as functions of luminance level, luminance ratio, interstimulus interval and adapting white for aperture and object colors," *Vision Research*, vol. 39, no. 23, pp. 3873–3891, November 1999.

[56] R.G.W. Hunt, *Measuring Color*. Surrey, UK: Fountain Press, 3rd edition, 1998.

[57] M.D. Fairchild, "Considering the surround in device-independent color imaging," *Color Research and Application*, vol. 20, no. 6, pp. 352–363, December 1995.

[58] G. Wyszecki and W.S. Stiles, *Color Science: Concepts and Methods, Quantitative Data and Formulas*. New York, USA: John Wiley & Sons, 2nd edition, 2000.

[59] International Organization for Standardization, "Colorimetry – Part 4: CIE $L^*a^*b^*$ colour space," ISO 11664-4, 2008.

[60] Commission Internationale de l'Eclairage, "Industrial colour-difference evaluation," CIE 116-1995.

[61] Commission Internationale de l'Eclairage, "Improvement to industrial colour-difference evaluation," CIE 142, 2001.

[62] R.S. Berns, *Billmeyer and Saltzman's Principles of Color Technology*. New York, USA: John Wiley & Sons, 3rd edition, 2000.

[63] International Organization for Standardization, "Colorimetry – Part 1: Standard colorimetric observers," ISO 11664-1, 2007.

[64] G. Sharma, "LCDs versus CRTs – color-calibration and gamut considerations," *Proceedings of the IEEE*, vol. 90, no. 4, pp. 605–622, April 2002.

[65] R. Palum, "How many photons are there?," in *Proceedings of the Image Processing, Image Quality, Image Capture Systems Conference*, Portland, OR, USA, April 2002, pp. 203–206.

[66] E.J. Giorgianni and T.E. Madden, *Digital Color Management*. Reading, MA, USA: Addison-Wesley, 1998.

[67] G.C. Higgins and R.N. Wolfe, "The relation of definition to sharpness and resolving power in a photographic system," *Journal of the Optical Society of America*, vol. 45, no. 2, pp. 121–125, February 1955.

[68] E. Granger and K. Cupery, "An optical merit function (SQF), which correlates with subjective image judgments," *Photographic Science and Engineering*, vol. 16, no. 3, pp. 221–230, May/June 1972.

[69] E. Jin, B. Keelan, J. Chen, J. Phillips, and Y. Chen, "Softcopy quality ruler method: Implementation and validation," *Proceedings SPIE*, vol. 7242, pp. 724206:1–14, January 2009.

[70] E. Jin and B. Keelan, "Slider-adjusted softcopy ruler for calibrated image quality assessment," *Journal of Electronic Imaging*, vol. 19, no. 1, pp. 011009, January 2010.

[71] S. Chang, B. Yu, and M. Vetterli, "Adaptive wavelet thresholding for image denoising and compression," *IEEE Transactions on Image Processing*, vol. 9, no. 9, pp. 1532–1546, September 2000.

[72] K.A. Parulski and R. Reisch, "Digital camera image storage formats," in *Single-Sensor Imaging: Methods and Applications for Digital Cameras*, R. Lukac (ed.), Boca Raton, FL, USA: CRC Press / Taylor & Francis, September 2008, pp. 351–379.

[73] G. Sharma and J. Trussell, "Digital color imaging," *IEEE Transactions on Image Processing*, vol. 6, no. 7, pp. 901–932, July 1997.

[74] A.B. Watson (ed.), *Digital Images and Human Vision*. Cambridge, MA, USA: MIT Press, 1993.

[75] W.B. Pennebaker and J.L. Mitchell, *JPEG Still Image Data Compression Standard*. New York, USA: Van Nostrand Reinhold, 1993.

[76] M.G. Ramos and S.S. Hemami, "Perceptually based scalable image coding for packet networks," *Journal of Electronic Imaging*, vol. 7, no. 3, pp. 453–463, July 1998.

[77] N. Jayant, J. Johnston, and R. Safranek, "Image compression based on models of human vision," in *Handbook of Visual Communications*, K. Smith, S. Moriarty, G. Barbatsis, and K. Kenney (eds.), San Diego, CA, USA: Academic Press, 1995, pp. 73–125.

[78] S.J. Daly, C.T. Chen, and M. Rabbani, "Adaptive block transform image coding method and apparatus," U.S. Patent 4 774 574, September 1988.

[79] G. Boreman, *Modulation Transfer Function in Optical and Electro-Optical Systems*. Bellingham, WA, USA: SPIE Press, vol. TT52, 2001.

# 6

## Joint White Balancing and Color Enhancement

**Rastislav Lukac**

## 6.1 Introduction

Despite recent advances in digital color imaging [1], [2], producing digital photographs that are faithful representations of the original visual scene remains still challenging due to a number of constraints under which digital cameras operate. In particular, differences in characteristics between image acquisition systems and the human visual system constitute an underlying problem in digital color imaging. Digital cameras typically use various color filters [3], [4], [5] to reduce the available color information to the light of certain wavelengths, which is then sampled by the image sensor, such as a *charge-coupled device*

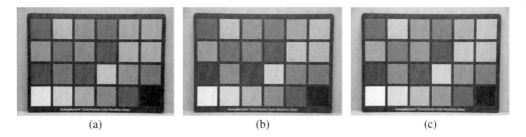

(a)                                    (b)                                    (c)

**FIGURE 6.1** (See color insert.)

The Macbeth color checker taken under the (a) daylight, (b) tungsten, and (c) fluorescent illumination.

(CCD) [6], [7] or a *complementary metal oxide semiconductor* (CMOS) sensor [8], [9]. Unfortunately, achieving a precise digital representation of the visual scene is still not quite possible and extensive processing [10], [11], [12] of acquired data is needed to reduce an impact of shortcomings of the imaging system.

For example, as shown in Figure 6.1, the coloration of captured images often appears different from the visual scene, depending on the illumination under which the image is taken. Although various light sources usually differ in their spectral characteristics [13], [14], [15], the human visual system is capable of discounting the dominant color of the environment, usually attributed to the color of a light source. This ability ensures the approximately constant appearance of an object for different illuminants [16], [17], [18], [19] and is known under the name of *chromatic adaptation*. However, digital cameras have to rely on the image processing methods, termed as *white balancing* [20], [21], [22], [23], to produce images with appearances close to what is observed by the humans. Some cameras use a sensor to measure the illumination of the scene; an alternative is to estimate the scene illuminant by analyzing the captured image data. In either case, fixed parameters are typically used for the whole image to perform the adjustment for the scene illuminant.

This chapter presents a framework [24], [25] that enhances the white balancing process by combining the global and local spectral characteristics of the captured image. This framework takes advantage of the refined color modeling and manipulation concepts in order to adjust the acquired color information adaptively in each pixel location. The presented design methodology is reasonably simple, highly flexible, and provides numerous attractive solutions that can produce visually pleasing images with enhanced coloration and improved color contrast.

To facilitate the subsequent discussions, Section 6.2 presents the fundamentals of color vision and digital color imaging, including the adaptation mechanisms of the human visual system, basic numerical representations of color signals, and various practical color modeling concepts. Section 6.3 focuses on the white balancing basics and briefly surveys several popular methods. Section 6.4 introduces the framework for joint white balancing and color enhancement. Motivation and design characteristics are discussed in detail, and some example solutions designed within this framework are showcased in different application scenarios and analyzed in terms of their computational complexity. Section 6.5 describes methods that are commonly used for color image quality evaluation and quantitative manipulation. Finally, conclusions are drawn in Section 6.6.

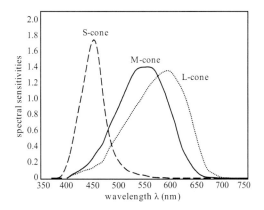

**FIGURE 6.2**

Estimated effective sensitivities of the cones. Peak sensitivities correspond to the light with wavelengths of approximately 420 nm for S-cones, 534 nm for M-cones, and 564 nm for L-cones.

## 6.2 Color Vision Basics

*Color* is a psycho-physiological sensation [26] used by the observers to sense the environment and understand its visual semantics. It is interpreted as a perceptual result of the light interacting with the spectral sensitivities of the photoreceptors.

Visible light is referred to as electromagnetic radiation with wavelengths $\lambda$ ranging from about 390 to 750 nanometers (nm). This small portion of the electromagnetic spectrum is referred to as the *visible spectrum*. Different colors correspond to electromagnetic waves of different wavelengths; the sensation of violet appears at 380 to 450 nm, blue at 450 to 475 nm, cyan at 476 to 495 nm, green at 526 to 606 nm, yellow at 570 to 590 nm, orange at 590 to 620 nm, and red at 620 to 750 nm. The region below 390 nm is called ultra-violet, the region above 750 nm is called infra-red.

The *human visual system* is based on two types of photoreceptors localized on the retina of the eye; the *rods* sensitive to light, and the *cones* sensitive to color [16]. The rods are crucial in *scotopic vision*, which refers to vision in low light conditions where no color is usually seen and only shades of gray can be perceived. In *photopic vision*, which refers to vision in typical light conditions, the rods become saturated and thus the perception mechanism completely relies on less-sensitive cones. Both rods and cones contribute to vision for certain illumination levels, creating thus gradual changes from scotopic to photopic vision.

It is estimated that humans are capable of resolving about 10 million color sensations [20]. The perception of color depends on the response of three types of cones, commonly called S-, M-, and L-cones for their respective sensitivity to short, middle, and long wavelengths. The different sensitivity of each type of cones is attributed to the spectral absorption characteristics of their photosensitive pigments. Two spectrally different lights that have the same L-, M-, and S-cone responses give the same color sensation; this phenomenon is termed as *metamers* [16], [17].

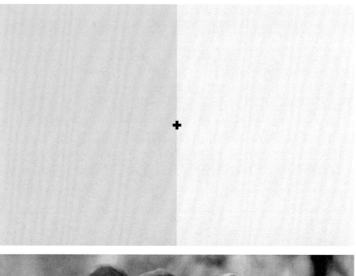

**FIGURE 6.3** (See color insert.)

Chromatic adaptation. Stare for twenty seconds at the cross in the center of the top picture. Then focus on the cross in the bottom picture; the color casts should disappear and the image should look normal.

### 6.2.1 Adaptation Mechanisms of the Human Visual System

The human visual system is capable of changing its sensitivity in order to adapt to prevailing lighting conditions (Figure 6.3), thus maximizing its ability to extract information from the actual environment. It operates over a very large range of illumination, although the range of light that can be processed and interpreted simultaneously is rather limited [17]. Some quick changes are achieved through the dilation and constriction of the pupil in the eye, thus directly controlling the amount of light that can enter the eye [18]. The actual sensitivity changes with the response of the receptive cells on the retina of the eye. This is a slower process and may take a few minutes until the visual system is fully adjusted to the actual lighting conditions. *Temporal adaptation* and *steady-state adaptation* refer, respectively, to the performance of the human visual system during the adaptation process and in the situations when this process has been completed [17].

*Light* and *dark adaptation* [17], [18] are the two special cases which refer to the process of adjusting for the transition from a bright environment to a dark environment and from a fully adapted dark environment to a bright environment, respectively. *Transient adaptation* [19] refers to the situations typical for high-contrast visual environments, where the eye has to adapt from low to high light levels and back in short intervals. This readaptation from one luminous background to another reduces the equivalent (perceived) contrast; this loss can be expressed using the so-called *transient adaptation factor*.

*Chromatic adaptation* [17], [18] characterizes the ability of the human visual system to discount the dominant color of the environment, usually attributed to the color of a light source, and thus approximately preserve the appearance of an object under various illuminants. Different types of the receptive cells in the eye are sensitive to different bands in the visible spectrum. If the actual lighting situation has a different color temperature compared to the previous lighting conditions, the cells responsible for sensing the light in a band with increased (or reduced) amount of light relative to the total amount of light from different bands, will reduce (or increase) their sensitivity relative to the sensitivity of the other cells. This effectively performs *automatic white balancing* of the *human visual system*. Assuming that chromatic adaptation is an independent gain regulation of the three sensors in the human visual system, these relations of the spectral sensitivities can be expressed as [17]:

$$l_2(\lambda) = k_l l_1(\lambda), \quad m_2(\lambda) = k_m m_1(\lambda), \quad s_2(\lambda) = k_s s_1(\lambda), \tag{6.1}$$

where $l_1(\lambda)$, $m_1(\lambda)$, and $s_1(\lambda)$ denote the the spectral sensitivities of the three receptors at one state of adaptation and $l_2(\lambda)$, $m_2(\lambda)$, and $s_2(\lambda)$ for a different state. The coefficients $k_l$, $k_m$, and $k_s$ are inversely related to the relative strengths of activation.

Chromatic adaptation models, such as the one presented above, are essential in describing the appearance of a stimulus and predicting whether two stimuli will match if viewed under disparate lighting conditions where only the color of the lighting has changed. Various computational models of chromatic adaptation are surveyed in References [17] and [18].

### 6.2.2 Numeral Representation of Color

Color cannot be specified without an observer and therefore it is not an inherent feature of an object. The *color stimulus* $S(\lambda)$ depends on both the *illumination* $I(\lambda)$ and the *object reflectance* $R(\lambda)$; this relationship can be expressed as follows [20], [27]:

$$S(\lambda) = I(\lambda)R(\lambda),$$
$$= \left( \sum_{j-1}^{m} \alpha_j I_j(\lambda) \right) \left( \sum_{k-1}^{n} \beta_k R_k(\lambda) \right),$$
$$= \sum_{j=1}^{m} \sum_{k=1}^{n} \alpha_j \beta_k I_j(\lambda) R_k(\lambda), \tag{6.2}$$

where $I_j(\lambda)$ and $R_k(\lambda)$ terms denote known basis functions. Given any object, it is possible to theoretically manipulate the illumination so that it produces any desired color stimulus.

Figure 6.4 shows the spectral power distribution of sunlight, tungsten, and fluorescent illuminations. The effect of these illumination types on the appearance of color is demonstrated in Figure 6.1.

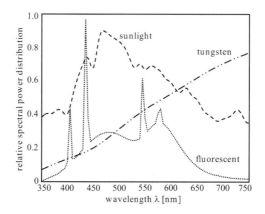

**FIGURE 6.4**

Spectral power distribution of common illuminations.

When an object with stimulus $S(\lambda)$ is observed, each of the three cone types responds to the stimulus by summing up the reaction at all wavelengths [20]:

$$X = \int_{390}^{750} l(\lambda)I(\lambda)R(\lambda)\,d\lambda,$$

$$Y = \int_{390}^{750} m(\lambda)I(\lambda)R(\lambda)\,d\lambda, \qquad (6.3)$$

$$Z = \int_{390}^{750} s(\lambda)I(\lambda)R(\lambda)\,d\lambda,$$

where $l(\lambda)$, $m(\lambda)$, and $s(\lambda)$ denote the spectral sensitivities of the L-, M-, and S-cones depicted in Figure 6.2.

Since three signals are generated based on the extent to which each type of cones is stimulated, any visible color can be numerically represented using three numbers called tristimulus values as a three-component vector within a three-dimensional coordinate system with color primaries lying on its axes [28]. The set of all such vectors constitutes the *color space*.

Numerous color spaces and the corresponding conversion formulas have been designed to convert the color data from one space to some other space, which is more suitable for completing a given task and/or addresses various design, performance, implementation, and application aspects of color image processing. The following discusses two representations relevant to the scope of this chapters.

### 6.2.2.1   Standardized Representations

The Commission Internationale de l'Éclairage (CIE) introduced the standardized CIE-RGB and CIE-XYZ color spaces. The well-known Red-Green-Blue (RGB) space was derived based on color matching experiments, aimed at finding a match between a color obtained through an additive mixture of *color primaries* and a color sensation [29]. Its standardized version [30] is used in most of today's image acquisition and display devices. It provides a reasonable resolution, range, depth, and stability of color reproduction while being efficiently implementable on various hardware platforms.

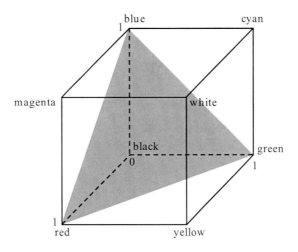

**FIGURE 6.5**

RGB color space.

The RGB color space can be derived from the XYZ space, which is another standardized space. Since the XYZ space is *device independent*, it is thus very useful in situations where consistent color representation across devices with different characteristics is required. Although the CIE-XYZ color space is rarely used in image processing applications, other color spaces can be derived from it through mathematical transforms.

A relationship between these two spaces can be expressed as follows [28], [31]:

$$
\begin{bmatrix} X \\ Y \\ Z \end{bmatrix} = \begin{bmatrix} 0.49 & 0.31 & 0.20 \\ 0.17697 & 0.81240 & 0.01063 \\ 0 & 0.01 & 0.99 \end{bmatrix} \begin{bmatrix} R \\ G \\ B \end{bmatrix}, \tag{6.4}
$$

where the $Y$ component corresponds to the luminance, whereas $X$ and $Z$ do not correspond to any perceptual attributes.

The RGB space is *additive*; as shown in Figure 6.5, any color can be obtained by combining the three primaries through their weighted contributions [29]. Equal contributions of all three primaries give a shadow of gray. The two extremes of gray are black and white, which correspond, respectively, to no contribution and the maximum contributions of the color primaries. When one primary contributes greatly, the resulting color is a shade of that dominant primary. Any pure *secondary color* is formed by maximum contributions of two primary colors: cyan is obtained using green and blue, magenta using red and blue, and yellow using red and green. When two primaries contribute greatly, the result is a shade of the corresponding secondary color.

The RGB space models the output of physical devices, and therefore it is considered *device dependent*. To consistently detect or reproduce the same RGB color vector, some form of color management is usually required. This relates to the specification of the white point, gamma correction curve, dynamic range, and viewing conditions [32]. Given an XYZ color vector with components ranging from zero to one and the reference white being the same as that of the RGB system, the conversion to sRGB values starts as follows:

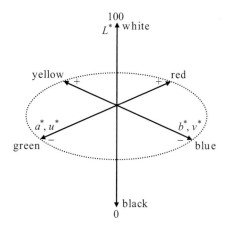

**FIGURE 6.6**

Lab / Luv color space.

$$\begin{bmatrix} R \\ G \\ B \end{bmatrix} = \begin{bmatrix} 3.2410 & -1.5374 & -0.4986 \\ -0.9692 & 1.8760 & 0.0416 \\ 0.0556 & -0.2040 & 1.0570 \end{bmatrix} \begin{bmatrix} X \\ Y \\ Z \end{bmatrix}. \tag{6.5}$$

To avoid values outside the nominal range, which are usually not supported in RGB encoding, both negative values and values exceeding one are clipped to zero and one, respectively. This step is followed by gamma correction:

$$f(\tau) = \begin{cases} 1.055\tau^{1/2.4} - 0.055 & \text{if } \tau > 0.00304, \\ 12.92\tau & \text{otherwise,} \end{cases} \tag{6.6}$$

where $\tau$ denotes the uncorrected color component. Finally, gamma-corrected components $f(R)$, $f(G)$, and $f(B)$ are multiplied by 255 to obtain their corresponding values in standard eight bits per channel encoding.

#### 6.2.2.2 Perceptual Representations

Perceptually uniform CIE $u,v$, CIE-Luv, and CIE-Lab representations are derived from XYZ values. In the case of CIE $u$ and $v$ values, the conversion formulas

$$u = \frac{4X}{X + 15Y + 3Z}, \quad v = \frac{9Y}{X + 15Y + 3Z} \tag{6.7}$$

can be used to form a *chromaticity diagram* that corresponds reasonably well to the characteristics of human visual perception [33].

Converting the data from CIE-XYZ to CIE-Luv and CIE-Lab color spaces [31] requires a reference point in order to account for adaptive characteristics of the human visual system [29]. Denoting this point as $[X_n, Y_n, Z_n]$ of the reference white under the reference illumination, the CIE-Lab values are calculated as

$$\begin{aligned} L^* &= 116f(Y/Y_n) - 16, \\ a^* &= 500\left(f(X/X_n) - f(Y/Y_n)\right), \\ b^* &= 200\left(f(Y/Y_n) - f(Z/Z_n)\right), \end{aligned} \tag{6.8}$$

whereas the CIE-Luv values are obtained as follows:

$$L^* = 116f(Y/Y_n) - 16,$$
$$u^* = 13L^*(u - u_n), \qquad (6.9)$$
$$v^* = 13L^*(v - v_n),$$

where $u_n = 4X_n/(X_n + 15Y_n + 3Z_n)$ and $v_n = 9Y_n/(X_n + 15Y_n + 3Z_n)$ correspond to the reference point $[X_n, Y_n, Z_n]$. The terms $u$ and $v$, calculated using Equation 6.7, correspond to the color vector $[X, Y, Z]$ under consideration. Since the human visual system exhibits different characteristics in normal illumination and low light levels, $f(\cdot)$ is defined as follows [31]:

$$f(\gamma) = \begin{cases} \gamma^{1/3} & \text{if } \gamma > 0.008856, \\ 7.787\gamma + 16/116 & \text{otherwise.} \end{cases} \qquad (6.10)$$

Figure 6.6 depicts the perceptual color space representation. The component $L^*$, ranging from zero (black) to 100 (white), represents the lightness of a color vector. All other components describe color; neutral or near neutral colors corresponds to zero or close to zero values of $u^*$ and $v^*$, or $a^*$ and $b^*$. Following the characteristics of *opponent* color spaces [34], $u^*$ and $a^*$ coordinates represent the difference between red and green, whereas $v^*$ and $b^*$ coordinates represent the difference between yellow and blue.

Unlike the CIE-XYZ color space, which is perceptually highly nonuniform, the CIE-Luv and CIE-Lab color spaces are suitable for quantitative manipulations involving color perception since equal Euclidean distances or mean square errors expressed in these two spaces equate to equal perceptual differences [35]. The CIE-Lab model also finds its application in color management [36], whereas the CIE-Luv model can assist in the registration of color differences experienced with lighting and displays [33].

### 6.2.3  Color Image Representation

An RGB color image $\mathbf{x}$ with $K_1$ rows and $K_2$ columns represents a two-dimensional matrix of three-component samples $\mathbf{x}_{(r,s)} = [x_{(r,s)1}, x_{(r,s)2}, x_{(r,s)3}]$ occupying the spatial location $(r,s)$, with $r = 1, 2, ..., K_1$ and $s = 1, 2, ..., K_2$ denoting the row and column coordinates. In the color vector $\mathbf{x}_{(r,s)}$, the terms $x_{(r,s)1}$, $x_{(r,s)2}$, and $x_{(r,s)3}$ denote the R, G, and B component, respectively. In standard eight-bit RGB representation, these components can range from 0 to 255. The large value of $x_{(r,s)k}$, for $k = 1, 2, 3$, indicates high contribution of the $k$th primary in the color vector $\mathbf{x}_{(r,s)}$. The process of displaying an image creates a graphical representation of the data matrix where the pixel values represent particular colors.

Each color vector $\mathbf{x}_{(r,s)}$ is uniquely defined by its magnitude

$$\sqrt{x_{(r,s)1}^2 + x_{(r,s)2}^2 + x_{(r,s)3}^2} \qquad (6.11)$$

and direction

$$\mathbf{x}_{(r,s)} / \sqrt{x_{(r,s)1}^2 + x_{(r,s)2}^2 + x_{(r,s)3}^2}, \qquad (6.12)$$

which indirectly indicate *luminance* and *chrominance* properties of RGB colors, and thus are important for human perception [37]. Note that the magnitude represents a scalar value, whereas the direction, as defined above, is a vector. Since the components of this vector are normalized, such vectors form the unit sphere in the vector space.

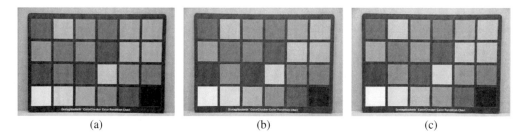

(a)              (b)              (c)

**FIGURE 6.7**

Luminance representation of the images shown in Figure 6.1 for the (a) daylight, (b) tungsten, and (c) fluorescent illumination.

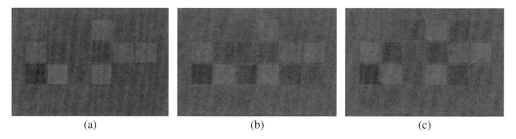

(a)              (b)              (c)

**FIGURE 6.8 (See color insert.)**

Chrominance representation of the images shown in Figure 6.1 for the (a) daylight, (b) tungsten, and (c) fluorescent illumination.

In practice, the luminance value $L_{(r,s)}$ of the color vector $\mathbf{x}_{(r,s)}$ is usually obtained as follows:

$$L_{(r,s)} = 0.299x_{(r,s)1} + 0.587x_{(r,s)2} + 0.114x_{(r,s)3}. \tag{6.13}$$

The weights assigned to individual color channels reflect the perceptual contributions of each color band to the luminance response of the human visual system [16].

The chrominance properties of color pixels are often expressed as the point on the triangle, which intersects the RGB color primaries in their maximum value. The coordinates of this point are obtained as follows [38]:

$$\frac{x_{(r,s)k}}{x_{(r,s)1} + x_{(r,s)2} + x_{(r,s)3}}, \quad \text{for } k = 1, 2, 3, \tag{6.14}$$

and their sum is equal to unity. Both above vector formulations of the chrominance represent the parametrization of the chromaticity space, where each chrominance line is entirely determined by its intersection point with the chromaticity sphere or chromaticity triangle in the three-dimensional vector space.

Figures 6.7 and 6.8, generated using Equations 6.13 and 6.14, respectively, show the luminance and chrominance representations of the images presented in Figure 6.1. As it can be seen, changing the illumination of the scene has a significant impact on the camera response, altering both the luminance and chrominance characteristics of the captured images.

### 6.2.4 Color Modeling

Any of the above definitions can be used to determine whether the two color vectors will appear similar to the observer under some viewing conditions. By looking at this problem from the other side, various color vectors matching the characteristics of the reference color vector can be derived based on some predetermined criterion. For example, simple but yet powerful color modeling concepts follow the observation that natural images consist of small regions that exhibit similar color characteristics. Since this is definitely the case of color chromaticity, which relates to the direction of the color vectors in the vector space, it is reasonable to assume that two color vectors $\mathbf{x}_{(r,s)}$ and $\mathbf{x}_{(i,j)}$ have the same chromaticity characteristics if they are collinear in the RGB color space. Based on the definition of dot product $\mathbf{x}_{(r,s)}.\mathbf{x}_{(i,j)} = \|\mathbf{x}_{(r,s)}\|\|\mathbf{x}_{(i,j)}\|\cos\left(\langle\mathbf{x}_{(r,s)},\mathbf{x}_{(i,j)}\rangle\right)$, where $\|\mathbf{x}_{(\cdot,\cdot)}\|$ denotes the length of $\mathbf{x}_{(\cdot,\cdot)}$ and $\langle\mathbf{x}_{(r,s)},\mathbf{x}_{(i,j)}\rangle$ denotes the angle between three-component color vectors $\mathbf{x}_{(r,s)}$ and $\mathbf{x}_{(i,j)}$, the following can be derived [39]:

$$\langle\mathbf{x}_{(r,s)},\mathbf{x}_{(i,j)}\rangle = 0 \Leftrightarrow \frac{\sum_{k=1}^{3} x_{(r,s)k} x_{(i,j)k}}{\sqrt{\sum_{k=1}^{3} x_{(r,s)k}^2}\sqrt{\sum_{k=1}^{3} x_{(i,j)k}^2}} = 1. \qquad (6.15)$$

The above concept can be extended by incorporating the magnitude information into the modeling assumption. Using color vectors $\mathbf{x}_{(i,j)}$ and $\mathbf{x}_{(r,s)}$ as inputs, the underlying modeling principle of identical color chromaticity enforces that their linearly shifted variants $[\mathbf{x}_{(r,s)} + \gamma\mathbf{I}]$ and $[\mathbf{x}_{(i,j)} + \gamma\mathbf{I}]$ are collinear vectors:

$$\langle\mathbf{x}_{(r,s)} + \gamma\mathbf{I}, \mathbf{x}_{(i,j)} + \gamma\mathbf{I}\rangle = 0 \Leftrightarrow \frac{\sum_{k=1}^{3}(x_{(r,s)k} + \gamma)(x_{(i,j)k} + \gamma)}{\sqrt{\sum_{k=1}^{3}(x_{(r,s)k} + \gamma)^2}\sqrt{\sum_{k=1}^{3}(x_{(i,j)k} + \gamma)^2}} = 1, \qquad (6.16)$$

where $\mathbf{I}$ is a unity vector of proper dimensions and $x_{(\cdot,\cdot)k} + \gamma$ is the $k$th component of the linearly shifted vector $[\mathbf{x}_{(\cdot,\cdot)} + \gamma\mathbf{I}] = [x_{(\cdot,\cdot)1} + \gamma, x_{(\cdot,\cdot)2} + \gamma, x_{(\cdot,\cdot)3} + \gamma]$. A number of solutions can be obtained by modifying the value of $\gamma$, which is a design parameter that controls the influence of both the directional and the magnitude characteristics.

By reducing the dimensionality of the vectors to two components, some popular color correlation-based models can be obtained, such as the *normalized color-ratio* model [40]:

$$\frac{x_{(r,s)k} + \gamma}{x_{(i,j)k} + \gamma} = \frac{x_{(r,s)2} + \gamma}{x_{(i,j)2} + \gamma}, \qquad (6.17)$$

or the *color-ratio* model [41] given by

$$\frac{x_{(r,s)k}}{x_{(i,j)k}} = \frac{x_{(r,s)2}}{x_{(i,j)2}}, \qquad (6.18)$$

which both enforce hue uniformity. Another modeling option is to imply uniform image intensity by constraining the component-wise magnitude differences, as implemented by the *color-difference* model [42]:

$$x_{(r,s)k} - x_{(i,j)k} = x_{(r,s)2} - x_{(i,j)2}. \qquad (6.19)$$

The above modeling concepts have proved to be very valuable in various image processing tasks, such as demosaicking, denoising, and interpolation. It will be shown later that similar concepts can also be used to enhance the white balancing process.

**FIGURE 6.9** (See color insert.)

Examples of the images taken in different lighting conditions without applying white balancing.

## 6.3 White Balancing

As previously noted, the humans observe the approximately constant appearance of an object in different lighting conditions. This phenomenon of chromatic adaptation, also termed as *color constancy*, suggests that the human visual system transforms recorded stimuli into representations of the scene reflectance that are largely independent of the scene illuminant [16].

*Automatic white balancing* of the visual data captured by a digital camera attempts to mimic the chromatic adaption functionality of the human visual system and usually produces an image that appears more neutral compared to its uncorrected version. This correction can be performed using chromatic-adaption transforms to ensure than the appearance of the resulting image is close to what the human observer remembers based upon his or her own state of adaptation. Figure 6.9 shows several images captured in different lighting conditions prior to performing white balancing.

Digital cameras use image sensors and various color filters to capture the scene in color. Similar to the process described in Equation 6.3, which characterizes the response of the human eye, the photosensitive elements in the camera respond to the stimulus $S(\lambda) = I(\lambda)R(\lambda)$ by summing up the reaction at all wavelengths, which gives

$$R_{\text{sensor}} = \int_{390}^{750} r(\lambda)I(\lambda)R(\lambda)\,\mathrm{d}\lambda,$$

$$G_{\text{sensor}} = \int_{390}^{750} g(\lambda)I(\lambda)R(\lambda)\,\mathrm{d}\lambda, \tag{6.20}$$

$$B_{\text{sensor}} = \int_{390}^{750} b(\lambda)I(\lambda)R(\lambda)\,\mathrm{d}\lambda,$$

where $r(\lambda)$, $g(\lambda)$, and $b(\lambda)$ denote the spectral sensitivities of the sensor cells with the red, green, and blue filters, respectively. The discrete version of the above equation is expressed as follows:

$$R_{\text{sensor}} = \sum_{j=1}^{m} r(\lambda_j) I(\lambda_j) R(\lambda_j) \Delta\lambda,$$

$$G_{\text{sensor}} = \sum_{j=1}^{m} g(\lambda_j) I(\lambda_j) R(\lambda_j) \Delta\lambda, \qquad (6.21)$$

$$B_{\text{sensor}} = \sum_{j=1}^{m} b(\lambda_j) I(\lambda_j) R(\lambda_j) \Delta\lambda.$$

Automatic white balancing can be seen as the process that aims at minimizing the effect of $I(\lambda)$ to ensure that $R_{\text{sensor}}$, $G_{\text{sensor}}$, and $B_{\text{sensor}}$ correlate with the object reflectance $R(\lambda)$ only [20], [43]. To solve this complex problem, various assumptions and simplifications are usually made, as discussed below.

### 6.3.1 Popular Solutions

In automatic white balancing, the adjustment parameters can be calculated according to some reference color present in the captured image, often using the pixels from white-like areas, which are characterized by similar high contributions of the red, green, and blue primaries in a given bit representation. This so-called *white-world* approach works based on the assumption that the bright regions in an image reflect the actual color of the light source [44]. If the scene does not contain natural white but it exhibits a sufficient amount of color variations, better results can be obtained using all pixels in the image [15], [23]. In this case, the underlying assumption is that the average reflectance of the scene is achromatic; that is, the mean value of the red, green, and blue channel is approximately equal. Unfortunately, this so-called *gray-world* approach can fail in images with a large background or large objects of uniform color. To obtain a more robust solution, the two above assumptions can be combined using the quadratic mapping of intensities [45].

A more sophisticated approach can aim at preselecting the input pixels to avoid being susceptible to statistical anomalies and use iterative processing to perform white balancing [46]. Another option is to use the illuminant voting scheme [47], which is a procedure repeated for various reflectance parameters in order to determine the corresponding illumination parameters that obtain the largest number of votes by solving a system of linear equations. In the so-called color-by-correlation method [43], the prior information about color distributions for various illuminants, as observed in the calibration step, is correlated with the color distribution from the actual image to determine the likelihood of each possible illuminant. Another solution combines various existing white balancing approaches to estimate the scene illuminant more precisely [48].

### 6.3.2 Mathematical Formulation

Once the prevailing illuminant is estimated, the white-balanced image **y** with the pixels $\mathbf{y}_{(r,s)} = [y_{(r,s)1}, y_{(r,s)2}, y_{(r,s)3}]$ is usually obtained as follows:

$$y_{(r,s)k} = \alpha_k x_{(r,s)k}, \text{ for } k = 1, 2, 3, \qquad (6.22)$$

where $\alpha_k$ denotes the gain associated with the color component $x_{(r,s)k}$. The process alters only the red and blue channels of the input image $\mathbf{x}$ with the pixels $\mathbf{x}_{(r,s)} = [x_{(r,s)1}, x_{(r,s)2}, x_{(r,s)3}]$ obtained via Equation 6.20, and keeps the green channel unchanged, that is, $\alpha_2 = 1$.

In the case of the popular gray-world algorithm and its variants, the channel gains are calculated as follows:

$$\alpha_1 = \frac{\bar{x}_2}{\bar{x}_1}, \quad \alpha_3 = \frac{\bar{x}_2}{\bar{x}_3}, \tag{6.23}$$

where the values

$$\bar{x}_k = \frac{1}{|\zeta|} \sum_{(i,j) \in \zeta} x_{(i,j)k}, \text{ for } k = 1, 2, 3, \tag{6.24}$$

constitute the global reference color vector $\bar{\mathbf{x}} = [\bar{x}_1, \bar{x}_2, \bar{x}_3]$ obtained by averaging, in a component-wise manner, the pixels selected according to some predetermined criterion. This criterion can be defined, for instance, to select white pixels, gray pixels, or all pixels in the input image $\mathbf{x}$ in order to obtain the set $\zeta$, with $|\zeta|$ denoting the number of pixels inside $\zeta$.

---

## 6.4 Joint White Balancing and Color Enhancement

This section presents a framework that can simultaneously perform *white balancing* and *color enhancement* using refined pixel-adaptive operations [24], [25]. Such adaptive processing, which combines both the global and local spectral characteristics of the input image, can be obtained, for instance, using the *color modeling* concepts presented in Section 6.2.4 or a simple *combinatorial approach*.

### 6.4.1 Spectral Modeling-Based Approach

More specifically, instead of applying the fixed gains $\alpha_k$ in all pixel locations to produce the white-balanced image $\mathbf{y}$ with pixels $\mathbf{y}_{(r,s)}$, the adaptive color adjustment process is enforced through the pixel-adaptive parameters $\alpha_{(r,s)k}$ associated with the pixel $\mathbf{x}_{(r,s)}$. Following the multiplicative nature of the adjustment process in Equation 6.22, the color-enhanced white-balanced pixels $\mathbf{y}_{(r,s)}$ can be obtained as follows:

$$y_{(r,s)k} = \alpha_{(r,s)k} x_{(r,s)k}, \text{ for } k = 1, 2, 3, \tag{6.25}$$

where $\alpha_{(r,s)k}$ are pixel-adaptive gains updated in each pixel position using the new reference vector $\hat{\mathbf{x}}_{(r,s)} = [\hat{x}_{(r,s)1}, \hat{x}_{(r,s)2}, \hat{x}_{(r,s)3}]$, which depends on the actual color pixel $\mathbf{x}_{(r,s)}$ and the global reference vector $\bar{\mathbf{x}}$.

Similar to Equation 6.23, the actual $\alpha_{(r,s)k}$ values in Equation 6.25 are formulated as the ratios of proper color components from the reference vector used to guide the adjustment process. Namely, as discussed in Reference [49] based on the rationale behind Equation 6.18, the pixel-adaptive gains are obtained as $\alpha_{(r,s)k} = \hat{x}_{(r,s)2}/\hat{x}_{(r,s)k}$ for

$\hat{x}_{(r,s)k} = x_{(r,s)2}\bar{x}_k/\bar{x}_2$, $\hat{x}_{(r,s)2} = x_{(r,s)k}\bar{x}_2/\bar{x}_k$, and $k \in \{1,3\}$, resulting in

$$\alpha_{(r,s)1} = \frac{x_{(r,s)1}\bar{x}_2^2}{x_{(r,s)2}\bar{x}_1^2}, \quad \alpha_{(r,s)3} = \frac{x_{(r,s)3}\bar{x}_2^2}{x_{(r,s)2}\bar{x}_3^2}, \tag{6.26}$$

where care needs to be taken to avoid dividing by zero.

In situations where $x_{(r,s)k}$ and $x_{(r,s)2}$ represent two opposite extremes in a given bit representation, the corresponding gain coefficient $\alpha_{(r,s)k}$, for $k \in \{1,3\}$, will have too large or too small value and Equation 6.25 can produce saturated colors. To address this problem, a positive constant $\gamma$ can be added to both the nominator and the denominator of ratio formulations in Equation 6.26 as follows:

$$\alpha_{(r,s)1} = \frac{x_{(r,s)1}\bar{x}_2^2 + \gamma}{x_{(r,s)2}\bar{x}_1^2 + \gamma}, \quad \alpha_{(r,s)3} = \frac{x_{(r,s)3}\bar{x}_2^2 + \gamma}{x_{(r,s)2}\bar{x}_3^2 + \gamma}, \tag{6.27}$$

which is equivalent to $\alpha_{(r,s)k} = (\hat{x}_{(r,s)2} + \gamma)/(\hat{x}_{(r,s)k} + \gamma)$, for $k \in \{1,3\}$. The parameter $\gamma$ controls the adjustment process; using $\gamma = 20$ usually gives satisfactory result in most situations, further increasing $\gamma$ above this value reduces the adjustment effect down to the original white-unbalanced coloration for some extreme settings since $\alpha_{(r,s)1} = \alpha_{(r,s)3} = 1$ for $\gamma \to \infty$. Since shifting linearly the components of the reference vector may significantly change the original gain values, an inverse shift should be added to the adjusted color components to compensate for this effect, which gives the following adjustment formula [25]:

$$y_{(r,s)k} = -\gamma + \alpha_{(r,s)k}(x_{(r,s)k} + \gamma), \text{ for } k = 1,2,3. \tag{6.28}$$

The above formula takes the form of the normalized color ratios in Equation 6.17. The components of the reference vector $\hat{x}_{(r,s)}$, obtained previously using standard color ratios, can also be redefined using the normalized color ratios as $\hat{x}_{(r,s)k} = -\gamma + (x_{(r,s)2} + \gamma)(\bar{x}_k + \gamma)/(\bar{x}_2 + \gamma)$ and $\hat{x}_{(r,s)2} = -\gamma + (x_{(r,s)k} + \gamma)(\bar{x}_2 + \gamma)/(\bar{x}_k + \gamma)$ for $\alpha_{(r,s)k}$ with $k \in \{1,3\}$. This modification gives

$$\alpha_{(r,s)1} = \frac{(x_{(r,s)1} + \gamma)(\bar{x}_2 + \gamma)^2}{(x_{(r,s)2} + \gamma)(\bar{x}_1 + \gamma)^2}, \quad \alpha_{(r,s)3} = \frac{(x_{(r,s)3} + \gamma)(\bar{x}_2 + \gamma)^2}{(x_{(r,s)2} + \gamma)(\bar{x}_3 + \gamma)^2}, \tag{6.29}$$

where too small values of $\gamma$ can result in images with excessive color saturation.

To increase the numerical stability of pixel-adaptive gains, the local color ratios $x_{(r,s)k}/x_{(r,s)2}$ in Equations 6.26 and 6.27 and $(x_{(r,s)k} + \gamma)/(x_{(r,s)2} + \gamma)$ in Equation 6.29 can be constrained as

$$\left(\frac{x_{(r,s)k}}{x_{(r,s)2}}\right)^\rho \quad \text{and} \quad \left(\frac{x_{(r,s)k} + \gamma}{x_{(r,s)2} + \gamma}\right)^\rho, \quad \text{for } k \in \{1,3\}, \tag{6.30}$$

where $\rho$ is a tunable parameter, with a typical operating range set as $-1 \leq \rho \leq 1$. Using $\rho = 0$ eliminates the contributions of the actual pixel from the gain calculations. Increasing the value of $\rho$ above zero enhances the color appearance of the image, whereas reducing the value of $\rho$ below zero suppresses the image coloration. Figures 6.10 and 6.11 demonstrate this behavior using Equations 6.28 and 6.29.

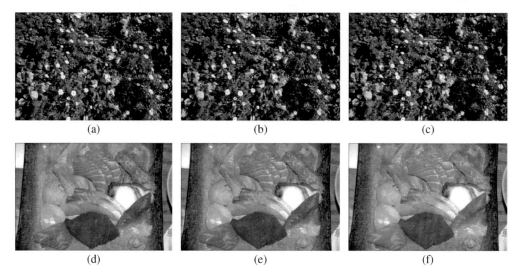

**FIGURE 6.10 (See color insert.)**

Joint white balancing and color enhancement using the spectral modeling approach: (a,d) $\rho = 0$, (b,e) $\rho = 0.25$, and (c,f) $\rho = 1.0$.

**FIGURE 6.11 (See color insert.)**

Joint white balancing and color enhancement using the spectral modeling approach: (a,d) $\rho = -0.5$, (b,e) $\rho = 0$, and (c,f) $\rho = 0.5$.

As suggested in the above discussion, various color modeling concepts can be employed within the presented framework. For example, in addition to the color ratio-based calculations, the reference vector $\hat{\mathbf{x}}_{(r,s)}$ can be obtained as $\hat{x}_{(r,s)k} = x_{(r,s)2} + \bar{x}_k - \bar{x}_2$ and $\hat{x}_{(r,s)2} = x_{(r,s)k} + \bar{x}_2 - \bar{x}_k$ for $\alpha_{(r,s)k}$ with $k = \in \{1,3\}$ using the color difference concept presented in Equation 6.19. The enumeration of all available options or the determination of the optimal design configuration according to some criteria is beyond the scope of this chapter.

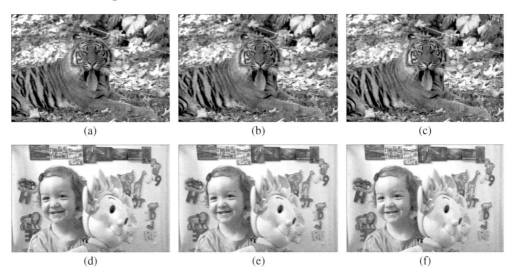

**FIGURE 6.12 (See color insert.)**

Joint white balancing and color enhancement using the combinatorial approach: (a,d) $w = 0.5$, (b,e) $w = 0$, and (c,f) $w = -0.5$.

### 6.4.2 Combinatorial Approach

Spectral modeling is not the only available option to combine the global and local reference color vectors in order to obtain pixel-adaptive gains. This combination function can be as simple as a *weighting scheme* with a tunable parameter assigned to control the contribution of the local gains and its complementary value assigned to control the contribution of the global gains [24], [50]:

$$\alpha_{(r,s)k} = w \frac{x_{(r,s)2}}{x_{(r,s)k}} + (1-w)\frac{\bar{x}_2}{\bar{x}_k}, \quad \text{for } k \in \{1,3\}, \tag{6.31}$$

where $w$ is a tunable parameter. Substituting $\alpha_{(r,s)k}$ in Equation 6.25 with the above gain expression leads to the following:

$$y_{(r,s)k} = wx_{(r,s)2} + (1-w)\frac{\bar{x}_2}{\bar{x}_k}x_{(r,s)k}, \quad \text{for } k = 1,2,3, \tag{6.32}$$

which is another formulation of pixel-adaptive white balancing.

The amount of color adjustment produced in Equation 6.32 consists of the white balancing term $(1-w)x_{(r,s)k}\bar{x}_2/\bar{x}_k$ and another term $wx_{(r,s)2}$, which can be seen as a color enhancement term for certain values of $w$. Namely, for $w = 1$ no white balancing is performed, as the method produces achromatic pixels by assigning the value of $x_{(r,s)2}$ from the input image to all three components of the output color vector $\mathbf{y}_{(r,s)}$. Reducing the value of $w$ toward zero enhances the amount of coloration from achromatic colors to those generated by white balancing. Further decreasing $w$, that is, setting $w$ to negative values results in color enhancement, with saturated colors produced for too low values of $w$. This behavior is demonstrated in Figure 6.12.

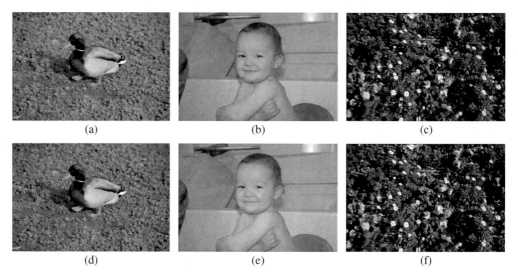

|        |        |        |
|--------|--------|--------|
| (a)    | (b)    | (c)    |
| (d)    | (e)    | (f)    |

**FIGURE 6.13** (See color insert.)

Color enhancement of images with inaccurate white balancing: (a–c) input images and (d–f) their enhanced versions.

### 6.4.3   Experimental Results

Following the common practice, the presented framework keeps the green channel unchanged using $\alpha_{(r,s)2} = 1$. Color adjustment effects are thus produced through the red and blue channel gains $\alpha_{(r,s)1}$ and $\alpha_{(r,s)3}$, which are updated in each pixel location of the captured image; that is, for $r = 1, 2, ..., K_1$ and $s = 1, 2, ..., K_2$.

The uncorrected color images, such as those shown in Figure 6.9, were produced by demosaicking the scaled raw single-sensor data captured using the Nikon D50 camera with a native resolution of $2014 \times 3039$ pixels. In order to do the best possible adjustment for the scene illuminant, the global reference vector $\bar{\mathbf{x}}$ was calculated using the pixels in white regions whenever possible.

Figures 6.10 to 6.12 show the results of joint white balancing and color enhancement for some example solutions. As it can be seen, these solutions produce a wide range of color enhancement. Moreover, as demonstrated in Figures 6.11 and 6.12, the color enhancement effects can range from image appearances with suppressed coloration to appearances with rich and vivid colors. The desired amount of color enhancement, according to some criteria, can be easily obtained through parameter tuning.

The next experiment focuses on the effect of color enhancement in situations with inaccurate white-balancing parameters. Figure 6.13 shows both the images with simulated (and perhaps exaggerated) color casts due to incorrect white balancing and the corresponding images after color enhancement. As expected, the color cast is more visible in the images with enhanced colors. In general, any color imperfection, regardless whether it is global (such as color casts) or local (such as color noise) in its nature, gets amplified by the color enhancement process. This behavior is typical also for other color enhancement operations in the camera processing pipeline, including color correction and image rendering (gamma correction, tone curve-based adjustment).

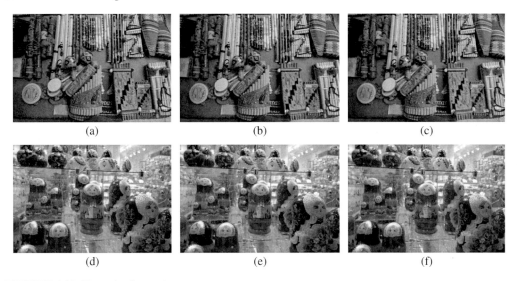

**FIGURE 6.14 (See color insert.)**

Color adjustment of the final camera images: (a,d) suppressed coloration, (b,e) original coloration, and (c,f) enhanced coloration.

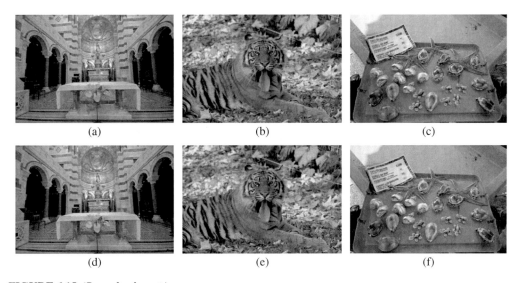

**FIGURE 6.15 (See color insert.)**

Color enhancement of the final camera images: (a–c) output camera images and (d–f) their enhanced versions.

Finally, Figures 6.14 and 6.15 show the results when the presented framework operates on the final processed camera images. To adapt to such a scenario, one can reasonably assume that the images are already white balanced; this implies that the components of the global reference vector $\bar{\mathbf{x}}$ should be set to the same value. Also in this case, the framework allows for a wide range of color enhancement, which suggests that the visually pleasing images with enhanced colors and improved color contrast can be always found through parameter tuning.

### 6.4.4 Computational Complexity Analysis

Apart from the actual performance, the computational efficiency is another realistic measure of practicality and usefulness of any algorithm. Several example solutions designed within the presented framework are analyzed below in terms of normalized operations, such as additions (ADDs), subtractions (SUBs), multiplications (MULTs), divisions (DIVS), and comparisons (COMPs).

For the color image with $K_1$ rows and $K_2$ columns, the solution defined in Equations 6.25 and 6.26 requires $2K_1K_2 + 2$ MULTs and $2K_1K_2 + 2$ DIVs to calculate the pixel-adaptive gains, and $2K_1K_2$ MULTs to apply these gains to the image. Implementing Equations 6.28 and 6.29 requires $4K_1K_2 + 4$ ADDs, $2K_1K_2 + 2$ MULTs, and $2K_1K_2 + 2$ DIVs to calculate the pixel-adaptive gains and additional $2K_1K_2$ SUBs and $2K_1K_2$ MULTs to apply these gains to the image. The tunable adjustment in Equation 6.30, implemented as a look-up table, will slightly add to the overall complexity. In the case of Equation 6.32, the adjustment process requires $2K_1K_2$ ADDs, 1 SUB, $4K_1K_2 + 2$ MULTs, and 2 DIVs. In addition to the above numbers, $3K_1K_2 - 3$ ADDs and 3 DIVs are needed when the global reference vector $\bar{x}$ in Equation 6.24 is calculated using all the pixels, or $3|\zeta| - 3$ ADDs, 3 DIVs, and $3K_1K_2$ COMPs when $\bar{x}$ is calculated using the selected pixels. As can be seen, all the example solutions have a very reasonable computational complexity, which should allow their efficient implementation in various real-time imaging applications.

Moreover, as reported in Reference [25], the execution of various solutions designed within the presented framework on a laptop equipped with an Intel Pentium IV 2.40 GHz processor, 512 MB RAM box, Windows XP operating system, and MS Visual C++ 5.0 programming environment can take (on average) between 0.4 and 0.85 seconds to process a $2014 \times 3039$ color image. Additional performance improvements are expected through extensive software optimization. Both the computational complexity analysis and the reported execution time suggest that the presented framework is cost-effective.

Finally, it should be noted that this chapter focuses on the color manipulation operations performed on full-color data. In the devices that acquire images using the image sensor covered by the color filter array, the full-color image is obtained using the process known as *demosaicking* or *color interpolation*. However, as discussed in References [24], [25], and [50], the presented framework is flexible and can be applied before or even combined with the demosaicking step. Thus, additional computational savings can be achieved.

### 6.4.5 Discussion

Numerous solutions can be designed within the presented framework by applying the underlying principle of mixing the global and local spectral characteristics of the captured image. In addition to using the actual pixel to produce the color enhancement effect, the framework allows extracting the local image characteristics using the samples inside a local neighborhood of the pixel being adjusted. These neighborhoods can be determined by dividing the entire image either into non-overlapping or overlapping blocks, and then selecting a block that, for example, contains the pixel location or minimizes the distance between its center and the pixel location being adjusted. However, such block-based processing may significantly increase the computational complexity compared to true pixel-adaptive methods, such as the example solutions presented in this chapter.

The presented framework can operate in conjunction with the white balancing gains recorded by the camera. This can be done through the replacement of ratios $\bar{x}_2/\bar{x}_k$, for $k \in \{1,3\}$, with the actual gains indicated by the camera for each of the respective color channels. If the gain associated with the green channel is not equal to one, then the white balancing gains should be normalized by the gain value associated with the green channel prior to performing the replacement.

In summary, the presented framework, which integrates white balancing and pixel-adaptive color enhancement operations, can be readily used in numerous digital camera and color imaging applications. The example solutions designed within this framework are flexible; they can be used in conjunction with various existing white balancing techniques and allow for both global or localized control of the color enhancement process. The framework is computationally efficient, has an excellent performance, and can produce visually pleasing color images with enhanced coloration and improved color contrast. The framework can be seen as an effective solution that complements traditional color correction, color saturation, and tone curve adjustment operations in the digital imaging pipeline or a stand-alone color enhancement tool suitable for image editing software and online image processing applications.

## 6.5   Color Image Quality Evaluation

The performance of color image processing methods, in terms of image quality, is usually evaluated using the subjective evaluation approach and the objective evaluation approach.

### 6.5.1   Subjective Assessment

Since most of the images are typically intended for human inspection, the human opinion on visual quality is very important [51]. Statistically meaningful results are usually obtained by conducting a large-scale study where human observers view and rate images, presented either in specific or random order and under identical viewing conditions, using a predetermined rating scale [52]. For example, considering the scope of this chapter, the score can range from one to five, reflecting poor, fair, good, very good, and excellent white balancing, respectively. Using the same score scale, the presence of any distortion, such as color artifacts and posterization effects, introduced to the image through the color adjustment process can be evaluated as very disruptive, disruptive, destructive but not disruptive, perceivable but not destructive, and imperceivable.

### 6.5.2   Objective Assessment

In addition to subjective assessment of image quality, automatic assessment of visual quality should be performed whenever possible, preferably using the methods that are capable of assessing the perceptual quality of images or can produce the scores that correlate well with subjective opinion.

In the CIE-Luv or CIE-Lab color space, perceptual differences between two colors can be measured using the Euclidean distance. This approach is frequently used in various simulations to evaluate the difference between the original image and its processed version. It is also common that the image quality is evaluated with respect to the known ground-truth data of some object(s) present in the scene, typically a target, such as the Macbeth color checker shown in Figure 6.1.

If the image is in the RGB format, the procedure includes the conversion of each pixel from the RGB values to the corresponding XYZ values using Equation 6.4, and then from these XYZ values to the target Luv or Lab values using the conversion formulas described in Section 6.2.2.2. Operating on the pixel level, the resulting perceptual difference between the two color vectors can be expressed as follows [29]:

$$\Delta E_{uv}^* = \sqrt{(L_1^* - L_2^*)^2 + (u_1^* - u_2^*)^2 + (v_1^* - v_2^*)^2}, \tag{6.33}$$

$$\Delta E_{ab}^* = \sqrt{(L_1^* - L_2^*)^2 + (a_1^* - a_2^*)^2 + (b_1^* - b_2^*)^2}. \tag{6.34}$$

The perceptual difference between the two images is usually calculated as an average of $\Delta E$ errors obtained in each pixel location. Psychovisual experiments have shown that the value of $\Delta E$ equal to unity represents a just noticeable difference (JND) in either of these two color models [53], although higher JND values may be needed to account for the complexity of visual information in pictorial scenes.

Since any color can be described in terms of its lightness, chroma, and hue, the CIE-Lab error can be equivalently expressed as follows [29], [33]:

$$\Delta E_{ab}^* = \sqrt{\left(\Delta L_{ab}^*\right)^2 + \left(\Delta C_{ab}^*\right)^2 + \left(\Delta H_{ab}^*\right)^2}, \tag{6.35}$$

where $\Delta L_{ab}^* = L_1^* - L_2^*$ denotes the difference in lightness,

$$\Delta C_{ab}^* = C_1^* - C_2^* = \sqrt{(a_1^*)^2 + (b_1^*)^2} - \sqrt{(a_2^*)^2 + (b_2^*)^2} \tag{6.36}$$

denotes the difference in chroma, and

$$\Delta H_{ab}^* = \sqrt{(a_1^* - a_2^*)^2 + (b_1^* - b_2^*)^2 - \left(\Delta C_{ab}^*\right)^2} \tag{6.37}$$

denotes a measure of hue difference. Note that the hue angle can be expressed as $h_{ab}^* = \arctan(b/a)$.

A more recent $\Delta E$ formulation weights the chroma and hue components by a function of chroma [54]:

$$\Delta E_{94}^* = \sqrt{\left(\frac{\Delta L_{ab}^*}{k_L S_L}\right)^2 + \left(\frac{\Delta C_{ab}^*}{k_C S_C}\right)^2 + \left(\frac{\Delta H_{ab}^*}{k_H S_H}\right)^2}, \tag{6.38}$$

where $S_L = 1$, $S_H = 1 + 0.015 \sqrt{C_1^* C_2^*}$, $S_C = 1 + 0.045 \sqrt{C_1^* C_2^*}$, and $k_L = k_H = k_C = 1$ for reference conditions. More sophisticated formulations of the $\Delta E$ measure can be found in References [55], [56], and [57], which employs spatial filtering to obtain an improved measure for pictorial scenes.

## 6.6 Conclusion

This chapter presented a framework that can simultaneously perform white balancing and color enhancement. This framework takes advantage of pixel-adaptive processing, which combines the local and global spectral characteristics of the captured visual data in order to produce the output image with the desired color appearance. Various specific solutions were presented as example methods based on relatively simple but yet powerful spectral modeling or combinatorial principles.

The framework is flexible, as it permits a wide range of functions that can follow the underlying concept of adaptive white balancing using a localized reference color vector that blends the spectral characteristics of the estimated illumination and the actual pixel. An additional degree of freedom in the design can be achieved through various tunable parameters that may be associated with such functions. The variety of such functions and efficient design methodology makes the presented framework very valuable in modern digital imaging and multimedia systems that attempt to mimic the human visual perception and use color as a cue for better image understanding and improved processing performance.

## Acknowledgment

The framework presented in this chapter was invented during the author's employment at Epson Canada Ltd. at Toronto, ON, Canada.

## References

[1] R. Lukac, *Single-Sensor Imaging: Methods and Applications for Digital Cameras*. Boca Raton, FL, USA: CRC Press / Taylor & Francis, September 2008.

[2] R. Lukac, *Computational Photography: Methods and Applications*. Boca Raton, FL, USA: CRC Press / Taylor & Francis, October 2010.

[3] J. Adams, K. Parulski, and K. Spaulding, "Color processing in digital cameras," *IEEE Micro*, vol. 18, no. 6, pp. 20–30, November 1998.

[4] R. Lukac and K.N. Plataniotis, "Color filter arrays: Design and performance analysis," *IEEE Transactions on Consumer Electronics*, vol. 51, no. 4, pp. 1260–1267, November 2005.

[5] K. Hirakawa and P.J. Wolfe, "Spatio-spectral sampling and color filter array design," in *Single-Sensor Imaging: Methods and Applications for Digital Cameras*, R. Lukac (ed.), Boca Raton, FL, USA: CRC Press / Taylor & Francis, September 2008, pp. 137–151.

[6] P.L.P. Dillon, D.M. Lewis, and F.G. Kaspar, "Color imaging system using a single CCD area array," *IEEE Journal of Solid-State Circuits*, vol. 13, no. 1, pp. 28–33, February 1978.

[7] B.T. Turko and G.J. Yates, "Low smear CCD camera for high frame rates," *IEEE Transactions on Nuclear Science*, vol. 36, no. 1, pp. 165–169, February 1989.

[8]  A.J. Blanksby and M.J. Loinaz, "Performance analysis of a color CMOS photogate image sensor," *IEEE Transactions on Electron Devices*, vol. 47, no. 1, pp. 55–64, January 2000.

[9]  D. Doswald, J. Haflinger, P. Blessing, N. Felber, P. Niederer, and W. Fichtner, "A 30-frames/s megapixel real-time CMOS image processor," *IEEE Journal of Solid-State Circuits*, vol. 35, no. 11, pp. 1732–1743, November 2000.

[10] K. Parulski and K.E. Spaulding, "Color image processing for digital cameras," in *Digital Color Imaging Handbook*, G. Sharma (ed.), Boca Raton, FL, USA: CRC Press, December 2002, pp. 728–757.

[11] R. Lukac, "Single-sensor digital color imaging fundamentals," in *Single-Sensor Imaging: Methods and Applications for Digital Cameras*, R. Lukac (ed.), Boca Raton, FL, USA: CRC Press / Taylor & Francis, September 2008, pp. 1–29.

[12] J.E. Adams and J.F. Hamilton, "Digital camera image processing chain design," in *Single-Sensor Imaging: Methods and Applications for Digital Cameras*, R. Lukac (ed.), Boca Raton, FL, USA: CRC Press / Taylor & Francis, September 2008, pp. 67–103.

[13] K. Barnard, L. Martin, A. Coath, and B. Funt, "A comparison of computational color constancy algorithms – Part I: Experiments with image data," *IEEE Transactions on Image Processing* vol. 11, no. 9, pp. 985–996, September 2002.

[14] G.D. Finlayson, "Three-, two-, one-, and six-dimensional color constancy," in *Color Image Processing: Methods and Applications*, R. Lukac and K.N. Plataniotis (eds.), Boca Raton, FL, USA: CRC Press / Taylor & Francis, October 2006, pp. 55–74.

[15] J. Lee, Y. Jung, B. Kim, and S. Ko, "An advanced video camera system with robust AF, AE, and AWB control," *IEEE Transactions on Consumer Electronics*, vol. 47, no. 3, pp. 694–699, August 2001.

[16] G. Sharma, "Color fundamentals for digital imaging," in *Digital Color Imaging Handbook*, G. Sharma (ed.), Boca Raton, FL, USA: CRC Press / Taylor & Francis, December 2002, pp. 1–113.

[17] E. Reinhard, E.A. Khan, A.O. Akyuz, and G.M. Johnson, *Color Imaging: Fundamentals and Applications*. Wellesley, MA, USA: AK Peters, July 2008.

[18] G.M. Johnson and M.D. Fairchild, "Visual psychophysics and color appearance," in *Digital Color Imaging Handbook*, G. Sharma (ed.), Boca Raton, FL, USA: CRC Press / Taylor & Francis, December 2002, pp. 173–238.

[19] "Lighting design glossary," in *Lighting Design and Simulation Knowledgebase*, Available online, http://www.schorsch.com/en/kbase/glossary/adaptation.html.

[20] E. Lam and G.S.K. Fung, "Automatic white balancing in digital photography," in *Single-Sensor Imaging: Methods and Applications for Digital Cameras*, R. Lukac (ed.), Boca Raton, FL, USA: CRC Press / Taylor & Francis, September 2008, pp. 267–294.

[21] N. Sampat, S. Venkataraman, and R. Kremens, "System implications of implementing white balance on consumer digital cameras," *Proceedings of SPIE*, vol. 3965, pp. 362–368, January 2000.

[22] C.C. Weng, H. Chen, and C.S. Fuh, "A novel automatic white balance method for digital still cameras," in *Proceedings of the IEEE International Symposium on Circuits and Systems*, Kobe, Japan, May 2005, vol. 4, pp. 3801–3804.

[23] N. Kehtarnavaz, N. Kim, and M. Gamadia, "Real-time auto white balancing for digital cameras using discrete wavelet transform-based scoring," *Journal of Real-Time Image Processing*, vol. 1, no. 1, pp. 89–97, October 2006.

[24] R. Lukac, "Automatic white balancing of a digital image," U.S. Patent 7 889 245, February 2011.

[25] R. Lukac, "New framework for automatic white balancing of digital camera images," *Signal Processing*, vol. 88, no. 3, pp. 582–593, March 2008.

[26] R. Gonzalez and R.E. Woods, *Digital Image Processing*. Reading, MA, USA: Prentice Hall, 3rd edition, August 2007.

[27] E. Giorgianni and T. Madden, *Digital Color Management*. Reading, MA, USA: Addison-Wesley, 2nd edition, January 2009.

[28] G. Wyszecki and W.S. Stiles, *Color Science: Concepts and Methods, Quantitative Data and Formulas*. New York, USA: Wiley-Interscience, 2nd edition, August 2000.

[29] H.J. Trussell, E. Saber, and M. Vrhel, "Color image processing," *IEEE Signal Processing Magazine*, vol. 22, no. 1, pp. 14–22, January 2005.

[30] M. Stokes, M. Anderson, S. Chandrasekar, and R. Motta, "A standard default color space for the internet – sRGB," Technical report, Available online, http://www.w3.org/Graphics/Color/sRGB.html.

[31] R.W.G. Hunt, *Measuring Colour*. Chichester, UK: Wiley, 4th edition, December 2011.

[32] S. Susstrunk, R. Buckley, and S. Swen, "Standard RGB color spaces," in *Proceedings of the Seventh Color Imaging Conference: Color Science, Systems, and Applications*, Scottsdale, AZ, USA, November 1999, pp. 127–134.

[33] S. Susstrunk, "Colorimetry," in *Focal Encyclopedia of Photography*, M.R. Peres (ed.), Burlington, MA, USA: Focal Press / Elsevier, 4th edition, 2007, pp. 388–393.

[34] R.G. Kuehni, *Color Space and Its Divisions: Color Order from Antiquity to the Present*. Hoboken, NJ, USA: Wiley-Interscience, March 2003.

[35] C.A. Poynton, *A Technical Introduction to Digital Video*. Toronto, ON, Canada: Prentice Hall, January 1996.

[36] A. Sharma, "ICC color management: Architecture and implementation," in *Color Image Processing: Methods and Applications*, R. Lukac and K.N. Plataniotis (eds.), Boca Raton, FL, USA: CRC Press / Taylor & Francis, October 2006, pp. 1–27.

[37] R. Lukac, B. Smolka, K. Martin, K.N. Plataniotis, and A.N. Venetsanopulos, "Vector filtering for color imaging," *IEEE Signal Processing Magazine*, vol. 22, no. 1, pp. 74–86, January 2005.

[38] J. Gomes and L. Velho, *Image Processing for Computer Graphics*. New York, USA: Springer-Verlag, May 1997.

[39] R. Lukac and K.N. Plataniotis, "Single-sensor camera image processing," in *Color Image Processing: Methods and Applications*, R. Lukac and K.N. Plataniotis (eds.), Boca Raton, FL, USA: CRC Press / Taylor & Francis, October 2006, pp. 363–392.

[40] R. Lukac and K.N. Plataniotis, "Normalized color-ratio modeling for CFA interpolation," *IEEE Transactions on Consumer Electronics*, vol. 50, no. 2, pp. 737–745, May 2004.

[41] D.R. Cok, "Signal processing method and apparatus for producing interpolated chrominance values in a sampled color image signal." U.S. Patent 4 642 678, February 1987.

[42] J. Adams, "Design of practical color filter array interpolation algorithms for digital cameras," *Proceedings of SPIE*, vol. 3028, pp. 117–125, February 1997.

[43] G.D. Finlayson, S.D. Hordley, and P.M. Hubel, "Color by correlation: A simple, unifying framework for color constancy," *IEEE Transactions on Pattern Analysis and Machine Intelligence*, vol. 23, no. 11, pp. 1209–1221, November 2001.

[44] N. Kehtarnavaz, H. Oh, and Y. Yoo, "Development and real-time implementation of auto white balancing scoring algorithm," *Real-Time Imaging*, vol. 8, no. 5, pp. 379–386, October 2002.

[45] E.Y. Lam, "Combining gray world and Retinex theory for automatic white balance in digital photography," in *Proceedings of the International Symposium on Consumer Electronics*, Macau, China, June 2005, pp. 134–139.

[46] J. Huo, Y. Chang, J. Wang, and X. Wei, "Robust automatic white balance algorithm using gray color points in images," *IEEE Transactions on Consumer Electronics*, vol. 52, no. 2, pp. 541–546, May 2006.

[47] G. Sapiro, "Color and illuminant voting," *IEEE Transactions on Pattern Analysis and Machine Intelligence*, vol. 21, no. 11, pp. 1210–1215, November 1999.

[48] S. Bianco, F. Gasparini, and R. Schettini, "Combining strategies for white balance," *Proceedings of SPIE*, vol. 6502, id. 65020D, January 2007.

[49] R. Lukac, "Refined automatic white balancing," *IET Electronics Letters*, vol. 43, no. 8, pp. 445–446, April 2007.

[50] R. Lukac, "Joint automatic demosaicking and white balancing," U.S. Patent 8 035 698, October 2011.

[51] A.K. Moorthy, K. Seshadrinathan, and A.C. Bovik, "Image and video quality assessment: Perception, psychophysical models, and algorithms," in *Perceptual Imaging: Methods and Applications*, R. Lukac (ed.), Boca Raton, FL, USA: CRC Press / Taylor & Francis, 2012.

[52] K.N. Plataniotis and A.N. Venetsanopoulos, *Color Image Processing and Applications*. New York, USA: Springer-Verlag, 2nd edition, December 2010.

[53] D.F. Rogers and R.E. Earnshaw, *Computer Graphics Techniques: Theory and Practice*. New York, USA: Springer-Verlag, October 2001.

[54] CIE publication No 116, Industrial colour difference evaluation. Central Bureau of the CIE, 1995

[55] M.R. Luo, G. Cui, and B. Rigg, "The development of the CIE 2000 colour difference formula: CIEDE2000," *Color Research and Applications*, vol. 26, no. 5, pp. 340–350, October 2001.

[56] M.R. Luo, G. Cui, B. Rigg, "Further comments on CIEDE2000," *Color Research and Applications*, vol. 27, no. 2, pp. 127–128, April 2002.

[57] B. Wandell, "S-CIELAB: A spatial extension of the CIE L*a*b* DeltaE color difference metric." Available online, http://white.stanford.edu/~brian/scielab/.

# 7

## Perceptual Thumbnail Generation

**Wei Feng, Liang Wan, Zhouchen Lin, Tien-Tsin Wong, and Zhi-Qiang Liu**

## 7.1   Introduction

As shrunk versions of pictures, thumbnails are commonly used in image preview, organization, and retrieval [1], [2], [3], [4]. In the age of digital media, a thumbnail plays a similar role for an image as an abstract does for an article. Just consider today's digital cameras; it is hard to find one with less than six Megapixels. Viewers expect to observe both the global composition and important perceptual features of the original full-size picture by inspecting its thumbnail, which should thus provide a faithful impression about the image content and quality.

In practice, thumbnails mainly serve in the following two aspects. First, they provide viewers a convenient way to preview high-resolution pictures on a small display screen. For instance, many portable devices, such as digital cameras and cell phones, provide prompt image preview function by small-size LCD displays [5]. Second, thumbnails enable the user to quickly browse a large number of images without being distracted by unimportant details. For example, almost all visual search engines, such as Google Images and Bing Images, use thumbnails to organize their returned search results. As the proliferation of digital images has been increasing in recent years, generating perceptually faithful thumbnails becomes very important.

From the historical viewpoint, the concept of thumbnail is gradually changed due to the developing practical requirements in real-world applications. At the very beginning, for a given full-size picture, a thumbnail was generated simply by filtering and subsampling [6], [7]. Then, to emphasize some important portions in original pictures, region of interest (ROI)-based thumbnails were proposed based on visual attention models [1], [8]. Compared to conventional downsampled thumbnails that equally treat all image parts and features in the original size, ROI-based thumbnails highlight visually important regions only, thus helping viewers to focus on essential parts, but inevitably losing the perceptual faith of global composition in original pictures. In recent years, moreover, people start to realize the necessity and importance of *perceptual thumbnails* [5], [9]. Specifically, perceptual thumbnails highlight the quality-related and perceptually important features in original pictures, with the global structural composition preserved. From perceptual thumbnails, viewers can easily assess the image quality without checking the original full-size versions, thus facilitating faithful image preview on small displays and fast browsing of a large number of pictures. In contrast, without perceptual thumbnails, evaluating the quality of images taken by digital cameras requires that the user has to repeatedly zoom in, zoom out at different scales, and shift across different parts of the pictures at higher resolutions, which is usually very time consuming, inconvenient, and ineffective. Moreover, the operation of zooming in may also make viewers lose the perception of the global composition of pictures.

Clearly, both conventional downsampled thumbnails and ROI-based thumbnails are not qualified to serve as perceptual thumbnails. As shown in Figure 7.1, some apparent perceptual features in the original pictures, for instance, blur and noise, may easily be lost in the downsampled thumbnails, no matter what interpolation algorithm is used. As mentioned above, the conventional downsampling process may significantly reduce the resolution of perceptual features in downsampled thumbnails. To discover a phenomenon in an image, the resolution of the phenomenon must be larger than some threshold that represents the perceiving capability of viewers. When the reduced resolution of naively downsampled perceptual features becomes less than viewers' perceiving thresholds, these features cannot be noticed. Therefore, in perceptual thumbnails, some important quality-related features should be highlighted and the resolution of perceptual features should not be affected by the downsampling process. Besides, ROI-based thumbnails cannot be used as perceptual thumbnails since this approach destroys the global composition of the original pictures. To this end, successful perceptual thumbnail generation should meet the four requirements discussed below.

- *Low cost*: As thumbnails are widely used in prompt preview of high-resolution pictures and fast browsing of a large number of images, they must be generated in an economic way in terms of their computation and storage. Hence, fast and simple algorithms for highlighting perceptual features are desirable.

- *Unsupervision*: Since one may want to quickly produce a large number of thumbnails in a browsing task, they should be generated automatically. Both preliminary training and on-the-fly parameter tuning should be avoided.

- *Highlighting multiple perceptual features*: To faithfully reflect the content and quality of original full-size pictures in thumbnails, multiple quality-related features, such as blur and noise, should be highlighted according to their respective strengths.

- *Preserving the original composition*: It is important for perceptual thumbnails to provide viewers with a clear impression about the global structure (composition) of original pictures. For instance, the user may want to quickly decide whether or not to take a second shot of the same scene by checking the thumbnails on small display screens in their cameras.

This chapter focuses on introducing the recent development of perceptual thumbnail generation techniques that satisfy the above four requirements. Within a unified framework, particular attention is paid to preserving and highlighting two types of quality-related perceptual features, that is, blur and noise.

Section 7.2 briefly introduces the techniques for traditional thumbnail generation and specifically discusses signal-level downsampled thumbnail generation and ROI-based thumbnail generation methods. Signal-level thumbnail generation concerns the development of different filtering or interpolation techniques in order to reduce aliasing artifacts in thumbnails. In contrast, ROI-based thumbnail generation aims at preserving salient image regions and essential structures during image resizing.

The next three sections focus on the technical details of perceptual thumbnails. Namely, Section 7.3 first described the commonly noticed visual cues/features that indicate image quality and need to be preserved in perceptual thumbnails. Then, it introduces a general framework for perceptual thumbnail generation, focused on blur and noise as two commonly encountered types of low-level perceptual features. This framework can be extended to include other types of perceptual features, such as bloom and overexposure. At last, this section briefly presents several state-of-the-art perceptual thumbnail generation methods that are based on the proposed general working flow. Sections 7.4 and 7.5 elaborate how to highlight spatially variant blur and homogeneous noise in thumbnails, respectively. Although blur and noise estimation has been studied for decades in image processing and computer vision, most previous deblurring and denoising methods tried to get an accurate estimate of blur and noise strength for the goal of recovering blur-free and noise-free images. Perceptual thumbnail generation, on the contrary, aims to highlight blur and noise at a noticeable scale in thumbnails. This raises the requirements of appropriate but inexact algorithms for blur and noise estimation, as well as alias-free blur and noise visualization. Therefore, these two sections also discuss how to quickly estimate blur and noise from the original images, and effectively visualize them in the downsampled thumbnails.

Section 7.6 presents a number of experimental results that comprehensively compare the performance of different techniques in perceptual thumbnail generation. Current perceptual thumbnail generation only handles blur and noise, the two low-level visual cues. Finally, this chapter concludes with Section 7.7, which summarizes the state-of-the-arts of perceptual thumbnail generation and discusses some possible ways to highlight other types of perceptual cues, including red-eye effect and bloom, in thumbnails.

## 7.2 Traditional Thumbnail Generation

Traditional thumbnail generation techniques can be classified as signal-level thumbnail generation and ROI-based thumbnail generation. Signal-level thumbnail generation techniques [7], [10], [11], [12], [13] create the thumbnail from a high-resolution original image by prefiltering and subsampling. The generated thumbnail preserves the original composition. ROI-based thumbnail generation techniques [2], [3], [14], [15], [16], on the other hand, aim at detecting and retaining important regions in the resulting thumbnail.

### 7.2.1 Signal-Level Thumbnail Generation

Signal-level thumbnail generation relies on image resampling techniques, which usually employ both filtering and subsampling [17]. Due to different filtering schemes employed, the quality of the resulting thumbnails may differ greatly. Among the existing image resampling techniques, the simplest method is decimation, that is, keeping some pixels of the original image and throwing others away. Without prefiltering the original image, decimation typically leads to significant aliasing artifacts.

To reduce such aliasing artifacts, lowpass filters can be used before decimation. For example, box filter computes an average of pixel values within the pixel neighborhood. Bilinear filter performs linear filtering first in one direction, and again in the other direction. The bicubic filter [10] is often chosen over the bilinear filter since the resampled image is smoother and has fewer aliasing artifacts. Other than second-order polynomials defined in bilinear filter and third-order polynomials defined in bicubic filter, high-order piecewise polynomials [12] can also be used, although the improvement is marginal. More sophisticated B-spline approximation is reported in the literature for image resampling [7], [11]. There exist other suitable linear methods, including Lanczos [6] and Mitchell [18] filters. The aforementioned filters are space-invariant (isotropic) in nature. In the context of texture mapping, perspective mapping leads to space-variant (anisotropic) footprint. For this application, elliptic Gaussian-like filters have been employed [19], [20], [21].

Another trend of image resampling is to perform edge-adaptive interpolation in order to generate sharp edges in the resampled image. In Reference [22], the authors developed an edge-directed image interpolation (NEDI) algorithm. They estimated local covariance from a low-resolution image and used the low-resolution covariance to adapt the image interpolation at a higher resolution. Reference [23] developed a method to optimally determine the local quadratic signal class for the image. Reference [24] proposed an edge-guided

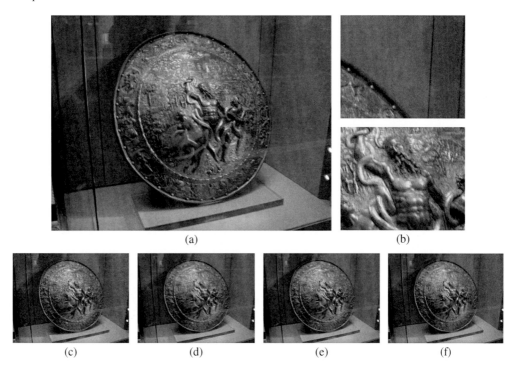

**FIGURE 7.1** (See color insert.)

Signal-level thumbnail generation: (a) original image, (b) two magnified regions cropped from the original image, and (c–f) thumbnails. Thumbnails are downsampled using (c) decimation, (d) bilinear, (e) bicubic, and (f) Lanczos filtering.

nonlinear interpolation algorithm, which estimates two observation sets in two orthogonal directions, and then fuses the two observation sets by the linear minimum mean square-error estimation technique. Another work [25] relied on a statistical edge dependency relating edge features of two different resolutions, and solved a constrained optimization. It should be noted that edge-adaptive resampling techniques are specially designed for image upsampling, and they are usually not suitable for image downsampling.

Figure 7.1 illustrates several commonly used image resampling methods. The original image has a resolution of $2048 \times 1536$ pixels, and is downsampled to a resolution of $320 \times 240$ pixels. As shown in Figure 7.1c, decimation results in severe aliasing artifacts. Bilinear filtering, shown in Figure 7.1d, removes artifacts. Bicubic filtering, shown in Figure 7.1e, produces a smoother thumbnail by further reducing the aliasing artifacts. The thumbnail generated using Lanczos filtering, shown in Figure 7.1f, has even fewer artifacts; however, this method is much slower than others. Obviously, signal-level thumbnail generation techniques can prevent aliasing artifacts and preserve the original image composition. However, they are not designed to preserve perceptual image quality during image resampling. In fact, they probably lose such features that are useful to the user to identify the image quality. Visual inspection of the magnified regions in Figure 7.1b reveals that the original high-resolution image suffers from noticeable blur and noise. However, the three thumbnails in Figures 7.1d to 7.1f all appear rather clean and clear. A qualitative analysis of the loss in the blur and noise will be presented in Section 7.3.

  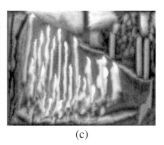

(a)                              (b)                              (c)

**FIGURE 7.2** (See color insert.)

ROI-based thumbnail generation: (a) high-resolution image, (b) automatic cropping, and (c) seam carving.

## 7.2.2 ROI-Based Thumbnail Generation

Unlike signal-level thumbnail generation, which preserves the original image composition, ROI-based thumbnail generation (Figure 7.2) just displays important image regions in the small-scale thumbnail. In Reference [1], the downsampling distortion was reduced by treating the high-resolution original image with edge-preserving smoothing, lossy image compression, or static codebook compression. Reference [8] used an image attention model based on region of interest, attention value, and minimal perceptible size to incorporate user attention in the process of adapting the high-resolution image on small-size screens. For screens with different scale and aspect ratio, different important regions were cropped from the original image for downsampling. Similarly, Reference [14] detected key components of the high-resolution image based on a saliency map, and then cropped images prior to downsampling (Figure 7.2c). Although the cropping-based methods can render important image portions in a more recognizable way, the user will lose the global overview of the original image.

Another research direction relating to thumbnail generation is image retargeting, which considers geometric constraints and image contents in the resizing process. Reference [26] maximized salient image objects by minimizing less important image portions in-between. Since the spatial spaces between image objects may be narrowed, the relative geometric relation between the objects can be altered in the thumbnail. Reference [15] presented a seaming carving approach (Figure 7.2c), which removes or inserts an eight-connected path of pixels on an image from top to bottom, or from left to right in one pass. This method, though quite simple, is rather time consuming and may damage the image structure severely. To address the problems in seaming carving, various retargeting techniques [16], [27], [28], [29] have been proposed. Although the techniques can be quite different, their objectives are similar: preserving the important image content, reducing visual artifacts, and preserving internal image structures. A comprehensive evaluation of state-of-art methods can be found in Reference [30].

Additionally, special concerns have been reported to address the situation when the high-resolution image contains texts. For example, Reference [3] created thumbnails for web pages by combining text summaries and image thumbnails. Reference [2] cropped and scaled image and text segments from the original image and generated a more readable layout for a particular display size. Following the aforementioned two works, many subsequent research works can be found, such as References [31], [32], and [33].

The ROI-based thumbnail generation techniques have been proven successful in preserving important image contents in the low-resolution thumbnail. However, none of them account for perceptual quality features in the high-resolution image. As a result, such techniques may remove these desired features during image downsizing.

## 7.3 Perceptual Thumbnail Generation

This section first describes several common visual/perceptual cues that indicate image quality. Then, it presents a general framework for perceptual thumbnail generation, specifically focusing on highlighting blur and noise in thumbnails. Finally, a brief overview of the state-of-the-art methods in perceptual thumbnail generation is presented.

### 7.3.1 Quality-Indicative Visual Cues

Given a picture shot with a digital camera, the user often demands to inspect whether or not the picture is well shot by quickly checking the thumbnail image. Such a quality inspection process can certainly benefit from properly highlighting some perceptual features in the thumbnail. However, in order to do so, one first needs to find what features or effects in the original picture should be highlighted in the thumbnail. According to practical photography experiences, there are generally two categories of quality-related perceptual features: low-level image features and high-level semantic features.

#### 7.3.1.1 Low-Level Image Features

There are some features that degrade the quality of an image but have less to do with the image content. Such features include blur, noise, bloom, overexposure, underexposure, and so on. In most cases, these features are quality indicative and should be noticed by the user during the photo-taking process, thus requiring to be preserved and highlighted by the downsampling operation when creating the thumbnails.

- *Blur*: Except for some special cases, a blurry picture is usually inferior to a sharp picture of the same scene. The user often accidently moves his or her hands during photo shooting, thus creating blur, which may also occur when there are moving objects in the scene or when some subjects are out of focus of the camera. Image blur, such as spatially varying blur and homogeneous blur, caused by different reasons may exhibit different spatial properties. In most cases, if the user notices unexpected blur in a picture through the preview screen of a digital camera, the user may choose to take a second shot on the same scene. Hence, blur is a very important quality-indicative effect in images and certainly should be highlighted in perceptual thumbnails. An example of image blur is shown in Figure 7.3a.

- *Noise*: Image noise is the random variation of color or brightness information produced by image sensors during exposure and analog-to-digital conversion. It often looks like color grains and scatters over the entire image. There are two types of

**FIGURE 7.3 (See color insert.)**

Quality-indicative visual cues: (a) blur, (b) noise, (c) bloom, and (d) red-eye defects.

image noise: fixed pattern noise and random noise [34]. Fixed pattern noise is generally visible when using long exposure time or when shooting under low-light conditions with high ISO speed. It has the same pattern under similar lighting conditions.
Random noise occurs at all exposure times and light levels. It appears as randomly
fluctuated intensities and colors, and is different even under the identical lighting
condition. Banding noise is a special type of random noise that appears as horizontal or vertical strikes. It is generated by the camera when reading data from image
sensors, and is more noticeable at high ISO speed. Generally, any noise degrades the
quality of a picture and therefore is also quality indicative. Figure 7.3b demonstrates
how image noise appears in a picture.

- *Bloom*: Bloom is the phenomenon that a bright light source appears as a bright halo and leads to color bleeding into nearby objects. This phenomenon occurs when the light source is so strong that the sensor pixels become saturated. For example, in Figure 7.3c, shooting direct sunlight will generate such effect around the boundary of sunlight area. Since the appearance of bloom will significantly destroy the local details of a scene, bloom is also quality indicative.

- *Overexposure and underexposure*: Overexposure and underexposure refer to the effect of losing details in highlight and shadow regions, respectively. This is usually caused by inappropriate setting of camera's shutter speed and exposure. In high-resolution photography, except for some special effects, overexposure and underexposure should be avoided in order to maintain high image quality.

### 7.3.1.2   High-Level Semantic Features

High-level perceptual features are highly related to the image content [35], as they either reflect some unexpected phenomena of the particular objects or even relate to some aesthetic aspects of the image. Some typical high-level perceptual features are listed below. High-level perceptual features are more difficult to be detected and highlighted in thumbnails than low-level perceptual features.

- *Defected eyes*: When shooting a human subject, the user usually cares whether the eyes were opened at the time of pressing the camera shutter and do not appear as red due to the reflected light from the flash (Figure 7.3d). To automatically detect and correct the red-eye effects in images, various methods have been designed [36], [37].

- *Simplicity*: The most discriminative factor that differentiates professionals from amateur photographers is whether the photo is simple [35]. For a picture, being simple usually means that it is easy to separate the subject from the background. There are various ways for professionals to achieve this goal, including making background out of focus, choosing significant color contrast, and using large lighting contrast between the subject and the background. Recently, simplicity was also used as an important criterion for image quality assessment [35].

- *Realism*: Similar to simplicity, realism is another high-level feature that reflects aesthetic image quality [35]. Professional photographers often carefully choose the lighting conditions and the color distribution, and make use of filters to enhance the color difference. They are also very deliberate in the picture composition of the subject and background. All these are for the purpose of keeping the foreground of pictures as realistic as possible.

In the context of thumbnail generation, some features (for example, simplicity and realism) might remain in the low-resolution thumbnail to some extent, while other features (for example, blur, noise, bloom, overexposure/underexposure, and defected eyes) are less noticeable due to the downsampling process. This chapter focuses on studying how to preserve two typical low-level image features, that is, blur and noise, in thumbnails. As already mentioned, these two phenomena may significantly degrade image quality and commonly exist in digital pictures.

## 7.3.2 A General Framework

Before presenting the general framework of producing perceptual thumbnails, the loss of blur and noise features by conventional image downsampling is first analyzed. Suppose the original image is represented as vector $x$. Denote the antialiasing lowpass filter as $T$ and subsampling operation as $S$. The conventional thumbnail can then be expressed as follows:

$$y = S \circ T(x), \tag{7.1}$$

where $\circ$ stands for function composition. When the original image suffers from blur and noise, it can be formulated as

$$x = B(\hat{x}) + n, \tag{7.2}$$

where $B$ and $n$ represent the blur and noise, respectively, and $\hat{x}$ is the ideal image without blur and noise degradation. Substituting Equation 7.2 into Equation 7.1 gives

$$y = S \circ T \circ B(\hat{x}) + S \circ T(n). \tag{7.3}$$

The following examines the term $S \circ T \circ B(\hat{x})$ of the above equation. Except for some extreme blurs, the lowpass filter $B$, usually representing accidental blur, has a much larger bandwidth than the lowpass filter $T$. In another words, compared to the influence of filter $T$, the influence of filter $B$ can be much less noticeable in small-resolution thumbnails. Consequently, the conventional downsampled thumbnails will appear rather sharp than at their original scales.

In the second term $S \circ T(n)$ of the same equation, the noise $n$, mostly composed of lots of high-frequency information, is filtered by $T$. Thus, it will become much less apparent than in the original resolution. For instance, if $n$ is a Gaussian noise of zero mean and moderate variance, $T(n)$ will be close to zero under typical downsampling factors. As a result, the conventional thumbnail $y$ will appear rather clean.

To highlight the blur and noise effects in $y$ with proper strength, the perceptual thumbnail can be formulated as follows:

$$y' = B(y) + n. \tag{7.4}$$

Intuitively, the conventional thumbnail should be blurred according to the blur strength in the original image, and the noise information should be superimposed at the same time. By this way, the user can easily inspect and discover the blur and noise level in the image by viewing the thumbnail alone.

Using the above *extract-and-superimpose* strategy, the general framework of perceptual thumbnail generation, which is capable of highlighting blur and noise features, can now be summarized. As illustrated in Figure 7.4, the high-resolution image is first downsampled to get the conventional thumbnail. Then, the blur and noise information present in the original high-resolution image is estimated and extracted. Afterward, the detected amount of blur is properly added into the thumbnail, followed by rendering the estimated noise in order to produce the refined perceptual thumbnail. Note that the presented framework for perceptual thumbnail generation is general enough and can be easily extended to include other types of perceptual features.

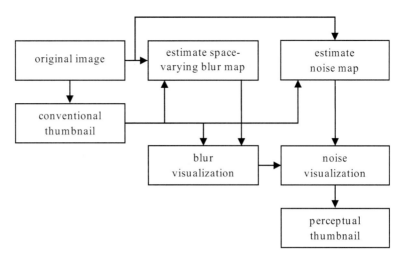

**FIGURE 7.4**

The framework of perceptual thumbnail generation, which particularly highlights two typical perceptual features, that is, blur and noise. This framework uses the general extract-and-superimpose strategy that can be easily extended to include other types of perceptual features.

### 7.3.3 State-of-the-Art Methods

Although perceptual thumbnail generation is relatively a new topic, there are several related preliminary works that have been reported in the literature. For instance, the image preview method introduced in Reference [5] divides the problem of perceptual thumbnail generation into structure enhancement and perceptual feature visualization tasks. The first task highlights the salient structure and suppresses the subtle details in the image by nonlinearly modulating the image gradient field. The second task estimates the strength of blur and noise in the original image and then superimposes blur and noise with appropriate degrees in the conventional thumbnail. Similarly, Reference [4] studied how to estimate and visualize blur and noise features in the thumbnail. The perceptual thumbnail were generated by directly blurring the conventional thumbnail and adding noise to it. The noise estimation method was further improved to achieve fast performance [9]. Other related efforts in this direction have been exploited in References [38] and [39].

Current perceptual thumbnail generation is highly related to the problems of image deblurring [40], [41], [42] and denoising [43], [44]. These two problems are rather difficult, and the relevant algorithms are usually complicated and time consuming. In contrast, instead of reconstructing the ideal high-resolution image, the goal of perceptual thumbnail generation is to visualize blur and noise in a proper way with reasonable complexity. This relieves the need of recovering actual blur kernels and estimating the noise accurately. Instead, a rough yet fast algorithm for blur and noise estimation is preferable. This is indeed the focus of the existing works related to perceptual thumbnail generation. Although most of these works consider blur and noise in perceptual thumbnails, they adopt quite different algorithms for blur/noise estimation and rendering. The following describes the two representative methods proposed in References [4] and [5].

## 7.4 Highlighting Blur in Thumbnails

This section suggests how to highlight the reduced blurriness in perceptual thumbnails. Specifically, for the blurred regions in the original high-resolution picture, the perceptual blur highlighting method should magnify the reduced blurriness in their downsampled counterparts, while for the sharp regions at original scales, the goal is to maintain their clearance in thumbnails. Hence, the two major steps in perceptual blur highlighting are blur strength extraction from the original pictures and faithful visualization of extracted blurriness in thumbnails.

### 7.4.1 Blur Estimation

In the literature, there exist many methods for blur estimation; these methods provide global blur metrics and evaluate the overall blurriness of an image [45], [46], [47]. A suitable global blur metric is helpful to assess images with uniform blurs, such as the motion blur due to shaking cameras. This is, however, insufficient for producing perceptual thumbnails, since the pictures may contain spatially varying blurs. For example, the out-of-focus regions may appear different blur strength at different depth. This raises the demands for local blur estimation. Although there exist studies focused on spatially varying blur determination [48], [49], [50], most of them rely on solving various optimization problems, which is is usually time consuming and inconvenient, and thus these methods are not suitable for efficiently generating perceptual thumbnails. To overcome this drawback, two techniques [5], [9] for roughly determining a local blur strength map from a high-resolution image will now be described.

#### 7.4.1.1 Gradient-Based Blur Estimation

Gradient-based blur estimation [5] relies on an important observation that the variation of a blurry edge region is much more gradual than that of a sharp edge region. Thus, the strength of regional blurriness can be quantitatively measured according to the variance of gradient angles/directions within the corresponding edge region. Specifically, the smaller the regional variance of gradient angles is, the blurrier the edge region would be. To verify this fact, Figure 7.5 shows the simulation test where each of the six high-resolution images were first blurred by Gaussian filters with increased standard deviations and then the variances of gradient directions in all edge regions were measured. As can be seen, as the strength of Gaussian filters increases, the average of directional variances in the gradient field decreases significantly.

Let $A_i$ denote the gradient angle of the $i$-th edge pixel, which can be obtained as

$$A_i = \arctan\left(\frac{\Delta_t(i)}{\Delta_s(i)}\right), \tag{7.5}$$

where $\Delta_t(i)$ and $\Delta_s(i)$ are the first-order differences of pixel $i$ along two spatial dimensions. Then, the gradient-based space-varying blur metric can be defined as follows:

$$B_i = \alpha \exp\left(-\beta \operatorname{var}(A_i)^\tau\right), \tag{7.6}$$

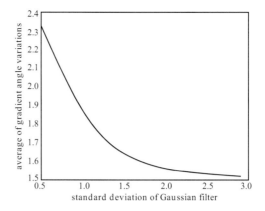

**FIGURE 7.5**

Inverse relationship of blur strength and regional variance of gradient directions.

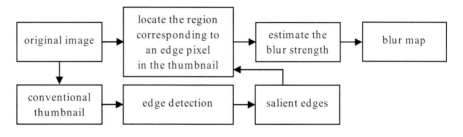

**FIGURE 7.6**

Gradient-based blur estimation flowchart.

where var$(\cdot)$ is the variance operator and returns the direction variance within the neighborhood of pixel $i$. Parameters $\alpha$, $\beta$, and $\tau$ control the influence of local gradient directional variance to the estimated blur strength. With this equation, the estimated blur strength is increasing with respect to the standard deviation of the Gaussian filter, as will be shown later.

In the context of perceptual thumbnail generation, there is no need to estimate the blur metric for the entire image at the original high resolution. In fact, one needs only to partially measure the blur strength in the low-resolution thumbnail image. Specifically, the blur around edge regions is visually much more noticeable and important to the viewers than in other regions [5], even when the entire image is very blurry. Hence, to speed up the computation, gradient-based blur estimation and highlight can be performed only around the edge regions in the thumbnail image by referring to its original version at the high resolution. As a result, as illustrated in Figure 7.6, gradient-based blur estimation is a three-step process. First, edge detection is performed on the conventional thumbnail to get salient edge pixels. Second, for each edge pixel in the thumbnail, its corresponding region at the original high-resolution picture is found. This small region defines the pixel neighborhood that is used to evaluate the gradient angel variance of current edge pixel. Finally, the blur metric is computed according to Equation 7.6. In this way, a blur strength map for all edge pixels in the thumbnail image can be obtained.

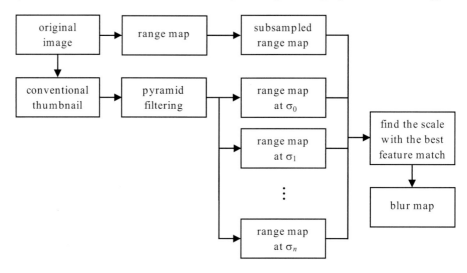

**FIGURE 7.7**

Scale-space-based blur estimation flowchart.


### 7.4.1.2  Scale-Space-Based Blur Estimation

Scale-space-based blur estimation [9] generates a space-varying blur map. Its basic idea, in contrast to gradient-based estimation, is first to smooth the conventional thumbnail using various Gaussian kernels with different scales and then to find the best filter scale such that the smoothed region in the filtered thumbnail image is most similar to the corresponding region in the high-resolution image. The algorithmic flow of scale-space-based blur estimation is illustrated in Figure 7.7. In order to generate a space-varying blur map, filter scale matching is performed for the entire thumbnail image. Similar to gradient-based blur estimation, this method does not differentiate between motion blur and out-of-focus blur, and just estimates the pixel-level blur metric. However, unlike gradient-based blur estimation, the scale-space method estimates the blur strength of all pixels in the thumbnail image.

In more detail, the conventional thumbnail $y$ is blurred by a series of Gaussian kernels $g$ with increasing scales $\{\sigma_j\}$, generating an indexed set of filtered thumbnails $f_j$ as follows:

$$f_j = g(\sigma_j) * y, \tag{7.7}$$

where $g(\cdot)$ represents Gaussian filter. As shown in Figure 7.7, for each blurred thumbnail, a local feature is extracted at each pixel and is compared to the features extracted from the high-resolution image. Here, the local feature of a pixel is defined as the maximum absolute difference between the appearances of the pixel and its surrounding eight neighbors. This generates a series of range maps $\{r_j\}$, where $j$ is the index of evaluated Gaussian filter scales [9]. Similarly, a range map for the high-resolution image is generated and subsampled to the thumbnail scale by further taking the maximum range value in a high-resolution neighborhood conforming to the subsampling factor. Specifically, denote the subsampled range map to be $r_o$. The blur map index at pixel $i$ can then be computed as

$$m_i = \min_j \{j \mid r_j(i) \le \gamma r_o(i)\}, \tag{7.8}$$

where the parameter $\gamma$ controls the amount of highlighted blur in thumbnails. Empirically, besides the conventional thumbnail, ten additional scales were used in experiments, starting from $\sigma_1 = 0.5$ and ending at $\sigma_{11} = 2.4$ with an increment of $0.2111$. Note that in the above equation, the blur map returns the index of evaluated Gaussian filters. For illustration purposes, the corresponding blur scale is visualized in Figure 7.7.

### 7.4.2 Blur Visualization

The second step of perceptual blur highlighting is *faithful blur visualization*. After performing either gradient-based blur estimation [5] or scale-space-based blur estimation [9], the extracted space-varying blur maps control Gaussian kernels with varying scales, which are finally superimposed into the downsampled thumbnail $y$ pixel by pixel and result in a faithfully blurred thumbnail $y_b$. Note that although the actual image blur in the original resolution may not be exactly conformed to Gaussian kernels, using Gaussian filters in the blur visualization for perceptual thumbnails is empirically a reasonable and effective choice, as the goal is to produce a visually faithful and noticeable blur effect.

Particularly, there are two types of methods to superimpose the space-varying Gaussian blurs in thumbnails. The first method is *incremental superimposition*. In the case of gradient-based blur estimation [5], the estimated pixel-level blur metric can be used as the standard deviation of a Gaussian filter. Note that the blur is estimated only at edge pixels of the conventional thumbnail. To get a reliable blurring effect, the blur strength at edge pixels is diffused to their neighborhood with self-adjusted filter scales. The blur metric for one pixel near an edge region is approximated as a weighted average of blur strengths of edge pixels nearby. The second method is *independent superimposition*. For the case of scale-space-based blur estimation [9], the estimated blur value corresponds to a specific Gaussian kernel with a particular scale. Since blur estimation is conducted at the entire thumbnail image, pixel values can be directly selected from the blurred thumbnails as follows:

$$y_b(i) = f_{m_i}(i), \tag{7.9}$$

where $m_i$ is the detected index of Gaussian filters for pixel $i$, and $f_{m_i}$ is the smoothed neighborhood of pixel $i$ in the thumbnail image.

Note that both incremental and independent superimposition have respective pros and cons. For instance, independent superimposition tends to produce artifacts in the blurred thumbnails, since there is nothing to protect the spatial coherence of blur strength by independent blur rendering. On the other hand, incremental superimposition has no such problem; however, the exact blur strength rendered into the thumbnail image is not implicitly clear when using incremental superimposition.

### 7.4.3 Evaluation

A simulation test with uniform blur is conducted to study how the blurriness is preserved across different blur strengthes. Six sharp high-resolution images are chosen as the test dataset. Each high-resolution image is blurred by a series of Gaussian kernels with increasing standard deviations starting from 0.5. The conventional thumbnails that match different blur strengths are prepared as well. Finally, the average blur strength is obtained for each blurred image using gradient-based blur estimation and scale-space-based blur estimation.

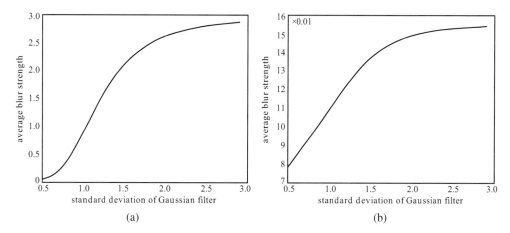

**FIGURE 7.8**

Relationship between detected blur strength and the ground truth Gaussian filter scales: (a) gradient-based blur estimation and (b) scale-space-based blur estimation.

Figure 7.8 shows the results of the average of the estimated blur standard deviations for the considered dataset. For gradient-based blur estimation, the estimated blur standard deviation is the computed blur metric. For scale-space-based blur estimation, the standard deviation of matched filter is retrieved according to the estimated index of tested Gaussian kernels. In both cases, the estimated standard deviation values increase as the blur gets more severe. In addition, they increase rapidly for small blur scales and become saturated for large blur scales.

## 7.5 Highlighting Noise in Thumbnails

Highlighting image noise in the thumbnail also involves two major steps, that is, estimating the noise in a high-resolution image and visualizing the noise in the image thumbnail.

### 7.5.1 Noise Estimation

Traditional image denoising [43], [44], [51] aims at reconstructing a noise-free image from its input noisy version. It is critical to have an accurate estimate of noise, otherwise the recovered image is still degraded. In perceptual thumbnail generation, the goal is to visualize noise and the exact precise form of the noise is not necessary for displaying noise in thumbnails. This relieves the requirement for noise estimation in two aspects. First, some prior knowledge about noise distribution can be exploited. Second, fast and inexact noise estimation methods [5], [9] can be used.

#### 7.5.1.1 Region-Based Noise Estimation

Image noise, as discussed in Section 7.3, is generated by image sensors during image acquisition. It is distributed over the entire image irrespective of the image content. How-

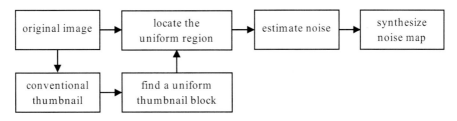

**FIGURE 7.9**

Region-based noise estimation flowchart.

ever, image noise is more visually apparent to the viewer in intensity-uniform regions than texture-intensive regions. It is because image noise is a high-frequency signal. Texture regions also have lots of high-frequency details; therefore, differentiating image noise from high-frequency texture details is usually difficult. Uniform regions, on the contrary, contain much less details, and hence the high-frequency information in uniform regions are mainly from image noise. Taking this prior knowledge, a region-based noise estimation was developed in Reference [5]. Its basic idea is to detect noise in a small uniform region of the high-resolution image, and then to synthesize a noise map in the thumbnail resolution based on the estimated noise region. Figure 7.9 illustrates the procedure.

Region-based noise estimation starts with the conventional thumbnail $y$ and divides it into non-overlapped regions $\{\Omega_k(y)\}$. The most uniform region $\Omega_u(y)$ is selected as the one with the minimum intensity variance, as follows:

$$\Omega_u(y) = \min_k\{k|var(\Omega_k(y))\}. \tag{7.10}$$

The uniform region $\Omega_u(x)$ in the high-resolution image can then be determined from $\Omega_u(y)$ via upsampling. To estimate the noise in $\Omega_u(x)$, a wavelet-based soft thresholding [51] is used. It first obtains empirical wavelet coefficients $c_l$ by pyramid filtering the noisy region $\Omega_u(x)$. Next, the soft thresholding nonlinearity is applied to the empirical wavelet coefficients:

$$\hat{c}_l = sgn(c_l)(|c_l| - \tau)_+, \tag{7.11}$$

where $(a)_+ = a$ if $a \geq 0$, and $(a)_+ = 0$ if $a < 0$. The threshold $\tau$ is specially chosen as $\tau = 1.6\sigma_l\sqrt{2\log(N)/N}$, where $N$ is the number of pixels in the noisy uniform region $\Omega_u(x)$. The noise standard deviation is estimated as $\sigma_l = c_m/0.6745$, where $c_m$ is the median absolute value of the normalized wavelet coefficients. Finally, the noise-free uniform region $\Omega_u(\hat{x})$ is recovered by inverting the pyramid filtering and the noise region is determined as

$$n_u(x) = \Omega_u(x) - \Omega_u(\hat{x}). \tag{7.12}$$

Note that the estimated noise region $n_u(x)$ may not be in the thumbnail resolution. Hence, a noise map $n(x)$ in the thumbnail resolution needs to be created. As noise distributes uniformly in the high-resolution image, one can use a simple yet efficient method to obtain $n(x)$. Namely, for each pixel in $n(x)$, its value is randomly selected from the estimated noise region $n_u(x)$. Although the resulted noise map $n(x)$ may not have exact match in the high-resolution image, it offers the viewer a quite similar visual experience.

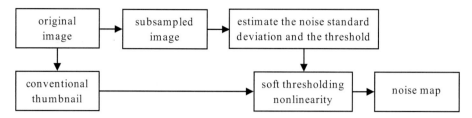

**FIGURE 7.10**

Multirate noise estimation flowchart.

### 7.5.1.2    Multirate Noise Estimation

The noise estimation method presented in Reference [4] also relies on wavelet-based soft thresholding [51]. Instead of a full wavelet transform, a single high-pass filtered signal is used to generate the noise. This signal is determined as

$$x_h = x - g(\sigma) * x, \tag{7.13}$$

where $g(\sigma)$ is a Gaussian filter with standard variation $\sigma = 1$. The soft threshold nonlinearity, as defined in Equation 7.11, is applied to the filtered signal. Then the noise map is given by

$$\rho(i) = \begin{cases} x_h(i), & \text{if } x_h(i) < \tau, \\ \tau, & \text{otherwise.} \end{cases} \tag{7.14}$$

where the threshold $\tau$ is estimated from $x_h$ without performing wavelet transform, and thus differing from region-based noise estimation. Here, the noise estimate $\rho$ corresponds to the high-resolution original image. According to Reference [52], subsampling the noise estimate $\rho$ by a scaling factor $t$ generates the noise map $n$ at the thumbnail resolution with the same standard deviation. Therefore, the noise map $n$ can be computed as

$$n(i) = \rho(ti). \tag{7.15}$$

In this noise estimation algorithm, estimating $\rho$ can be rather slow due to the high resolution of the original image. A fast version of this algorithm was developed in Reference [9] by exchanging the order of subsampling and noise generation. More specifically, multirate signal transformations [52] was used to estimate the noise at the low thumbnail resolution. Multirate noise estimation is feasible since there are enough pixels at the thumbnail resolution for the determination of the noise standard deviation. In addition, the soft threshold nonlinearity commutes with the subsampling operator. Due to these two facts, subsampling can be applied to the high-resolution image and the noise estimate is performed on the subsampled low-resolution image. Note that the threshold $\tau$ is estimated from the low-resolution image subsampled from the high-resolution image. Figure 7.10 illustrates the process of multirate noise estimation. Readers are referred to Reference [9] for a more detailed analysis of multirate noise estimation.

### 7.5.2    Noise Visualization

Visualizing noise is rather simple. As the noise is assumed to be additive to the ideal signal, the noise-preserved thumbnail is formed by adding the estimated noise map $n$ to

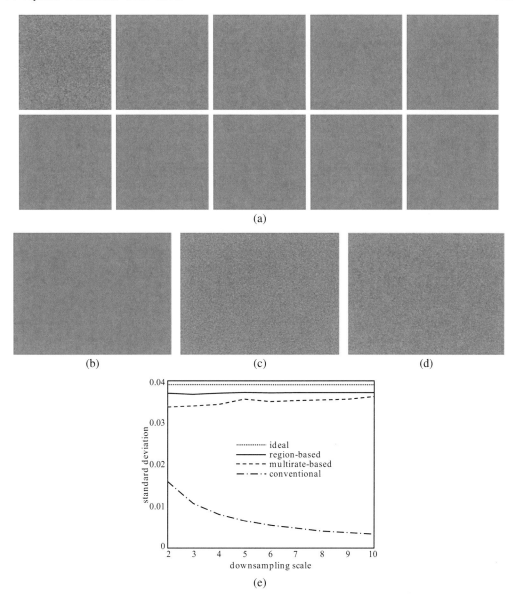

**FIGURE 7.11**

Evaluation of noise estimation: (a) $100 \times 100$ region at downsampling scales ranging from one to ten, with one corresponding to the original noise, (b) conventional thumbnail at the downsampling scale of ten, (c) region-based thumbnail, (d) multirate estimation-based thumbnail, and (e) plots of standard deviation of pixel intensities against the downsampling scales.

the thumbnail. Considering image blur, the noise visualization is performed after blur visualization as follows:

$$y_n(i) = y_b(i) + mag * n(i), \tag{7.16}$$

where the parameter *mag* controls the strength of noise to be visualized.

### 7.5.3 Evaluation

A simulation using a noise image is conducted to evaluate the performance of the two noise estimation methods. To roughly simulate the observed noise in photographs [53], the noise image is generated by adding a Gaussian noise with the standard deviation $\sigma = 10/255 = 0.392$ to a gray image with constant values. In Figure 7.11a, the $2048 \times 1536$ noise image (with one patch shown in the left-top position) is downsampled by different subsampling factors. At each subsampling scale, a $100 \times 100$ region is cropped for illustration. It is obvious that increasing the subsampling scale reduces the strength of noise gradually. In the thumbnail of a resolution $205 \times 154$ shown in Figure 7.11b, the noise is almost invisible. The two noise estimation methods, with the outputs shown in Figures 7.11c and 7.11d, are applied at each downsampling scale to estimate the noise map. Figure 7.11e shows the results of the standard deviations for the estimated noise maps with respect to the subsampling factors. The conventional thumbnail has a decreased standard deviation. The decreasing slope verifies the observation that the larger the subsampling factor is, the less noisy the thumbnail appears. Both noise estimation methods, on the other hand, have nearly horizontal lines and quite close to the ideal curve. The results indicate that the two methods are able to preserve the noise standard deviations faithfully across scale.

## 7.6 Experimental Results

This section presents some experimental results and comparisons of perceptual thumbnail generation with the ability of highlighting blur and noise. All results reported here are produced using $\alpha = 1$, $\beta = 3.0$, $\tau = 2.0$, and $\gamma = 1.5$.

First, the experimental results on blur highlighting are reported. Note that a good blur highlighting algorithm should be able to reflect present blur and should not to introduce obvious blur when the picture is sharp. Figure 7.12 demonstrates a blurry picture due to camera shake. The conventional thumbnail with $423 \times 317$ pixels appears rather clear and sharp, although the magnified regions in the original high-resolution image with $2816 \times 2112$ pixels look blurry. Both gradient-based and scale-space-based blur estimation can reflect blur in the modulated thumbnails. Visual inspection of the corresponding blur maps reveals that both methods have effects around edge regions. The difference is that the gradient-based method visualizes blur around salient edges only, while the scale-space-based method is effective in most edges, including weak edges. In terms of visual experience, the scale-space-based method introduces larger blur that is thus more visible to viewers; however, it appears to produce block artifacts.

Now that the existing blur highlighting methods are effective for blurry pictures, their performance is exploited in the case of sharp pictures, for which the methods are expected to introduce as small blur as possible. Refer to Figures 7.13 and 7.14. The two high-resolution pictures with $2048 \times 1536$ pixels are shot without motion blur or camera shake. One can see in the magnified image region that the edges are rather strong and sharp. Hence, the perceptual thumbnail at a resolution of $342 \times 256$ pixels should look similar to the conventional thumbnail, which is also sharp and clear. A close inspection of the

**FIGURE 7.12** (See color insert.)

Highlighting blur in the thumbnail when the original image is blurred: (a) conventional thumbnail at a resolution of $423 \times 317$ pixels, (b) two magnified regions from the original image with $2816 \times 2112$ pixels, (c) perceptual thumbnail obtained using gradient-based blur estimation, (d) corresponding blur map, (e) perceptual thumbnail obtained using the scale-space-based blur estimation, and (f) corresponding blur map.

thumbnails in Figures 7.13c, 7.13d, 7.14b, and 7.14c reveals that either method introduces some blur in the resulted thumbnails. From side-by-side comparisons, it can be observed that the gradient-based method suffers less from the unexpected blur, and the scale-space-based method may introduce obvious distracting blur in some regions.

**FIGURE 7.13 (See color insert.)**

Highlighting blur in the thumbnail when the original image is sharp: (a) conventional thumbnail with $341 \times 256$ pixels, (b) magnified region from the original image with $2048 \times 1536$ pixels, (c) perceptual thumbnail via gradient-based blur estimation, and (d) perceptual thumbnail via scale-space-based blur estimation.

**FIGURE 7.14 (See color insert.)**

Highlighting blur in the thumbnail when the original image with $1536 \times 2048$ pixels is sharp: (a) conventional thumbnail with $256 \times 341$ pixels, (b) perceptual thumbnail via gradient-based blur estimation, and (c) perceptual thumbnail via scale-space-based blur estimation.

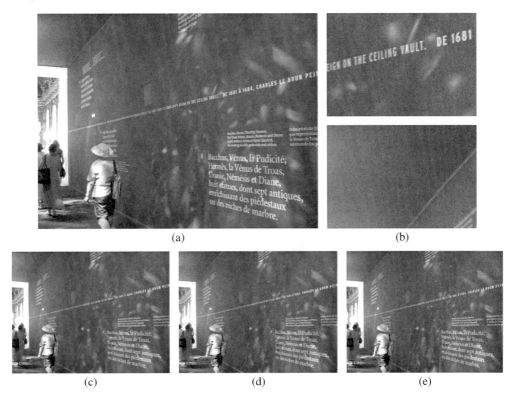

**FIGURE 7.15**

Highlighting noise in the thumbnail: (a) original noisy image, (b) two magnified regions from the original image, (c) conventional thumbnail, (d) perceptual thumbnail obtained using region-based noise estimation, and (e) perceptual thumbnail obtained using multirate noise estimation.

The following discusses the experimental results for noise highlighting. Similar to the requirements of blur highlighting methods, a good noise highlighting method shall be able to visualize image noise for noisy images, and not to introduce apparent noise for noise-free images. Figures 7.15 and 7.16 compare the conventional thumbnail and the perceptual thumbnail from region-based noise estimation and multirate noise estimation. The two high-resolution images ($2272 \times 1704$ and $2048 \times 1536$ pixels) suffer from severe noise; however, such noise cannot be observed when viewing their conventional thumbnails. On the contrary, the two noise estimation methods are able to generate thumbnails that reflect noise in a similar way to the original high-resolution images.

In Figure 7.17, noise estimation is performed for a noise-free high-resolution image with $2048 \times 1536$ pixels. The perceptual thumbnail should be similar to the conventional thumbnail which looks clean. Comparing the results reveals that the thumbnail from region-based noise estimation is almost the same as the conventional thumbnail. The thumbnail from multirate noise estimation has details enhanced in texture-intensive regions. Here, both noise estimation methods used Equation 7.16 with the parameter $mag = 2$. A smaller value of this parameter can be used to reduce the level of detail enhancement.

Finally, Figure 7.18 shows an example that combines both blur highlighting and noise highlighting. The high-resolution image with $2048 \times 1536$ pixels suffers from image blur

**FIGURE 7.16**

Highlighting noise in the thumbnail: (a) original noisy image, (b) two magnified regions from the original image, (c) conventional thumbnail, (d) thumbnail obtained using region-based noise estimation, and (e) thumbnail obtained using multirate noise estimation.

and noise that take effect in the entire image. The thumbnails generated by using bilinear, bicubic, and Lanczos filtering methods shown in Figures 7.18c to Figures 7.18e appear rather clean and clear; by viewing these thumbnails the user cannot tell that the high-resolution image is noisy and blurry. This is not the case when the thumbnails are created by adding blur only according to the gradient-based and scale-space-based blur estimation methods. Comparing the resulted thumbnails with the conventional ones, image blur is reflected in the resulted thumbnails. The final thumbnails are generated by superimposing the estimated noise. As can be seen in Figures 7.18h and 7.18k, the perceptual thumbnails well reflect the blur and noise present in the high-resolution image.

## 7.7    Conclusion

This chapter discussed perceptual thumbnail generation, a practical problem available in digital photography, image browsing, image searching, and web applications. Differing from the conventional image thumbnails, the perceptual thumbnails allows inspecting the

(a)                                                                    (b)

(c)                                    (d)                                    (e)

**FIGURE 7.17**

Highlighting noise in the thumbnail: (a) original noise-free image, (b) two magnified regions from the original image, (c) conventional thumbnail, (d) thumbnail obtained using region-based noise estimation, and (e) thumbnail obtained using multirate noise estimation.

image quality at the small thumbnail resolution. Perceptual thumbnails serve in an intuitive way by displaying the perception-related visual cues in the thumbnail resolution.

Perceptual thumbnail generation is still a rather new problem in the field of computer vision and media computing. Existing methods mainly focus on two low-level visual cues, that is, blur and noise. Although blur and noise estimation and removal have been extensively studied, the goal of most techniques is to obtain the blur kernel and noise model and in turn to reconstruct the ideal image. Perceptual thumbnail generation does not have such strict requirements for blur and noise estimation. A rough and fast estimation is sufficient and necessary in this case. Existing techniques on perceptual thumbnail generation exploit how to assess the local spatial-varying blur strength and the noise in a fast way.

One potential future task for perceptual thumbnail generation is to highlight other visual cues, like the red-eye effect. There are many works on detecting and removal red-eye effects [36], [37]. However, visualizing red eyes in the thumbnail is not as straightforward as highlighting blur and noise. Superimposing the detected red eyes with the thumbnail may result in an unnatural preview. To highlight red eyes in the thumbnail, a possible solution is to mark its region in the thumbnail. For example, one may draw a red ellipse around the defected eyes. Image bloom is another phenomenon which is associated with similar problems, that is, how to effectively visualize it in thumbnails.

**FIGURE 7.18 (See color insert.)**

Highlighting blur and noise in the thumbnail: (a) original image with both blur and noise, (b) two magnified regions from the original image, (c) conventional thumbnails obtained using bilinear filtering, (d) conventional thumbnails obtained using bicubic filtering, (e) conventional thumbnails obtained using Lanczos filtering, (f) gradient-based estimation blur map, (g) perceptual thumbnail obtained using gradient-based blur estimation, (h) perceptual thumbnail obtained using gradient-based blur estimation and region-based noise estimation, (i) scale-space-based estimation blur map, (i) perceptual thumbnail obtained using scale-space-based blur estimation, and (k) perceptual thumbnail obtained using scale-space-based blur estimation and multirate noise estimation.

## Acknowledgment

This research is partially supported by project #MMT-p2-11 of the Shun Hing Institute of Advanced Engineering, The Chinese University of Hong Kong.

## References

[1] C. Burton, J. Johnston, and E. Sonenberg, "Case study: An empirical investigation of thumbnail image recognition," in *Proceedings of the IEEE Symposium on Information Visualization*, Atlanta, Georgia, USA, October 1995, pp. 115–121.

[2] K. Berkner, E. Schwartz, and C. Marle, "Smart nails: Display- and image-dependent thumbnails," in *Proceedings of the SPIE Conference on Document Recognition and Retrieval*, San Jose, CA, USA, December 2003, pp. 54–65.

[3] A. Woodruff, A. Faulring, R. Rosenholtz, J. Morrison, and P. Pirolli, "Using thumbnails to search the web," in *Proceedings of the ACM Conference on Human Factors in Computing Systems*, Seattle, WA, USA, March 2001, pp. 198–205.

[4] R. Samadani, S.H. Lim, and D. Tretter, "Representative image thumbnails for good browsing," in *Proceedings of the IEEE International Conference on Image Processing*, San Antonio, TX, USA, September 2007, vol. 2, pp. 193–196.

[5] L. Wan, W. Feng, Z.C. Lin, T.T. Wong, and Z.Q. Liu, "Perceptual image preview," *Multimedia Systems*, vol. 14, no. 4, pp. 195–204, September 2008.

[6] W. Burger and M.J. Burge, *Principles of Digital Image Processing: Core Algorithms*. London, UK: Springer, 2009.

[7] A. Muñoz, T. Blu, and M. Unser, "Least-squares image resizing using finite differences," *IEEE Transactions on Image Processing*, vol. 10, no. 9, pp. 1365–1378, September 2001.

[8] L.Q. Chen, X. Xie, X. Fan, W.Y. Ma, H.J. Zhang, and H.Q. Zhou, "A visual attention model for adapting images on small displays," *ACM Multimedia Systems Journal*, vol. 9, no. 4, pp. 353–364, October 2003.

[9] R. Samadani, T.A. Mauer, D.M. Berfanger, and J.H. Clark, "Image thumbnails that represent blur and noise," *IEEE Transactions on Image Processing*, vol. 19, no. 2, pp. 363–373, Feburary 2010.

[10] R.G. Keys, "Cubic convolution interpolation for digital image processing," *IEEE Transaction on Acoustics, Speech, and Signal Processing*, vol. 29, no. 6, pp. 1153–1160, December 1981.

[11] M. Unser, "Splines: A perfect fit for signal and image processing," *IEEE Transaction on Signal Processing*, vol. 16, no. 6, pp. 22–38, November 1999.

[12] E.H.W. Meijering, W.J. Niessen, and M.A. Viergever, "Piecewise polynomial kernels for image interpolation: A generalization of cubic convolution," in *Proceedings of the IEEE International Conference of Image Processing*, Kobe, Japan, October 1999, vol. 3, pp. 647–651.

[13] X. Li, *Edge Directed Statistical Inference with Applications to Image Processing*. PhD thesis, Princeton University, Princeton, NJ, USA, 2000.

[14] B. Suh, H. Ling, B.B. Bederson, and D.W. Jacobs, "Automatic thumbnail cropping and its effectiveness," in *Proceedings of the 16th Annual ACM Symposium on User Interface Software and Technology*, Vancouver, Canada, November 2003, pp. 95–104.

[15] S. Avidan and A. Shamir, "Seam carving for content-aware image resizing," *ACM Transaction on Graphics*, vol. 26, no. 3, pp. 10:1–9, July 2007.

[16] Y.S. Wang, C.L. Tai, O. Sorkine, and T.Y. Lee, "Optimized scale-and-stretch for image resizing," *ACM Transactions on Graphics*, vol. 27, no. 5, pp. 118:1–8, December 2008.

[17] P. Thévenaz, T. Blu, and M. Unser, *Image Interpolation and Resampling*. New York, USA: Academic Press, 2000.

[18] D.P. Mitchell and A.N. Netravali, "Reconstruction filters in computer graphics," *SIGGRAPH Computer Graphics*, vol. 22, no. 4, pp. 221–228, August 1988.

[19] N. Greene and P. Heckbert, "Creating raster omnimax images from multiple perspective views using the elliptical weighted average filter," *IEEE Computer Graphics and Applications*, vol. 6, no. 6, pp. 21–27, June 1986.

[20] P. Heckbert, *Fundamentals of Texture Mapping and Image Warping*. Master's thesis, University of California, Berkeley, CA, USA, 1989.

[21] J. McCormack, R. Perry, K.I. Farkas, and N.P. Jouppi, "Feline: Fast elliptical lines for anisotropic texture mapping," in *Proceedings of the 26th Annual Conference on Computer Graphics and Interactive Techniques*, Los Angeles, CA, USA, August 1999, pp. 243–250.

[22] X. Li and M. Orchard, "New edge-directed interpolation," *IEEE Transactions on Image Processing*, vol. 10, no. 10, pp. 1521–1527, October 2001.

[23] D. Muresan and T. Parks, "Adaptively quadratic (AQUA) image interpolation," *IEEE Transactions on Image Processing*, vol. 13, no. 5, pp. 690–698, May 2004.

[24] L. Zhang and X. Wu, "An edge-guided image interpolation algorithm via directional filtering and data fusion," *IEEE Transactions on Image Processing*, vol. 15, no. 8, pp. 2226–2238, August 2006.

[25] R. Fattal, "Image upsampling via imposed edge statistics," *ACM Transactions on Graphics*, vol. 26, no. 3, pp. 95:1–8, July 2007.

[26] V.R. Sctlur, *Optimizing Computer Imagery for More Effective Visual Communication*. PhD thesis, Northwestern University, Evanston, IL, USA, 2005.

[27] Y. Pritch, E. Kav-Venaki, and S. Peleg, "Shift-map image editing," in *Proceedings of the International Conference on Computer Vision*, Kyoto, Japan, September 2009, pp. 151–158.

[28] P. Kráhenbúhl, M. Lang, A. Hornung, and M. Gross, "A system for retargeting of streaming video," *ACM Transactions on Graphics*, vol. 28, no. 5, pp. 126:1–126:10, December 2009.

[29] M. Rubinstein, A. Shamir, and S. Avidan, "Multi-operator media retargeting," *ACM Transactions on Graphics*, vol. 28, no. 3, pp. 23:1–11, July 2009.

[30] M. Rubinstein, D. Gutierrez, O. Sorkine, and A. Shamir, "A comparative study of image retargeting," *ACM Transactions on Graphics*, vol. 29, no. 6, pp. 160:1–10, December 2010.

[31] H. Lam and P. Baudisch, "Summary thumbnails: Readable overviews for small screen web browsers," in *Proceedings of the SIGCHI Conference on Human Factors in Computing Systems*, Portland, OR, USA, April 2005, pp. 681–690.

[32] A. Cockburn, C. Gutwin, and J. Alexander, "Faster document navigation with space-filling thumbnails," in *Proceedings of the SIGCHI Conference on Human Factors in Computing Systems*, Montreal, QC, Canada, April 2006, pp. 1–10.

[33] S. Baluja, "Browsing on small screens: Recasting web-page segmentation into an efficient machine learning framework," in *Proceedings of the 15th International Conference on World Wide Web*, Edinburgh, Scotland, May 2006, pp. 33–42.

[34] A. Wrotniak, "Noise in digital cameras." Available online, http://www.wrotniak.net/photo/tech/noise.html, 2008.

[35] Y. Ke, X. Tang, and F. Jing, "The design of high-level features for photo quality assessment," in *Proceedings of the IEEE Conference on Computer Vision and Pattern Recognition*, New York, USA, June 2006, vol. 1, pp. 419–426.

[36] L. Zhang, Y. Sun, M. Li, and H. Zhang, "Automated red-eye detection and correction in digital photographs," in *Proceedings of the International Conference on Image Processing*, Singapore, October 2004, vol. 4, pp. 2363–2366.

[37] F. Volken, J. Terrier, and P. Vandewalle, "Automatic red-eye removal based on sclera and skin tone detection," in *Proceedings of the IS&T Third European Conference on Color in Graphics, Imaging and Vision*, Sydney, Australia, July 2006, pp. 359–364.

[38] N. El-Yamany, *Faithful Quality Representation of High-Resolution Imagess at Low Resolutions*. PhD thesis, Southern Methodist University, Dallas, TX, USA, 2010.

[39] M. Trentacoste, R. Mantiuk, and W. Heidrich, "Quality-preserving image downsizing," in *Proceedings of the 3rd ACM Conference and Ehibition on Computer Graphics and Interactive Techniques in Asia*, Seoul, Korea, December 2010, p. 74.

[40] D. Kundur and D. Hatzinakos, "Blind image deconvolution," *IEEE Signal Processing Magazine*, vol. 13, no. 3, pp. 43–64, May 1996.

[41] Q. Shan, J. Jia, and A. Agarwala, "High-quality motion deblurring from a single image," *ACM Transactions on Graphics*, vol. 27, no. 3, pp. 73:1–10, August 2008.

[42] N. Joshi, C.L. Zitnick, R. Szeliski, and D.J. Kriegman, "Image deblurring and denoising using color priors," in *Proceedings of the IEEE Conference on Computer Vision and Pattern Recognition*, Miami, FL, USA, June 2009, pp. 1550–1557.

[43] A.S. Wilsky, "Multiresolution Markov models for signal and image processing," *Proceedings of the IEEE*, vol. 90, no. 8, pp. 1396–1458, August 2002.

[44] M. Zhang and B.K. Gunturk, "Multiresolution bilateral filtering for image denoising," *IEEE Transactions on Image Processing*, vol. 17, no. 12, pp. 2324–2333, December 2008.

[45] P. Marziliano, F. Dufaux, S. Winkler, and T. Ebrahimi, "A no-reference perceptual blur metric," in *Proceedings of the International Conference on Image Processing*, Lausanne, Switzerland, December 2002, vol. 3, pp. 57–60.

[46] J. Caviedes and F. Oberti, "A new sharpness metric based on local kurtosis, edge and energy information," *Signal Processing: Image Communication*, vol. 19, no. 2, pp. 147–161, February 2004.

[47] R. Ferzli and L. Karam, "No-reference objective wavelet based noise immune image sharpness metric," in *Proceedings of the IEEE International Conference on Image Processing*, Genoa, Italy, September 2005, vol. 1, pp. 405–408.

[48] M.C. Chiang and T.E. Boult, "Local blur estimation and super-resolution," in *Proceedings of the Conference on Computer Vision and Pattern Recognition*, June 1997, pp. 821–826.

[49] J.H. Elder and S.W. Zucker, "Local scale control for edge detection and blur estimation," *IEEE Transaction on Pattern Analysis and Machine Intelligence*, vol. 20, no. 7, pp. 699–716, July 1998.

[50] S. Bae and F. Durand, "Defocus magnification," *Computer Graphics Forum*, vol. 26, no. 3, pp. 571–579, September 2007.

[51] D.L. Donoho, "De-noising by soft-thresholding," *IEEE Transactions on Information Theory*, vol. 41, no. 3, pp. 613–627, May 1995.

[52] P. Vaidyanathan, *Multirate Systems and Filter Banks*. Upper Saddle River, NJ, USA: Prentice Hall, 1993.

[53] S.H. Lim, "Characterization of noise in digital photographs for image processing," *Proceedings of SPIE*, vol. 6069, pp. 219–228, April 2006.

# 8

## Patch-Based Image Processing: From Dictionary Learning to Structural Clustering

Xin Li

## 8.1   Historical Background and Overview

Where does patch come from? The earliest appearance of *patch* as a technical term in the open literature can be found in Reference [1], which focuses on textural features for image classification. In those old days, pixels are termed as resolution cells and small-area patches typically refer to a window or patch of $3 \times 3$ pixels. Later, Reference [2] introduced an image enhancement approach based on local statistics, such as mean and variance, calculated on a patch-by-patch basis in order to drive the filter to remove additive or multiplicative noise. Computing optical flow [3] or disparity field [4] from a pair of images was also based on the assumption that the gradient of image intensity field can be locally calculated within a patch, where patch is often interchanged with the term *window* or *block*. Fast advances of communication and computing technologies stimulated the research into image compression, especially the development of Joint Photographic Experts Group (JPEG) coding standard [5]. Under the context of JPEG compression, an $8 \times 8$ discrete cosine transform (DCT) [6] became the favorite but also contributed to the notorious block artifacts; from this perspective, decomposing an image into non-overlapping patches or blocks represents a simple yet effective way of overcoming the curse of dimensionality [7].

The establishment of wavelet theory [8], also known as filter-bank theory in the field of electrical engineering [9] and multi-resolution analysis in the field of computer science [10], rapidly shattered the hope of lapped or overlapped-block transforms [11]. This new wavelet theory, connected with the intuitive Laplacian pyramid [12] but with a lot more mathematical rigor, has found numerous applications from image coding [13] to seismic signal analysis [14]. The sparsity or heavy-tailed distribution of wavelet coefficients has become the signature of *transient events* — thanks to the good localization property of wavelet bases in space and frequency, singularities (such as edges, lines, and corners) in natural images only produce a small number of exceptions in the wavelet space. Mathematically, Besov-space functions [15] and statistical modeling of wavelet coefficients [16], [17] became the new frontiers of image modeling. The concept of patch or window was quickly absorbed into this new context; for instance, it is often convenient to estimate the local variance of wavelet coefficients within a local patch or window [18], [19], [20].

It was the vision community who discovered that wavelet-based image models, when applied to the task of texture synthesis, might not be as splendid as what they have delivered for lossy image compression. The breakthrough on texture synthesis was presented in Reference [21], where nonparametric sampling techniques convincingly demonstrated the capability of synthesizing a large class of textures and arguably outperformed competing parametric models in the wavelet space [22]. Since then, patch-based image models, powered by the magic of nonparametric sampling, have shown increasingly more vitality. This resulted in many applications, such as *image inpainting* with exemplar-based [23], [24], fragment-based [25], and deterministic annealing based [26] approaches, *image denoising* with nonlocal-mean [27], patch-based [28], and block-matching three-dimensional (BM3D) [29] techniques, *super-resolution imaging* with exemplar-based [30], epitome-based [31], sparsity-based [32] approaches, and *object recognition* with scale-invariant feature transform (SIFT) [33], local binary pattern (LBP) [34], and speeded-up robust features (SURF) [35] techniques. Most recently, the unification of patch models and sparse representation [36], [37] finally gave the birth to a new meaning of patch-based image models, commonly referred to as nonlocal sparsity.

This chapter reviews the basic principles of underlying patch-based image models, demonstrates their potential in engineering applications, and explores their possible scientific connections with human vision models. Section 8.2 defines two classes — descriptive (analysis) and generative (synthesis) — of patch-based image models and discusses their relationship. Section 8.3 presents first-generation patch-based image models and emphasizes their root in the Hilbert space. Section 8.4 introduces second-generation patch-based image models and articulates their connection with set theoretic estimation techniques [38]. The key for transition from first-generation to second-generation models will be to think outside of Hilbert space or to *"think globally, fit locally"* [39]. Section 8.5 presents extensive and reproducible experimental results to demonstrate the superiority of second-generation patch models in image denoising, compression artifact removal, and inverse halftoning. Section 8.6 attempts to make some overarching connections with the known theories of contour integration [40] and redundancy exploitation [41] in neuroscience. Under the hypothesis that sparsity has been adopted by the human visual system as an efficient coding strategy for adapting to the statistics of natural stimuli in the physical world, the ultimate unsettled question appears to be: *where does the sparsity come from?*

## 8.2    Patch-Based Image Models: Descriptive versus Generative

Why are image models or representations needed? As argued in Reference [42], one important reason is that they provide an abstraction of the large amount of data contained in an image. Image models enable to systematically develop algorithms for accomplishing a particular image-related task. The nature of the specific task (for example, recognition versus compression) suggests which of the two classes of image models – *descriptive* and *generative*, often associated with the tasks of image analysis and synthesis, respectively – should be used. Descriptive models focus on the extraction of distinctive features from a given image such that they can facilitate the task of classifying the image into one of several categories [43]. Whether a plausible image can be reconstructed from those features is irrelevant to descriptive models. By contrast, generative models preserve the image information and their synthesis capability lends them more desirable for the task of compression and restoration than that of classification and recognition. However, recent trend in sparsity-based recognition [44] has challenged the above common perception about generative models – they could achieve highly competent performance in recognition tasks even though the capability of synthesis is not necessary.

Common to both descriptive and generative models is a challenge called the curse of dimensionality. Since an image is usually decomposed of $10^4$ to $10^8$ pixels, developing models directly in that high-dimensional space is often practically infeasible. This is where the concept of *patch* becomes useful; instead of working with the whole image, one can focus his attention to a small area of so-called image patches. Why is this a good idea? For the class of descriptive models, local patches often experience much less distortion than global images and therefore it becomes easier to define the similarity between two local patches. Such an idea is essential to the success of a class of local keypoint descriptors, such as SIFT [33], local binary pattern (LBP) [34], SURF [35] and affine SIFT (ASIFT) [45]. Despite the outstanding performance and increasing influence of those local descriptors, the interest here is focused on the class of patch-based generative models.[1]

For the class of generative models, the discussion should articulate more about the overlapping among image patched. In a complete or non-redundant representations, such as block-based DCT coding [5], no overlapping between adjacent patches is allowed – a naive argument often goes like that patch overlapping introduces redundancy and therefore is bad for the task of lossy image compression. Such claim is partially true or can only be justified from an encoder perspective where satisfying the bit budget constraint is eminent. It will be shown in the remainder of this chapter that overcomplete or redundant representations, not just allowing but promoting patch overlapping, are highly desirable not only in various engineering applications related to image restoration but also in scientific investigation of how to mathematically model the human visual system. As shown in Section 8.6, there exists physiological evidence supporting the significant overlap among local receptive fields of adjacent neurons [47], which clearly indicates that redundancy has been exploited by the

---

[1]It will be argued at the end of this chapter that under the context of bio-inspired computing, local descriptive models have to be extended to admit global synthesis in order to reflect the organizational principles underlying neural systems [46].

law of evolution in nature [41]. As discussed in Section 8.3, first-generation patch-based models basically echo the message of taking a local view toward a complex phenomenon, which has been at the foundation of classical physics [48]. As the complexity of a natural system increases (for instance, from physical to chemical to biological) [49], the fundamental assumption with locality principle is likely to fall apart, which motivates discussion of second-generation patch models in Section 8.4.

## 8.3 First-Generation Patch Models: Dictionary Construction and Learning

This section reviews first-generation patch models, which are designed with the local view and under the influential paradigm of Hilbert space. This design principle can be considered to historically follow two competing yet complementary paths: the *construction* of a signal-independent dictionary (or basis functions) and the *learning* of a dictionary from training data (and therefore signal dependent). Both lines of attacks have turned out fruitful and it has often been found that two approaches end up with dictionaries of similar characteristics and algorithms of comparable performance.

Mathematically, it is a standard practice of decomposing a given signal $f$ under a collection of basis functions $\{\phi_1, ..., \phi_n\}$ as $f = \sum_{i=1}^{N} a_i \phi_i$, where $a_i = < f, \phi_i >$ denotes the inner product between the signal of interest and basis functions. The challenge here often surrounds the choice of dictionary or basis functions; for example, what dictionary could give the most parsimonious representation of $f$? For stationary Gaussian process, the optimal choice of dictionary has been long known as Karhunen-Loève transform (KLT) [50], also known as principal component analysis [51]. However, since the basis functions of KLT are calculated from the covariance matrix, they are signal dependent (therefore consume extra bandwidth which is undesirable for image compression) and computationally demanding (therefore limit its potential in practical applications). Consequently, alternative approaches toward dictionary construction have been proposed in the literature; the story of discrete-cosine-transform (DCT) [52] is summarized below.

The motivation behind the definition of DCT is that it provides a good approximation to the optimal KLT for a class of so-called AR(1) process [53]. It has been experimentally shown [52] that DCT achieves better energy compaction property than Fourier or Walsh-Hadamard transform and almost identical performance to KLT for an AR(1) source with correlation coefficient of $\rho = 0.9$. Additionally, the discovery of fast algorithms for computing DCT (for example, Reference [6]) greatly facilitates the adoption of DCT by JPEG [5], which is the first lossy image compression standard, and several video coding standards, including Moving Picture Experts Group (MPEG) [54] and H.263 [55] standards. However, since each block or image patch is processed independently from others, both JPEG and MPEG produce notorious block artifacts at the low bit rates. Even in the new image compression standard JPEG2000 [13], the concept of tile is no different from that of block except that the choice of dictionary varies – from DCT to discrete-wavelet transform (DWT). Therefore, patch-based models were reinvented in JPEG2000 with different motivation of supporting object-based or region-of-interest coding [56].

Another line of research on patch-based models deals with learning signal-dependent basis functions from training data. The discovery of independent component analysis (ICA) algorithms [57], [58], [59] supplied different technical communities a cool and new tool that rapidly found it to be applied to a variety of signal processing, computer vision, and machine learning tasks. It is worth noting that the independent components of natural images were found to be edge-like filters [60], which resemble the receptive fields of simple cells in visual cortex [61]. However, ICA has failed to deliver is a precise characterization of statistical *independence*; that is, it is often difficult to justify the independency assumption except few idealistic scenarios, such as blind source separation, where source mixing admits a linear model [62]. By contrast, recent advances in dictionary learning, such as K-SVD [63], [64], [65], have shown that a learned dictionary can support both reconstruction and recognition tasks.

Regardless of construction or learning, an open question that has not been sufficiently addressed in the open literature is the size of dictionary $N$. In engineering applications such as data compression, $N$ is often chosen to be the same as the size of the original image because that is when the system is neither underdetermined not overdetermined. However, it has been suggested that a larger $N$ or an overcomplete representation might achieve higher sparsity. Depending on how the sparsity of transform-coefficients $a$ is measured, one could face convex (for example, $l_1$ or $l_2$-norm) and nonconvex (for example, $l_0$-norm) optimization problems in determining $a_i$'s for a fixed dictionary. This problem has been known as basis pursuit [66], [67], and there has been growing interest in the development of new pursuit algorithms in recent years (for example, K-SVD [63], [68]). Nevertheless, the question of choosing dictionary size $N$ remains open and the difficulty with answering this question partially motivates exploring second-generation patch-based models, as elaborated below.

## 8.4 Second-Generation Patch Models

### 8.4.1 From Local Transience to Nonlocal Invariance

The difficulty of choosing a proper dictionary size has to be appreciated from a point of view that raises the skepticism at a more fundamental level, that is, the Hilbert-space formulation or the locality assumption. To facilitate the illustration, Figure 8.1 depicts a toy example of two-dimensional spiral data. Namely, Figure 8.1a shows the learned basis vectors (highlighted by "+") by K-SVD algorithm [63] as the dictionary size increases from $M = 5$ to $M = 40$. It is not difficult to verify that the sparsity (as measured by the percentage of nonzero coefficients) does improve along with the dictionary size; but what remains troublesome is that the topology of learned dictionary has little to do with the actual probability density function (pdf) of the spiral data. However, by switching the tool, in this case from K-SVD to an expectation maximization (EM)-based learning of finite mixture model [69], the approximation of spiral pdf by a collection of Gaussian mixture models does appear plausible, as shown in Figure 8.1b.

The above example suggests that Hilbert-space functions are not the only possible formulation to model signals. The introduction of artificial structures, such as inner product

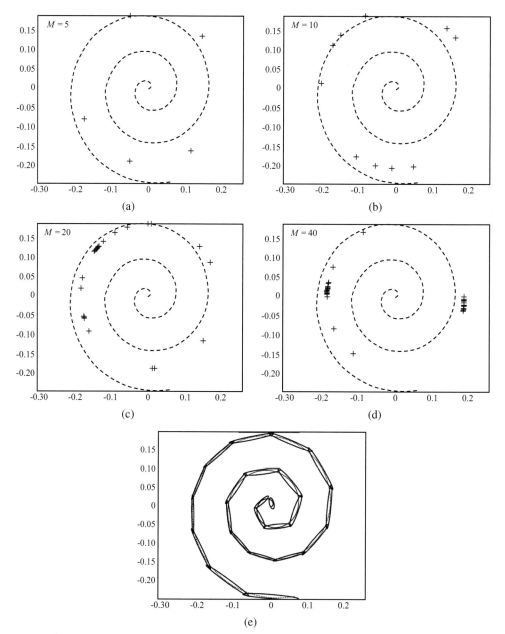

**FIGURE 8.1**

Two ways of discovering the manifold structure underlying a signal: (a–d) dictionary learning [63] with atoms in dictionary **A** highlighted using the '+' symbol, (e) structural clustering [69] with each segment denoting a different cluster $p(\mathbf{x}|\theta_{\mathbf{m}})$.

and basis function, has been better justified, instead of taking them for granted. In fact, it is well-known from the mathematical literature [70] that Hilbert space is at the bottom of the hierarchy of mathematical spaces. On top of it, there exist several other choices, such as metric space and topological space, which contain a larger set of functions than what Hilbert space does. Moreover, locality assumption is likely to be only a part of the story because the definition of local neighborhood is conditioned on the way to represent a signal. Adjacent points in the patch space might correspond to two image patches spatially distant from each other; for example, consider self-repeating texture patterns. This summary motivates the following study of second-generation patch models.

Instead of committing to Hilbert-space formulation, it is proposed here to consider pdf of image patches as a low-dimensional manifold embedded in the patch space and ask: *what is a good tool for studying manifold*? The idea of analyzing the topology of a manifold by studying differential functions on that manifold was first conceived in Reference [71]. Along this line of reasoning, one can see that an alternative approach toward modeling complex signals is to model their associated dynamical systems or more precisely a *nonlinear* map that operates on the signal manifold. Note that the nonlinearity is the key for departing from the standard practice of linear system analysis. More important, the focus here will be put on the nonlinearity arising from the bilateral representation of image signals [72], that is, two patches could become neighbors if they are close either in the spatial domain or in the range space. The following nonlinear filter (NLF) in the continuous space has been found general enough to incorporate both local and nonlocal filters in the literature:

$$P_{NLF}(f(x,y)) = \frac{\int_\Omega w(x,y;x',y')f(x',y')dx'dy'}{\int_\Omega w(x,y;x',y')dx'dy'}, \tag{8.1}$$

where $w(x,y;w',y')$ is a generic weighting function related to the structural similarity between two patches at $(x,y)$ and $(x',y')$.

It should be noted that despite the linear weighting form at the surface, the weighting coefficients are *recursively* defined with respect to the patch similarity and therefore intrinsically it is nonlinear. In fact, there have only been primarily two types of nonlinear operations in the open literature: *memoryless* (related to nonlinear point-operations such as thresholding [73] and sigmoid function [74]) and *with memory* (related to hysteresis [75] and clustering [76], [77], [78]). The nonlinearity of Equation 8.1 could arise from either hard thresholding or the soft definition (connected with clustering) of weighting coefficients. A more tantalizing question is: how can this new perspective of nonlinear map be useful to processing or modeling image signals? It will be shown in this section that it does have practical applications in set theoretic image restoration, whereas extensive experimental results will be reported in the next section.

Set theoretic image estimation [38] or alternating projection-based image restoration [79] has been extensively studied by the image processing community. What novel element do the second-generation patch-based models bring about? It is argued here that the *a priori* knowledge about the image source emphasizes the *global invariant* property (in contrast to the local transient property characterized by first-generation patch models). Such global view is particularly effective for the class of image structures, including regular edges and textures, because they are often approximately translation-invariant. The translation invariance of edges and textures is essential to the justification of quantifying weight coefficients by structural similarity in Equation 8.1; a heuristic idea that has been widely adopted by various filtering techniques from bilateral filter [72] to nonlocal-mean denoising [27].

It has been shown [80] that nonlinear filter based projection operator $P_{NLF}$ as defined by Equation 8.1 is a non-expansive mapping [81]. Such result generalizes previous one for first-generation patch models [82], which makes it convenient to develop a new class of image restoration algorithms with nonlocal regularization techniques. In traditional paradigm of image estimation via alternating projections [83], [84], the target of interest is pursued as the intersection of two constraint sets: observation constraint set specified by the degradation model (for example, blurring or quantization) and regularization constraint set carrying out *a priori* knowledge. By replacing local regularization constraint set by the nonlocal one, a new class of projection-based image restoration algorithms can be obtained.

> **ALGORITHM 8.1** Projection-based image restoration with nonlocal regularization. Input is the degraded image $\mathbf{g}$, output is the estimated image $\hat{\mathbf{f}}$.
>
> 1. Initialization: Set $\hat{\mathbf{f}}^{(0)}$ to be an initial estimate.
> 2. Main loop:
>    - Projection onto the observation constraint set via $\mathbf{f}^{k+1/2} = P_{obs}\mathbf{f}^k$, where $P_{obs}$ is dependent on a specific degradation model.
>    - Projection onto the prior constraint set via $\mathbf{f}^{k+1} = P_{NLF}\mathbf{f}^{k+1/2}$, where $P_{NLF}$ is given by Equation 8.1.

Inspired by the success of BM3D denoising [29], the following experiment is based on a specific implementation of $P_{NLF}$. The basic idea is to select a set of exemplars from the collection of patches (for example, those at a fixed sampling lattice) and perform $k$ nearest-neighbor ($k$NN) search for each exemplar. Then, a transform-domain thresholding strategy is applied to the three-dimensional data array formed by each two-dimensional exemplar and its $k$ nearest neighbors. The non-expansive property of such BM3D-thresholding based $P_{NLF}$ has been rigorously shown in Reference [85], which theoretically justifies the good numerical behavior of Algorithm 8.1. However, it is noted that the regularization constraint set arising from nonlocal similarity is non-convex because the cost functional of data clustering (for example, $k$NN and $k$-means) is seldom convex. The nonconvexity of sparsity optimization is the other significant departure of second-generation patch models from their first-generation counterpart.

### 8.4.2 From Convex to Nonconvex Sparsity Optimization

Though less popular, the problem of nonconvex minimization has been extensively studied. The most well-known optimization technique for nonconvex problems is likely to be simulated annealing [86] inspired by the analogy between combinatorial optimization and statistical mechanics. It was argued in Reference [86] that "*as the size of optimization problems increases, the worst-case analysis of a problem will become increasingly irrelevant, and the average performance of algorithms will dominate the analysis of practical applications.*" Such observation has motivated the introduction of an artificial temperature parameter to simulate the annealing process in nature. Simulated annealing was often implemented by Metropolis algorithm based on local random perturbation; more powerful Monte Carlo nonlocal/clustering methods [87], [88] have been developed afterward. An important insight behind clustering algorithms is that collective updating could eliminate the problem of "*critical slowing down*" [89].

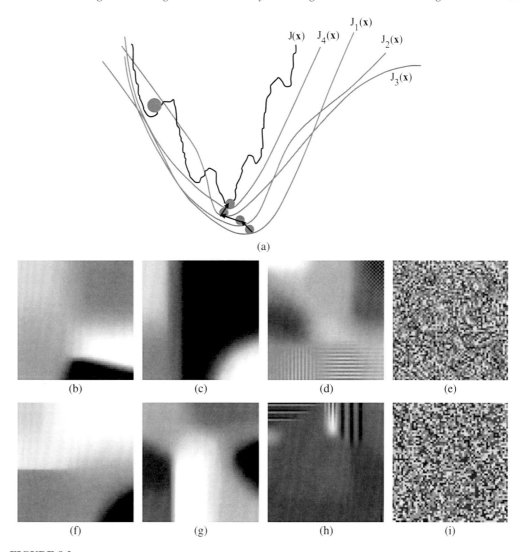

**FIGURE 8.2**

Illustration of deterministic annealing. (a) The minimum of a nonconvex function $J$ may be found by a series of convex approximations $J_i$. (b–i) Sample stable states (local minima) corresponding to varying temperatures.

For image restoration applications, one can argue that it is also the *average* performance that matters because any restoration algorithm is supposed to work for a class of images instead of an individual one. In fact, it is often difficult to come up with the worst-case example in the first place (that is, *what kind of image is the hardest to recover?*). What seems an unknown connection between nonlocal Monte Carlo methods and nonlocal sparse representations is the role played by *clustering*. More specifically, the observation constraint appears to suggest that it is not necessary to do annealing in a stochastic but deterministic fashion because observation data **g** offers a valuable hint for clustering. Indeed, deterministic annealing (DA), often known as graduated nonconvexity (GNC) in the vision literature [90], has been proposed as a statistical physics inspired approach toward data clustering [91], [92].

The basic idea behind DA is simple and appealing; one can modify the nonconvex cost function $J(\mathbf{f})$ in such a way that the global optimum can be approximated through a sequence of convex cost functions $J_p^*(\mathbf{f})$ (the auxiliary variable $p$ parameterizes the annealing process), as shown in Figure 8.2a. When the iteration starts from a small parameter favoring a smooth cost function (that is, high temperature), it is relatively easy to find a favorable local optimum. As the cost function becomes more jagged after several iterations, the local optimum will be gradually driven toward the global minimum as the temperature decreases [90]. In addition to the above standard interpretation, a twist can be added from the saddle-point thinking here; functions $J_p^*(\mathbf{f})$ do not need to be convex *everywhere* but only locally around the point of interest. In other words, the strategy of convex approximation can be made *data dependent*; therefore, as the point of interest $\mathbf{f}$ moves to $\mathbf{f'}$, a different sequence of locally convex cost functions would be in action.

To illustrate how DA works, the following Monte Carlo experiment has been designed. Starting with random noise realization, iterative filtering is applied via $\mathbf{f}^{n+1} = P_\theta \mathbf{f}^n$, where $P_\theta$ denotes a nonlocal filter with a temperature parameter $\theta$ (the threshold $T$ in nonlinear shrinkage or the Lagrangian multiplier $\lambda$ in nonlocal regularization [93]). As long as the $P_\theta$ is non-expansive [80], it can be observed that $lim_{n\to\infty} \mathbf{f}^n$ would converge to a fixed point; without any observation constraint, such fixed point could wander in the phase space depending on the order parameter $\theta$. By varying the $\theta$ value, varying structures in physical systems (for example, crystal versus glass) can be observed; analogously in image restoration, the constellations of an image manifold is discovered: *smooth regions*, *regular edges*, *and textures* as shown in Figure 8.2b, where the hard thresholding stage of BM3D denoising [29] is used as $P_T$. It is also worth to note that the phenomenon of *phase transition* (as the temperature varies) has not been observed for conventional local image models, such as Markov random field (MRF) [94]. In other words, clustering has played a subtle role in connecting nonlocal sparsity with nonconvex optimization; even though such connection has not been fully understood yet (some preliminary results can be found in Reference [85]).

---

## 8.5 Experimental Results

This section presents reproducible experimental results to demonstrate the superiority of second-generation patch models. Due to the space limitation, only experiments related to image denoising, compression artifacts removal, and inverse halftoning are included. The source codes and various simulations, including deblurring and super-resolution, are available online.[2] Additional results can be found in References [37], [80], [95], and [96].

### 8.5.1 Image Denoising

The simplest restoration task is likely to be the removal of additive white Gaussian noise from a degraded image $\mathbf{y} = \mathbf{x} + \mathbf{w}$, where $w \sim N(0, \sigma_w^2)$. Such problem is often called image denoising in the literature and has been extensively studied in the past decades. The

---

[2]http://www.csee.wvu.edu/~xinl/demo/nonlocal.html

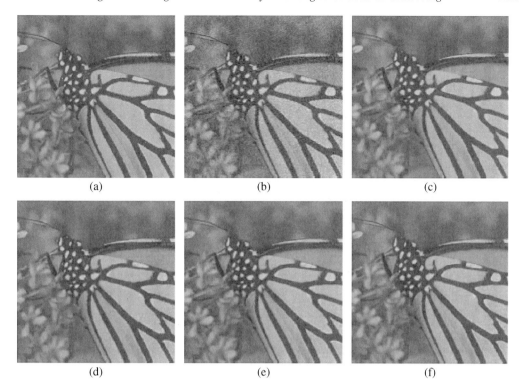

**FIGURE 8.3**

Denoising performance comparison for *Monarch* image: (a) original image, (b) noisy image with $\sigma_w = 20$, (c) SA-DCT output [97] with $PSNR = 30.06$ dB and $SSIM = 0.9125$, (d) K-SVD output [68] with $PSNR = 30.09$ dB and $SSIM = 0.9102$, (e) BM3D output [29] with $PSNR = 30.45$ dB and $SSIM = 0.9194$, and (f) CSR output [37] with $PSNR = 30.57$ dB and $SSIM = 0.9190$.

following experiment compares four leading image denoising algorithms with reproducible source codes; these algorithm use concepts of shape-adaptive DCT (SA-DCT)[3] [97], K-SVD[4] [68], BM3D[5] [29], and clustering-based sparse representation (CSR)[6] [37]. More specifically, SA-DCT is among the very best techniques built upon local sparsity models. K-SVD and BM3D represent the approaches of obtaining nonlocal sparse representations via dictionary learning and structural clustering, respectively. Finally, CSR combines both dictionary learning and structural clustering.

The two test images used in the experiment are *Monarch* (edge-class) and *Straw* (texture-class); the two popular objective quality metrics employed here are the peak signal-to-noise ratio (PSNR) and the structural similarity (SSIM) index [98]. Figures 8.3 and 8.4 allow the comparison of of denoised images by different algorithms, both subjectively and objectively. It can be observed that nonlocal methods noticeably outperform the local one and the gain becomes more significant for the class of texture structures. This is not surprising because nonlocal similarity is stronger in texture images. Another observation is that

---

[3]http://www.cs.tut.fi/~foi/SA-DCT/
[4]http://www.cs.technion.ac.il/~ronrubin/software.html
[5]http://www.cs.tut.fi/~foi/GCF-BM3D/
[6]http://www.csee.wvu.edu/~xinl/CSR.html

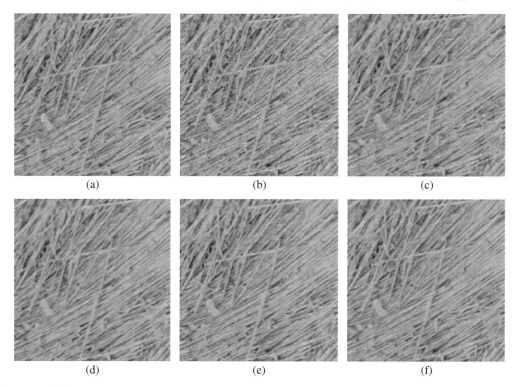

**FIGURE 8.4**

Denoising performance comparison for *Straw* image: (a) original image, (b) noisy image with $\sigma_w = 20$, (c) SA-DCT output [97] with $PSNR = 26.22$ dB and $SSIM = 0.8760$, (d) K-SVD output [68] with $PSNR = 26.91$ dB and $SSIM = 0.8894$, (e) BM3D output [29] with $PSNR = 27.09$ dB and $SSIM = 0.8968$, and (f) CSR output [37] with $PSNR = 27.35$ dB and $SSIM = 0.9003$.

both the ideas of dictionary learning and structural clustering can lead to improved denoising performance and their gain is complementary. Combining those two ideas often lead to significant performance improvements for the class of texture images; from a manifold point of view, dictionary learning and structural clustering can be interpreted as adapting to the local geometry and global topology, respectively. More detailed comparison and analysis related to image denoising can be found in References [29], [68], and [97]. Recent advances in this field are presented in References [36], [99], [100], and [101].

It is interesting to note the potential benefit of exploiting more general geometric invariance than translation-invariance. For instance, if it is known that an image is rotation-invariant, one can construct a dictionary $\mathscr{D}$ consisting of patches generated from not only translation but also rotations and reflections (conceptually they can be viewed as a kind of bootstrap resampling techniques [102]). It has been found for the class of images satisfying more strict global invariant properties, the improvements over BM3D achieved by bootstrapping are impressive both subjectively and objectively (see Figure 8.5). Visual quality improvements primarily come from better recovery of global symmetric structures. Note that the objective SSIM metric does not always correlate with subjective assessment (SSIM is based on local comparisons only). In other words, the measurement of global symmetry in an image is still beyond the reach of existing image quality assessment techniques.

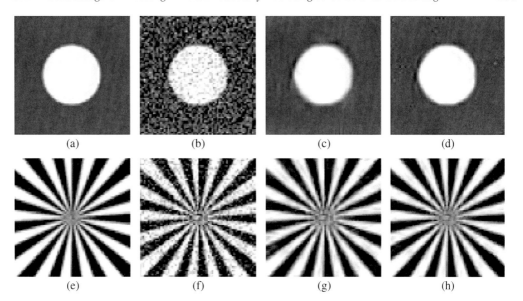

**FIGURE 8.5**

Performance comparison of BM3D and exemplar-based EM-like (EBEM) denoising methods: (a) original *Disk* image, (b) noisy image with $\sigma_w = 50$, (c) BM3D output with $PSNR = 29.36$ dB and $SSIM = 0.9243$, and (d) EBEM output [100] with $PSNR = 31.38$ dB and $SSIM = 0.9214$; (e) original *Rays* image, (f) noisy image with $\sigma_w = 50$, (g) BM3D output with $PSNR = 20.99$ dB and $SSIM = 0.9497$, and (h) EBEM output [100] with $PSNR = 23.05$ dB and $SSIM = 0.9670$.

### 8.5.2 Compression Artifact Removal

Lossy image coding is probably the most successful application of wavelets as of today. From the embedded zerotree wavelet (EZW) method [16] to the new JPEG2000 standard [13], nonlinear approximation of Besov-space functions has well served as the models for photographic images thanks to the good localization property of wavelet bases in space and frequency. However, from a set theoretic perspective toward lossy image coding [103], the original image belongs to a convex set characterized by the quantization cell, that is, $\mathbf{g} = Q[\mathbf{Tf}]$ where $\mathbf{T}$ denotes the transform and $Q$ is a uniform quantizer. Outputting the center of quantization cell corresponds to a maximum-likelihood (ML) decoding strategy whose optimality becomes questionable at low bit-rates (for example, the notorious ringing artifacts around sharp edges produced by wavelet coders). An improved decoding strategy is to pursue maximum *a posteriori* (MAP) estimation of $\mathbf{f}$ based on the observation data $\mathbf{g}$ and the a priori knowledge about $\mathbf{f}$. The nonlocal sparsity regularization conveyed by second-generation patch models distinguish them from previous attacks based on first-generation patch models (see, for example, References [84], [104], and [105]).

The above idea has been tested using the set partitioning in hierarchical trees (SPIHT)[7] algorithm [106], which is a widely used benchmark codec. For a uniform quantizer with stepsize of $\Delta$, the observation constraint set is defined as $C_{obs} = \{f(m,n)|y(m,n) - \Delta/2 \leq x(m,n) < y(m,n) + \Delta/2\}$. The corresponding projection operator is defined as $P_{obs}x = y +$

---

[7] http://www.cipr.rpi.edu/research/SPIHT/spiht3.html

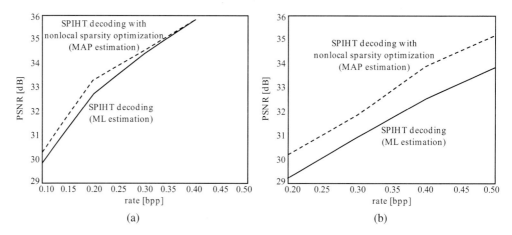

**FIGURE 8.6**

PSNR performance improvement of MAP decoding for images (a) *Lena* and (b) *Barbara*.

$\Delta/2$ for $x > y + \Delta/2$, $P_{obs}x = y$ for $x \in [y - \Delta/2, y + \Delta/2]$, and $P_{obs}x = y - \Delta/2$ for $x < y - \Delta/2$. In Reference [107], it has been opted to implement the projection onto prior constraint set $P_{prior}$ by a BM3D-based nonlocal filter with threshold $\delta$ (it plays the role of Lagrangian multiplier). A deterministic annealing protocol similar to Reference [108] has been adopted as $\delta_n = \delta_0 - 0.02n, n = 1 - 10$, where $\delta_0$ is the initial threshold set proportional to the average of block-wise variations within **y**. Figure 8.6 shows the PSNR gain achieved over SPIHT [106] by MAP decoding for two typical $512 \times 512$ images *Lena* and *Barbara*. It is not surprising that the gain is much more dramatic for *Barbara* image because it contains significant portions of texture patterns. Visual inspection of decoded images in Figures 8.7

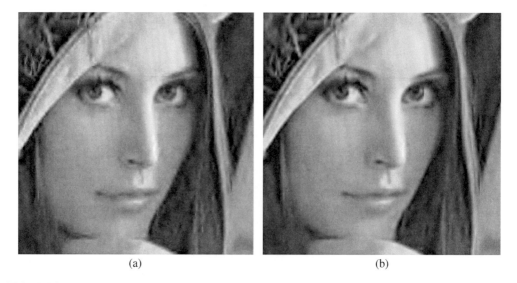

**FIGURE 8.7**

Performance comparison of SPIHT decoding for *Lena* image at the bit rate of 0.20 bpp: (a) ML estimation with $PSNR = 32.70$ dB and $SSIM = 0.8669$, (b) MAP estimation with $PSNR = 33.29$ dB and $SSIM = 0.8784$.

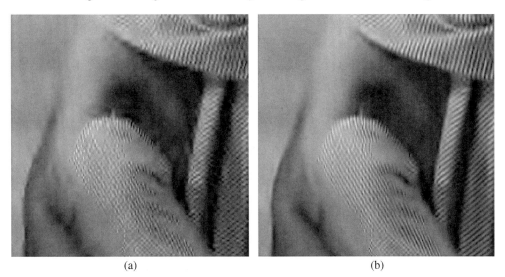

(a)  (b)

**FIGURE 8.8**

Performance comparison of SPIHT decoding for *Barbara* image at the bit rate of 0.20 bpp: (a) ML estimation with $PSNR = 26.19$ dB and $SSIM = 0.7572$, (b) MAP estimation with $PSNR = 27.33$ dB and $SSIM = 0.7942$.

and 8.8 confirms that texture regions are much better reconstructed by MAP estimation than ML scheme. For *Lena* image, the gain is more observable for the low bit rates than high bit rates. Such finding can be explained by the prioritization of bits in scalable coder, such as SPIHT. The first batch of bits typically contain important structural information at the global level whose self-similarity is justified. As bit rate increases, more bits are assigned to less regular textural information locally (for example, hair region); therefore, the gain of exploiting the nonlocal sparsity diminishes.

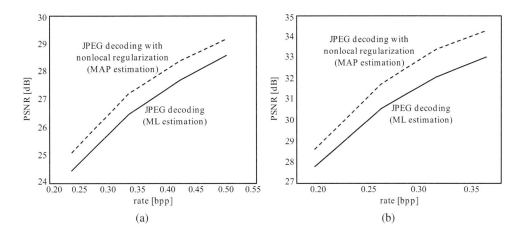

(a)  (b)

**FIGURE 8.9**

PSNR performance improvement of JPEG system due to graduated nonconvexity-based nonlocal regularization for images (a) *Cameraman* and (b) *House*.

(a)                                                        (b)

**FIGURE 8.10**

Performance comparison of JPEG decoding for *Cameraman* image at the bit rate of 0.42 bpp: (a) without nonlocal regularization with $PSNR = 27.71$ dB and $SSIM = 0.8355$, (b) with nonlocal regularization with $PSNR = 28.39$ dB and $SSIM = 0.8570$.

(a)                                                        (b)

**FIGURE 8.11**

Performance comparison of JPEG decoding for *House* image at the bit rate of 0.32 bpp: (a) without nonlocal regularization with $PSNR = 32.07$ dB and $SSIM = 0.8452$, (b) with nonlocal regularization with $PSNR = 33.22$ dB and $SSIM = 0.8636$.

The above discussion has also justified the merit of nonlocal sparsity regularization on deblocking JPEG-coded images. It will be more convincing if the reconstructed images can be compared on the same test image with previous deblocking techniques such as those in References [84], [104], and [105]. Unfortunately, such comparison is difficult due to the lack of source codes or executables of previous works. For the $512 \times 512$ *Lena* image and

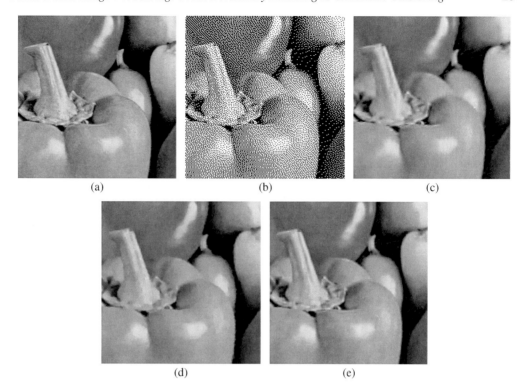

(a)               (b)               (c)

(d)                     (e)

**FIGURE 8.12**

Inverse halftoning comparison: (a) $200 \times 200$ portion of original *Peppers* image, (b) halftoned image after Floyd-Steinberg diffusion, (c) reconstructed image by wavelet-based inverse halftoning [116] with $PSNR = 31.03$ dB, (d) reconstructed image by TV restoration [112] with $PSNR = 30.92$ dB, and (e) reconstructed image by nonlocal regularization with $PSNR = 32.64$ dB.

at the bit rate of 0.24 bpp, nonlocal regularization can easily achieve the gain of 1.3 dB without parameter optimization, while both projection-based [84] and wavelet-based [105] methods reported the gain of around 0.8 dB. Moreover, as can be seen in Figures 8.10 to 8.11, visual quality improvements achieved by nonlocal sparsity regularization are also convincing.

### 8.5.3 Inverse Halftoning

The last application of second-generation patch models is inverse halftoning, which refers to the conversion of a halftoned image to a continuous-tone version, as demonstrated in Figures 8.12 and 8.13. This application has been chosen here for the purpose of emphasizing the point that heavy-tailed signal models, local or nonlocal, are capable of handling a wide range of noise, including halftoning noise. In the past decade, a flurry of competing inverse halftoning techniques have been developed with varying choice of regularization strategies, example techniques include linear filtering [109], Markov random field (MRF) based method [110], wavelet thresholding [111], nonlinear filtering [112], hybrid least-mean-square (LMS) and minimum mean square error (MMSE) estimation [113], and look-up-table method [114]. Despite the diversity of those strategies, they are unanimously

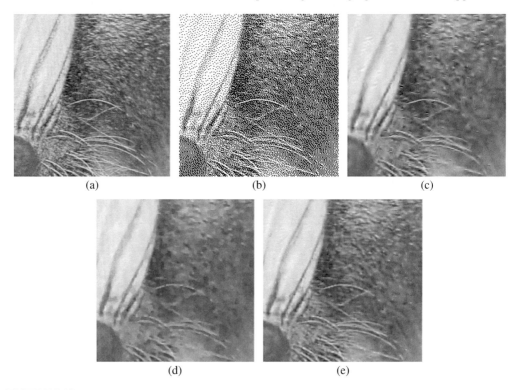

**FIGURE 8.13**

Inverse halftoning comparison: (a) $200 \times 200$ portion of original *Baboon* image, (b) halftoned image after Floyd-Steinberg diffusion, (c) reconstructed image by wavelet-based inverse halftoning [116] with $PSNR = 23.03$ dB, (d) reconstructed image by TV restoration [112] with $PSNR = 21.60$ dB, and (e) reconstructed image by nonlocal regularization $PSNR = 23.36$ dB.

local and produce comparable performance. Just like lossy image coding, halftoning can be also interpreted from a set theoretic perspective [115]. Therefore, it is convenient to obtain an inverse halftoning algorithm with nonlocal sparse representations; one only needs to modify the projection operator associated with the observation constraint set. The initial estimate is simply set to be a lowpass filtered version (the $3 \times 3$ blurring kernel is $[121; 242; 121]/16$). A similar deterministic annealing procedure to compression artifact removal is adopted as $\delta_n = 10 - n$, for $n = 1, 2, ..., 10$.

The nonlocal inverse halftoning algorithm has been compared with two local schemes, that is, wavelet-based inverse halftoning via deconvolution (WInHD)[8] [116] and inverse halftoning based on total variation (TV),[9] which is a technique similar to nonlinear filtering [112]. Experiments are performed on test images *Peppers* (edge-class) and *Baboon* (texture-class). For the *Peppers* image, the best reported PSNR results in the literature are in the range of $30 - 31.5$ dB, while the proposed nonlocal technique can achieve 32.64 dB. Moreover, as shown in Figure 8.12, the nonlocal method avoids producing unpleasant ringing artifacts around sharp edges.

---

[8]http://www.dsp.rice.edu/files/software/winhd.zip
[9]http://www.math.ucla.edu/~gilboa/PDE-filt/tv_denoising.html

Texture structures are often more challenging to handle for the task of inverse halftoning because they tend to have similar characteristics to halftoning noise locally. The importance of preserving texture structures has not been openly addressed in the literature of inverse halftoning before. Therefore, including experimental results for texture images in this chapter may be useful for those who are interested in the inverse halftoning problem in order to maintain a healthy skepticism toward the current practice — in computational science, what kind of test data should be used in the experiments is often an issue as important as the development of new algorithms. Figure 8.13 contains the objective and subjective quality comparison among original and reconstructed images by different inverse halftoning schemes. It is noted that for this specific image, TV model is a poor match again as the reconstructed image by wavelet method suffers from the loss of high-frequency information around the hair and beard region. By contrast, the proposed nonlocal scheme appears to obtain the most visually pleasant reconstruction even though the PSNR gain is modest; this is partially due to the fact that the fur pattern is somewhat irregular and therefore not exactly translation invariant.

## 8.6 Connections with Redundancy Exploitation and Contour Integration

In addition to support engineering applications, image models play a major role in shedding new insights into the understanding of sensory coding exploited by the human visual system. First-generation patch models can be viewed as being connected with the redundancy reduction hypothesis [117]. In the original proposal (influenced by communication theory [118]), the objective of sensory coding was thought of as reducing the redundancy in the stimulus; such view has been shared by many source coding researchers who pursue the most parsimonious representation of a source. The sparsity of image representation – when cast under the framework of sensory coding – has led to several influential works on computational modeling of the primary visual cortex (also called V1) [119], [120]. The hindsight after decade of research has suggested that lossy image compression and visual perception might be two related but disparate problems. For example, encoder is often more important than decoder in lossy image compression; however, it is the decoding strategy of neural systems that conceals the secret of sensory memory and adaptation [121], [122].

More important, as technology advances, anatomical and physiological studies of cortex in most mammals have shown that the channel capacity (as measured by the number of neurons) increases (instead of decreases) when moving to the higher levels. Such observation has even forced to abandon the original redundancy reduction hypothesis and replace it by a so-called redundancy exploitation hypothesis [41]. In this modified proposal, the role of redundancy is argued to support more accurate prediction which includes both reconstruction (a low-level vision task) and recognition (a high-level vision task) as special cases. In fact, a hypothesis unifying the inference tasks at low-level and high-level is highly desirable especially from the perspective of avoiding semantic gaps [123]. The equal importance of both feedforward and feedback connections in visual cortex also suggests that redundancy or statistical regularity of sensory stimuli is exploited by the human visual system in a seamless fashion, no matter what level of inference task it deals with.

So how do the second-generation patch models fit the paradigm of redundancy exploitation? As already discussed, the major innovation behind second-generation patch models lies in the exploitation of nonlocal sparsity arising from translation invariant property of regular edges and textures. In other words, the clustering of similar patches manifests another important source of statistical regularity (complementary to their local smoothness). Note that the transition from local smoothness to nonlocal (structural) similarity also matches the fact of increasing redundancy at higher level of visual cortex because storing more sophisticated similarity relationship requires more neurons (conceptually similar to the Quine-McCluskey method in logic design [124] and the theory of structural mapping [125]). Exactly how does the human visual system identify two similar patches (up to not just translation but more general projective transformations) is the task of perceptual organization and still largely remains a mystery. However, the positive role played by clustering to reconstruction (the focal point of this chapter) as well as recognition (for example, recent advances in sparsity-based robust face recognition [44]) has clearly shown the promising direction of better understanding redundancy exploitation hypothesis.

Along this line of reasoning, it can be envisioned that a logically natural next step of extending second-generation patch models is to integrate them with the models of contour integration [40], [126]. Along the contour of an object, how does the human visual system collectively organize local patches together to form the sensation of a single contour? To address this question, it seems that one can either generalize the definition of similar patches from translation invariance to rotational invariance [34] or explore the feasibility of integrating straight contours (lines). The latter hypothesis seems to be more biologically plausible in view of what is known about the anatomy and physiology of frog vision [127]. Various studies on optical illusion (for example, Kanizsa triangle [128]) also seem to support the existence of neurons devoted to the reconstruction or imagination of global long contours. For frogs, their capability of recognizing preys and predators heavily relies on an efficient sensory coding strategy developed through evolution, that is, the image is not analyzed pixel-by-pixel but in terms of a matrix of overlapping contexts. The key lesson learned from studying frog vision is the importance of relationship between a local receptive field and its neighboring ones (for example, Figures 17.2 to 17.6 in Reference [47]); maybe similar organizational principles have been recycled by the law of evolution across different scales (that is, figure-ground separation in ventral pathway and texture segmentation in dorsal pathway [129]).

## 8.7 Conclusion

This chapter has reviewed the history of patch-based image models, focusing on the class of generative ones. The evolution from first-generation patch models (dictionary construction and learning) to second-generation patch models (structural clustering and sparsity optimization) offers refreshing insights on how locality and convexity have served as double-bladed sword in mathematical modeling of photographic images. On one hand, local variation and nonlocal similarity turn out to be the two sides of the same coin as they jointly characterize the statistical regularity of image source; on the other hand, nonconvexity associated with the clustering procedure dictates to maintain a healthy skepticism toward the increasing popularity of convex optimization such as $l_1$-based techniques. It is strongly believed that it is more fruitful to mathematically model image source by thinking outside the box of Hilbert space. A differential function on a manifold will, in a typical case, directly reflect the underlying topology. The nonlinear projection operators introduced in this chapter can be viewed as the first step toward such topological thinking of patch-based image models. The abstraction of nearness in a topological space or patch similarity is an issue that deserves further study and could pave the path toward the next-generation modeling of images – what is more important than the concept of patch is the organizational principle underlying patches.

In addition to theoretic reasoning, this chapter has also strived hard to make the presented experimental results fully reproducible, which is as important as mental reproducibility in scientific research. Especially in view of the physical origin of image data, a good theoretic model or representation does not necessarily lead to a better image processing algorithm. The risk of model-data mismatch appears to be among the most less-acknowledged challenges to image processing community. As the legendary physicist R. Feynmann once said: *"It doesn't matter how beautiful your theory is, it doesn't matter how smart you are. If it doesn't agree with experiment, it's wrong."* Computational science, including image processing, lives by its *"disdain for authority and reliance on experimentation"* (Chris Quigg). As the only way of falsifying a theory or model, image data and processing algorithms could play a role bigger than serving engineering applications alone. It is the last mission of this chapter to make an open call for promoting the current state of reproducible research not only in image processing but also across its adjacent fields, such as computer vision and graphics, machine learning, and speech processing. Only by stepping on each other's shoulders (rather than feet) can further open the view at the horizon – a new vista of understanding geometry through topology might be there!

## References

[1] R. Haralick, K. Shanmugam, and I. Dinstein, "Textural features for image classification," *IEEE Transactions on Systems, Man and Cybernetics*, vol. 3, no. 6, pp. 610–621, November 1973.

[2] J. Lee, "Digital image enhancement and noise filtering by use of local statistics," *IEEE Transactions on Pattern Analysis and Machine Intelligence*, vol. 2, no. 2, pp. 165–168, March 1980.

[3] B. Horn and B. Schunck, "Determining optical flow," *Artificial Intelligence*, vol. 17, no. 1–3, pp. 185–203, August 1981.

[4] B. Lucas and T. Kanade, "An iterative image registration technique with an application to stereo vision," in *Proceedings of the International Joint Conference on Artificial Intelligence*, Vancouver, BC, Canada, August 1981, vol. 2, pp. 674–679.

[5] J.L.M. William and B. Pennebaker, *JPEG: Still Image Data Compression Standard*. New York, USA: Kluwer, 1992.

[6] W. Chen, C. Smith, and S. Fralick, "A fast computational algorithm for the discrete cosine transform," *IEEE Transactions on Communications*, vol. 25, no. 9, pp. 1004–1009, September 1977.

[7] R. Bellman, *Dynamic Programming*. Princeton, NJ, USA: Princeton University Press, 1957.

[8] I. Daubechies, *Ten Lectures on Wavelets*. Philadelphia, PA, USA: SIAM, 1992.

[9] M. Vetterli and J. Kovacevic, *Wavelets and Subband Coding*. New Jersey, NJ, USA: Prentice Hall, Englewood Cliffs, 1995.

[10] S. Mallat, *A Wavelet Tour of Signal Processing*, 2nd Edition. San Diego, CA, USA: Academic Press, 1999.

[11] H. Malvar, "Lapped transforms for efficient transform/subband coding," *IEEE Transactions on Acoustics, Speech and Signal Processing*, vol. 38, no. 6, pp. 969–978, June 1990.

[12] P. Burt and E. Adelson, "The Laplacian pyramid as a compact image code," *IEEE Transactions on Communications*, vol. 31, no. 4, pp. 532–540, April 1983.

[13] D. Taubman and M. Marcellin, *JPEG2000: Image Compression Fundamentals, Standards, and Practice*. Norwell, MA, USA: Kluwer, 2001.

[14] A. Grossman and J. Morlet, "Decomposition of Hardy functions into square integrable wavelets of constant shape," *Fundamental Papers in Wavelet Theory*, vol. 15, no. 4, pp. 723–736, July 1984

[15] R.A. DeVore, B. Jawerth, and B.J. Lucier, "Image compression through wavelet transform coding," *IEEE Transactions on Information Theory*, vol. 38, no. 2, pp. 719–746, March 1992.

[16] J.M. Shapiro, "Embedded image coding using zerotrees of wavelet coefficients," *IEEE Transactions on Signal Processing*, vol. 41, no. 12, pp. 3445–3462, December 1993.

[17] E.P. Simoncelli and E.H. Adelson, "Noise removal via Bayesian wavelet coring," in *Proceedings of the IEEE International Conference on Image Processing*, Lausanne, Switzerland, September 1996, vol. 1, pp. 379–382.

[18] I.K.M.K. Mihcak and K. Ramchandran, "Local statistical modeling of wavelet image coefficients and its application to denoising," in *Proceedings of the IEEE International Conference on Acoustics, Speech, and Signal Processing*, Phoenix, AZ, USA, March 1999, vol. 6, pp. 3253–3256.

[19] X. Li and M. Orchard, "Spatially adaptive image denoising under overcomplete expansion," in *Proceedings of the IEEE International Conference on Image Processing*, Vancouver, BC, Canada, September 2000, vol. 3, pp. 300–303.

[20] S.G. Chang, B. Yu, and M. Vetterli, "Adaptive wavelet thresholding for image denoising and compression," *IEEE Transactions on Image Processing*, vol. 9, no. 9, pp. 1532–1546, September 2000.

[21] A. Efros and T. Leung, "Texture synthesis by non-parametric sampling," in *Proceedings of the International Conference on Computer Vision*, Kerkyra, Greece, September 1999, vol. 2, pp. 1033–1038.

[22] J. Portilla and E. Simoncelli, "A parametric texture model based on joint statistics of complex wavelet coefficients," *International Journal of Computer Vision*, vol. 40, no. 1, pp. 49–71, October 2000.

[23] P.P.A. Criminisi and K. Toyama, "Object removal by exemplar-based inpainting," in *Proceedings of the IEEE Conference on Computer Vision and Pattern Recognition*, Madison, WI, USA, June 2003, vol. 2, pp. 721–728.

[24] A. Criminisi, P. Perez, and K. Toyama, "Region filling and object removal by exemplar-based image inpainting," *IEEE Transactions on Image Processing*, vol. 13, no. 9, pp. 1200–1212, September 2004.

[25] I. Drori, D. Cohen-Or, and H. Yeshurun, "Fragment-based image completion," *ACM Transactions on Graphics*, vol. 22, no. 3, pp. 303–312, July 2003.

[26] X. Li, "On modeling interchannel dependency for color image denoising," *International Journal of Imaging Systems and Technology*, vol. 17, no. 3, pp. 163–173, June 2007.

[27] A. Buades, B. Coll, and J.M. Morel, "A non-local algorithm for image denoising," in *Proceedings of the IEEE Conference on Computer Vision and Pattern Recognition*, San Diego, CA, USA, June 2005, vol. 2, pp. 60–65.

[28] C. Kervrann and J. Boulanger, "Unsupervised patch-based image regularization and representation," in *Proceedings of the European Conference on Computer Vision*, Graz, Austria, May 2006, pp. IV:555–567.

[29] K. Dabov, A. Foi, V. Katkovnik, and K. Egiazarian, "Image denoising by sparse 3-D transform-domain collaborative filtering," *IEEE Transactions on Image Processing*, vol. 16, no. 8, pp. 2080–2095, August 2007.

[30] W.T. Freeman, T.R. Jones, and E.C. Pasztor, "Example-based super-resolution," *IEEE Computer Graphics and Applications*, vol. 22, no. 2, pp. 56–65, March/April 2002.

[31] V. Cheung, B.J. Frey, and N. Jojic, "Video epitomes," in *Proceedings of the IEEE Conference on Computer Vision and Pattern Recognition*, San Diego, CA, USA, June 2005, vol. 1, pp. 42–49.

[32] J. Yang, J. Wright, T. Huang, and Y. Ma, "Image super-resolution as sparse representation of raw image patches," *Proceedings of the IEEE Conference on Computer Vision and Pattern Recognition*, Anchorage, AK, USA, June 2008, pp. 1–8.

[33] D.G. Lowe, "Distinctive image features from scale-invariant keypoints," *International Journal of Computer Vision*, vol. 60, no. 2, pp. 91–110, November 2004.

[34] T. Ojala, M. Pietikainen, and T. Maenpaa, "Multiresolution gray-scale and rotation invariant texture classification with local binary patterns," *IEEE Transactions on Pattern Analysis and Machine Intelligence*, vol. 24, no. 7, pp. 971–987, July 2002.

[35] H. Bay, T. Tuytelaars, and L. Van Gool, "SURF: Speeded up robust features," in *Proceedings of the European Conference on Computer Vision*, Graz, Austria, May 2006, pp. 404–417.

[36] J. Mairal, F. Bach, J. Ponce, G. Sapiro, and A. Zisserman, "Non-local sparse models for image restoration," in *Proceedings of the IEEE International Conference on Computer Vision*, Kyoto, Japan, October 2009, pp. 2272–2279.

[37] W. Dong, X. Li, L. Zhang, and G. Shi, "Sparsity-based image via dictionary learning and structural clustering," in *Proceedings of the IEEE Conference on Computer Vision and Pattern Recognition*, Colorado Springs, CO, USA, June 2011, pp. 457–464.

[38] P.L. Combettes, "The foundations of set theoretic estimation," *Proceedings of the IEEE*, vol. 81, no. 2, pp. 182–208, February 1993.

[39] L.K. Saul, S.T. Roweis, and Y. Singer, "Think globally, fit locally: Unsupervised learning of low dimensional manifolds," *Journal of Machine Learning Research*, vol. 4, pp. 119–155, December 2003.

[40] D. Field, A. Hayes, and R. Hess, "Contour integration by the human visual system: Evidence for a local," *Vision Research*, vol. 33, no. 2, pp. 173–193, January 1993.

[41] H. Barlow, "Redundancy reduction revisited," *Network: Computation in Neural Systems*, vol. 12, no. 3, pp. 241–253, August 2001.

[42] T. Chen, "The past, present, and future of image and multidimensional signal processing," *IEEE Signal Processing Magazine*, vol. 15, no. 2, pp. 21–58, March 1998.

[43] C.E. Guo, S.C. Zhu, and Y.N. Wu, "Modeling visual patterns by integrating descriptive and generative methods," *International Journal of Computer Vision*, vol. 53, no. 1, pp. 5–29, June 2003.

[44] J. Wright, A. Yang, A. Ganesh, S. Sastry, and Y. Ma, "Robust face recognition via sparse representation," *IEEE Transactions on Pattern Analysis and Machine Intelligence*, vol. 31, no. 2, pp. 210–227, February 2009.

[45] J. Morel and G. Yu, "ASIFT: A new framework for fully affine invariant image comparison," *SIAM Journal on Imaging Sciences*, vol. 2, no. 2, pp. 438–469, April 2009.

[46] C. Mead, *Analog VLSI and Neural Systems*. Boston, MA, USA: Addison-Wesley Longman Publishing, 1989.

[47] S. Zeki, *A Vision of the Brain*. Cambridge, MA, USA: Oxford University Press, 1993.

[48] K. Wilson, "The renormalization group: Critical phenomena and the Kondo problem," *Reviews of Modern Physics*, vol. 47, no. 4, pp. 773–840, October 1975.

[49] W. Ashby, "Principles of the self-organizing system," in *Principles of Self-Organization*, Transactions of the University of Illinois Symposium, H. von Foerster and G.W. Zopf, Jr. (eds.), London, UK: Pergamon Pres, 1962, pp. 255–278.

[50] A. Jain, *Fundamentals of Digital Image Processing*. Englewood Cliffs, NJ, USA: Prentice-Hall, 1989.

[51] R. Duda, P. Hart, and D. Stork, *Pattern Classification*, 2nd Edition. New York, USA: Wiley, 2001.

[52] N. Ahmed, T. Natarajan, and K. Rao, "Discrete cosine transfom," *IEEE Transactions on Computers*, vol. 100, no. 1, pp. 90–93, January 1974.

[53] N. Jayant and P. Noll, *Digital Coding of Waveforms: Principles and Applications to Speech and Video*. Englewood Cliffs, NJ, USA: Prentice-Hall, 1984.

[54] J.L. Mitchell, W.B. Pennebaker, C.E. Fogg, and D.J. LeGall, *MPEG Video Compression Standards*. London UK: Chapman-Hall 1996.

[55] G. Cote, B. Erol, M. Gallant, and F. Kossentini, "H.263+: Video coding at low bit rates," *IEEE Transactions on Circuits and Systems for Video Technology*, vol. 8, no. 7, pp. 849–866, November 1998.

[56] C. Christopoulos, J. Askelof, and M. Larsson, "Efficient methods for encoding regions of interest in the upcoming JPEG2000 still image coding standard," *IEEE Signal Processing Letters*, vol. 7, no. 9, pp. 247–249, September 2000.

[57] A. Bell and T. Sejnowski, "An information-maximization approach to blind separation and blind deconvolution," *Neural Computation*, vol. 7, no. 6, pp. 1129–1259, November 1995.

[58] P. Comon, "Independent component analysis, a new concept?," *Signal Processing*, vol. 36, no. 3, pp. 287–314, April 1994.

[59] A. Hyvarinen, J. Karhunen, and E. Oja, *Independent Component Analysis*. New York, USA: John Wiley, 2001.

[60] A.J. Bell and T.J. Sejnowski, "The independent components of natural scenes are edge filters," *Vision Research*, vol. 37, no. 23, pp. 3327–3338, December 1997.

[61] J.G. Daugman, "Uncertainty relation for resolution in space, spatial frequency, and orientation optimized by two-dimensional visual cortical filters," *Journal of the Optical Society of America A*, vol. 2, no. 7, pp. 1160–1169, July 1985.

[62] A. Belouchrani, K. Abed-Meraim, J. Cardoso, and E. Moulines, "A blind source separation technique using second-order statistics," *IEEE Transactions on Signal Processing*, vol. 45, no. 2, pp. 434–444, February 1997.

[63] M. Aharon, M. Elad, and A. Bruckstein, "K-SVD: An algorithm for designing overcomplete dictionaries for sparse representation," *IEEE Transactions on Signal Processing*, vol. 54, no. 11, pp. 4311–4322, November 2006.

[64] K. Kreutz-Delgado, J. Murray, B. Rao, K. Engan, T. Lee, and T. Sejnowski, "Dictionary learning algorithms for sparse representation," *Neural Computation*, vol. 15, no. 2, pp. 349–396, February 2003.

[65] J. Mairal, F. Bach, J. Ponce, G. Sapiro, and A. Zisserman, "Supervised dictionary learning," *Arxiv Preprint*, arXiv:0809.3083, September 2008.

[66] Z. Chen, C. Micchelli, and Y. Xu, "A construction of interpolating wavelets on invariant sets," *Mathematics of Computation*, vol. 68, no. 228, pp. 1569–1587, March 1999.

[67] F.Bergeaud and S. Mallat, "Matching pursuit: Adaptative representations of images," *Computational and Applied Mathematics*, vol. 15, no. 2, pp. 97–110, 1996.

[68] M. Elad and M. Aharon, "Image denoising via sparse and redundant representations over learned dictionaries," *IEEE Transactions on Image Processing*, vol. 15, no. 12, pp. 3736–3745, December 2006.

[69] M.A.T. Figueiredo and A.K. Jain, "Unsupervised learning of finite mixture models," *IEEE Transactions Pattern Analysis and Machine Intelligence*, vol. 24, no. 3, pp. 381–396, March 2002.

[70] E. Kreyszig, *Introductory Functional Analysis with Applications*. New York, USA: John Wiley, 1989.

[71] M. Morse, *The Calculus of Variations in the Large*. New York, USA: American Mathematical Society, 1934.

[72] C. Tomasi and R. Manduchi, "Bilateral filtering for gray and color images," in *Proceedings of the International Conference on Computer Vision*, Bombay, India, January 1998, pp. 839–846.

[73] D. Donoho and I. Johnstone, "Ideal spatial adaptation by wavelet shrinkage," *Biometrika*, vol. 81, no. 3, pp. 425–455, 1994.

[74] J. Hertz, A. Krogh, and R.G. Palmer, *Introduction to the Theory of Neural Computation*. Boston, MA, USA: Addison-Wesley Longman Publishing, 1991.

[75] I. Mayergoyz, "Mathematical models of hysteresis," *Physical Review Letters*, vol. 56, no. 15, pp. 1518–1521, September 1986.

[76] T. Cover and P. Hart, "Nearest neighbor pattern classification," *IEEE Transactions on Information Theory*, vol. 13, no. 1, pp. 21–27, January 1967.

[77] J. Hartigan and M. Wong, "A K-means clustering algorithm," *Journal of the Royal Statistical Society C*, vol. 28, no. 1, pp. 100–108, 1979.

[78] A. Ng, M. Jordan, and Y. Weiss, "On spectral clustering: Analysis and an algorithm," *Advances in Neural Information Processing Systems*, vol. 2, pp. 849–856, 2002.

[79] D. Youla, "Generalized image restoration by the method of alternating orthogonal projections," *IEEE Transactions on Circuits and System*, vol. 9, no. 9, pp. 694–702, September 1978.

[80] X. Li, "Fine-granularity and spatially-adaptive regularization for projection-based image deblurring," *IEEE Transactions on Image Processing*, vol. 20, no. 4, pp. 971–983, April 2011.

[81] K. Goebel and W.A. Kirk, *Topics in Metric Fixed Point Theory*. New York, USA: Cambridge University Press, 1990.

[82] O.G. Guleryuz, "Nonlinear approximation based image recovery using adaptive sparse reconstructions and iterated denoising – Part I: Theory," *IEEE Transactions on Image Processing*, vol. 15, no. 3, pp. 539–554, March 2006.

[83] M.I. Sezan and H.J. Trussell, "Prototype image constraints for set-theoretic image restoration," *IEEE Transactions on Signal Processing*, vol. 39, no. 10, pp. 2275–2285, October 1991.

[84] Y. Yang, N. Galatsanos, and A. Katsaggelos, "Projection-based spatially adaptive reconstrcution of block-transform compressed images," *IEEE Transactions on Image Processing*, vol. 4, no. 7, pp. 896–908, July 1995.

[85] X. Li, "Image recovery from hybrid sparse representation: A determinisitc annealing approach," *IEEE Journal of Selected Topics in Signal Processing*, vol. 9, no. 5, pp. 953–962, September 2011.

[86] S. Kirkpatrick, C.D. Gelatt, and M.P. Vecchi, "Optimization by simulated annealing," *Science*, vol. 220, no. 4598, pp. 671–680, May 1983.

[87] R. Swendsen and J. Wang, "Nonuniversal critical dynamics in Monte Carlo simulations," *Physical Review Letters*, vol. 58, no. 2, pp. 86–88, 1987.

[88] U. Wolff, "Collective Monte Carlo updating for spin systems," *Physical Review Letters*, vol. 62, no. 4, pp. 361–364, 1989.

[89] J.S. Liu, *Monte Carlo Strategies in Scientific Computing*. New York, USA: Springer, 2001.

[90] A. Blake and A. Zisserman, *Visual Reconstruction*. Cambridge, MA, USA: MIT Press, 1987.

[91] K. Rose, E. Gurewwitz, and G. Fox, "A deterministic annealing approach to clustering," *Pattern Recognition Letters*, vol. 11, no. 9, pp. 589–594, September 1990.

[92] K. Rose, "Deterministic annealing for clustering, compression, classification, regression, and related optimization problems," *Proceedings of the IEEE*, vol. 86, no. 11, pp. 2210–2239, November 1998.

[93] A. Elmoataz, O. Lezoray, and S. Bougleux, "Nonlocal discrete regularization on weighted graphs: A framework for image and manifold processing," *IEEE Transactions on Image Processing*, vol. 17, no. 7, pp. 1047–1060, July 2008.

[94] S. Geman and D. Geman, "Stochastic relaxation, Gibbs distributions, and the Bayesian restoration of images," *IEEE Transactions on Pattern Analysis and Machine Intelligence*, vol. 6, no. 6, pp. 721–741, November 1984.

[95] X. Li, W. Dong, L. Zhang, and G. Shi, "Image deblurring with local adaptive sparsity and nonlocal robust regularization," in *Proceedings of the IEEE International Conference on Image Processing*, Brussels, Belgium, September 2011, pp. 1881–1884.

[96] W. Dong, X. Li, L. Zhang, and G. Shi, "Image reconstruction with locally adaptive sparsity and nonlocal robust regularization," *Inverse Problems*, submitted, 2011.

[97] A. Foi, V. Katkovnik, and K. Egiazarian, "Pointwise shape-adaptive DCT for high-quality denoising and deblocking of grayscale and color images," *IEEE Transactions on Image Processing*, vol. 16, no. 5, pp. 1395–1411, May 2007.

[98] Z. Wang, A.C. Bovik, H.R. Sheikh, and E.P. Simoncelli, "Image quality assessment: From error visibility to structural similarity," *IEEE Transactions on Image Processing*, vol. 13, no. 4, pp. 600–612, April 2004.

[99] P. Chatterjee and P. Milanfar, "Clustering-based denoising with locally learned dictionaries," *IEEE Transactions on Image Processing*, vol. 18, no. 7, pp. 1438–1451, July 2009.

[100] X. Li, "Exemplar-based EM-like image denoising via manifold reconstruction," in *Proceedings of the IEEE International Conference on Image Processing*, Hong Kong, September 2010, pp. 73–76.

[101] V. Katkovnik, A. Foi, K. Egiazarian, and J. Astola, "From local kernel to nonlocal multiple-model image denoising," *International Journal of Computer Vision*, vol. 86, no. 1, pp. 1–32, January 2010.

[102] B. Efron and R. Tibshirani, *An Introduction to the Bootstrap*. Boca Raton, FL, USA: CRC Press, 1994.

[103] N. Thao and M. Vetterli, "Set theoretic compression with an application to image coding," *IEEE Transactions on Image Processing*, vol. 7, no. 7, pp. 1051–1056, July 1998.

[104] A. Zakhor, "Iterative procedures for reduction of blocking effects in transform image coding," *IEEE Transactions on Circuits Systems Video Technology*, vol. 2, no. 1, pp. 91–95, January 1992.

[105] Z. Xiong, M. Orchard, and Y. Zhang, "A deblocking algorithm for JPEG compressed images using overcomplete wavelet representations," *IEEE Transactions on Circuit and Systems for Video Technology*, vol. 7, no. 2, pp. 433–437, April 1997.

[106] A. Said and W.A. Pearlman, "A new fast and efficient image codec based on set partitioning in hierarchical trees," *IEEE Transactions on Circuits and Systems for Video Technology*, vol. 6, no. 3, pp. 243–250, June 1996.

[107] X. Li, "Collective sensing: A fixed-point approach in the metric space," in *Proceedings of the SPIE Conference on Visual Communication and Image Processing*, Huangshan, China, July 2010, pp. 7744–7746.

[108] O.G. Guleryuz, "Nonlinear approximation based image recovery using adaptive sparse reconstructions and iterated denoising – Part II: Adaptive algorithms," *IEEE Transactions on Image Processing*, vol. 15, no. 3, pp. 555–571, March 2006.

[109] P. Wong, "Adaptive error diffusion and its application in multiresolution rendering," *IEEE Transactions on Image Processing*, vol. 5, no. 7, pp. 1184–1196, July 1996.

[110] R.L. Stevenson, "Inverse halftoning via MAP estimation," *IEEE Transactions on Image Processing*, vol. 6, no. 4, pp. 574–583, April 1997.

[111] Z. Xiong, K. Ramchandran, and M. Orchard, "Inverse halftoning using wavelets," *IEEE Transactions on Image Processing*, vol. 8, no. 10, pp. 1479–1483, October 1999.

[112] T. Kite, B. Evans, and A. Bovik, "Modeling and quality assessment of halftoning by error diffusion," *IEEE Transactions on Image Processing*, vol. 9, no. 5, pp. 909–922, May 2000.

[113] P.C. Chang, C.S. Yu, and T.H. Lee, "Hybrid LMS-MMSE inverse halftoning technique," *IEEE Transactions on Image Processing*, vol. 10, no. 1, pp. 95–103, January 2001.

[114] M. Mese and P.P. Vaidyanathan, "Look-up table (LUT) method for inverse halftoning," *IEEE Transactions on Image Processing*, vol. 10, no. 10, pp. 1566–1578, October 2001.

[115] N. Thao, "Set theoretic inverse halftoning," in *Proceedings of the IEEE International Conference on Image Processing*, Washington, DC, USA, October 1997, vol. 1, pp. 783–786.

[116] R. Neelamani, R.D. Nowak, and R.G. Baraniuk, "Winhd: Wavelet-based inverse halftoning via deconvolution," *Rejecta Mathematica*, vol. 1, no. 1, pp. 84–103, July 2009.

[117] H.B. Barlow, "Possible principles underlying the transformations of sensory messages," in *Sensory Communication*, W.A. Rosenblith (ed.), Cambridge, MA, USA: MIT Press, 1961, pp. 217–234.

[118] C.E. Shannon, "A mathematical theory of communication," *Bell System Technical Journal*, vol. 37, pp. 379–423, 623–656, 1948.

[119] D.J. Field, "What is the goal of sensory coding?," *Neural Computation*, vol. 6, no. 4, pp. 559–601, July 1994.

[120] B. Olshausen and D. Field, "Emergence of simple-cell receptive field properties by learning a sparse code for natural images," *Nature*, vol. 381, no. 6583, pp. 607–609, June 1996.

[121] F. Rieke, *Spikes: Exploring the Neural Code*. Cambridge, MA, USA: MIT Press, 1999.

[122] P. Dayan and L. Abbott, *Theoretical Neuroscience*. Cambridge, MA, USA: MIT Press, 2001.

[123] M. Ehrig, *Ontology Alignment: Bridging the Semantic Gap.* New York, USA: Springer-Verlag, 2007.

[124] C. Roth, Jr., *Fundamentals of Logic Design.* Stamford, CT, USA: Brooks/Cole Publishing, 2009.

[125] D. Gentner, "Structure-mapping: A theoretical framework for analogy," *Cognitive Science: A Multidisciplinary Journal,* vol. 7, no. 2, pp. 155–170, April–June 1983.

[126] Z. Li, "A neural model of contour integration in the primary visual cortex," *Neural Computation,* vol. 10, no. 4, pp. 903–940, May 1998.

[127] H. Maturana, J. Lettvin, W. McCulloch, and W. Pitts, "Anatomy and physiology of vision in the frog (*Rana pipiens*)," *The Journal of General Physiology,* vol. 43, no. 6, p. 129, July 1960.

[128] J. Sterman, "Learning in and about complex systems," *System Dynamics Review,* vol. 10, no. 2–3, pp. 291–330, Fall 1994.

[129] L. Ungerleider and J. Haxby, "'What' and 'where' in the human brain," *Current Opinion in Neurobiology,* vol. 4, no. 2, pp. 157–165, April 1994.

# 9

## Perceptually Driven Super-Resolution Techniques

**Nabil Sadaka and Lina Karam**

## 9.1   Introduction

Nowadays, high-resolution (HR) image/video applications are reaching every home and soon will become a necessity rather than a mere luxury. Whether it is through a high-definition (HD) television, computer monitor, HD camera, or even a handheld phone, HR multimedia applications are becoming essential components of consumers' daily lives.

Moreover, the demand for HR images transcends the need for offering a better-quality picture to the common viewer. HR imagery is continuing to gain popularity and dominate many industries that require accurate image analysis. For example, in surveillance applications, high-resolution images are needed for a better performance of target detection, face recognition, and text recognition [1], [2]. Furthermore, medical imaging applications require HR images for accurate assessment and detection of small lesions [3], [4].

The capture and delivery of HR multimedia content is a complex and problematic process [5]. In any typical imaging system or multimedia delivery chain, the quality of HR media can be impaired by processes of acquisition, transmission, and display. The resolution of the imaging systems is physically limited by the pixel density in image sensors. On one hand, increasing the number of pixels in an image sensor via reducing the pixel size is limited by the existence of shot noise and associated high cost. On the other hand, increasing the chip size is deemed ineffective due to the existence of a large capacitance that slows the charge transfer rate and limits the data transfer [6]. Thus, a major drawback of HR image acquisition in many of the aforementioned applications is the inadequate resolution of the sensors, either because of cost or hardware limitation [7]. In addition to hardware limitations, HR imagery can be blurred during acquisition due to atmospheric turbulence and the camera-to-object motion [8]. Transmission of HR media requires extremely high bandwidth, which is usually not available in real-time scenarios. Thus, high compression rates are imposed on the media content, resulting in annoying compression artifacts and high frame dropping rates [9]. Moreover, large displays use interpolation techniques to scale video content to fit a target screen size, thus introducing blurred details and system noise enlargement. Hence, advanced signal processing, such as super-resolution (SR) imaging, offers a cost-effective way to increase image resolution by overcoming sensor hardware limitations and reducing media impairments.

When choosing a solution for the SR problem, there is always a trade-off between computational efficiency and HR image quality. Existing SR approaches suffer from extremely high computational requirements due to the high dimensionality in terms of large number of unknowns to be estimated when solving the SR inverse problem. Recently, SR techniques are at the heart of HR image/video technologies in various application areas, such as consumer electronics, entertainment, digital communications, military, and biomedical imaging. Most of these applications require real-time or near real-time processing due to limitations on computational power with the essential demand for high image quality requirements. Efficient SR methods represent vital solutions to the digital multimedia acquisition, delivery, and display sectors. Iterative SR solutions, that are commonly Bayesian or regularized optimization-based approaches, can converge to a high-quality SR solution but suffer from high computational complexity. Moreover, non-iterative SR solutions, such as the kernel-based SR approaches, are inherently efficient in nature but can result in limited HR reconstruction quality. As a consequence, a new class of selective SR estimators [10], [11], [12], [13], [14], [15] have been introduced to reduce the dimensionality or the computational complexity of popular SR algorithms while maintaining the desired visual quality of the reconstructed HR image. Generally, these selective SR algorithms detect only a subset of active pixels that are super-resolved iteratively based on the pixels' local significance measure to the final SR result. The subset of active pixels selected for processing can be detected using non-perceptual measures based only on numerical image characteristics or perceptual measures based on the human visual system (HVS) perception of low-level features of the image content. In References [10] and [11], a non-perceptual selective approach is presented where the local gradients of the estimated HR image at each iteration are used to detect the pixels with significant spatial activities (that is, pixels at which the gradient is above a certain threshold). The drawback of this approach lies in manually tweaking the gradient threshold for each image differently to attain the best desired visual SR quality.

Unlike the hard thresholding applied to the non-perceptual gradient measures in References [10] and [11], selective perceptually based SR approaches presented in References [12], [13], [14], and [15] adaptively determine the set of significant pixels using an automated perceptual decision mechanism without any manual tuning. These perceptual detection mechanisms are based on perceptual contrast sensitivity threshold modeling, which detects details with perceived contrast over a uniform background, and visual attention modeling, which detects salient low-level features in the image. Such perceptually driven SR methods proved to significantly reduce the computational complexity of iterative SR algorithms without any perceptible loss in the desired enhanced image/video quality. Perceptually based SR imaging is a new research area, where very few approaches have been proposed so far. Existing solutions usually focus on enhancing the perceptual quality of the super-resolved image and ignore the essential factor of computational efficiency. For example, a perceptual SR algorithm presented in Reference [16] minimizes a set of perceptual error measures that preserve the fidelity and enhances the smoothness and details of the reconstructed image. Unfortunately, this approach is not very efficient since it is still an iterative SR solution which uses a conjugate gradient or constrained least squares solution, where every pixel in the SR error function is processed inclusively.

This chapter focuses on efficient selective SR techniques driven by perceptual modeling of the human visual system. Section 9.2 presents a survey of relevant SR techniques. The multi-frame SR observation model is described in Section 9.3. Section 9.4 provides a general background on multi-frame SR formulation and comparisons. Section 9.5 presents the selective perceptual SR framework employed to enhance the computational efficiency without degrading the HR quality. The contrast sensitivity threshold model used for detecting active pixels for selective SR processing is described and a mechanism for detecting the perceptually significant pixels, that are selected for SR processing, is presented. Section 9.6 presents an efficient attentive-selective perceptual SR framework based on salient low-level features detection concepts. Several saliency detection algorithms based on visual attention modeling are reviewed. Section 9.7 describes the application of the perceptually driven SR framework to increase the computational efficiency of a maximum *a posteriori* (MAP)-based SR method and a two-step fusion-restoration SR method, respectively. It also includes simulation results. Finally, this chapter concludes with Section 9.8.

## 9.2 Main Super-Resolution Techniques

Super-resolution (SR) techniques are considered to be effective image enhancement and reconstruction solutions that approach the resolution enhancement problem from two dif-

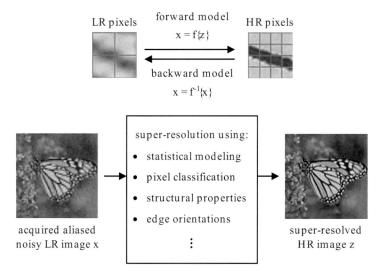

**FIGURE 9.1**

Single-frame SR block diagram. The forward model, $f\{z\}$, is a mathematical description of the image degradation process exploiting the relationship between HR neighboring pixels. The inverse problem or backward
model $f^{-1}\{x\}$ is estimating the HR image from the low-quality captured image.

ferent perspectives. The so-called single-frame SR methods require one under-sampled or
degraded image to reconstruct a higher resolution enhanced image whereas the multi-frame
SR methods combine multiple degraded frames or views of a scene to estimate a higher resolution enhanced image. This section presents a survey of relevant single/multi-frame SR
techniques and introduces essential SR background material.

### 9.2.1   Single-Frame Super-Resolution

   Single-frame SR approaches are widely applied solutions to the resolution enhancement
problem, especially in cases where only one degraded low-resolution (LR) observation
is available [17]. Single-frame SR approaches can also be referred to as image interpolation or reconstruction techniques and thus can be used interchangeably. Recovering a
high-resolution image from an under-sampled (according to Nyquist limits [18]) and noisy
observation is a highly ill-posed inverse problem. Thus, prior image models relating neighboring pixels or prior models learned through similar image patches are needed extra information that can aid in solving the inverse problem. A common approach among these
SR techniques is that they take advantage of the relation between neighboring pixels of the
same image to estimate the values of missing pixels. Figure 9.1 illustrates a general single-
frame SR process where $f\{z\}$ and $f^{-1}\{x\}$ denote the forward and backward degradation
process of the imaging system, respectively.

   A well-known problem with common kernel-based super-resolution, such as bilinear
and bicubic interpolation [19], is the blurring and blockiness effects of sharp edges. The
blurring of sharp edges results from the inaccurate kernel resizing to adapt to the edge
sharpness. The edge blockiness, also known as staircase effects, is mainly due to the failure of the filter to adapt to various edge orientations. Reference [20] proposed an edge-

directed interpolation method that uses local covariance estimates of the input LR frame to adapt interpolation coefficients to arbitrarily oriented edges of the reconstructed HR image. This method is motivated by the geometric regularity property [21] of an ideal step edge. Recently, Reference [22] presented a new edge-directed interpolation approach that uses multiple low resolution training windows to reduce the covariance mismatch between the LR and HR pixels. Reference [23] proposed a smart interpolation method that uses anisotropic diffusion [24], a technique aiming at reducing noise and enhancing contrast without significantly degrading edges and objects in an image. References [25], [26], [27], [28], and [29] proposed SR methods based on the optimal recovery principle, which model the local image regions using ellipsoidal signal classes and adapts the interpolating kernels accordingly. Another model-based SR approach [30] is based on multi-resolution analysis in the wavelet domain, where the statistical relationships between coefficients at coarser scales are modeled using hidden Markov trees to predict the coefficients on the finest scale. Finally, learning-based SR approaches recover missing high-frequency information in an image through matching using a large image database [31], [32], [33], [34].

Many single-frame SR methods are computationally efficient. However, super-resolving from a single low-resolution image is known to be a highly ill-posed inverse problem due to the low number of observations relative to the large number of missing pixels/unknowns. Thus, the gain in quality in the single-frame SR approach is limited by the minimal number of information provided to recover missing details in the reconstructed HR signal.

### 9.2.2 Multi-Frame Super-Resolution

Multi-frame SR techniques offer a better solution to the resolution enhancement problem by exploiting extra information from several neighboring frames in a video sequence. In a multi-frame acquisition system, the subpixel motion between the camera and object allows the frames of the video sequence to contain oversampled similar information that make the reconstruction of an HR image possible.

This chapter focuses on multi-frame SR techniques that enhance the resolution of images by combining information from multiple low-resolution (LR) frames of the same scene to estimate a high-resolution (HR) unaliased and sharp/deblurred image under realistic bandwidth requirements and cost-effective hardware [6]. Interest in multi-frame super-resolution re-emerged in the recent years mainly due to the use of multi-frame image sequences, which can take advantage of additional spatio-temporal information available in the video content, and the increase of hardware computational power and advancement of display technologies, which make SR applications possible. Figure 9.2 presents a block diagram showing a multi-frame SR estimation process using multiple degraded LR frames, with subpixel shifts, to estimate one HR reference frame. The figure also shows the LR pixels registered relative to a common HR reference grid to further visualize the number of pixels used in the process of estimating the missing pixels in the SR image.

Since Reference [35] first introduced the multi-frame image restoration and registration problem/solution, several multi-frame SR approaches have been proposed in the past two decades. These multi-frame SR approaches are described and broadly categorized according to their methods of solution into Bayesian, regularized norm minimization, fusion-restoration (FR), and non-iterative fusion-interpolation (FI) approaches. Bayesian maxi-

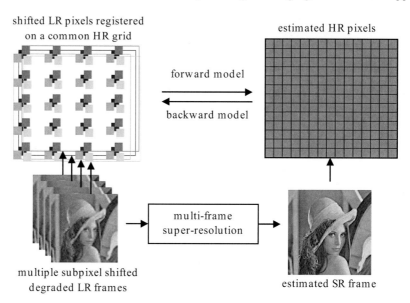

**FIGURE 9.2**

Multi-frame SR block diagram.

mum *a posteriori* (MAP) solutions have gained great attention and proved to be effective due to the inclusion of *a priori* knowledge about the HR estimate and inherent probabilistic formulation of the relation between the LR observations and the HR image [36], [37], [38], [39]. Regularized norm minimization solutions aim at minimizing an error term with some smoothness regularization assumption on the HR estimate [40], [41], [42]. It was shown in Reference [42] that the $l_1$-norm minimization approach is the most robust to errors in the system and that the bilateral total variation regularization function gave the best performance in terms of robustness and edge preservation. The two-step fusion-restoration (FR) SR estimation methods [40], [42] were devised in order to decrease the computational requirements and to increase the robustness to outliers. The FR methods employ a non-iterative fusion step, which includes a one-step registration process, and a restoration step, which simultaneously deblurs and denoises the fused image by minimizing an error function with or without a specific regularization term. The non-iterative fusion-interpolation (FI) SR approaches [43], [44] are composed of a fusion step and a non-iterative reconstruction step using kernel-based interpolation, which is inherently efficient in nature; however, these non-iterative methods can result in lower visual quality than the iterative methods as discussed later in this chapter. All the previous methods consider the subpixel motion between the observations to recover missing information by solving the SR problem.

## 9.3 Multi-Frame Super-Resolution Observation Model

Multi-frame reconstruction techniques, as described earlier in Section 9.2.2, reduce the sensitivity of the SR inverse problem solution to some given LR observation measures

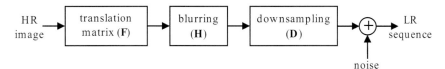

**FIGURE 9.3**

Multi-frame SR observation model.

by exploiting extra information from several neighboring frames in a video sequence. A necessary assumption for multi-frame SR solutions to work is the existence of subpixel shifts between the observed LR frames. Due to these fractional pixel shifts, registered LR samples will not always fall on a uniformly spaced HR grid, thus providing oversampled information necessary for solving the SR inverse problem. The LR pixels in the data acquisition model are defined as a weighted sum of appropriate HR pixels. The weighting function, also known as the system degradation matrix, models the blurring caused by the point spread function (PSF) of the optics. An additive noise term can be added to compensate for any random errors and reading sensor noise in the acquisition model. Assuming that the resolution enhancement factor is constant and the LR frames are acquired by the same camera, it is logical to consider the same PSF and statistical noise for all the LR observations. Taking into consideration all these assumptions, a common observation model for the SR problem is formulated.

Consider $K$ low-resolution frames $\mathbf{y}_k$, for $k = 1, 2, ..., K$, each arranged in lexicographical form of size $N_1 N_2 \times 1$ pixels. Let $L_1$ and $L_2$ be the resolution enhancement factors in the horizontal and vertical directions, respectively. For simplicity, it is assumed here that $L = L_1 = L_2$. The values of the pixels in the $k$th low-resolution frame of the sequence can be expressed in matrix notation as

$$\mathbf{y}_k = \mathbf{W}_k \mathbf{z} + \mathbf{n}, \tag{9.1}$$

where $\mathbf{z}$ represents the lexicographically ordered undegraded HR image of size $N \times 1$ for $N = L^2 N_1 N_2$, and $\mathbf{n}$ is the additive noise modeled as an independent and identically distributed (i.i.d.) Gaussian random variable with variance $\sigma_\eta^2$. In Equation 9.1, the degradation matrix is represented by $\mathbf{W}_k = DHF_k$, where $F_k$ is the warping matrix of size $N \times N$, the term $H$ denotes the blurring matrix of size $N \times N$ representing the common PSF function, and $D$ is the decimation matrix of size $N_1 N_2 \times N$.

Let $\mathbf{y}$ be the observation vector composed of all the LR vectors $\mathbf{y}_k$, for $k = 1, 2, ..., K$, concatenated vertically. The pointwise notation counterpart of Equation 9.1 for the $m$th element of $\mathbf{y}_k$ is given by

$$y_m = \sum_{r=1}^{N} w_{m,r} z_r + n_m, \tag{9.2}$$

where $n_m$ represents the additive noise and $w_{m,r}$ represents the contribution of $z_r$ (which is the $r$th HR pixel in $\mathbf{z}$), to $y_m$ (which is the $m$th LR pixel in the observation vector $\mathbf{y}$). Figure 9.3 illustrates the SR observation model for acquiring the LR frames.

## 9.4 Super-Resolution Problem Formulation

The formulation of the existing solutions for the SR problem usually falls into three main categories:

- The Bayesian maximum likelihood (ML) methods and maximum *a posteriori* (pMAP) methods. The former ones produce a super-resolution image that maximizes the probability of the observed LR input images under a given model, whereas the latter ones aim at stabilizing the ML solution under noisy conditions by making explicit use of prior HR information.

- The regularized-norm minimization methods produce an SR image by minimizing an error criteria ($l_p$-norm) with a regularization term. Efficient methods emerged from this category by swapping the order of the warping and blurring operators in the observation model to fuse the images into one HR grid followed by an iterative regularized optimization solution. An example is the fusion-restoration class of SR methods, as discussed in Section 9.7.2.

- The non-iterative kernel-based methods, also referred to as fusion-interpolation (FI) methods, merge all the observations on a common HR grid and solve for the best interpolation through adaptive kernel design techniques. This category is inherently computationally efficient since it is non-iterative in nature.

### 9.4.1 Bayesian Super-Resolution Formulation

Bayesian MAP-based estimators are common solutions for the SR problem since they offer fast convergence and high-quality performance [36], [37], [38], [39]. In Bayesian SR solutions all parameters or unknowns (that is, HR image, motion parameters, and noise) and observable variables (that is, the LR observations) are assumed to be unknown stochastic quantities with assumed probability distributions based on subjective beliefs. In the following formulation of the MAP solution, the motion parameters are assumed to be known for simplicity. For example, in cases of compressed video content the motion vectors can be retrieved from the headers of the bitstream (note that motion estimation is a mature field of research with many proposed accurate methods). In order to estimate the HR image $\mathbf{z}$, a Bayesian MAP estimator is formed given the low-resolution frames $\mathbf{y}_k$, for $k = 1, 2, ..., K$, and appropriate prior. The HR estimate $\hat{\mathbf{z}}$ can be computed by maximizing the *a posteriori* probability $\Pr(\mathbf{z}|\{\mathbf{y}_k\})$, or by minimizing the log-likelihood function as follows:

$$\hat{\mathbf{z}} = \arg\min_{\mathbf{z}} \log[\Pr(\mathbf{z}|\{\mathbf{y}_k\})]. \tag{9.3}$$

Using Bayes rule and assuming that the LR observations $\mathbf{y}_k$ are statistically independent of $\mathbf{z}$, the problem reduces to

$$\hat{\mathbf{z}} = \arg\min_{\mathbf{z}} \{-\log[\Pr(\{\mathbf{y}_k\}|\mathbf{z})] - \log[\Pr(\mathbf{z})]\}. \tag{9.4}$$

Now solving for an accurate HR estimate in Equation 9.4 is highly dependent on the prior HR image density $\Pr(\mathbf{z})$ and the conditional LR density $\Pr(\{\mathbf{y}_k\}|\mathbf{z})$ models. Note that when dropping the prior HR probability model in Equation 9.4, the MAP optimization problem reduces to an ML estimation problem that is highly unstable under small errors in the parameters of the acquisition model and noisy conditions [37]. It has been widely used in Bayesian SR formulation literature [36], [37], [38], [39], [45], [46], [47], [48], [49], [50], [51], [52] that the noise model is assumed to be a zero mean Gaussian. From Equation 9.4, and given that the elements of $\mathbf{n}_k$ are i.i.d Gaussian random variables, the conditional probability distribution can be modeled as follows:

$$\Pr(\mathbf{y}_k|\mathbf{z}) = \frac{1}{(2\pi)^{\frac{N}{2}} \sigma_\eta^N} \exp\left\{ -\frac{1}{2\sigma_\eta^2} \|\mathbf{y}_k - \mathbf{W}_k\mathbf{z}\|^2 \right\}, \qquad (9.5)$$

where $\sigma_\eta^2$ is the noise variance. The problem of determining which HR image prior model is the best for a particular HR reconstruction is still an open problem widely targeted by various existing literature [6], [53]. However, a common approach followed by existing MAP-based SR solutions is the assumption of smoothness constraints on the HR priors within homogeneous regions [36], [37], [54], [55]. These priors can generally be modeled as

$$\Pr(\mathbf{z}) \propto \exp\left[ \frac{-\lambda}{2} \|\mathbf{Q}\mathbf{z}\|^2 \right], \qquad (9.6)$$

where $\mathbf{Q}$ represents a linear high-pass operator that penalizes the estimates that are not smooth and $\lambda$ controls the variance of the prior distribution. In References [36] and [37], these piecewise smoothness priors in Equation 9.6 take the form of Huber-Markov random fields that are modeled as Gibbs prior functional according to Reference [56]. Then, the prior model can be written as follows:

$$\Pr(\mathbf{z}) = \frac{1}{2}\mathbf{z}^T \mathbf{C}_z^{-1}\mathbf{z}, \qquad (9.7)$$

where $\mathbf{C}_z$ is the covariance of the HR image prior model imposing piecewise smoothness constraints between neighboring pixels. Thus, with the smoothness prior model (Equation 9.7) and mutually independent additive Gaussian noise on the prior error model (Equation 9.5), the MAP SR estimation problem can be formulated by minimizing the following convex cost function with a unique global minimum:

$$f(\mathbf{z}) = \frac{1}{2\sigma_\eta^2} \sum_{k=1}^{K} \left(\mathbf{y}_k - \mathbf{W}_k\mathbf{z}^T\right)\left(\mathbf{y}_k - \mathbf{W}_k\mathbf{z}\right) + \frac{1}{2}\mathbf{z}^T \mathbf{C}_z^{-1}\mathbf{z}, \qquad (9.8)$$

where $\mathbf{W}_k$ is the degradation matrix for frame $k$, the term $\sigma_\eta^2$ denotes the noise variance, $\mathbf{z}$ is the HR frame in lexicographical form, and $\mathbf{y}_k$ are the observed LR frames in vector form. Thus, the MAP estimator can be reformulated as a least-squares error minimization problem, which in matrix notation is the $l_2$-norm square of the error vector, with a smoothness regularization constraint, given by

$$\hat{\mathbf{z}} = \arg\min_{\mathbf{z}} \left\{ C_\eta^{-1} \sum_{k=1}^{K} \|\mathbf{y}_k - \mathbf{W}_k\mathbf{z}\|_2^2 + \lambda\Gamma(\mathbf{z}) \right\}, \qquad (9.9)$$

where $\|.\|_2^2$ is the square of the $l_2$-norm and $C_\eta$ is the i.i.d. Gaussian noise covariance matrix equal to $\sigma_\eta^2 I$. The $\lambda$ in the second term is a regularization weighting factor, and $\Gamma(\mathbf{z})$ is a smoothness regularization constraint in function of the SR image prior.

### 9.4.2 Regularized-Norm Minimization Formulation

Here, an SR image is estimated by following a regularized-norm minimization paradigm [40], [41], [42], [57]. Then, in an underdetermined system of equations (Equation 9.1), estimating the HR image $\mathbf{z}$ given a sequence of LR observations $\mathbf{y}_k$, for $k = 1, 2, ..., K$, is commonly formulated as an optimization problem minimizing an error criteria and a regularization term. Thus, the SR optimization problem can be written as

$$\hat{\mathbf{z}} = \arg\min_{\mathbf{z}} \{f(\mathbf{z})\}, \qquad (9.10)$$

where the cost function $f(\mathbf{z})$ has the following form:

$$f(\mathbf{z}) = \frac{1}{\gamma} \sum_{k=1}^{K} E\left(\mathbf{y}_k, \mathbf{W}_k \mathbf{z}\right) + \lambda \Gamma(\mathbf{z}). \qquad (9.11)$$

In the above equation, $E(.)$ is the error term in function of $\mathbf{y}_k$ and $\mathbf{z}$. The weighting factors $\gamma$ and $\lambda$ are constant tuning parameters. The regularization term $\Gamma(.)$, which is in function of the HR image only, is designed to preserve important image content or structures, such as edges and objects, and also to increase the robustness of the solution to outliers and errors in the system. Unlike MAP-based algorithms, such as the one presented in Section 9.4.1, there is no *a priori* assumption made about the distribution function of the reconstructed HR image. Reference [42] proved that an $l_1$-norm imposed on the error residual is the most robust solution against outliers. Also in Reference [42], different regularization terms are considered for best performance in terms of robustness and edge preservation. Therefore, a cost function formulation for the SR problem can be expressed as follows:

$$f(\mathbf{z}) = \frac{1}{\gamma} \sum_{k=1}^{K} \|\mathbf{y}_k - \mathbf{W}_k \mathbf{z}\|_p^p + \lambda \Gamma(\mathbf{z}), \qquad (9.12)$$

where $\|.\|_p^p$ is the $l_p$-norm raised to the power $p = \{1, 2\}$. The weighting factor $\gamma$ is set to $\{1, \sigma_\eta^2\}$ for $p = \{1, 2\}$, respectively. Previous SR solutions are based on the warp-blur observation model following Equation 9.1. Assuming a circularly symmetric blurring matrix and a translation or rotation type of motion, then the motion and blur matrices in Equation 9.1 can be swapped. Consequently, the observation model, referred to by the blur-warp model, can be deduced from Equation 9.1 by defining the degradation matrix as $\mathbf{W}_k = DF_k H$, where $F_k$ is the warping matrix of size $N \times N$, the term $H$ denotes the blurring matrix of size $N \times N$ representing the common PSF function, and $D$ is the decimation matrix of size $N_1 N_2 \times N$. The question as to which of the two models (blur-warp or warp-blur) should be used in SR solutions is addressed in Reference [58]. Following this blur-warp observation model, a fast implementation of the regularized-norm minimization solution, referred to as the fusion-restoration (FR) approach, can be achieved by solving for a blurred estimate $\mathbf{z}_b = H\hat{\mathbf{z}}$ of the HR image followed by an interpolation and deblurring iterative step. Often, the blurred HR estimate is a non-iterative approach composed of

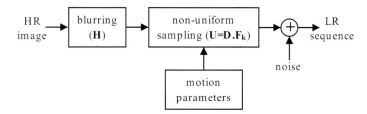

**FIGURE 9.4**

The blur-warp observation model for multi-frame SR.

registering all the LR observations relative to the HR grid and estimating the HR pixel by using an average or median operator of the LR pixels at each HR location [40], [41], [42]. The formulation of this FR approach will be discussed in Section 9.7.2.

### 9.4.3 Non-Iterative Kernel-Based Formulation

In the category of non-iterative kernel-based SR solutions, following the blur-warp observation model, all the LR observations are registered and merged on a common HR grid and non-iterative kernel-based solutions solve for the best estimate through adaptive kernel design techniques. The blur-warp acquisition model can be visualized as a non-uniform sampler ($\mathbf{U} = DF_k$) applied on the blurred HR estimate $\mathbf{z}_b$ as shown in Figure 9.4.

A locally adaptive approach using such SR estimators is described in References [43] and [44], where the fused LR samples are processed using a moving observation window to estimate the interpolation kernel and an estimation window to apply the designed kernel on the spanned LR observed samples to estimate the missing HR pixels. Figure 9.5 illustrates the non-iterative estimation approach using locally adaptive convolution kernel processing.

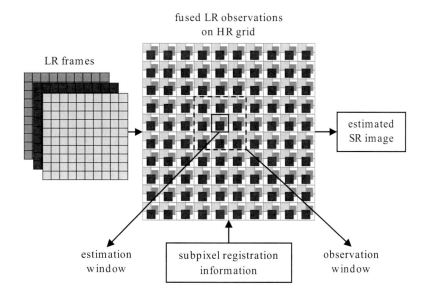

**FIGURE 9.5**

Non-iterative kernel-based SR.

From Figure 9.5, consider that the observation window is of size $W_x \times W_y$ pixels on the HR grid and spans $P = KW_xW_y/L^2$ LR pixels denoted by vector $\mathbf{G}_i$, and the estimation window is of size $D_x \times D_y$ pixels on the HR grid and spans $D_xD_y$ estimated SR pixels denoted by vector $\mathbf{D}_i$, where $i$ is the location of the respective window index. Estimating a local set of SR pixels can then be achieved by simply filtering the vector $\mathbf{G}_i$ by its locally designed kernel coefficients $\mathbf{W}_i$, following $\mathbf{D}_i = \mathbf{W}_i.\mathbf{G}_i$. In Reference [43], the interpolation kernel coefficients are designed by minimizing the mean square error of $\mathbf{D}_i - \mathbf{W}_i\mathbf{G}_i$. Solving for the optimal weights of the matrix $\mathbf{W}_i$ for each partition reduces to an adaptive Wiener filter solution for the considered observed LR pixels in the window. In References [44] and [59], the weights of the Wiener filter are defined as

$$\mathbf{W}_i = \mathbf{R}_i^{-1}\mathbf{P}_i, \qquad (9.13)$$

where $\mathbf{R}_i = E\left\{\mathbf{G}_i\mathbf{G}_i^T\right\}$ and $\mathbf{P}_i = E\left\{\mathbf{G}_i\mathbf{D}_i^T\right\}$. Thus, the determination of the weighting coefficients in Equation 9.13 requires the unknown HR image that can be either modeled parametrically or by training data. To avoid training, a parametric modeling approach can be adopted as described in Reference [43]. This category of SR estimation is inherently computationally efficient since it is non-iterative in nature.

### 9.4.4   Super-Resolution Solutions in Practice

Existing SR approaches suffer from extremely high computational requirements due to the high dimensionality in terms of large number of unknowns to be estimated in the solution of the SR inverse problem. Generally, MAP-based SR methods are computationally expensive, but can converge to a high-quality solution. Even for solutions with fast convergence rates, commonly used Bayesian approaches are conditioned on the number of different LR observations and the HR image prior statistical model that can lead to very high computational requirements even for small image estimates. To reduce the computations required for the regularized norm minimization SR solutions, the fusion-restoration (FR) methods register and merge all the LR observations on one HR grid before using an iterative regularized minimization reconstruction process. However, these solutions are still computationally intensive due to the high dimensionality of the problem with good reconstruction quality. Moreover, the non-iterative fusion-interpolation (FI) SR approaches, also referred to as kernel-based SR, are inherently less computationally intensive but suffer from limited reconstruction quality depending on their assumed statistical model. Learning-based SR approaches recover missing high-frequency information in an image through matching using a large image database of training sets. The drawbacks of these methods is the requirement of large representative training sets that are targeted toward example-based and specific applications. For example, the MAP-based SR approach in Reference [37] requires a total of approximately $395 \times 10^6$ multiplication and addition operations to estimate an HR image with $256 \times 256$ pixels from 16 LR frames of size $64 \times 64$ pixels with translational motion and noise. Also, for the same problem, the FR-based SR method in Reference [42] requires approximately a total of $200 \times 10^6$ multiplication and addition operations. Additionally, the inherently faster non-iterative FI-based SR approach in Reference [43], with parameters set as $W_x = W_y = 12$, $D_x = D_y = 4$, and $\rho = 0.75$, requires $72 \times 10^6$ multiplication and addition operations but suffers from a limited reconstruction quality.

**FIGURE 9.6**

Super-resolved $256 \times 256$ HR *Cameraman* image obtained using sixteen $64 \times 64$ low-resolution images with noise standard deviation $\sigma_n = 4$: (a) original image, (b) bicubic interpolation with $PSNR = 22.44$ dB, (c) baseline MAP-SR with $PSNR = 25.67$ dB, (d) baseline FR-SR with $PSNR = 25.73$ dB, and (e) non-iterative FI-SR with $PSNR = 24.11$ dB.

Figure 9.6 compares the visual quality of the HR estimates generated using various SR solutions described in References [37], [42], and [43]. As it can be seen, the MAP-based [37] and the FR-based [42] result in a noticeably sharper image and approximately 2 dB increase in PSNR compared to the FI-based SR approach [43]. Given these results, the following sections present a new trend of efficient selective quality-preserving SR techniques that are driven by perceptual modeling of low-level features.

## 9.5 Perceptually Driven Efficient Super-Resolution

The high dimensionality of the SR image reconstruction problem demands high computational efficiency to be deemed of any practical value. As described in Section 9.4, many powerful iterative solutions were proposed to reduce the complexity and increase the stability of solving a very large system of linear equations. Although these SR methods are theoretically justifiable and presented reliable results in terms of image quality and robustness, still at each iteration all the pixels are processed on an HR grid inclusively and thus they still suffer from high dimensionality in solving the inverse problem. As a consequence,

(a)         (b)         (c)

(d)         (e)         (f)

**FIGURE 9.7**

Super-resolved $256 \times 256$ HR *Cameraman* image obtained using sixteen $64 \times 64$ low-resolution images with noise standard deviation $\sigma_n = 10$: (a) original image, (b) bicubic interpolation, (c) baseline MAP SR, (d) SELP-MAP SR, (e) gradient-detector with block-based mean thresholds (G-MAP) SR, and (f) entropy-detector with block-based mean thresholds (E MAP) SR.

a selective perceptually driven efficient (SELP) SR framework, that takes into account the preservation of perceptual quality and the reduction of computational efficiency, has been introduced to reduce the dimensionality of existing iterative SR techniques. The perceptual SR approach is selective in nature; only a limited set of perceptually significant pixels, detected through human perceptual and/or attentive modeling, are super-resolved.

Early attempts on selective SR processing, as presented in References [10] and [11], used a non-perceptual gradient-based approach in order to detect active pixels that are significant for SR processing. Although the gradient-based approach (and other similar high-frequency detection-based approaches) resulted in savings, it suffered from a significant drawback, consisting of using a different threshold on the gradient for different images in order to be able to detect the pixels of interest and achieve good performance. This is not practical as it requires tweaking the gradient threshold for each image differently. Furthermore, general high-frequency detection methods, such as gradient or entropy based, do not incorporate any perceptual weighting and cannot automatically adapt to an image's local high-frequency content that is perceptually relevant to the human visual system (HVS). Thus, in the SELP SR method [12], [13], a set of perceptually significant pixels is determined adaptively using an automatic perceptual detection mechanism that proves to work over a broad set of images without any manual tuning.

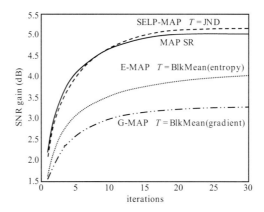

**FIGURE 9.8**

SNR comparison of the baseline MAP SR, SELP-MAP SR, G-MAP SR with block mean thresholds, and E-MAP SR approaches with block mean thresholds using sixteen $64 \times 64$ low-resolution images with noise standard deviation $\sigma_n = 10$ for the $256 \times 256$ *Cameraman* image.

The problem of devising automatic detection thresholds that can adapt to local image content perceptually is of major importance. Figures 9.7 and 9.8 demonstrate the superior performance of perceptual automatic thresholds, such as the SELP SR algorithm of References [12] and [13], and the one discussed in the following section, over the simple non-perceptual gradient and entropy-based approaches. To avoid the impractical manual tweaking of the detection thresholds for each image differently and to eliminate global thresholding that does not adapt to local image features and which was shown in Reference [11] to require manual tweaking to lead to good performance, the local non-perceptual detection thresholds are computed as the mean value corresponding to the magnitude of the gradient or entropy of each block of $8 \times 8$ pixels. That is, the detection mechanism in Reference [11] is replaced by locally thresholding the magnitude of the gradient (G-MAP), and locally thresholding the entropy (E-MAP).

Figure 9.7 clearly demonstrates the superior perceptual quality of the SELP SR method (Figure 9.7d) as compared with the non-perceptual methods, including the gradient-based (Figure 9.7e) and the entropy-based (Figure 9.7f) methods. Compared to the gradient-based and the entropy-based MAP SR methods, the SELP-MAP method results in a better reconstruction of edges, as it can be seen around regions such as the tripod of the camera and the face of the cameraman. Furthermore, Figure 9.8 illustrates the significant increase in SNR gains per iteration using the SELP-MAP method compared to the gradient-based and entropy-based MAP SR methods. It also shows that the SELP-MAP and the baseline MAP SR algorithms have similar performance.

### 9.5.1 Selective Perceptual Super-Resolution Framework

A block diagram of the SELP-SR estimation framework is shown in Figure 9.9. Initially, a rough estimate $z_0$ of the high-resolution image is obtained by either interpolating one of the LR images in the observed sequence or by fusing all the LR frames on one HR grid (also referred to as the shift-and-add technique). Other techniques, such as learning-based approaches [31], [32], [33], can also be used to produce initial estimates. At each iteration

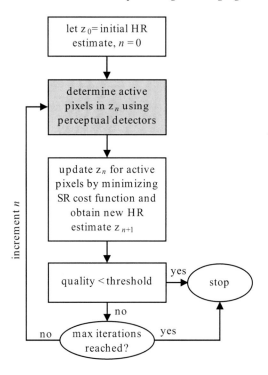

**FIGURE 9.9**

Flowchart of the selective perceptual SR framework.

and for each estimated HR image, active pixels are detected based on the human visual detection model described in Section 9.5.2. Then, only these perceptually active pixels are updated by the SR estimation algorithm, which is based generally on minimizing an SR cost function. As a result, at each iteration, only a subset of pixels (that is, the active pixels) is selected for the SR processing phase. The algorithm stops whenever the change between the current HR estimate and the previous one is less than a small number $\varepsilon$, or whenever the total number of specified iterations is reached. Only the selected perceptually active pixels need to be included in computing the change in the estimated SR frames between iterations; the other pixels remain unchanged. The SELP SR framework is very flexible in that any iterative SR estimation algorithm can be easily integrated in the SELP system.

### 9.5.2 Perceptual Contrast Sensitivity Threshold Model

Neurons in the primary visual cortex of the human visual system (HVS) are sensitive to various stimuli of low-level features in a scene, such as color, orientation, contrast, and luminance intensity [60]. The luminance sensitivity also referred to as light adaptation is the discrimination of luminance variations at every light level. Moreover, the contrast sensitivity is the response to local variations of luminance to the surrounding luminance [60]. Limits on the human visual sensitivity to low-level stimuli, such as light and contrast, are converted to masking thresholds that are used in perceptual modeling. Masking thresholds are levels above which a human can start distinguishing a significant stimulus or distortion [61]. Thus, the human visual detection model discriminates between image components based on contrast sensitivity of local information to their surroundings. In References [12] and [13], the SELP SR scheme attempts to exclude less significant information from SR processing by exploiting the masking properties of the human visual system through generating contrast sensitivity detection thresholds. The contrast sensitivity threshold is the measure of the smallest contrast, or the so-called just noticeable difference (JND), that yields a visible signal over a uniform background.

Digital natural images can be represented using linear weighted combinations of cosine functions through the discrete cosine transform (DCT). This is exploited in the lossy JPEG standard and many other image compression algorithms, including lossless ones. The models presented in References [62] and [63] derive contrast sensitivity thresholds and contrast masking thresholds for natural images in the DCT domain. The contrast sensitivity model considers the screen resolution, the viewing distance, the minimum display luminance $L_{min}$, and the maximum display luminance $L_{max}$. The contrast sensitivity thresholds are computed locally in the spatial domain using a sliding window of size $N_{blk} \times N_{blk}$. The obtained thresholds per block will be used to select the pixels to be super-resolved for each HR estimate. The model, described in this section, involves first computing the contrast sensitivity threshold $t_{128}$ for a uniform block having a mean grayscale value equals to 128, and then obtaining the threshold for any block having arbitrary mean intensity using the approximation model presented in Reference [63].

The contrast sensitivity threshold $t_{128}$ of a block in the spatial domain is computed as

$$t_{128} = \frac{TM_g}{L_{max} - L_{min}}, \tag{9.14}$$

where $M_g$ is the total number of grayscale levels (that is, $M_g = 255$ for eight-bit images), and $L_{min}$ and $L_{max}$ are the minimum and maximum display luminances, respectively. The threshold luminance $T$ is evaluated based on the parametric model derived in Reference [62] using a parabolic approximation, where $T = \min(10^{g_{0,1}}, 10^{g_{1,0}})$ and the terms $g_{0,1}$ and $g_{1,0}$ are defined as follows:

$$g_{0,1} = \log_{10} T_{min} + K(\log_{10}[1/(2N_{blk}w_y)] - \log_{10}[f_{min}])^2, \tag{9.15}$$

$$g_{1,0} = \log_{10} T_{min} + K(\log_{10}[1/(2N_{blk}w_x)] - \log_{10}[f_{min}])^2, \tag{9.16}$$

where $w_x$ and $w_y$ denote the horizontal width and vertical height of a pixel in degrees of visual angle, respectively. The term $T_{min}$ denotes the luminance threshold at the frequency $f_{min}$, where the threshold is minimum, and $K$ determines the steepness of the parabola. The parameters $T_{min}$, $f_{min}$, and $K$ are the luminance-dependent parameters of the parabolic model and are computed as follows [62]:

$$T_{min} = \begin{cases} \left(\frac{L}{L_T}\right)^{\alpha_T} \frac{L_T}{S_0} & \text{for } L \leq L_T, \\ L/S_0 & \text{for } L > L_T, \end{cases} \tag{9.17}$$

$$f_{min} = \begin{cases} f_0 \left(\frac{L}{L_f}\right)^{\alpha_f} & \text{for } L \leq L_f, \\ f_0 & \text{for } L > L_f, \end{cases} \tag{9.18}$$

$$K = \begin{cases} K_0 \left(\frac{L}{L_K}\right)^{\alpha_K} & \text{for } L \leq L_K, \\ K_0 & \text{for } L > L_K. \end{cases} \tag{9.19}$$

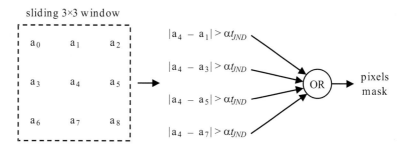

**FIGURE 9.10**

Generation process of the SELP mask.

In Reference [62], the values of the constants in Equations 9.17 to 9.19 are set as $L_T = 13.45\ cd/m^2$, $S_0 = 94.7$, $\alpha_T = 0.649$, $\alpha_f = 0.182$, $f_0 = 6.78\ cycles/deg$, $L_f = 300\ cd/m^2$, $K_0 = 3.125$, $\alpha_K = 0.0706$, and $L_K = 300\ cd/m^2$. For a background value of 128, the local background luminance is computed as

$$L = L_{min} + 128\frac{L_{max} - L_{min}}{M_g}, \tag{9.20}$$

where $L_{min}$ and $L_{max}$ denote the minumum and maximum luminance of the display, respectively. Once the threshold $t_{128}$ at a grayscale value 128 is calculated using Equation 9.14, the just noticeable difference (JND) thresholds for the other grayscale values are approximated using a power function [63] as follows:

$$t_{JND} = t_{128}\left(\frac{\sum_{n_1=0}^{N_{blk}-1}\sum_{n_2=0}^{N_{blk}-1}I_{n_1,n_2}}{N_{blk}^2(128)}\right)^{a_T}, \tag{9.21}$$

where $I_{n_1,n_2}$ is the intensity level at pixel location $(n_1, n_2)$ and $a_T$ is a correction exponent that controls the degree to which luminance masking occurs and is set to $a_T = 0.649$, as given in Reference [63]. Note that if the block has a mean of 128, then $t_{JND}$ of Equation 9.21 reduces to $t_{128}$ as expected.

A DELL UltraSharp 1905 FP liquid crystal display is used to show the images. For a screen resolution of $1280 \times 1024$, and for a measured luminance of $L_{min} = 0\ cd/m^2$ and $L_{max} = 175\ cd/m^2$, the threshold $t_{128}$ is computed to be equal to 3.3092 for $N_{blk} = 8$.

### 9.5.3 Perceptually Significant Pixels Detection

The perceptual mask determines the significant pixels that need to be processed at each iteration of the SR algorithm. For each estimated HR image, the perceptual mask is generated based on comparisons with the computed JND thresholds $t_{JND}$ (Section 9.5.2) over a local block of size $8 \times 8$ pixels. Note that $t_{128}$ is computed only once according to Equation 9.14 since it is a constant. Also, for all eight-bit images, the remaining $t_{JND}$ values, corresponding to the 255 possible mean intensity values (other than 128), can be precomputed and stored in a lookup table (LUT) in memory. For each image block, the mean of the block is computed and the corresponding $t_{JND}$ is simply retrieved from the LUT.

(a)                              (b)

**FIGURE 9.11**

Illustration of the detected set of active pixels: (a) the blurred and noisy version of the $512 \times 512$ *Monarch* image and (b) the corresponding map $\mathbf{M}_p$ with white intensities denoting the active pixels selected using the perceptual mask.

As shown in Figure 9.10, after $t_{JND}$ is obtained, the center pixel of a sliding $3 \times 3$ window is compared to its four cardinal neighbors. If any absolute difference between the current center pixel and any of its four cardinal neighbors is greater than $t_{JND}$, then the corresponding pixel mask is set to one indicating an active pixel. The remaining pixel mask locations are set to zero corresponding to non-active pixels. The sliding window will scan all the pixels of the HR image estimate to generate the perceptual mask. In general, the active locations consist of pixels in regions where edges are visible to the HVS. As a result, the perceptual mask signals a subset of pixels that are perceptually relevant for SR processing, resulting in significant computational savings. The SELP framework saves a significant number of computations in the SR processing stage with minimal overhead of operations per pixel for generating the perceptual mask. Figure 9.11 shows an example of the perceptually active pixels that are obtained by applying the SELP mask to a blurred and noisy version of the $512 \times 512$ *Monarch* image.

## 9.6 Attentively Driven Efficient Super-Resolution

To further enhance the SELP SR framework, previous work [14], [15] showed that not all the detail pixels detected by the SELP algorithm are needed to preserve the overall visual quality of an HR image. It is known that the human visual system scans a visual scene through a small window, restricted by the foveal region, having high central resolution and degrading resolution toward the peripheries. A large field of view is processed by a number of fixation points, attended to with high visual acuity, connected by fast eye movements referred to as saccades. The ordered selection of these regions of interest is predicted according to a visual attention model by studying the eye movement sensitivity to top-down mechanisms, such as image understanding, and bottom-up salient image features, such as contrast of color intensity, edge orientations, and object motion [64], [65].

Given the fact that the attended regions are processed at high visual acuity, artifacts present in these regions are better perceived by the HVS than artifacts present in non-attended areas. In consequence, the observer's judgment of image quality is prejudiced by distortions present in salient regions as shown in Reference [66]. Following this logic, saliency maps generated by visual attention modeling, can play a fundamental role in reducing the number of processed pixels of selective SR approaches. Hence, an attentive selective perceptual (AT-SELP) SR estimators, that exploit the human visual attention for processing the visual content, are introduced in References [14], [15], and [67]. Moreover, different low-level features influenced by visual attention models presented in References [68], [69], and [70] are studied to illustrate the efficiency and quality of the attentive SR framework.

### 9.6.1  Attentive-Selective Perceptual Super-Resolution Framework

A block diagram of the AT-SELP SR estimation framework is shown in Figure 9.12. Similarly as in Figure 9.9, at each iteration and for each estimated HR image, active pixels are detected based on the human visual detection model described in Sections 9.5.2 and 9.6.2. Then, only these perceptually active pixels are updated by the SR estimation algorithm which is generally based on minimizing an SR cost function as in Equation 9.10. As a result, at each iteration, only a subset of pixels (that is, the active pixels) is selected for the SR processing phase.

The first phase of the SR algorithm processes the perceptually active pixels determined by a contrast sensitivity mask $\mathbf{M}_p$, as explained in Section 9.5.3. Then, in the second phase of the SR estimation, only the subset of active pixels that is determined to be salient by the selective attention mask $\mathbf{M}_a$ is further iterated upon. The process of updating the HR estimates of the perceptual/attentive active pixels continues until a maximum number of iterations is reached or the system stabilizes; that is, until $|\mathbf{M}_a.(\mathbf{z}_{n+1} \quad \mathbf{z}_n)|/|\mathbf{M}_a.\mathbf{z}_n| < c$ in the attentive active region and $|\mathbf{M}_p.(\mathbf{z}_{n+1} - \mathbf{z}_n)|/|\mathbf{M}_p.\mathbf{z}_n| < s.\varepsilon$ in the perceptual non-attentive active region. Here, $s$ is a scaling factor greater than 1 and $\varepsilon$ is a predetermined threshold that represents the desired accuracy or quality of the SR algorithm. Only the selected perceptually active and salient attentive pixels need to be included in computing the change in the estimated SR frames between iterations; the other pixels do not change their values. It is necessary in the first phase of the algorithm to super-resolve the perceptually significant information of the non-attended regions to a certain acceptable quality level ($s.\varepsilon$) that will not attract and bias the HVS perception of the background quality, thus, leading to an acceptable homogeneous quality of the entire image. As a result, the salient regions are reconstructed with a higher visual acuity while maintaining a trade-off between smoothing the flat regions dominated by noise and sharpening the perceptually relevant edges.

### 9.6.2  Saliency-Based Mask Detection

Any saliency-based visual attention model, such as those in References [68], [69], and [70], can be adopted to detect the attentive mask $\mathbf{M}_a$ in the AT-SELP SR framework to further reduce the set of active pixels selected by the contrast sensitivity detector, thus reducing the computational complexity of the SR algorithms. Saliency maps $\mathbf{S}$ combine several low-level image features that compete to attract the human attention, providing

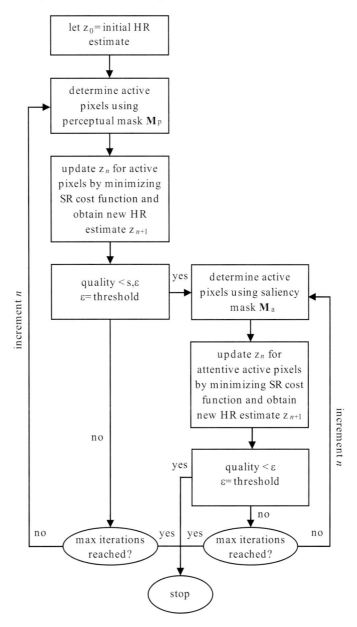

**FIGURE 9.12**

Flowchart of the AT-SELP SR framework.

measures of the level of attention at every point in the visual scene. After computing the saliency information at each point in the scene, the attentive mask $\mathbf{M}_a$ is generated by choosing the pixel locations corresponding to the highest $\tau\%$ of the saliency map values. Existing saliency map generation techniques inspired by a hierarchical visual attention model (IT) [68], a foveated gaze attentive model (GAF) [69], and a frequency-tuned attention model (FT) [70] are detailed in this section for further comparisons and analysis in the AT-SELP SR framework.

### 9.6.2.1 Hierarchical Saliency Model

Reference [68] computes a hierarchical saliency map $\mathbf{S}_{IT}$ using center-surround differences of intensity and orientation between different dyadic Gaussian scales. The input image $I$ is subsampled into a dyadic Gaussian pyramid of four levels $\sigma$, obtained by progressively filtering and downsampling each direction separately. The intensity information is the image intensity values at each pyramid level, $I_\sigma$. The orientation information is calculated by convolving the intensity pyramid with Gabor filters as follows:

$$O(\sigma, \theta) = \|I_\sigma * G_0(\theta)\| + \|I_\sigma * G_{\pi/2}(\theta)\|, \tag{9.22}$$

where $G_\psi(\theta)$ is a Gabor filter with phase $\psi = [0, \pi/2]$ and orientation $\theta = [0, \pi/4, \pi/2, 3\pi/4]$. The foveated visual perception and the antagonistic "center-surround" process are implemented as across-scale differences between fine levels $c = \{1, 2\}$ corresponding to center pixels, and coarse levels $s = c + \delta$ with chosen $\delta = \{1, 2\}$ corresponding to surround pixels. The across-scale difference $\ominus$ is calculated by interpolation to the finer scale followed by point-by-point subtraction. Feature maps that signify the sensitivity of the HVS to differences in intensity and orientation are calculated as follows:

$$F_I(c, s) = |I(c) \ominus I(s)|, \tag{9.23}$$
$$F_O(c, s, \theta) = |O(c, \theta) \ominus O(s, \theta)|. \tag{9.24}$$

At this point, four intensity feature maps and sixteen orientation feature maps are created. Each group of feature maps is combined into two conspicuity maps through across-scale addition $\oplus$ by downsampling each map to the second scale of the pyramid followed by point-by-point addition. A map normalization operator $\aleph(.)$ is applied to scale the values of different ranges into a common fixed range $[0, M]$. The conspicuity maps for intensity and orientation are calculated as follows:

$$C_I = \oplus_{c=1}^2 \oplus_{s=c+1}^{c+2} \aleph(I(c, s)), \tag{9.25}$$
$$C_O = \sum_\theta \oplus_{c=1}^2 \oplus_{s=c+1}^{c+2} \aleph(O(c, s, \theta)). \tag{9.26}$$

A saliency map is then calculated by averaging over the two normalized conspicuities as

$$S = \frac{1}{2}(\aleph(C_I) + \aleph(C_O)). \tag{9.27}$$

Figure 9.13a shows the attentive mask $\mathbf{M}_a(IT)$ generated from the hierarchical saliency map with $\tau = 20\%$. The resulting attended regions are used in the AT-SELP method to promote the detected subset of attended pixels for further iterations or enhancement.

### 9.6.2.2 Foveated Saliency Model

A foveated visual attention model in Reference [69] is based on the analysis of the statistics of low-level features at the point of gaze in a scene. The saliency map $\mathbf{S}_{GAF}$ is generated by a foveated combination of low-level image features, such as mean luminance, contrast, and bandpass outputs of both luminance and contrast. For an image patch of size $M = M_1 \times M_2$, the mean luminance $\bar{I}$ is calculated as follows:

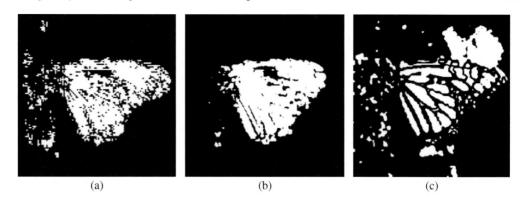

(a)          (b)          (c)

**FIGURE 9.13**

Illustration of the detected set of active pixels (20% of highest saliency values) for the blurred and noisy version of the $512 \times 512$ *Monarch* image shown in Figure 9.11 by generating and applying: (a) the hierarchical attentive mask $\mathbf{M}_a(IT)$, (b) the foveated attentive mask $\mathbf{M}_a(GAF)$, and (c) the frequency-tuned attentive mask $\mathbf{M}_a(FT)$. White intensities denote active pixels.

$$\bar{I} = \frac{1}{\sum_{i=1}^{M} w_i} \sum_{i=1}^{M} I_i w_i, \tag{9.28}$$

where $I_i$ is the grayscale value of the pixel at patch location $i$, and $w_i$ is the raised cosine function given by

$$w_i = 0.5[\cos(\frac{\pi r_i}{R}) + 1], \tag{9.29}$$

where $r_i = \sqrt{(x_i - x_c)^2 + (y_i - y_c)^2}$ is the radial distance of pixel location $(x_i, y_i)$ from the center of the patch $(x_c, y_c)$ and $R$ is patch radius. Then, the root-mean-squared contrast $C$ is computed as follows:

$$C = \sqrt{\frac{1}{\sum_{i=1}^{M} w_i} \sum_{i=1}^{M} w_i \frac{(I_i - \bar{I})^2}{(\bar{I})^2}}. \tag{9.30}$$

The bandpass of patch luminance features are computed by using Gabor kernels designed for different eccentricity values. Eccentricity values $e$, measured in degrees of visual angle, represent the distance needed to reach a particular patch. Then the bandpass-luminance for each patch is computed as $G_{lum} = \max |G_{lum}(e) * I(e)|$, where $G_{lum}(e)$ is the designed Gabor filter at eccentricity $e$. Similarly, the bandpass of patch contrast is computed using $G_{grad} = \max |G_{grad}(e) * |\nabla(I(e))||$, where $G_{grad}(e)$ is the designed Gabor filter at eccentricity $e$ of the image gradient $|\nabla(I(e))|$. More details on the Gabor filters design can be found in Reference [69]. The saliency map is then calculated by averaging all the computed image features after scaling them to a common fixed range $[0, M]$. Figure 9.13b shows the attentive mask $\mathbf{M}_a(GAF)$ generated from the foveated saliency map with $\tau = 20\%$.

### 9.6.2.3 Frequency-Tuned Saliency Model

Reference [70] proposed a frequency-tuned saliency approach, using difference of Gaussians (DoG) of luminance intensity, to generate saliency map information $\mathbf{S}_{FT}$. The bandpass filters using DoG are designed by properly selecting the standard deviation of the

Gaussian filters targeting saliency generation. In Reference [70], the saliency map information ($\mathbf{S}_{FT}$) is generated by taking the magnitude of the differences between the image mean vector and Gaussian blurred version using a $5 \times 5$ separable kernel as follows:

$$\mathbf{S}_{FT} = |I_\mu - I_{w_{hc}}|, \tag{9.31}$$

where $|.|$ is the norm operator, $I_\mu$ is the arithmetic mean pixel value of the image, and $I_{w_{hc}}$ is the Gaussian blurred version of the original image. Figure 9.13c shows an example of the attentive mask $\mathbf{M}_a(FT)$ to a blurred and noisy version of the $512 \times 512$ *Monarch* image. Comparing the masks shown in Figures 9.13a to 9.13c and the perceptual mask ($\mathbf{M}_p$) shown in Figure 9.11 reveals that the significantly lower number of active pixels is selected for SR processing using the attentive masking phase.

The existing visual attention approaches [68], [69] introduce a significant computational overhead for computing the saliency maps. These would be useful for the AT-SELP framework only if the saliency maps were already precomputed as part of another task (such as visual quality assessment for example). Such available saliency maps can then be exploited to improve the efficiency of the SR process. However, if these saliency maps were not available, generating those using the existing visual attention methods would not improve the SR process due to the high computational overhead that is introduced when computing these saliency maps. In order to tackle this issue and make the AT-SELP process feasible when saliency maps are not available, an efficient low-complexity JND-based saliency detector targeted toward efficient selective SR was recently proposed in Reference [67].

## 9.7 Perceptual SR Estimation Applications

The perceptually driven SR frameworks (SELP, AT-SELP), previously discussed in Sections 9.5 and 9.6, are very flexible in that any iterative SR estimation algorithm can be easily integrated in the perceptually selective system. Hence, application of the selective perceptual (SELP) and the attentive selective (AT-SELP) SR framework to an iterative MAP-based SR method [37] and a FR-based SR method [42] are described below in Sections 9.7.1 and 9.7.2, respectively. The attentively or perceptually driven SR schemes result in significant computational savings while maintaining the perceptual SR image quality, as compared to the iterative baseline SR schemes [37], [42].

### 9.7.1 Perceptual MAP-Based SR Estimation

Existing Bayesian MAP-based SR estimators present high-quality estimation but suffer from high computational requirements [36], [37], [38], [39]. In order to illustrate the significant reduction in computations for MAP SR techniques, the popular gradient-descent optimization-based algorithm from Reference [37] is integrated into the attentive-perceptual selective SR framework (SELP, AT-SELP).

In Reference [37], a uniform detector sensitivity is assumed over the span of the detector degradation model, then the point spread function (PSF) $H$ in the observation model

(Section 9.3) is represented using an averaging filter. Following the Bayesian MAP-based SR formulation in Section 9.4.1, the regularization smoothness constraint term (in the cost function of Equation 9.9) is represented in Reference [37] by $\Gamma(\mathbf{z}) = \frac{1}{2}\mathbf{z}^T \mathbf{C}_z^{-1}\mathbf{z}$, where $\mathbf{C}_z$ is the covariance of the HR image prior model imposing a piecewise smoothness relationship between neighboring pixels in $\mathbf{z}$ as follows [37]:

$$\mathbf{C}_z^{-1} = \frac{1}{\lambda} \sum_{i=1}^{N} d_{i,k} \left( \sum_{j=1}^{N} d_{i,j} z_j \right), \tag{9.32}$$

where $\lambda$ is scaling factor controlling the effect of rapidly changing features in $\mathbf{z}$, and $z_j$ is the pixel at the $j$th location of the lexicographically ordered vector $\mathbf{z}$. The coefficients $d_{i,j}$, which express *a priori* assumption about the relationship between neighboring pixels in $\mathbf{z}$, are defined as follows [37]:

$$d_{i,j} = \begin{cases} 1 & \text{for } i = j, \\ -1/4 & \text{for } i \neq j : z_j \in \text{cardinal neighbors of } z_i \end{cases} \tag{9.33}$$

In the AT-SELP-MAP SR scheme, the gradient descent minimization procedure is applied selectively and only to pixels that are determined to be perceptually significant by the attentive-perceptual detectors (Sections 9.5.3 and 9.6.2). Following the steepest descent solution in the perceptually driven framework, the HR image is estimated using Equation 9.8 as follows [37]:

$$\hat{\mathbf{z}}_{n+1} = \hat{\mathbf{z}}_n - \beta_n \mathbf{M}\{\frac{1}{\sigma_\eta^2} \sum_{k=1}^{K} \mathbf{W}^T (\mathbf{W}\mathbf{z}_n - \mathbf{y}) + \frac{1}{2}\mathbf{C}_z^{-1}\mathbf{z}_n\}, \tag{9.34}$$

where $\mathbf{M}$ represents the attentive-perceptual mask for selecting active pixels at every iteration of the SR process. Note that the initial HR image estimate is an interpolated version of one of the LR frames. The elements in $\mathbf{M}$ take binary values, where ones indicate active pixels that should be included in the SR update at the current iteration, whereas zero mask values indicate that the corresponding pixels are non-active and will not be processed in the current iteration.

### 9.7.2 Perceptual Fusion-Restoration SR Estimation

The efficient perceptually driven framework can be used in conjunction with the fusion-restoration (FR) SR algorithms. Reference [42] proposed a two-step algorithm by using first a non-iterative data fusion step followed by an iterative gradient-descent deblurring-interpolation step. This algorithm models the relative motion between low-resolution frames as translational and the point spread function (PSF) as an $L_1 \times L_2$ Gaussian lowpass filter with a standard deviation equals to 1. Following the regularized-norm SR formulation in Section 9.4.2, a fast fusion-restoration implementation of the minimization solution (Equation 9.12) can be achieved by solving for a blurred estimate $\mathbf{z}_b = H\hat{\mathbf{z}}$ of the HR image followed by an interpolation and deblurring iterative step. An initial blurred version of the HR estimate $\mathbf{z}_b$ is produced in the data fusion step by registration followed by a median operator of the LR frames on the HR grid, referred to as the "median shift and add" operator.

As for the regularization term in Equation 9.12, a bilateral total variation regularization that preserves edges is adopted as follows [42]:

$$\Gamma_{BTV}(\mathbf{z}) = \sum_{l=-R}^{R} \sum_{m=0}^{R} \alpha^{|m|+|l|} \left\| \mathbf{z} - S_x^l S_y^m \mathbf{z} \right\|_1, \tag{9.35}$$

where $S_x^l$ and $S_y^m$ shift the HR image $\mathbf{z}$ by $l$ and $m$ pixels in the horizontal and vertical directions, respectively, and $R \geq 1$ represents several scales of shifting values. The weight $\alpha$ is applied as a decaying factor for convergence purposes, and is chosen between $0 \leq \alpha \leq 1$. The SR problem in Equation 9.12 reduces to deblurring and interpolating for the missing pixels in the initial HR estimate $\mathbf{z}_b$ that is formulated as a regularized $l_1$-norm minimization problem [42]. In the AT-SELP-FR SR scheme, the steepest descent minimization is applied selectively only to the active pixels identified by the perceptual detection schemes:

$$\hat{\mathbf{z}}_{n+1} = \hat{\mathbf{z}}_n - \beta_n \mathbf{M} \left[ H^T A^T sign(AH\hat{\mathbf{z}}_n - A\mathbf{z}_b) + \right.$$

$$\left. \lambda \sum_{l=-R}^{R} \sum_{m=0}^{R} \alpha^{|m|+|l|} \left[ I - S_y^{-m} S_x^{-l} \right] sign(\hat{\mathbf{z}}_n - S_x^l S_y^m \hat{\mathbf{z}}_n) \right], \tag{9.36}$$

where $\beta_n$ is the step size in the direction of the gradient and $\lambda$ is a regularization weighting factor. Matrix $A$ is an $N \times N$ diagonal matrix with diagonal values equal to the square root of the number of measurements that contribute to make each element of $\mathbf{z}_b$. Also, $S_x^{-l}$ and $S_y^{-m}$ define a shifting effect in the opposite directions of $S_x^l$ and $S_y^m$. The term $\mathbf{M}$ represents the perceptually attentive masking that selects the active pixels that are processed at each iteration, thus reducing the computations required in Reference [42].

### 9.7.3 Perceptually Driven SR Results

The performance of the efficient perceptually driven SR framework is assessed using a set of test images shown in Figure 9.14. The images used in this experiment include *Cameraman*, *Lena*, and *Clock* from the USC image database [71], and *Monarch* from the LIVE image database [72], as shown in Figure 9.14. These images, all of the size of $256 \times 256$ pixels, differ in their characteristics. For example, the *Clock* and *Monarch* images contain many smooth regions, while the *Cameraman* and *Lena* images have more edges and texture variations. Moreover, the *Lena* image can be used to demonstrate the application of SR in face recognition applications.

|        (a)        |        (b)        |        (c)        |        (d)        |

**FIGURE 9.14**

Original $256 \times 256$ test images: (a) *Cameraman*, (b) *Lena*, (c) *Clock*, and (d) *Monarch*.

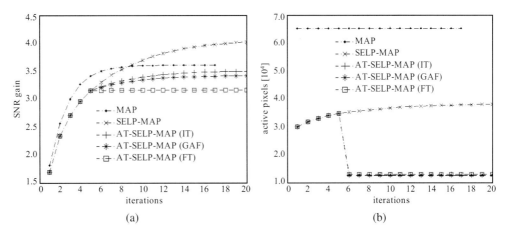

(a)　　　　　　　　　　　(b)

**FIGURE 9.15**

Comparison between the baseline MAP, SELP-MAP, and the AT-SELP-MAP SR estimators using sixteen $64 \times 64$ LR images, resizing factor $L = 4$, and noise variance $\sigma_\eta^2 = 16$ for the ninth frame of the $256 \times 256$ *Cameraman* sequence: (a) SNR gain per iteration and (b) number of processed pixels per iteration.

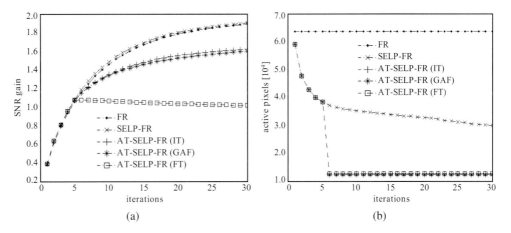

(a)　　　　　　　　　　　(b)

**FIGURE 9.16**

Comparison between the baseline FR, SELP-FR, and the AT-SELP-FR SR estimators using sixteen $64 \times 64$ LR images, resizing factor $L = 4$, and noise variance $\sigma_\eta^2 = 16$ for the ninth frame of the $256 \times 256$ *Cameraman* sequence: (a) SNR gain per iteration and (b) number of processed pixels per iteration.

A sequence of LR images is generated from a single HR image by passing the HR image through the SR degradation model described in Section 9.3 to generate a sequence of blurred, shifted, and noisy LR images. Then, for a resolution factor of four, the degradation process is applied by randomly shifting the reference $256 \times 256$ HR image in the horizontal and vertical directions, blurring the shifted images with a lowpass filter of size $4 \times 4$, and subsampling the result by a factor of four in each direction to generate sixteen $64 \times 64$ LR frames. The blur filters are modeled as an average filter and as a Gaussian filter with standard deviation of one for the MAP-based [37] and the FR-based [42] SR observation models, respectively. Then, an additive Gaussian noise of variance $\sigma_\eta^2 = 16$ is added to the resulting LR sequence. Figure 9.3 depicts the simulated sequence generation process.

The attentive-selective perceptual MAP-based super-resolution and attentive-selective perceptual fusion-restoration super-resolution schemes referred to as AT-SELP-MAP [15] and AT-SELP-FR [14], respectively, are compared with their existing non-selective counterparts MAP-SR [37] and FR-SR [42], as well as the selective SR counterparts SELP-MAP and SELP-FR [12], [13]. The simulation parameters of the compared MAP-based SR methods are set to $\lambda = 100$, $\varepsilon = 0.0001$, and $s = 100$, and a maximum of twenty iterations are performed. On the other hand, the simulation parameters of the compared FR-based SR methods are set to $R = 2$, $\alpha = 0.6$, $\lambda = 0.08$, and $\beta = 8$, $\varepsilon = 0.0001$, $s = 120$, and a maximum of thirty iterations are performed. The parameter $\tau$ of the saliency map detection is set to 20% to identify the attentive active pixels. Experimentally, percentage values between 20 and 40% offered promising results in terms of quality and computational efficiency trade-off. These numbers are plausible, since shifts of attention based on competing saliency results needs 30 to 70 ms [73]. Then, in a 30.33 ms time interval given to view one frame, assuming 30 fps video sequence, one can only perceive 20 to 40% of the salient regions. For the saliency mask detection schemes, different saliency maps generated from existing visual attention models presented in References [68], [69], [70] are also integrated and tested in the proposed AT-SELP SR framework.

Figure 9.15 gives a quantitative comparison among the baseline MAP, SELP-MAP, and proposed AT-SELP-MAP SR methods using different saliency detection techniques applied to super-resolve the ninth frame of the simulated *Cameraman* LR sequence. Similarly, Figure 9.16 shows a quantitative comparison among the baseline FR, SELP-FR, and proposed AT-SELP-FR SR methods. Error measures, represented by the SNR gain per iteration, and the computational complexity, signified by the number of processed pixels per iteration, are presented in Figures 9.15, 9.16a, and 9.16b, respectively. It can be seen that in the case of baseline MAP and FR SR methods all the pixels are processed at each iteration for all the images (that is, for a $256 \times 256$ image, the total number of processed pixels is 65536 at each iteration). As shown in Figures 9.15b and 9.16b for the SELP-MAP and SELP-FR SR methods, respectively, the number of processed pixels per iteration varies from one image to the other depending on the visual content. In Figures 9.15b and 9.16b, the visual attention processing takes effect around the fifth iteration, thus further reducing the detected active pixels processed. Due to the attentive selectivity, the AT-SELP SR framework presents considerable savings in terms of the number of processed pixels, leading to a significant reduction in computational complexity. Comparing the SNR measures in Figures 9.15a and 9.16a, it can be easily seen that the $\mathbf{M}_a(IT)$ mask integrated into the AT-SELP SR framework has the best error performance among the other implemented saliency mask detectors. Also, the efficient and simple saliency mask generator $\mathbf{M}_a(FT)$ adopted from Reference [70] does not enhance the overall quality of the SR estimate due to selecting objects and features in the image that are not relevant to SR processing applications. Furthermore, the overall quantitative assessment is also shown in terms of PSNR measures in Tables 9.1 and 9.2.

Since the employed PSNR and SNR gain measures do not necessarily reflect the resulting visual quality [74], Figures 9.17 and 9.18 show the SR outputs obtained for the ninth frame of the simulated *Cameraman* sequence for all compared MAP-based and FR-based SR methods, respectively. Despite the fact that the AT-SELP-SR scheme processes significantly less pixels per iteration, it produces a comparable visual quality to the existing

**TABLE 9.1**

PSNR values in dBs of MAP-based SR methods for all test sequences using magnification factor $L = 4$ and noise variance $\sigma_\eta^2 = 16$.

| Method | Test Sequence | | | | |
| --- | --- | --- | --- | --- | --- |
| | Cameraman | Lena | Clock | Monarch | Average |
| MAP | 25.67 | 25.22 | 28.57 | 21.01 | 25.12 |
| SELP-MAP | 26.08 | 25.51 | 29.18 | 21.73 | 25.63 |
| AT-SELP-MAP (IT) | 25.56 | 25.16 | 28.64 | 21.20 | 25.14 |
| AT-SELP-MAP (GAF) | 25.48 | 25.19 | 28.63 | 21.22 | 25.13 |
| AT-SELP-MAP (FT) | 25.23 | 24.98 | 28.47 | 20.85 | 24.88 |

**TABLE 9.2**

PSNR values in dBs of FR-based SR methods for all test sequences using magnification factor $L = 4$ and noise variance $\sigma_\eta^2 = 16$.

| Method | Test Sequence | | | | |
| --- | --- | --- | --- | --- | --- |
| | Cameraman | Lena | Clock | Monarch | Average |
| FR | 25.73 | 24.74 | 28.58 | 21.78 | 25.21 |
| SELP-FR | 25.74 | 24.75 | 28.52 | 21.81 | 25.21 |
| AT-SELP-FR (IT) | 25.48 | 24.66 | 28.36 | 21.59 | 25.02 |
| AT-SELP-FR (GAF) | 25.46 | 24.63 | 28.29 | 21.58 | 24.99 |
| AT-SELP-FR (FT) | 24.90 | 24.38 | 27.98 | 21.01 | 24.57 |

selective and non-selective [37], [42] SR schemes. Furthermore, Tables 9.3 and 9.4 show similar pixel savings achieved by the different visual attention methods when integrated in the AT-SELP SR framework. It is shown in Table 9.3 that up to 66% in pixel savings can be achieved by the AT-SELP-MAP SR over the baseline MAP SR, while around 31% in pixel savings is achieved over the efficient SELP-MAP SR. Table 9.4 shows that the proposed AT-SELP-FR framework saves around 71% in SR processing over the non-selective baseline-FR and around 44% in SR processing over the efficient SELP-FR SR.

## 9.8   Conclusion

This chapter provided an overview of existing SR techniques, with a focus on perceptually driven multi-frame super-resolution (SR) techniques. The MAP-based SR methods, that are based on Bayesian formulations of the SR problem, strongly depend on prior HR information. The MAP SR approaches were shown to have a good reconstruction quality but with high computational requirements. The FR-based SR methods provide relatively efficient implementations of the regularized-norm minimization methods by reducing matrix operations and are more robust to errors.

This chapter also presented a new class of perceptually driven selective (SELP, AT-SELP) SR solutions that can reduce the computational complexity of iterative SR problems while

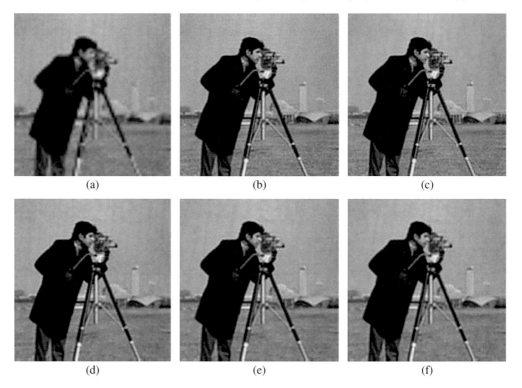

(a)    (b)    (c)

(d)    (e)    (f)

**FIGURE 9.17**

Super-resolved ninth frame of $256 \times 256$ HR *Cameraman* image obtained using sixteen $64 \times 64$ low-resolution images with $\sigma_\eta^2 = 16$ and $\tau = 20\%$: (a) bicubic interpolation, (b) baseline MAP-SR, (c) SELP-MAP SR, (d) AT-SELP-MAP SR using hierarchical $\mathbf{M}_a(IT)$, (e) AT-SELP-MAP SR using foveated $\mathbf{M}_a(GAF)$, and (f) AT-SELP-MAP SR using frequency-tuned $\mathbf{M}_u(FT)$.

**TABLE 9.3**

Percentage of pixel savings of MAP-based SR methods for all test sequences using magnification factor $L = 4$ and noise variance $\sigma_\eta^2 = 16$.

| Method | Test Sequence | | | | |
| --- | --- | --- | --- | --- | --- |
| | *Cameraman* | *Lena* | *Clock* | *Monarch* | Average |
| MAP | — | — | — | — | — |
| SELP-MAP | 34.93 | 16.37 | 46.61 | 29.10 | 31.75 |
| AT-SELP-MAP (IT) | 68.01 | 62.76 | 67.61 | 68.34 | 66.68 |
| AT-SELP-MAP (GAF) | 67.85 | 61.47 | 67.00 | 68.80 | 66.28 |
| AT-SELP-MAP (FT) | 67.41 | 62.25 | 67.68 | 67.97 | 66.33 |

maintaining the desired estimated HR media quality. It was shown that general high-frequency and edge detection methods, such as gradient-based or entropy-based methods, that do not incorporate any perceptual weighting, cannot automatically adapt to local visual characteristics that are perceptually relevant to the human visual system (HVS). Thus, the problem of devising automatic detection thresholds that can adapt to the local perceptually relevant image content is addressed. Due to human visual attention, not all the detail pixels

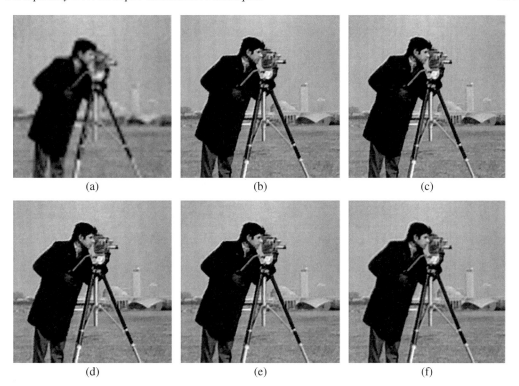

**FIGURE 9.18**

Super-resolved ninth frame of $256 \times 256$ HR *Cameraman* image obtained using sixteen $64 \times 64$ low-resolution images with $\sigma_\eta^2 = 16$ and $\tau = 20\%$: (a) bicubic interpolation, (b) baseline FR-SR, (c) SELP-FR SR, (d) AT-SELP-FR SR using hierarchical $\mathbf{M}_a(IT)$, (e) AT-SELP-FR SR using foveated $\mathbf{M}_a(GAF)$, and (f) AT-SELP-FR SR using frequency-tuned $\mathbf{M}_a(FT)$.

**TABLE 9.4**

Percentage of pixel savings of FR-based SR methods for all test sequences using magnification factor $L = 4$ and noise variance $\sigma_\eta^2 = 16$.

| Method | Test Sequence | | | | |
|---|---|---|---|---|---|
| | *Cameraman* | *Lena* | *Clock* | *Monarch* | Average |
| FR | — | — | — | — | — |
| SELP-FR | 44.47 | 36.56 | 62.77 | 34.47 | 44.57 |
| AT-SELP-FR (IT) | 71.17 | 70.23 | 74.20 | 69.37 | 71.24 |
| AT-SELP-FR (GAF) | 71.75 | 70.57 | 74.38 | 69.08 | 71.45 |
| AT-SELP-FR (FT) | 71.33 | 70.63 | 74.37 | 68.53 | 71.22 |

detected by the contrast sensitivity threshold model are needed to preserve the overall visual quality of an HR image. Toward this goal, an efficient attentive-selective perceptual (AT-SELP) SR framework jointly driven by human visual perception and saliency-based visual attention models, was presented. The effectiveness of the proposed attentive-perceptual framework in reducing the amount of SR processing while maintaining the desired visual quality was verified by investigating different saliency map techniques that combine several low-level features, such as center-surround differences in intensity and orientation, patch luminance and contrast, and bandpass outputs of patch luminance and contrast. The SELP and AT-SELP SR frameworks were shown to be easily integrated into a MAP-based SR algorithm as well as a FR-based SR estimator. Simulation results showed significant reduction on average in computational complexity with comparable visual quality in terms of subjectively perceived quality as well as PSNR, SNR, or MAE gains.

## References

[1] T. Akgun, Y. Altunbasak, and R. Mersereau, "Super-resolution reconstruction of hyperspectral images," *IEEE Transactions on Image Processing*, vol. 14, no. 11, pp. 1860–1875, November 2005.

[2] T. Celik, C. Direkoglu, H. Ozkaramanli, H. Demirel, and M. Uyguroglu, "Region-based super-resolution aided facial feature extraction from low-resolution sequences," in *Proceedings of the IEEE International Conference on Acoustics, Speech, and Signal Processing*, Philadelphia, PA, USA, March 2005, vol. 2, pp. 789–792.

[3] J. Kennedy, O. Israel, A. Frenkel, R. Bar-Shalom, and H. Azhari, "Super-resolution in PET imaging," *IEEE Transactions on Medical Imaging*, vol. 25, no. 2, pp. 137–147, February 2006.

[4] M. Robinson, S. Farsiu, J. Lo, and C. Toth, "Efficient restoration and enhancement of super-resolved X-ray images," in *Proceedings of the IEEE International Conference on Image Processing*, San Diego, CA, USA, October 2008, pp. 629–632.

[5] M. Mancuso and S. Battiato, "An introduction to the digital still camera technology," *ST Journal of System Research*, vol. 2, no. 2, pp. 1–9, December 2001.

[6] S.C. Park, M.K. Park, and M.G. Kang, "Super-resolution image reconstruction: A technical overview," *IEEE Signal Processing Magazine*, vol. 20, no. 3, pp. 21–36, May 2003.

[7] T. Komatsu, K. Aizawa, T. Igarashi, and T. Saito, "Signal-processing based method for acquiring very high resolution images with multiple cameras and its theoretical analysis," *IEE Proceedings on Communications, Speech and Vision*, vol. 140, no. 1, pp. 19–24, February 1993.

[8] B.C. Tom, N.P. Galatsanos, and A.K. Katsaggelos, "Reconstruction of a high resolution image from multiple low resolution images," in *Super-Resolution Imaging*, S. Chaudhuri (ed.), Boston, MA, USA: Kluwer Academic Publishers, 2001, pp. 73–105.

[9] Y. Wang, J. Ostermann, and Y.Q. Zhang, *Video Processing and Communications*. Boston, MA, USA: Prentice Hall, 2002.

[10] Z. Ivanovski, L. Karam, and G. Abousleman, "Selective Bayesian estimation for efficient super-resolution," in *Proceedings of the Fourth IEEE International Symposium on Signal Processing and Information Technology*, Rome, Italy, December 2004, pp. 433–436.

[11] Z. Ivanovski, L. Panovski, and L.J. Karam, "Efficient edge-enhanced super-resolution," in *Proceedings of the 3rd International Conference on Sciences of Electronic, Technologies of Information and Telecommunications*, Sousse, Tunisia, March 2005, pp. 35:1–5.

[12] R. Ferzli, Z. Ivanovski, and L. Karam, "An efficient, selective, perceptual-based super-resolution estimator," in *Proceedings of the IEEE International Conference on Image Processing*, San Diego, CA, USA, October 2008, pp. 1260–1263.

[13] L. Karam, N. Sadaka, R. Ferzli, and Z. Ivanovski, "An efficient, selective, perceptual-based super-resolution estimator," *IEEE Transactions on Image Processing*, 2011.

[14] N. Sadaka and L. Karam, "Efficient perceptual attentive super-resolution," in *Proceedings of the IEEE International Conference on Image Processing*, Cairo, Egypt, November 2009, pp. 3113–3116.

[15] N. Sadaka and L. Karam, "Perceptual attentive super-resolution," in *Proceedings of the International Workshop on Video Processing and Quality Metrics for Consumer Electronics*, Scottsdale, AZ, USA, January 2009.

[16] F. Liu, J. Wang, S. Zhu, M. Gleicher, and Y. Gong, "Visual-quality optimizing super resolution," *Computer Graphics Forum*, vol. 28, no. 1, pp. 127–140, March 2009.

[17] J. van Ouwerkerk, "Image super-resolution survey," *Image and Vision Computing*, vol. 24, no. 10, pp. 1039–1052, October 2006.

[18] A.K. Jain, *Fundamentals of Digital Image Processing*. Englewood Cliffs, NJ, USA: Prentice Hall, 1989.

[19] T. Lehmann, C. Gonner, and K. Spitzer, "Survey: Interpolation methods in medical image processing," *IEEE Transactions on Medical Imaging*, vol. 18, no. 11, pp. 1049–1075, November 1999.

[20] X. Li and M. Orchard, "New edge-directed interpolation," *IEEE Transactions on Image Processing*, vol. 10, no. 10, pp. 1521–1527, October 2001.

[21] S.G. Mallat, *A Wavelet Tour of Signal Processing*. New York, USA: Academic, 1998.

[22] W.S. Tam, C.W. Kok, and W.C. Siu, "Modified edge-directed interpolation for images," *Journal of Electronic Imaging*, vol. 19, no. 1, pp. 013011:1–20, January 2010.

[23] S. Battiato, G. Gallo, and F. Stanco, "Smart interpolation by anisotropic diffusion," in *Proceedings of the International Conference on Image Analysis and Processing*, Rome, Italy, September 2003, pp. 572–577.

[24] P. Perona and J. Malik, "Scale-space and edge detection using anisotropic diffusion," *IEEE Transactions on Pattern Analysis and Machine Intelligence*, vol. 12, no. 7, pp. 629–639, July 1990.

[25] D. Muresan and T. Parks, "Adaptively quadratic (AQUA) image interpolation," *IEEE Transactions on Image Processing*, vol. 13, no. 5, pp. 690–698, May 2004.

[26] D. Muresan and T. Parks, "Adaptive, optimal-recovery image interpolation," in *Proceedings of the IEEE International Conference on Acoustics, Speech, and Signal Processing*, Salt Lake City, UT, USA, May 2001, vol. 3, pp. 1949–1952.

[27] D. Muresan and T. Parks, "Optimal recovery approach to image interpolation," in *Proceedings of the IEEE International Conference on Image Processing*, Thessaloniki, Greece, October 2001, vol. 3, pp. 848–851.

[28] D. Muresan and T. Parks, "Demosaicing using optimal recovery," *IEEE Transactions on Image Processing*, vol. 14, no. 2, pp. 267–278, February 2005.

[29] D. Muresan and T. Parks, "Prediction of image detail," in *Proceedings of the IEEE International Conference on Image Processing*, Vancouver, BC, Canada, September 2000, vol. 2, pp. 323–326.

[30] K. Kinebuchi, D. Muresan, and T. Parks, "Image interpolation using wavelet based hidden Markov trees," in *Proceedings of the IEEE International Conference on Acoustics, Speech, and Signal Processing*, Salt Lake City, UT, USA, May 2001, vol. 3, pp. 1957–1960.

[31] W. Freeman, T. Jones, and E. Pasztor, "Example-based super-resolution," *IEEE Computer Graphics and Applications*, vol. 22, no. 2, pp. 56–65, March/April 2002.

[32] P. Gajjar and M. Joshi, "New learning based super-resolution: Use of DWT and IGMRF prior," *IEEE Transactions on Image Processing*, vol. 19, no. 5, pp. 1201–1213, May 2010.

[33] Z. Xiong, X. Sun, and F. Wu, "Robust web image/video super-resolution," *IEEE Transactions on Image Processing*, vol. 19, no. 8, pp. 2017–2028, August 2010.

[34] Y.W. Tai, W.S. Tong, and C.K. Tang, "Perceptually-inspired and edge-directed color image super-resolution," in *Proceedings of the IEEE Conference on Computer Vision and Pattern Recognition*, New York, USA, June 2006, vol. 2, pp. 1948–1955.

[35] R.Y. Tsai and T.S. Huang, "Multiframe image restoration and registration," in *Advances in Computer Vision and Image Processing*, R.Y. Tsai and T.S. Huang (eds.), Greenwich, CT, USA: JAI Press Inc., 1984, vol. 1, pp. 317–339.

[36] R.R. Shultz and R.L. Stevenson, "Extraction of high-resolution frames from video sequences," *IEEE Transactions on Image Processing*, vol. 5, no. 6, pp. 996–1011, June 1996.

[37] R.C. Hardie, K. Barnard, and E.E. Armstrong, "Joint MAP registration and high-resolution image estimation using a sequence of undersampled images," *IEEE Transactions on Image Processing*, vol. 6, no. 12, pp. 1621–1633, December 1997.

[38] B. Gunturk and M. Gevrekci, "High-resolution image reconstruction from multiple differently exposed images," *IEEE Signal Processing Letters*, vol. 13, no. 4, pp. 197–200, April 2006.

[39] B. Gunturk, Y. Altunbasak, and R. Mersereau, "Super-resolution reconstruction of compressed video using transform-domain statistics," *IEEE Transactions on Image Processing*, vol. 13, no. 1, pp. 33–43, January 2004.

[40] M. Elad and Y. Hel-Or, "A fast super-resolution reconstruction algorithm for pure translational motion and common space invariant blur," *IEEE Transactions on Image Processing*, vol. 10, no. 8, pp. 1187–1193, August 2001.

[41] A. Zomet, A. Rav-Acha, and S. Peleg, "Robust super-resolution," in *Proceedings of the IEEE Conference on Computer Vision and Pattern Recognition*, Kauai, HI, USA, December 2001, vol. 1, pp. 645–650.

[42] S. Farsiu, D. Robinson, M. Elad, and P. Milanfar, "Fast and robust multiframe super-resolution," *IEEE Transactions on Image Processing*, vol. 13, no. 10, pp. 1327–1344, October 2004.

[43] R. Hardie, "A fast image super-resolution algorithm using an adaptive Wiener filter," *IEEE Transactions on Image Processing*, vol. 16, no. 12, pp. 2953–2964, December 2007.

[44] B. Narayanan, R. Hardie, K. Barner, and M. Shao, "A computationally efficient super-resolution algorithm for video processing using partition filters," *IEEE Transactions on Circuits and Systems for Video Technology*, vol. 17, no. 5, pp. 621–634, May 2007.

[45] D. Chen and R. Schultz, "Extraction of high-resolution video stills from MPEG image sequences," in *Proceedings of the IEEE International Conference on Image Processing*, Los Alamitos, CA, USA, October 1998, vol. 2, pp. 465–469.

[46] B. Gunturk, Y. Altunbasak, and R. Mersereau, "Multiframe resolution-enhancement methods for compressed video," *IEEE Signal Processing Letters*, vol. 9, no. 6, pp. 170–174, June 2002.

[47] J. Mateos, A. Katsaggelos, and R. Molina, "Resolution enhancement of compressed low resolution video," in *Proceedings of the IEEE International Conference on Acoustics, Speech, and Signal Processing*, Istanbul, Turkey, June 2000, vol. 4, pp. 1919–1922.

[48] J. Mateos, A. Katsaggelos, and R. Molina, "Simultaneous motion estimation and resolution enhancement of compressed low resolution video," in *Proceedings of the IEEE International Conference on Image Processing*, Vancouver, BC, Canada, September 2000, vol. 2, pp. 653–656.

[49] S.C. Park, M.G. Kang, C. Segall, and A. Katsaggelos, "Spatially adaptive high-resolution image reconstruction of low-resolution DCT-based compressed images," in *Proceedings of the IEEE International Conference on Image Processing*, Rochester, NY, USA, September 2002, vol. 2, pp. 861–864.

[50] S.C. Park, M.G. Kang, C. Segall, and A. Katsaggelos, "Spatially adaptive high-resolution image reconstruction of DCT-based compressed images," *IEEE Transactions on Image Processing*, vol. 13, no. 4, pp. 573–585, April 2004.

[51] C. Segall, A. Katsaggelos, R. Molina, and J. Mateos, "Bayesian resolution enhancement of compressed video," *IEEE Transactions on Image Processing*, vol. 13, no. 7, pp. 898–911, July 2004.

[52] C. Segall, R. Molina, and A. Katsaggelos, "High-resolution images from low-resolution compressed video," *IEEE Signal Processing Magazine*, vol. 20, no. 3, pp. 37–48, May 2003.

[53] M.G. Kang and A. Katsaggelos, "General choice of the regularization functional in regularized image restoration," *IEEE Transactions on Image Processing*, vol. 4, no. 5, pp. 594–602, May 1995.

[54] R. Hardie, S. Cain, K. Barnard, J. Bognar, E. Armstrong, and E. Watson, "High resolution image reconstruction from a sequence of rotated and translated infrared images," in *Proceedings of SPIE*, vol. 3063, pp. 113–124, July 1997.

[55] R. Hardie, K. Barnard, J. Bognar, E. Armstrong, and E. Watson, "High-resolution image reconstruction from a sequence of rotated and translated frames and its application to an infrared imaging system," *Optical Engineering*, vol. 37, no. 1, pp. 247–260, January 1998.

[56] R. Schultz and R. Stevenson, "A Bayesian approach to image expansion for improved definition," *IEEE Transactions on Image Processing*, vol. 3, no. 3, pp. 233–242, May 1994.

[57] D. Mudugamuwa, X. He, D. Wei, and C.H. Ahn, "Super-resolution by prediction based sub-pel motion estimation," in *Proceedings of the International Conference Image and Vision Computing*, Dubai, United Arab Emirates, February 2009, pp. 282–287.

[58] Z. Wang and F. Qi, "On ambiguities in super-resolution modeling," *IEEE Signal Processing Letters*, vol. 11, no. 8, pp. 678–681, August 2004.

[59] M. Shao, K. Barner, and R. Hardie, "Partition-based interpolation for color filter array demosaicking and super-resolution reconstruction," *Optical Engineering*, vol. 44, no. 10, pp. 107003:1–14, October 2005.

[60] S. Winkler, *Digital Video Quality: Vision Models and Metrics*. New York, USA: Wiley and Sons, 2005, pp. 20–30.

[61] A. Bovik, *Handbook of Image and Video Processing*. San Diego, CA, USA: Academic Press, 2000, pp. 669–684.

[62] A. Ahumada and H. Peterson, "Luminance-model-based DCT quantization for color image compression," *Proceedings of SPIE*, vol. 1666, pp. 365–374, August 1992.

[63] A.B. Watson, "DCT quantization matrices visually optimized for individual images," *Proceedings of SPIE*, vol. 1913, pp. 202–216, February 1993.

[64] R. Raj, W. Geisler, R. Frazor, and A. Bovik, "Natural contrast statistics and the selection of visual fixations," in *Proceedings of the IEEE International Conference on Image Processing*, Atlanta, GA, USA, October 2006, pp. 1152–1155.

[65] U. Rajashekar, I. van der Linde, A. Bovik, and L. Cormack, "Foveated analysis and selection of visual fixations in natural scenes," in *International Conference on Image Processing*, Atlanta, GA, USA, October 2006, pp. 453–456.

[66] N. Sadaka, L. Karam, R. Ferzli, and G. Abousleman, "A no-reference perceptual image sharpness metric based on saliency-weighted foveal pooling," in *Proceedings of the IEEE International Conference on Image Processing*, San Diego, CA, USA, October 2008, pp. 369–372.

[67] N. Sadaka and L. Karam, "Efficient super-resolution driven by saliency selectivity," in *Proceedings of the IEEE International Conference on Image Processing*, Brussels, Belgium, September 2011.

[68] L. Itti, C. Koch, and E. Niebur, "A model of saliency-based visual attention for rapid scene analysis," *IEEE Transactions on Pattern Analysis and Machine Intelligence*, vol. 20, no. 11, 1254–1259, November 1998.

[69] U. Rajashekar, I. van der Linde, A. Bovik, and L. Cormack, "GAFFE: A gaze-attentive fixation finding engine," *IEEE Transactions on Image Processing*, vol. 17, no. 4, pp. 564-573, April 2008.

[70] R. Achanta, S. Hemami, F. Estrada, and S. Susstrunk, "Frequency-tuned salient region detection," in *Proceedings of the IEEE Conference on Computer Vision and Pattern Recognition*, Miami, FL, USA, June 2009, pp. 1597–1604.

[71] A.G. Weber, "The USC-SIPI image database version 5." Available online, http://sipi.usc.edu/database/.

[72] H.R. Sheikh, Z.Wang, L. Cormack, and A. Bovik, "Live image quality assessment database release 2." Available online, http://live.ece.utexas.edu/research/quality.

[73] L. Itti and C. Koch, "A saliency-based search mechanism for overt and covert shifts of visual attention," in *Vision Research*, vol. 40, no. 10–12, pp. 1489–1506, June 2000.

[74] B. Girod, "What's wrong with mean-squared error," in *Digital Images and Human Vision* A.B. Watson (ed.), Cambridge, MA, USA: MIT Press, 1993, pp. 207–220.

# 10

## Methods of Dither Array Construction Employing Models of Visual Perception

**Daniel L. Lau, Gonzalo R. Arce, and Gonzalo J. Garateguy**

## 10.1   Introduction

Digital halftoning refers to the process of converting a continuous-tone image or photograph into a binary pattern of black and white pixels for display on binary devices, such as ink-jet or electrophotographic printers. Early approaches to digital halftoning mimic the analog process of projecting a film negative, through a silk screen, onto high-contrast lithographic film. These early approaches produce periodic patterns of round dot clusters varying in size with tone such that light and dark shades of gray were represented by small and large dot clusters, respectively. By thinking of the size of round clusters as amplitude, these periodic clustered-dot methods are generally referred to halftoning by means of amplitude modulation.

By turning pixels on according to the average gray level of the entire cell, the resulting halftone loses fine details from the original; furthermore, significant computational resources are spent calculating the specific path by which pixels are turned on according to the pixel's row and column coordinate within the cell. So as a means of preserving finer image details as well as to reduce overall computational complexity, the pixels of the halftone cell can be independently thresholded according to a quantization level stored in a dither array of the same size and orientation of the cell. As such, a solid black line of small width (1–2 pixels) will be preserved (Figure 10.1a). The specific arrangement of consecutive thresholds within the dither array determines the resulting dot shape with early approaches to dither array construction recreating the spiraling path of Post-Script screening.

2

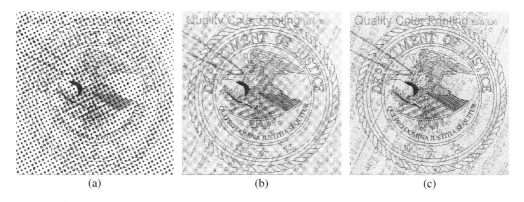

(a)  (b)  (c)

**FIGURE 10.1**

Demonstrations of halftoning by means of (a) a clustered-dot dither array, (b) Bayer's dither array, and (c) a blue-noise dither array.

Later approaches to halftoning involve the process of error diffusion, where a pixel is quantized to black or white, and its quantization error is passed on to neighboring, soon-to-be-processed pixels. The resulting patterns are composed of an aperiodic pattern of isolated dots, homogeneously distributed where light shades of gray are represented by a loose packing of dots and dark shades by a tight packing. Thinking of the spacing of dots as their frequency, methods like error diffusion that create varying shades of gray by manipulating the spacing between dots are generally referred to as halftoning by means of frequency modulation. Because of the high computational complexity of error diffusion, a means of producing frequency modulation patterns using dither arrays was developed.

## 10.2 State of the Art Work

While patterns of printed dot clusters, formed by early dither arrays, create consistent, smooth halftones in electrophotographic device unable to reliably print isolated dots, these same arrays create an apparent spatial resolution much lower than the native resolution of a reliable printing device such as an ink-jet printer. In such devices, higher image quality can be achieved through the dispersion of consecutive thresholds as first specified by Bayer [1] (Figure 10.1b). The drawback of Bayer's dither is that by distributing thresholds along a deterministic path, resulting halftones show a strong periodic structure that imparts an unnatural appearance on the image [2].

In order to eliminate the periodic textures of Bayer's dither while preserving the high fidelity of isolated dot patterns, various authors have proposed distributing consecutive thresholds in such a way that print dot patterns minimize the visibility of resulting binary texture to the human viewer. Such approaches require careful consideration of how the human visual system works. To this end, various lowpass filter models have been employed, such as in References [2], [3], [4], [5], [6], [7], [8], [9], and [10], to introduce so-called blue-noise dither arrays where consecutive thresholds are distributed in a pseudo-random fashion to create the illusion that the resulting halftone was created by means of error diffusion (Figure 10.1c). Error diffusion [11], [12], [13] is a neighborhood process of quantizing the current input pixel and then diffusing the resulting quantization error onto soon-to-be-processed input pixels. While involving significantly more computational resources than dither array halftoning, error diffusion patterns are visually superior to periodic dither arrays because the algorithm minimizes the low-frequency content of the halftone, creating patterns composed exclusively of high-frequency spatial content. As blue is the high-frequency component of visible white light, Reference [12] coined the term "blue-noise" to describe these patterns. The human visual system, being less sensitive to random patterns while finding isolated dots harder to see, would find the resulting texture less visible overall.

The process of constructing blue-noise dither arrays typically involves the steps of generating a set of binary dither patterns, with a separate pattern for each unique gray level, and then summing the patterns pixelwise to produce a multilevel threshold matrix. An eight-bit display, for example, will involve the construction of 256 unique dither patterns ranging from an all black pattern up to an all white pattern. In constructing the dither pattern for an intermediate gray level, an iterative process is used that swaps black and white pixels in order to minimize the pattern's low-frequency energy under a stacking constraint such that pixels set to white for one gray level are also white for all higher gray levels. Summing all of the component dither patterns pixelwise creates a threshold matrix such that a particular pixel threshold value equals to the percentage of times that pixel is set to one across the component dither pattern set. When halftoning, an output pixel is then set to 1 if the corresponding input pixel equals or exceeds the corresponding threshold in the dither array. For large images, the dither array is tiled end-to-end until each input pixel has an associated threshold value. For this reason, it is extremely important that the dither array show no discontinuity in texture when wrapping around from one side to the other.

In comparing various approaches to dither array construction, the majority of algorithms differ in the way that they measure low-frequency energy to determine which pixels to swap. In the case of Ulichney's void-and-cluster (VAC) algorithm [2], the current iteration's dither pattern is filtered by means of circular convolution with a human visual model consisting of a lowpass, Gaussian filter kernel. Having applied this filter, the resulting visual model of the halftone is then a measure of white pixel density such that low values indicate regions of the dither pattern with a low concentration of white pixels while large values indicate regions with a high concentration. Wanting to achieve a homogeneous distribution of ones and zeros, this algorithm selects the white pixel corresponding to the highest concentration of white pixels and sets it to black. The image is then refiltered to determine the current concentration of white pixels minus the one that was just toggled. From the resulting concentration, the algorithm then selects the black pixel corresponding to the lowest concentration of white pixels and sets it to white. Repeating this process of voiding and clustering, the algorithm stops after the black pixel set to white is the same pixel that was just turned to black during the void stage.

A popular alternative to void-and-cluster is the direct binary search (DBS) [14], which is an iterative process. For a particular iteration, a pixel is either toggled of swapped with one of its eight nearest surrounding neighbors, depending upon which modification leads to the largest drop in squared error between the visual model of the halftone and original continuous-tone input image. If none of the modifications lead to a reduction in error, then

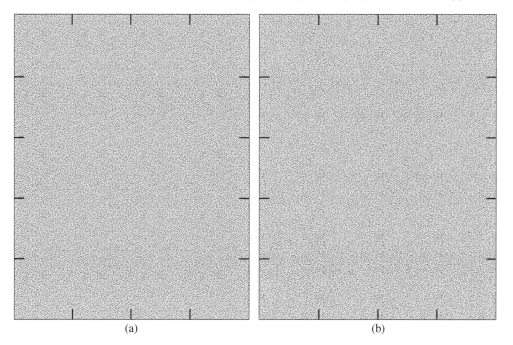

(a)                                                                                  (b)

**FIGURE 10.2**

Illustration of the tiling artifacts created by (a) a single repeating blue-noise dither of size $108 \times 108$ pixels and (b) ten randomly selected dither arrays of size $108 \times 108$ pixels with a common border region [18].

no modification is performed with processing continuing with the next pixel in the raster order. If, after a complete pass through the image, no modifications are performed, then the algorithm has converged and processing stops; otherwise, the algorithm performs another pass through the image. Because direct binary search is a steepest descent algorithm like void-and-cluster, the final result is susceptible to local minima, depending on the initial halftone as well as the particular HVS model.

To date, the various blue-noise dither array construction algorithms fail to fully achieve the smooth textures of error diffusion due to their reliance on steepest descent algorithms to minimize a visual cost measure, resulting in the algorithms falling into local minima. As such, this chapter introduces a postprocessing stage based on centroidal Voronoi tessellations (CVTs) to design dither arrays that, when thresholded with at a constant gray level, yield binary patterns that approximate the ideal blue-noise pattern regardless of the input gray level. This algorithm allows for a fine tuning in the distribution of minority pixels and a fine control over the quality of binary patterns at different gray levels. Centroidal Voronoi tessellations have been used in many applications ranging from data compression to optimal quadrature rules, optimal representation, quantization, clustering [15], and also halftoning. In halftoning, CVTs have been used before [16], [17] for the automatic generation of stipplings, which is a method traditionally performed by artists to represent continuous tone images as black an white images.

In Reference [17], an initial binary image, created using some halftoning algorithm is optimized through the use of weighted Voronoi digrams, using the original continuous tone image as the density function. This approach leads to good results but involves a high computational load since numerical integrations have to be performed at each iteration of the algorithm. Techniques to precalculate dot distributions, and then use look up tables to generate the halftone are also presented in Reference [17], but the quality of the halftones obtained degrades considerably with respect to the initial approach. In general the usefulness of CVTs for any application is based on its minimization properties and the capacity to create uniform distributions of dots. The algorithm presented here takes advantage of these properties to optimize the distributions of minority pixels in the binary patterns of the array. By introducing a modification of Lloyd's algorithm, this optimization can be controlled, yielding binary patterns of even quality across the complete grayscale.

Now regardless of the construction method, the use of the dither array technique for halftoning continuous tone images have a fundamental problem related with the unavoidable tiling process of multiple copies of the dither array. As demonstrated in Figure 10.2a, the tiling creates a globally periodic halftone that is visible to the human eye. To overcome this difficulty, Reference [18] explored the possibility of constructing multiple dither arrays with component arrays tiled end-to-end in a random ordering. As shown in Figure 10.2b, this process involves constructing dither arrays with a common border, such that each pattern satisfies a wrap-around property with itself and all other component masks. However, the authors offered no criteria by which to set the size of the common border region, where too small a region leads to texture discontinuities between tiles while too large a region leads to visible tiling artifacts in the bands of the common region.

Along similar lines as described above, Reference [19] proposed the use of multiple small blue-noise screens, designed using direct binary search that could be tiled in a random order to yield high-quality halftones with an overall screen that is periodic but not having periodic artifacts. A similar approach was then offered in Reference [20] which proposed the use of screen tile indices as a watermark. Finally, Reference [21] proposed the use of multiple small constant binary texture tiles, arranged in any manner, to design highly compressible halftones.

This chapter presents a generalization of the approach first offered in Reference [18], by which the size of the common border region between component arrays is optimized as a function of the gray level. In addition, this chapter also reviews the extension of the stochastic dither arrays to nonzero screen angles, taking advantage of the human eye's reduced sensitivity to diagonal correlation. This technique can be easily extended to color halftoning, where the possibility of tiling artifacts can be further reduced by using different screen angles for each color, as performed in traditional clustered-dot dithering.

In closing this chapter, the focus is shifted on the especially challenging problem of lenticular printing, which spatially multiplexes a series of component images to create a composite image as illustrated in Figure 10.3 [22]. These images are then transferred onto the backside of a lenticular lens array. When viewed through the lenses and depending on the incident angle through the array, the human eye sees a single de-interlaced component image. Oriented horizontally, the rotation of the lens array creates an animated image sequence. Oriented vertically, the lens array forms a stereoscopic pair that, given a sufficient number of component viewing angles, creates a full-color holographic display. Within the printing industry, lenticular imaging is most commonly associated with advertising, and in particular, most observers will associate lenticular printing with movie posters and other kiosk displays.

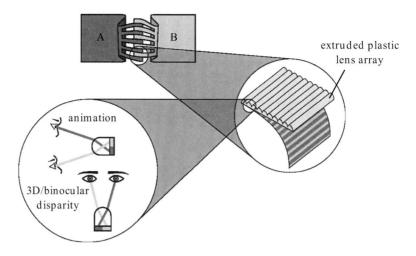

**FIGURE 10.3**

Illustration of the lenticular imaging process where: (top-center) images are divided into strips and are interlaced together into on graphic, (right) the graphic is printed directly onto the back of an extruded lens, and (bottom-right) the lenticule isolates and magnifies the interlaced image beneath, depending on the angle of observation, such that if the lenticule runs vertically, then a different image is delivered to each eye to create a 3D image. If the lenticule runs horizontal, then a rotation of the lens array creates an animation effect.

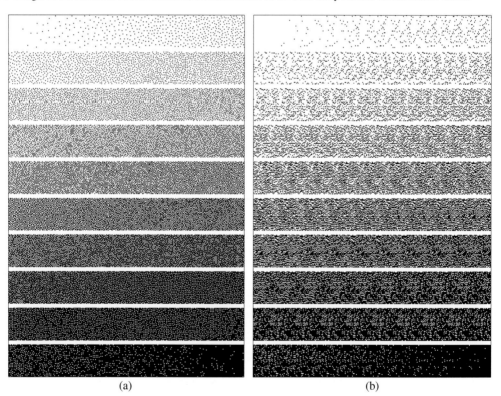

(a)                                                    (b)

**FIGURE 10.4**

The halftoned grayscale ramps (a) before and (b) after downsampling using a traditional void-and-cluster dither array.

Due to the manner in which the lens array de-interlaces the component images, the use of a blue-noise dither array to halftone the composite image creates a noisy uncorrelated appearance as demonstrated in Figure 10.4. So as an attempt to maintain a smooth homogeneous distribution of printed dots in the composite image while also achieving a smooth blue-noise appearance in the component de-interlace images, this chapter introduces a modified technique for constructing blue-noise dither arrays that constrains their construction to maintain a smooth texture in both their original form as well as in their downsampled renditions. Specifically, the construction of these *lenticular aware* dither arrays will be studied by means of the void-and-cluster algorithm [2] where the computational complexity associated with this technique of converting a continuous tone pixel value into a binary value is a pointwise operation, requiring no additional complexity than in a traditional halftoning application.

## 10.3 Dither Array Construction

The masking process is a point process by which each pixel of the continuous tone input image $f[\mathbf{n}]$ of $G$ gray levels is compared to a corresponding pixel in a threshold matrix $S[\mathbf{n}]$ of size $M \times N$, which is called a mask or screen. If the input is greater than the threshold at position $\mathbf{n}$, the binary output denoted by $b[\mathbf{n}]$ is set to one or otherwise to zero. In the masking processes, there is a distinct output binary pattern $I_i$ corresponding to each constant input of gray level $i$ with $i \in \{0, 1, ..., G-1\}$. The number of pixels set to one in each binary pattern corresponds to the gray level it represents. The parameter $I_0$ is the all zero matrix while $I_{G-1}$ is the all ones matrix. And $I_i$ has $\lfloor MNi/(G-1) + 0.5 \rfloor$ pixels set to one.

If a position in $I_{i_0}$ is set to one, at gray level $i_0$, that position remains set to one for any other $I_{i_1}$ with $i_1 > i_0$. Conversely, if a position in $I_{i_2}$ is set to zero, at gray level $i_2$, then $I_{i_1}$ is set to zero in the same position for any $i_1 < i_2$ (see Figure 10.5). These restrictions are called stacking constraints given by

$$\begin{aligned} \text{if } I_{i-1}[\mathbf{n}] = 1 &\Rightarrow I_i[\mathbf{n}] = 1, \\ \text{if } I_{i+1}[\mathbf{n}] = 0 &\Rightarrow I_i[\mathbf{n}] = 0. \end{aligned} \tag{10.1}$$

Having all the binary patterns corresponding to each of the input gray levels, the threshold matrix $S[\mathbf{n}]$ is given by

$$S[\mathbf{n}] = \sum_{i=0}^{G-1} I_i[\mathbf{n}], \tag{10.2}$$

and the output of the masking process is

$$b[\mathbf{n}] = \begin{cases} 1 & \text{if } f[\mathbf{n}] > S[\mathbf{n} \bmod[M,N]], \\ 0 & \text{if } f[\mathbf{n}] \leq S[\mathbf{n} \bmod[M,N]]. \end{cases} \tag{10.3}$$

The order in which binary patterns are designed has an important impact on its quality, since with advancing in the process more and more pixels are constrained leaving less room for the selection of new minority pixels.

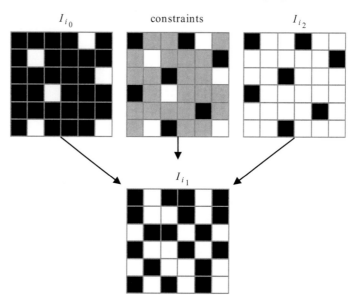

**FIGURE 10.5**

Stacking constraints for gray levels $i_0 < i_1 < i_2$. White pixels represent a 1, black pixels represent a 0, and gray pixels represent an undetermined value in $I_{i_1}$. These values are the ones that must be set to either 1 or 0 using the binary pattern design algorithm.

In the process of designing $I_{i_1}$ based on previously designed patterns $I_{i_0}$ and $I_{i_2}$ with $i_0 < i_1 < i_2$, the stacking constraint implies that there would be $\lfloor MNi_0/(G-1) + 0.5 \rfloor$ pixels constrained to be one in $I_{i_1}$ and $MN - \lfloor MNi_2/(G-1) + 0.5 \rfloor$ pixels constrained to be zero, leaving

$$\lfloor MNi_2/(G-1) + 0.5 \rfloor - \lfloor MNi_0/(G-1) + 0.5 \rfloor \tag{10.4}$$

unconstrained pixels to be determined at the design stage. The criterion to select the best design order would be to maximize the number of undetermined pixels at every stage, since this allows to generate better binary patterns using the appropriate algorithm. A way to maximize Equation 10.4 is by having $i_2$ and $i_0$ as far as possible at every iteration and that is accomplished if the design order pivots between the extremes of the grayscale, that is, $i = 0, G-1, 1, G-2, ..., (G-2)/2$. Following this ordering, the base patterns $I_{i_0}$ and $I_{i_2}$ always will be in opposite sides of the grayscale. The pixels considered as minority pixels are the ones set to one for $I_{i_0}$ and to zero for $I_{i_2}$.

As it is desirable to use the same algorithm to create binary patterns at either end of the grayscale range, regardless of the definition of a minority pixel, two sets of binary patterns defined as $\{I_i^b\}_{i=1}^{(G-2)/2}$ and $\{I_i^w\}_{i=1}^{(G-2)/2}$ (assuming $G$ even) are considered here. These binary patterns correspond to binary patterns in the mask, as $I_i = I_i^w$ for $i = 0, ..., (G-2)/2$ and $I_i = 1 - I_{G-1-i}^b$ for $i = (G-2)/2+1, ..., G-1$. The letters $b$ and $w$ stand for black (bits set to zero) and white (bits set to one) since the patterns in $\{I_i^b\}$ arises for input gray levels $i > (G-1)/2$ and the patterns in $\{I_i^w\}$ for $i < (G-1)/2$. Using these two sets, a minority pixel will always be equal to one in any given pattern and the same algorithm can be used either to design $I_i^w$ or $I_i^b$. The design order proposed is then

$\{I_1^b, I_1^w, I_2^b, I_2^w, \ldots, I_{(G-2)/2}^b, I_{(G-2)/2}^w\}$ and the stacking constraints reformulated in terms of $I_i^w$ and $I_i^b$ are

$$
\begin{aligned}
\text{if } I_{i-1}^b[\mathbf{n}] = 1 &\Rightarrow I_i^b[\mathbf{n}] = 1, \\
\text{if } I_{i-1}^w[\mathbf{n}] = 1 &\Rightarrow I_i^w[\mathbf{n}] = 1,
\end{aligned}
\tag{10.5}
$$

and

$$
\begin{aligned}
\text{if } I_i^b[\mathbf{n}] = 1 &\Rightarrow I_j^w[\mathbf{n}] = 0 \text{ for } j = i, \ldots, (G-2)/2, \\
\text{if } I_i^w[\mathbf{n}] = 1 &\Rightarrow I_j^b[\mathbf{n}] = 0 \text{ for } j = i, \ldots, (G-2)/2.
\end{aligned}
\tag{10.6}
$$

Once the two sets of binary patterns are designed, the mask is obtained by adding up both sets in two partial threshold matrices

$$
B = \sum_{i=1}^{(G-2)/2} I_i^b, \quad W = \sum_{i=1}^{(G-2)/2} I_i^w,
\tag{10.7}
$$

and then calculating

$$
S = \left(\frac{G-2}{2}\right) \mathbf{1}_{M \times N} - B + W,
\tag{10.8}
$$

where $\mathbf{1}_{M \times N}$ is a matrix of dimension $M \times N$ filled with ones.

Each binary pattern $I_i^b$ is designed based on the previously designed binary patterns $I_{i-1}^b$ and $I_{i-1}^w$ and for the case of $I_i^w$ is based on the patterns $I_i^b$ and $I_{i-1}^w$. Starting from $I_0^w$ and $I_0^b$ as the all zero matrices, the algorithm that determines where to locate new minority pixels in $I_{i-1}^b$ to generate $I_i^b$ consists of a proper number of minority pixels being added to $I_{i-1}^b$ such that the resulting pattern represents gray level $i$ (that is, has $N_i = \lfloor MNi/(G-1) + 0.5 \rfloor$ pixels set to one). If the number of pixels set to one in $I_{i-1}^b$ is $N_{i-1}$, then the number of pixels needed to create $I_i^b$ is $K_i = N_i - N_{i-1}$. These $K_i$ pixels are placed one by one using the same procedure as in void-and-cluster [2]. The binary pattern $I_{i-1}^b$ is circularly convolved with a human visual system filter model $B$ and then a new minority pixel is added in the position $\mathbf{n}^*$ corresponding to the minimum of the filtered pattern, subject to the stacking constraint

$$
\mathbf{n}^* = \underset{\mathbf{n}}{\arg\min} \left( I_{i-1}^b \otimes B \right)[\mathbf{n}]
\tag{10.9}
$$

such that

$$
I_{i-1}^w[\mathbf{n}] = 0, \quad I_{i-1}^b[\mathbf{n}] = 0.
\tag{10.10}
$$

Instead of the traditional Gaussian visual model $B = e^{-\frac{m^2+n^2}{2\sigma^2}}$, the following model

$$
B[\mathbf{n}] = e^{-\frac{(|m|^P + |n|^P)^{2/P}}{2\sigma^2}}
\tag{10.11}
$$

can be used and tuned to increase the likelihood of minority pixel placement along vertical or diagonal directions (see Figure 10.6).

The parameter $p$ determines the degree of isotropy and also influences the likelihood of minority pixel placement. By choosing values of $p < 2$, the likelihood of placing minority pixels in diagonal directions is increased since the $L^p$ norm has greater values than the $L^2$ norm in diagonal directions and thus smaller values of the filter. For $p > 2$, the filter takes a square shape, product of the approximation of the $L^p$ norm to the infinity norm

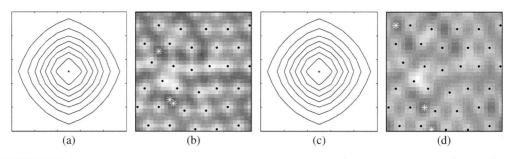

|     |     |     |     |
| --- | --- | --- | --- |
| (a) | (b) | (c) | (d) |

**FIGURE 10.6**

(a) Level curves of the filter with $p = 1.5$ and $\sigma = 2$, (b) filtered binary pattern of size $32 \times 32$, (c) level curves of the filter with $p = 3$ and $\sigma = 2$, and (d) filtered binary pattern of size $32 \times 32$. The $*$ symbols denote local minimums and $\bullet$ symbols indicate the locations of minority pixels in the pattern.

as $p$ grows. In this case, the likelihood of placing minority pixels along horizontal and vertical directions increases. This filter also has the particularity that along the vertical and horizontal axis it always decays as the Gaussian for any $p$; the setting $p = 2$ yields the two-dimensional Gaussian filter. Here, the values of $p$ and the variance $\sigma = \lambda_b$ are adjusted according to the gray level of the binary pattern being generated.

With regards to the parameters defining the visual model, Reference [23] established that the mean distance between minority pixels in an ideal blue-noise pattern is determined by the principal wavelength

$$\lambda_b = \begin{cases} 1/\sqrt{g} & \text{for } 0 < g \leq 1/4, \\ 2 & \text{for } 1/4 < g \leq 3/4, \\ 1/\sqrt{1-g} & \text{for } 3/4 < g \leq 1, \end{cases} \tag{10.12}$$

where $g = i/(G - 1)$ is the gray level represented by the pattern and $\lambda_b$ is the principal wavelength of the pattern. Furthermore, the same work studied the influence of the regular grid on the spectral characteristics of binary patterns and, in particular, the aliasing effects that appears at gray levels in the region $1/4 < g < 3/4$. From that study, it is known that to avoid disturbing artifacts and to maintain radial symmetry in the spectrum, some amount of clustering must be allowed in the region $1/4 < g < 3/4$ while for the region $g < 1/4$ and $g > 3/4$, the dots are mostly isolated and the principal wavelength can be used as an accurate approximation of the interpoint distance. A good method to design binary patterns according to this model is one that is capable of maintaining the interpoint distance as uniformly as possible for $g < 1/4$ and $g > 3/4$ and that also allows some clustering in the region $1/4 < g < 3/4$ while keeping radial symmetry.

## 10.4 Blue-Noise Patterns and CVTs

The algorithm proposed for constructing dither arrays in this section is a hybrid algorithm that smoothly switches the binary pattern design method between the regions $g < 1/4$, $g > 3/4$, and $1/4 < g < 3/4$. For the region $g < 1/4$ and $g > 3/4$, a variation of Lloyd's algorithm is proposed to improve the uniformity of interpoint distances that can be tuned according to the gray level and that is capable of optimizing a binary pattern without significantly disturbing the previously designed binary patterns. On the other hand, for the region $1/4 < g < 3/4$, the criteria to design binary patterns have to be based on the minimization of the low-frequency content and the preservation of the radial symmetry of the spectrum, rather than in the homogenization of the interpoint distances. As such, any of the commonly utilized dot placement algorithms can be selected. In particular, the void-and-cluster algorithm is used since it is one with less complexity and, as will be demonstrated later, leads to very good results if used in conjunction with the Lloyd optimization stage and with a specially tuned filter.

The proposed variant of Lloyd's algorithm, described in this section, involves iteratively tweaking the position of minority pixels within the component dither patterns using Voronoi tessellations where, in order to escape the local minima of void-and-cluster, the discrete-space dither patterns are mapped to continuous-space. The points are then mapped back to discrete-space after completing the optimization. This relaxation of discrete to continuous-space allows for a smooth and continuous optimization of the points distribution, which considerably improves the quality of the final binary pattern. As the discussion constantly goes from one space to the other, it is fundamental to clearly define the discrete and continuous spaces, introduce a proper representation of binary patterns in both spaces, and define the mappings between the two spaces.

This description begins by defining the discrete space of minority pixels as

$$D = \{\mathbf{n} = [m,n] \in \mathbb{Z}^2 \mid 0 < m \le M, 0 < n \le N\} \tag{10.13}$$

and the continuous space of minority pixels as

$$C = \{\mathbf{z} = (x,y) \in \mathbb{R}^2 \mid 0 < y \le M, 0 < x \le N\}, \tag{10.14}$$

where $\mathbf{n} = [m,n]$ are the discrete coordinates of a minority pixel and $\mathbf{z} = (x,y)$ are its continuous coordinates. Each binary pattern $I_i^b$ or $I_i^w$ in the discrete space will be represented by an $M \times N$ binary matrix with $N_i = \lfloor MNi/(G-1) + 0.5 \rfloor$ entries set to one. In addition, the binary pattern $DI_i^b = I_i^b - I_{i-1}^b$ are defined, that corresponds to the set of pixels swapped to one from $I_{i-1}^b$ to $I_i^b$, according to the stacking constraint (the definition of $DI_i^w$ is analogous). Each binary pattern $DI_i^b$ has a number of minority pixels $K_i$ such that $N_i = N_{i-1} + K_i$.

Based on the definition of $DI_i^b$ and $DI_i^w$, any binary pattern can be expressed as

$$I_i^b = \sum_{j=1}^{i} DI_j^b, \quad I_i^w = \sum_{j=1}^{i} DI_j^w. \tag{10.15}$$

On the other hand, in the continuous space, the corresponding representation of the binary pattern $I_i^b$ is the set of $N_i$ points $\mathscr{I}_i^b = \{\mathbf{z}_j\}_{j=1}^{N_i}$ such that $\mathbf{z}_j = (n_j, m_j)$, where $\mathbf{n}_j = [m_j, n_j]$ are the coordinates of each pixel set to one in $I_i^b$. The set $\mathscr{DI}_i^b = \{\mathbf{z}_{i,l}^b\}_{l=1}^{K_i}$ is defined as the set of points $\mathbf{z}_{i,l}^b = (n_l, m_l)$ corresponding to each pixel set to one in $DI_i^b$ (and analogously for $\mathscr{DI}_i^w = \{\mathbf{z}_{i,l}^w\}_{l=1}^{K_i}$). As in the discrete case, the binary patterns $\mathscr{I}_i^b$ and $\mathscr{I}_i^w$ can be expressed as a function of the sets $\mathscr{DI}_j^w$ and $\mathscr{DI}_j^b$ as follows:

$$\mathscr{I}_i^b = \bigcup_{j=1}^{i} \mathscr{DI}_j^b, \quad \mathscr{I}_i^w = \bigcup_{j=1}^{i} \mathscr{DI}_j^w. \tag{10.16}$$

Now, while the mapping to go from the discrete to the continuous space is clearly defined for each minority pixel as $\mathbf{z}_j = (n_j, m_j)$, the inverse mapping is not clearly defined since the continuous space has a infinite number of possible points in it whereas discrete space contains only a finite set of elements. Therefore in any mapping, different sets of points in the continuous space will have the same discrete representation. The inverse mapping used here is the following.

Given a set of points in the continuous space $\mathscr{I}_i = \{\mathbf{z}_j\}_{j=1}^{N_i}$, the associated matrix in the discrete space $I_i$ is defined by quantizing the coordinates of the points $\mathbf{z}_j = (x_j, y_j)$ such that

$$n_j = \begin{cases} \lfloor x_j + 0.5 \rfloor & \text{if } \lfloor x_j + 0.5 \rfloor \geq 1, \\ \lfloor x_j + 0.5 \rfloor + N & \text{if } \lfloor x_j + 0.5 \rfloor < 1, \end{cases} \tag{10.17}$$

$$m_j = \begin{cases} \lfloor y_j + 0.5 \rfloor & \text{if } \lfloor y_j + 0.5 \rfloor \geq 1, \\ \lfloor y_j + 0.5 \rfloor + M & \text{if } \lfloor y_j + 0.5 \rfloor < 1, \end{cases} \tag{10.18}$$

and the pixels in $I_i$ according to

$$I_i(\mathbf{n}) = \begin{cases} 1 & \text{if } \mathbf{n} = [m_j, n_j] \text{ for } j = 1, ..., N_i, \\ 0 & \text{otherwise.} \end{cases} \tag{10.19}$$

This mapping is denoted by $I_i = \mathscr{Q}(\mathscr{I}_i)$, and as long as the points in $\mathscr{I}_i$ are sufficiently separated $(d(\mathbf{z}_i, \mathbf{z}_j) > 1/2^{1/2} \ \forall \ i, j)$, then this mapping returns a matrix $I_i$ with the same number of minority pixels as the input set $\mathscr{I}_i$.

Given a set of $K$ points $\{\mathbf{z}_i\}_{i=1}^{K}$ in $\Omega \subset \mathbb{R}^2$ that will be called here as the set of generators, the Voronoi tessellation is defined as the collection of $K$ subsets $\{V_i\}_{i=1}^{K} \subset \Omega$ such that each $V_i$ is given by

$$V_i = \{\mathbf{z} \in \Omega | d(\mathbf{z}, \mathbf{z}_i) < d(\mathbf{z}, \mathbf{z}_j) \ \forall j \neq i\}. \tag{10.20}$$

Note that $V_i$ is the set of nearest neighbors to the point $\mathbf{z}_i$ under the distance metric $d(\cdot, \cdot)$ and also that this collection of subsets covers the whole set $\Omega$ except only by the borders of the Voronoi regions. Figure 10.7a depicts an example of a Voronoi tessellation of four points in which a wrap-around Euclidean distance was used. Evidently for the same set of points there are different Voronoi tessellations, depending on the distance metric used. A thorough review can be found in Reference [24] where many variations and applications of Voronoi tessellations defined for different distances are presented. The distance used here is the Euclidean distance that wraps around the set $\Omega$. This selection is due to the fact that Voronoi tessellations are used in the design process of binary patterns, and all binary patterns in a dither array must have the wrap-around property.

Besides the previous definitions, the definition of the centroid of a Voronoi region is also necessary. As a consequence of the distance metric used, it can be demonstrated [24] that all the Voronoi cells $V_i$ are convex polygons. Then, denoting its vertices by $\mathbf{v}_i = (x_i, y_i)$, the definition of the centroid $\mathbf{z}_i^* = (x_i^*, y_i^*)$ of a Voronoi cell $V_i$ is given by

$$x_i^* = \frac{1}{6A} \sum_{i=0}^{N-1} (x_i + x_{i+1})(x_i y_{i+1} - x_{i+1} y_i), \\ y_i^* = \frac{1}{6A} \sum_{i=0}^{N-1} (y_i + y_{i+1})(x_i y_{i+1} - x_{i+1} y_i), \tag{10.21}$$

where $A = \frac{1}{2} \sum_{i=0}^{N-1} (x_i y_{i+1} - x_{i+1} y_i)$ is the area of the cell and $\mathbf{v}_i = (x_i, y_i)$ are the coordinates of its vertices.

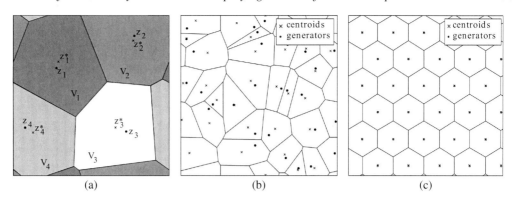

**FIGURE 10.7**

(a) Example of Voronoi tessellation using the wrap-around distance, where $V_i$ are the Voronoi cells, $\mathbf{z}_i^*$ the centroids of the cells, and $\mathbf{z}_i$ its generators; (b) Voronoi tessellation of a random distribution of generators (30 points); and (c) Centroidal Voronoi tessellation (30 points).

Centroidal Voronoi tessellations (CVTs) are a particular kind of Voronoi tessellations in which all the centroids of the cells $\mathbf{z}_i^*$ corresponds to its generators $\mathbf{z}_i$, that is, $\mathbf{z}_i^* = \mathbf{z}_i$ for all $i$. Figures 10.7b and 10.7c show examples of a regular and a centroidal Voronoi tessellation, respectively. As it can be appreciated, the centroids and generators do not coincide, in general, for a given distribution of points. There are various method to construct CVTs [15]; Algorithm 10.1 depicts popular Lloyd's method. In the following, a modification of this algorithm is used as a second stage of the binary pattern design, as a global optimization over the whole binary pattern constructed using the void-and-cluster algorithm.

**ALGORITHM 10.1** Lloyd's algorithm.

1. Set a starting distribution of points $\{\mathbf{z}_i\}_{i=1}^{K_i}$.
2. Calculate the Voronoi tessellation generated by the distribution.
3. Calculate the centroids $\mathbf{z}_i^*$ of each Voronoi cell $V_i$ according to Equation 10.21.
4. Replace the generators by the centroids $\mathbf{z}_i \leftarrow \mathbf{z}_i^*$.
5. Terminate if the maximum number of iterations is reached; otherwise do step 2.

From the binary pattern $I_i^b$ created by means of void-and-cluster, its continuous version is obtained as $\mathscr{I}_i^b = \bigcup_{j=1}^{i} \mathscr{D}\mathscr{I}_j^b$. The set $\mathscr{D}\mathscr{I}_j^b$ is the continuous version of the pixels added to go from $I_{j-1}^b$ to $I_j^b$, and each set is composed of individual points denoted by $\mathbf{z}_{j,l}$. The subindex $j$ identifies the stage in which the point was introduced, and the subindex $l$ identifies each individual point within that stage. In the following steps, the integer constraint over the coordinates of minority pixels is relaxed and they are treated as points in the continuous space. Eventually, they have to be remapped to the discrete space, but this step is performed after the optimization of interpoint distances had converged. Before explaining the modified Lloyd's algorithm used for the optimization of binary patterns, a set of constants $\mu_j$ needs to be defined; these constants are associated to each of the minority pixels which are used to control its mobility. The values of $\mu_j$ go from zero to one, with zero meaning that the minority pixel is not able to move at all, and one meaning complete freedom of movement.

**ALGORITHM 10.2** Modified Lloyd's algorithm.

1. Start with the set $\mathscr{I}_i^b = \bigcup_{j=1}^i \mathscr{D}\mathscr{I}_j^b$ formed from all previous $\mathscr{D}\mathscr{I}_j^b = \{\mathbf{z}_{j,l}^b\}_{l=1}^{K_j}$ patterns.

2. Assign a weight $\mu_j$ to each point $\mathbf{z}_{j,l}^b$ in the set of generators $\mathscr{I}_i^b$.

3. Compute the Voronoi tessellation $\{V_{j,l}^b\}$ associated with $\mathscr{I}_i^b$.

4. Calculate the centroid $\mathbf{z}_{j,l}^{*b}$ of each Voronoi cell $V_{j,l}^b$ in the tessellation.

5. Update $\mathbf{z}_{j,l}^b \leftarrow \mathbf{z}_{j,l}^b + (\mathbf{z}_{j,l}^{*b} - \mathbf{z}_{j,l}^b)\mu_j$.

6. Terminate if the maximum number of iterations $N_{max}$ is reached and output $\mathscr{I}_i^b = \bigcup_{j=1}^i \{\mathbf{z}_{j,l}^b\}_{l=1}^{K_j}$; otherwise go to step 3.

Taking as input the continuous version of the binary pattern $I_i^b$, the steps of the modified Lloyd's algorithm are listed in Algorithm 10.2. In contrast to the traditional void-and-cluster algorithm in which once the minority pixels are located they remain unchanged for the rest of the process, this two-step algorithm of void-and-cluster followed by Lloyd's algorithm permits the soft placement of minority pixels. It takes into account several binary patterns in which minority pixels are allowed to move with respect to their initial positions.

As an illustration of the improved spatial distribution of minority pixels afforded by Lloyd's algorithm, Figure 10.8 depicts three different binary patterns generated using void-and-cluster (VAC) [2], direct binary search (DBS) [14], and centroidal Voronoi tessellations (CVTs). The patterns are compared in terms of their radially averaged power spectral densities (RAPSD) defined in Reference [25] by dividing the Fourier domain into a series of non-overlapping annular rings and taking the average power within each ring. The power spectra are being generated by means of Bartlett's method of averaging periodograms. In Figure 10.8, the pattern resulting from a CVT is clearly the one with less power at low frequencies and the one that has the most isotropic distribution of points.

With regards to adding points in subsequent dither patterns under the stacking constraint, the Lloyd's process allows for the early added points to reconfigure themselves in order to yield a more uniform distribution in the emerging binary pattern. The amount of movement allowed to previous minority pixels is controlled by the constant $\mu_j$ in such a way that in each iteration smaller constants are assigned to older points. By doing so, binary patterns are altered for a few design stages after their creation, but eventually converge to a fixed distribution. The drawback of the continuous optimization is that in the process, the stacking constraints with respect to the complementary set of binary patterns $\{I_i^w\}$ can be violated. To account for that, after the optimization, the output is quantized $I_i^b = \mathcal{Q}(\mathscr{I}_i^b)$ and the stacking constraints in Equation 10.6 are checked. If some of the points do not meet the stacking constraints, they are removed one by one and relocated using the same procedure as in the first stage (see Equation 10.9). The resulting pattern $I_i^b$ is converted to its continuous version and used in the next iteration to generate $I_{i+1}^b$.

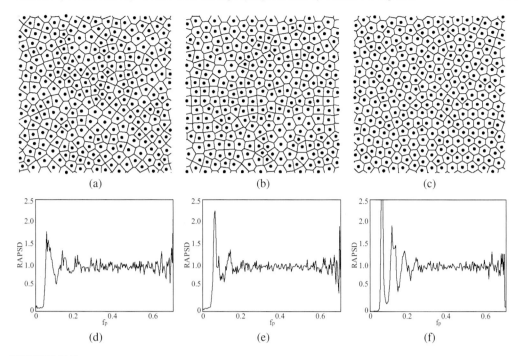

**FIGURE 10.8**

Centroidal Voronoi tessellations of patterns from (a) VAC, (b) DBS, and (c) CVT. Radially averaged power spectral densities (RAPSD) of patterns from (d) VAC, (e) DBS, and (f) CVT.

The set of parameters used at each component pattern change with the gray level where in the first stage of void-and-cluster, the parameters $p$ and $\sigma$ control the likelihood of minority pixel placement. In the second stage, the parameters $\mu_j$ and $N_{max}$ need to be set; $\mu_j$ parameters control the mobility of each point, and $N_{max}$ denotes the number of iterations of the modified Lloyd's algorithm. As explained in Section 10.3, there are two different regions of the grayscale in which the criteria used to design the binary patterns change. For $g < 1/4$ and $g > 3/4$, the goal is to optimize the interpoint distances. For $1/4 < g < 3/4$, the goal is to reduce the low frequency content and to maintain the radial symmetry in the spectral domain. In the first stage of the design process, it was found through extensive simulation that the set of parameters giving the best results are $p = 1.6$ and $\sigma = \lambda_b$ for $g < 1/4$ and $g > 3/4$, and $p = 2$ and $\sigma = 1.5$ for $1/4 < g < 3/4$. For $g < 1/4$ and $g > 3/4$, the process to design the mask is based on the binary pattern design algorithm presented. For $1/4 < g < 3/4$, the process is based on a searching process using the modified filter.

As the second stage aims to uniformly distribute the added points, more iterations of Lloyd's algorithm must be performed for lighter and darker gray levels than for midrange gray levels. The values used for the masks shown in this chapter are $N_{max} = 50$ for the first two binary patterns $I_1^b$ and $I_1^w$, and $N_{max} = 10$ for the remaining patterns. The selection of the number of iterations was determined experimentally and might change for different mask sizes. The optimal value has to be determined in each case. Regarding the parameter $\mu_j$, it was found that in lighter and darker regions of the grayscale bests results are obtained if only two binary patterns are optimized at a time while for midrange gray levels up to four binary patterns can be optimized. As such, the values of the constants are updated at each

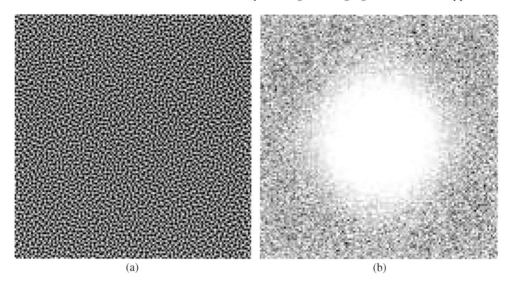

(a)                                                        (b)

**FIGURE 10.9**

The resulting eight-bit $128 \times 128$ blue-noise dither array in (a) the spatial domain and (b) the spectral domain.

iteration in an exponentially decreasing sequence, initializing all the constants at $\mu_j = 0.94$, and then updating their values as $\mu_j = \mu_j^{25}$ for $g < 1/32$ and $g > 31/32$, and $\mu_j = \mu_j^4$ for $1/32 < g < 1/4$ and $3/4 < g < 31/32$. For $1/4 < g < 3/4$, the parameter $\mu_j$ is set to zero since no Lloyd optimization is performed.

The mask design process starts by setting $I_i^b = \mathbf{0}$ and $I_i^w = \mathbf{0}$, and then the binary patterns up to $I_{(G-2)/4}^b$ and $I_{(G-2)/4}^w$ are designed using the two-step procedure of void-and-cluster followed by Lloyd's algorithm under that stacking constraint. For subsequent patterns in the range from $i = (G-2)/4 + 1$ to $i = (G-2)/2$, a searching process is used where, given $I_{i-1}^b$ and $I_{i-1}^w$, the pattern $I_i^b$ is designed by adding $K_i$ minority pixels to $I_{i-1}^b$ to complete the $N_i = \lfloor MNi/(G-1) + 0.5 \rfloor$ necessary for gray level $i$. The minority pixels are added one by one, subject to the stacking constraints in Equations 10.5 and 10.6 in positions $\mathbf{n}^*$ given by

$$\mathbf{n}^* = \operatorname*{argmin}_{\mathbf{n}} \left( I_{i-1}^b \otimes B \right) [\mathbf{n}],$$
$$\text{subject to } I_{i-1}^w [\mathbf{n}] = 0 \text{ and } I_{i-1}^b [\mathbf{n}] = 0. \tag{10.22}$$

Analogously, $I_i^w$ is designed adding $K_i$ minority pixels to $I_{i-1}^w$ in positions $\mathbf{n}^*$ given by

$$\mathbf{n}^* = \operatorname*{argmin}_{\mathbf{n}} \left( I_{i-1}^w \otimes B \right) [\mathbf{n}],$$
$$\text{subject to } I_{i-1}^w [\mathbf{n}] = 0 \text{ and } I_i^b [\mathbf{n}] = 0. \tag{10.23}$$

The process continues until the two last binary patterns $I_{(G-2)/2}^b$ and $I_{(G-2)/2}^w$ are generated.

It is important to note that only the continuous versions of the binary patterns are stored, updating the sets $\mathscr{D}\mathscr{I}_j^b$ at each iteration. In contrast to the algorithm used for $i < (G-2)/4$, there is no optimization of the interpoint distance through Lloyd's algorithm, and the placement of minority pixels is always performed in the discrete space in the range $i = (G-2)/4 + 1$ to $i = (G-2)/2$. However, their continuous space versions are stored to reconstruct the discrete patterns at the end according to Equation 10.24. The

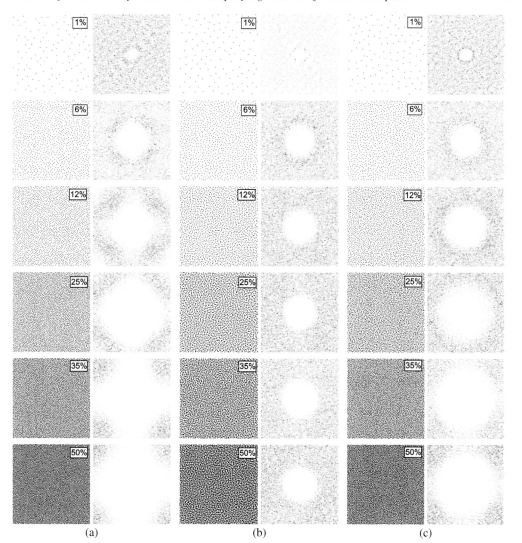

(a)    (b)    (c)

**FIGURE 10.10**

Comparison of binary patterns generated using (a) VAC, (b) DBS, and (c) CVT algorithm. The number shown over the binary patterns is the ink coverage percentage of the pattern and the column on the right of each binary pattern correspond to the absolute value of its *DFT* with the DC component removed.

matrices $B$ and $W$ are built according to Equation 10.7 and the mask is generated according to Equation 10.8, where the sets of binary patterns in the discrete space are recovered from their continuous versions by computing

$$I_i^b = \sum_{j=1}^{i} \mathcal{Q}(\mathcal{D}\mathcal{I}_j^b), \quad I_i^w = \sum_{j=1}^{i} \mathcal{Q}(\mathcal{D}\mathcal{I}_j^w). \tag{10.24}$$

Figure 10.9 depicts the resulting dither array along with the magnitude of its Fourier transform, created using this process. As expected, this array has a highpass power spectrum.

In order to compare the quality of binary patterns from masks designed using the Voronoi optimization algorithm, Figure 10.10 depicts a set of binary patterns from VAC, DBS, and

| (a) | (b) | (c) |

**FIGURE 10.11**

Halftoned images resulting from the masking process using (a) VAC mask, (b) DBS mask, and (c) Voronoi optimized mask. The size of the rendered image is $400 \times 602$ pixels and it is printed at 160 dpi.

CVT masks along with their Fourier transforms. These patterns correspond to only half of the grayscale, since as a consequence of the design order used, the characteristics of patterns in diametrically opposed gray levels are almost equal.

The binary patterns from the VAC mask (Figure 10.10a), designed using the VAC algorithm [2] with a fixed variance of the filter set to $\sigma = 2.5$, have much of their power in the high-frequency region, with power accumulation in diagonal directions for $g = 1/2$. As noted in Section 10.4, a good blue-noise mask must have some degree of clustering for $1/4 < g < 3/4$; the VAC mask does not present any clustering in that region, yielding a noisy appearance and the emergence of artifacts at midrange gray levels. The patterns from the DBS mask (Figure 10.10b), on the contrary, have a more stable appearance and a much more symmetric spectrum, but at the expense of a higher degree of clustering for $1/8 < g < 7/8$ with power accumulation at lower frequencies. Finally, the patterns from the CVT mask (Figure 10.10c) are in between the patterns from VAC and DBS, having less low-frequency content than DBS patterns, but at the same time preserving the radial symmetry in the frequency domain for all gray levels. This is understandable because the Voronoi optimization algorithm can be tuned to smoothly switch the criteria for binary pattern design in different regions of the grayscale. Through the appropriate selection of parameters, one can design patterns that are better approximations to the ideal blue-noise pattern. Figure 10.11 presents halftoned images using the VAC, DBS, and CVT masks.

## 10.5    Artifact-Free Dither Arrays

As shown in Figure 10.2, the problem with halftoning by means of a single, blue-noise dither array is that, through the process of tiling, the resulting halftone becomes periodic and at high print resolutions, this periodicity imparts a visually disturbing texture onto the image. Reference [18] addresses this problem by generating a set of unique dither patterns such that, regardless of their ordering, would show no apparent discontinuities in texture when tiled together. This phenomenon is here referred to as a *wrap-across* property. The resulting halftoning process would reduce the tiling artifacts of traditional dither array halftoning while maintaining the computational simplicity of a point-process.

In order to generate a set of dither arrays that satisfy the wrap-across property, the method of Reference [18] starts from a dither array generated according to the traditional technique. From this initial array, the method generates a set of subsequent dither arrays, where the boundary pixels are constrained to correspond exactly with those of the initial array. Specifically in this chapter, the initial sets of binary patterns $\{I_{i,1}^b, I_{i,1}^w : i = 0, 1, \ldots, (G-2)/2\}$ are generated for the first dither array $S_1$ using the algorithm described in previous sections. Given this initial array, a mask $M$ is defined as a binary image with the same row and column dimensions as the proposed component binary patterns. This matrix has the boundary pixels of $k$ rows at the top and bottom and $k$ columns at the right and left equal to one, indicating that these pixels must remain unchanged in the new dither array. The center pixels of $M$ are set to zero, indicating that these pixels need to be designed independently.

Having the mask $M$, the process begins by generating the set of binary patterns $\{I_{i,2}^b, I_{i,2}^w : i = 0, 1, \ldots, (G-2)/2\}$ corresponding to the second mask $S_2$. Starting from $I_{0,2}^b$ and $I_{0,2}^w$ as all zero matrices, the pattern $I_{1,2}^b$ is generated by adding $K_1 = \lfloor MN1/(G-1) + 0.5 \rfloor$ ones to $I_{0,2}^b$. The term $I_{1,2}^b$ is initialized as $I_{1,2}^b = I_{0,2}^b$ and the remaining ones are added as follows:

$$\text{if } M[\mathbf{n}] = 1 \quad \Rightarrow \quad I_{1,2}^b[\mathbf{n}] = I_{1,1}^b[\mathbf{n}]. \tag{10.25}$$

For $M[\mathbf{n}] = 0$, the remaining minority pixels to complete the $K_1$ pixels are added one by one in positions $\mathbf{n}^*$, where

$$\mathbf{n}^* = \underset{\mathbf{n}}{\operatorname{argmin}} \left( I_{1,2}^b \otimes B \right) [\mathbf{n}],$$
$$\text{subject to } I_{1,2}^w[\mathbf{n}] = 0, \ I_{1,2}^b[\mathbf{n}] = 0, \text{ and } M[\mathbf{n}] = 0. \tag{10.26}$$

In the next step, the continuous space version of the resulting pattern from the previous process $\mathscr{I}_{1,2}^b$ is optimized using the modified Lloyd's algorithm with the constants $\mu_j$ set to zero for the pixels in the common border and to one for the rest of the pixels. The term $I_{1,2}^b$ is obtained by mapping the result of the optimization to the discrete space as $I_{1,2}^b = \mathscr{Q}(\mathscr{I}_{1,2}^b)$.

The second binary pattern $I_{1,2}^w$ is generated in a similar way. First, the pattern $I_{1,2}^w$ is initialized as $I_{1,2}^w = I_{0,2}^w$ and the pixels in the common border are set as $I_{1,2}^w = I_{1,1}^w$ in positions where $M[\mathbf{n}] = 1$. The remaining pixels are added in positions $\mathbf{n}^*$, where

$$\mathbf{n}^* = \underset{\mathbf{n}}{\operatorname{argmin}} \left( I_{1,2}^w \otimes B \right) [\mathbf{n}],$$
$$\text{subject to } I_{1,2}^w[\mathbf{n}] = 0, \ I_{1,2}^b[\mathbf{n}] = 0, \text{ and } M[\mathbf{n}] = 0. \tag{10.27}$$

Then, the continuous version of $I_{1,2}^w$, $\mathscr{I}_{1,2}^w$ is optimized setting the constants $\mu_j$ to zero for pixels in the common border and to one for the rest. The discrete version of the binary pattern is obtained by doing $I_{1,2}^w = \mathscr{Q}(\mathscr{I}_{1,2}^w)$. If for some pixels the conditions in Equations 10.5 and 10.6 are not met, these pixels are removed one by one and placed according

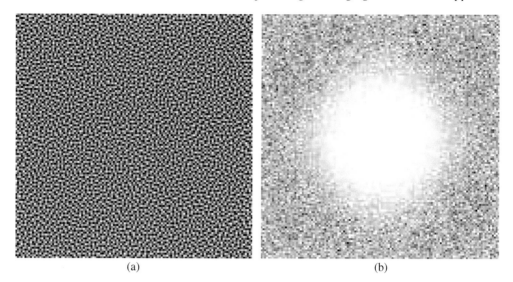

(a)                                                          (b)

**FIGURE 10.12**

The resulting eight-bit $128 \times 128$ complementary blue-noise dither array derived from dither array of Figure 10.9 in (a) the spatial domain and (b) the spectral domain.

to Equation 10.27. For the first two binary patterns, the number of iterations of Lloyd is set to $N_{max} = 50$. For the remaining patterns, the parameters are set as described previously. The remaining binary patterns $\{I_{i,2}^b, I_{i,2}^w : i = 2, 3, \ldots, (G-2)/2\}$ are designed in an analogous way as in Section 10.4 with the difference that at each iteration the binary patterns $I_{i,2}^b$ and $I_{i,2}^w$ are generated by first copying the common border pixels $I_{i,2}^b[\mathbf{n}] = I_{i,1}^b[\mathbf{n}]$ and $I_{i,2}^w[\mathbf{n}] = I_{i,1}^w[\mathbf{n}]$ in positions where $M[\mathbf{n}] = 1$. The remaining pixels to complete the $K_i$ necessary are added in positions where $M[\mathbf{n}] = 0$ according to Equations 10.22 and 10.23 and optimized in the same way as the first two patterns $I_{1,2}^w$ and $I_{1,2}^b$. The results of applying this algorithm to the dither array presented in Figure 10.9 are shown in Figure 10.12 with clear blue-noise characteristics.

In order to further minimize the likelihood of visible tiling artifacts deriving from the common boundary region, the goal should be to reduce $k$, which denotes the number of pixels, by defining the width of the boundaries in $M$. This approach is limited by the expected spacing between printed dots of the component dither patterns such that too small $k$ results in discontinuities in texture near the boundary between two unique dither patterns. Here, $k = \lceil \lambda_b \rceil$ is defined as a gray-level dependent variable equal to the principal wavelength, defined in Equation 10.12. This way takes advantage of what is known from Reference [10] about the correlation between printed dots of blue-noise patterns, ensuring that unconstrained thresholds of one component array are twice $\lambda_b$ apart from those of neighboring component arrays. Figure 10.13a shows the result of applying the adaptive border technique and Figure 10.13b shows the common pixels between two complementary dither arrays resulting from the approach of Reference [18]. The common pixels near the boundaries are greatly reduced for the adaptive technique, leading to a reduction in tiling artifacts as well. Figure 10.14 depicts the corresponding halftone pattern representing the 70% gray level using ten randomly selected masks with adaptive common border.

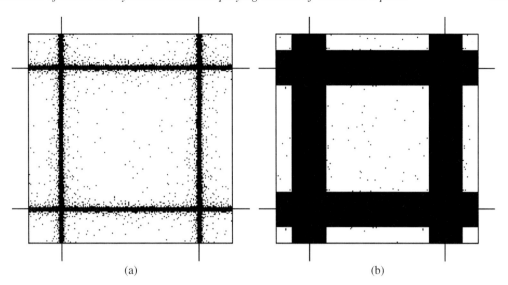

(a)                       (b)

**FIGURE 10.13**

The common pixels, marked in black, between the complementary dither arrays: (a) the result of adapting the common border according to $\lambda_b$, (b) the result of setting a common border to ten pixels.

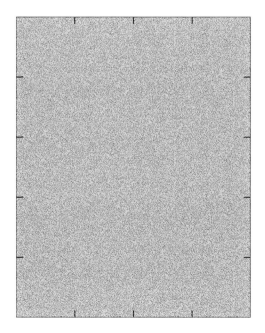

**FIGURE 10.14**

Illustration of the tiling artifacts created by ten randomly selected dither arrays with a minimized common border region.

As a further means of reducing the visibility of the repeating boundaries, one can rely on the human visual system's reduced sensitivity to diagonal correlation and generate dither arrays at rational screen angles. In particular, it is known that, at rational screen angles,

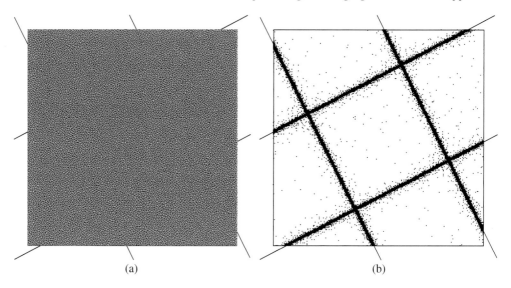

(a)                                              (b)

**FIGURE 10.15**

(a) The result of tiling nine dither arrays created with the adaptive border technique at an angle of $27^o$, and (b) common pixels, marked in black, between the base dither array and nine common border dither arrays.

the halftone cells of a traditional amplitude modulation screen form a regular tiling pattern. Combining these halftone cells (on the order of $12 \times 12$ or $16 \times 16$ pixels) into larger cells allows creating a regular tiling pattern whose cell period is on the order of a traditional blue-noise dither array ($128 \times 128$ or $256 \times 256$). Figure 10.15 illustrates the tiling pattern created by such an arrangement using a $27^o$ screen angle with a minimized border region and cell size of $128 \times 128$. Now, because the dither array can be aligned at any rational screen angle, an obvious extension of the proposed dither array scheme is to vary the screen angles between color primaries as is performed with traditional, clustered-dot ordered dithering. By doing so, the likelihood of tiling artifacts is further minimized, as could also be achieved by varying the size of the component dither arrays between colors. This is especially advantageous for small screen sizes, where the common border region accounts for a significant percentage of dither array area and hence, a large portion of the final halftone contains a repeating artifact.

To study the reduction in tiling artifacts achieved by the adaptive common border technique presented in Section 10.5, two patches of $3 \times 3$ dither arrays are compared. The first one denoted by $\mathbf{Y}$ corresponds to the tiling of nine randomly selected arrays generated with the adaptive technique, and the second one denoted by $\mathbf{W}$ correspond to the periodic tiling of one dither array $S_1$. Denoting by $S_i$ the common border dither arrays, the cells $\mathbf{Y}$ and $\mathbf{W}$ are given by

$$\mathbf{Y} = \begin{pmatrix} S_1 & S_2 & S_3 \\ S_4 & S_5 & S_6 \\ S_7 & S_8 & S_9 \end{pmatrix}, \quad \mathbf{W} = \begin{pmatrix} S_1 & S_1 & S_1 \\ S_1 & S_1 & S_1 \\ S_1 & S_1 & S_1 \end{pmatrix}. \tag{10.28}$$

To quantify the amount of reduction in tiling artifact, the autocorrelations of both cells $\mathbf{Y}$ and $\mathbf{W}$ are evaluated and the peaks present at multiples of the dither array size are compared. Calling $\mathbf{FY} = FFT(\mathbf{Y})$ and $\mathbf{FW} = FFT(\mathbf{W})$, where FFT stands for the

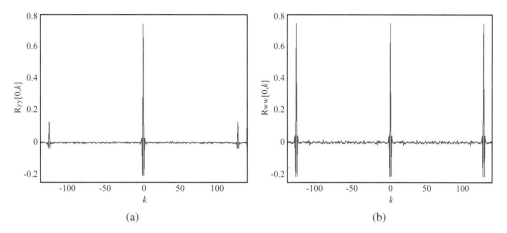

**FIGURE 10.16**

(a) Autocorrelation of **Y** evaluated at $[0,k]$. (b) Autocorrelation of **W** evaluated at $[0,k]$. The size of the dither arrays used were $128 \times 128$ pixels.

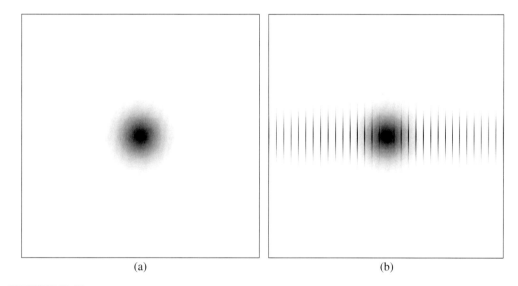

**FIGURE 10.17**

Gaussian lowpass filters used by VAC: (a) standard and (b) lenticular aware filter. Darker colors indicate larger filter values.

fast Fourier transform, the autocorrelations can be obtained as $R_{\mathbf{YY}} = IFFT(|\mathbf{FY}|^2)$ and $R_{\mathbf{WW}} = IFFT(|\mathbf{FW}|^2)$, with the DC component of **FY** and **FW** subtracted before calculating the inverse FFT (IFFT). Figure 10.16 shows the values of both autocorrelations evaluated at the horizontal axis $[0,k]$, since it is along horizontal and vertical directions where the peaks in the autocorrelation are found. It can be seen that the peaks at $k = 128$ and $k = -128$ are significatively reduced in the cell containing the adaptive border arrays with respect to the periodic cell.

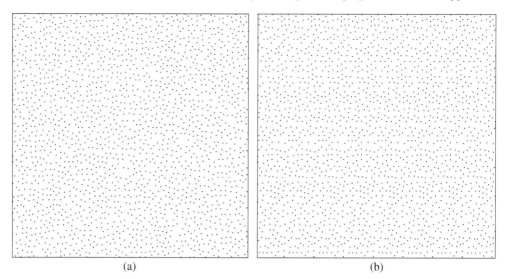

(a)                               (b)

**FIGURE 10.18**

(a) The lenticular aware dither pattern. (b) The corresponding dither pattern constructed by repeating a down-sampled rendition eight times.

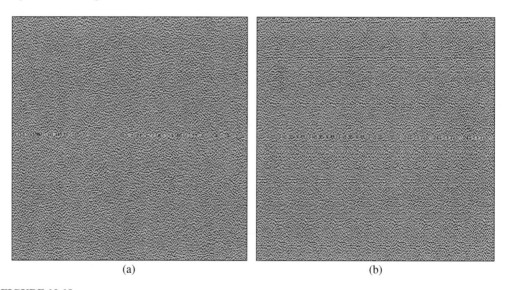

(a)                               (b)

**FIGURE 10.19**

(a) The lenticular aware dither array. (b) The corresponding dither array constructed by repeating a downsampled rendition eight times.

## 10.6 Lenticular Screening

Now, in order to modify the void-and-cluster algorithm to produce lenticular aware dither arrays, the lowpass filter is modified to maintain smooth textures in the downsampled image such that

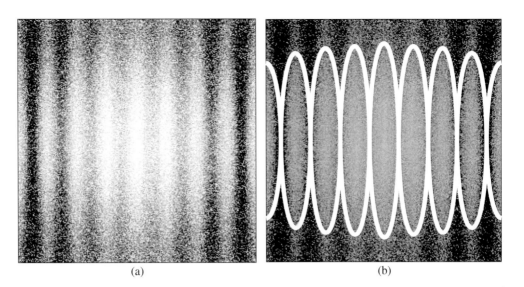

**FIGURE 10.20**

(a) The magnitude of the Fourier transform of the lenticular aware dither array. (b) The red-noise voids, indicated by white ovals, caused by the upsampling operation on the lowpass filters used by VAC.

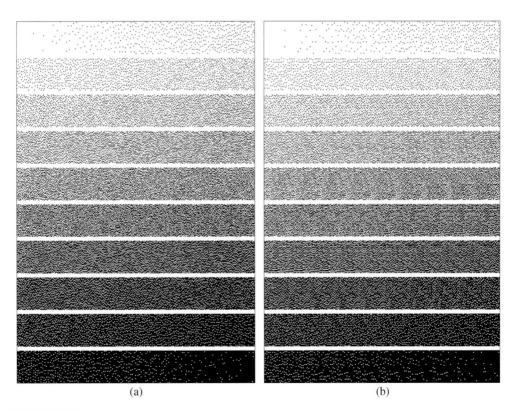

**FIGURE 10.21**

The halftoned grayscale ramps (a) before and (b) after downsampling using a lenticular aware dither array.

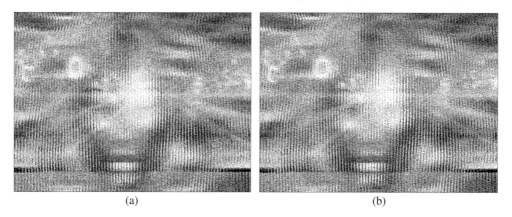

(a)            (b)

**FIGURE 10.22**

A composite lenticular image halftoned by means of (a) a traditional VAC dither array and (b) a lenticular aware VAC dither array.

$$
B[\mathbf{n}] = \begin{cases} \exp\left(-\frac{(|m|^{p}+(|n|/N)^{p})^{2/p}}{2\sigma^{2}}\right) & \text{for } n = 0, +-N, +-2N, +-3N, \ldots \\ \exp\left(-\frac{(|m|^{p}+|n|^{p})^{2/p}}{2\sigma^{2}}\right) & \text{otherwise,} \end{cases} \tag{10.29}
$$

where $N$ is the number of printed pixel columns under each lens of the lenticular image. In the case of having eight pixel columns under each lens, Figure 10.17 compares the new lenticular aware filter against a traditional VAC version. Applying this new filter, Figure 10.18 shows the dither pattern before and after downsampling, where no discontinuity in texture can be observed, although there are some unwanted hole artifacts. As for the resulting dither array, Figure 10.19 shows the final dither array and the corresponding dither array formed by first downsampling the original array by a factor of eight and then tiling the resulting array eight times to create a new $256 \times 256$ array. Figure 10.20a presents the magnitude of the raw dither array's Fourier transform that shows the effects of the upsampling operation on the lowpass filter. This results in multiple red-noise voids, which are indicated in Figure 10.20b by the white ovals superimposed over the Fourier transform plot.

Looking now at the dot patterns produced by this array, Figure 10.21 shows the corresponding grayscale ramp and image produced by thresholding with the lenticular aware dither array with neither dot pattern showing any significant discontinuity in texture except at extreme gray levels. To evaluate the effects of the new dither array on lenticular images, Figure 10.22 shows the halftones produced by a traditional and the new lenticular aware dither array for a lenticular image composed of eight component images with each image corresponding to single pixel-wide slices. In the composite images, neither halftone shows any discontinuity in texture caused by the dither array. However, in Figure 10.23, which shows one of the eight de-interlaced component images, one can see what looks like a white-noise halftone produced by the traditional dither array while the lenticular aware array creates a traditional blue-noise texture.

With regards to the change in texture across two neighboring component images, such as would be visible when progressing through an animated sequence, one can look at the textures produced when the composite image is downsampled by a factor one-off from its true rate, in this case downsampled by seven instead of eight. The resulting textures are

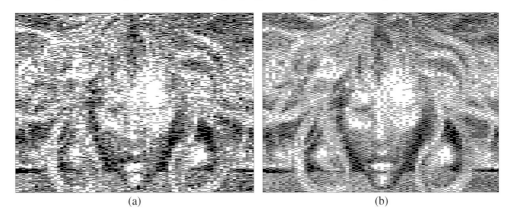

**FIGURE 10.23**

A de-interlaced lenticular image halftoned by means of (a) a traditional VAC dither array and (b) a lenticular aware VAC dither array.

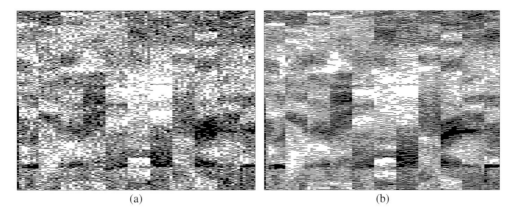

**FIGURE 10.24**

A de-interlaced lenticular image downsampled by a factor of seven (one-off from the true rate of eight) and halftoned by means of (a) a traditional VAC dither array and (b) a lenticular aware VAC dither array.

visible in Figure 10.24 where, again, the traditional dither array produces a white-noise texture while the new array maintains a much smoother texture. The vertical blind effect, visible in these two figures, are a result of the transition from the eight component image back to the first component image occurring at every eight sample, since one component image is shifted with each downsampled pixel. What these two figures suggest is that as a human subject's eye animates through the lenticular sequence, it will see larger amounts of twinkling in the traditional halftone versus the lenticular aware technique.

A further implication of Figure 10.24 is that it suggests that the size of the lenticular aware dither array does not need to be an integer multiple of the downsampling rate, and that one could make the dither array of size $KN - 1$, where $K$ is some positive integer. Such a dither array would have the advantage of producing a periodic texture in the de-interlaced image with a horizontal period of $(KN)^2$ lenses instead of $K$ lenses. It would still have a vertical period of approximately $KN$ pixels, but at the printer's native resolution, not the lens resolution. Of course, it is possible to construct non-square dither arrays

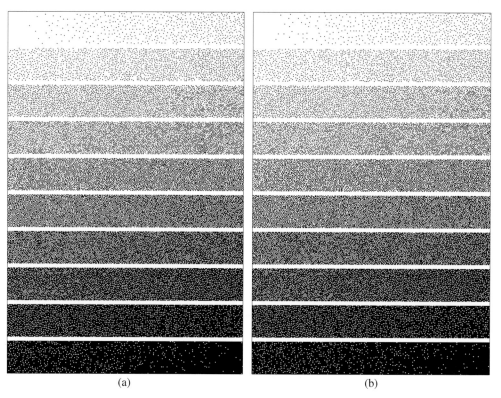

(a)                                              (b)

**FIGURE 10.25**

The halftoned grayscale ramps (a) before and (b) after downsampling using a lenticular aware dither array of size $255 \times 255$ pixels.

with large vertical periods. The important feature here is that one can achieve much larger horizontal periods in the de-interlaced images with modest periods in the composite image. Figure 10.25 demonstrates the reduction in tiling artifacts achieved through this process, where Figure 10.25b shows the grayscale ramp produced by a downsampled rendition of the lenticular aware dither array. In summary, the lenticular aware VAC process produces consistent halftone textures in both the composite and de-interlaced renditions of a lenticular image with reduced tiling artifacts achieved by using a non-integer multiple of the downsampling rate for the dither array width. Tiling artifacts in the composite image are of no significant consequence.

## 10.7    Conclusion

By introducing a concise description of the ideal spectral shape of visually pleasing dither patterns, Reference [12] introduced a means by which dither patterns could be quantitatively evaluated, and hence, one could compare multiple halftoning algorithms with the better technique being the one with a spectral profile closest in shape to this ideal model. While this *cost* played a fundamental role in the optimization of error-diffusion [26], [27],

[28], [29], its most significant contribution may be its leading to the construction of aperiodic dispersed-dot dither arrays. While sharing many similarities to periodic clustered-dot dithering arrays, blue-noise arrays have, so far, failed to achieve the same level of variability/flexibility with their only significant variation being the dither array size.

This chapter has attempted to bridge the gap between blue-noise dither arrays and traditional ordered dithering, offering a wider range of parameters by which the user can optimize threshold placement. Through the selection of different sets of parameters $\mu_j$, $N_{max}$, $p$, and $\sigma$ for each gray level, the textures and radial symmetry of the binary patterns obtained can be controlled with greater versatility. The amount of power accumulated along vertical or diagonal directions can be controlled by choosing $p > 2$ or $p < 2$ in the searching filter and the uniformity of the whole pattern regulated by performing more iterations of the modified Lloyd's algorithm. In addition, an even quality across the whole grayscale can be achieved by optimizing multiple binary patterns simultaneously at each iteration, choosing the rate at which the $\mu_j$ constants change with each iteration.

The reduction in tiling artifacts achieved by optimizing the common border of dither arrays has proved to be significant, as shown in Figure 10.16. The design of dither arrays at rational screen angles have further reduced the presence of artifacts allowing to produce nearly artifact-free screens. Beyond the reduction in tiling artifacts, the technique of tiling common border dither arrays creates new opportunities for data encryption and digital watermarking [19], [30], [31], in which a message can be encoded in the ordering chosen for the tiling of small binary patterns.

Looking at lenticular printing, it should be noted that even simply modifications to existing techniques can produce dramatic improvements over applying traditional halftoning algorithms. This chapter has presented such a modification by altering the lowpass filter of void-and-cluster algorithm. In doing so, the proposed dither array produces smooth, blue-noise textures in both the raw composite image as well as the de-interlaced component images. Beyond this improvement, this chapter has also introduced a novel means of evaluating halftone texture in lenticular images by looking at the halftone produced by downsample one-off from the true de-interlacing rate. This new means of texture evaluation attempts to analyze the change in dot patterns across neighboring component images as would be seen in an animated lenticular sequence.

## Acknowledgment

Figure 10.9 is image courtesy of the University of Delaware.

## References

[1] B. Bayer, "An optimum method for two level rendition of continuous-tone pictures," in *Proceedings of the IEEE International Conference on Communications*, Seattle, WA, USA, June

1973, pp. 11–15.

[2]  R.A. Ulichney, "The void-and-cluster method for dither array generation," *Proceedings of SPIE*, vol. 1913, pp. 332–343, February 1993.

[3]  J. Sullivan and L. Ray, "Digital halftoning with correlated minimum visual modulation patterns," U.S. Patent 5 214 517, May 1993.

[4]  J. Sullivan, L. Ray, and R. Miller, "Design of minimum visual modulation halftone patterns," *IEEE Transactions on Systems, Man and Cybernetics*, vol. 21, no. 1, pp. 33–38, January 1991.

[5]  T. Mitsa and K.J. Parker, "Digital halftoning technique using a blue noise mask," *Journal of the Optical Society of America*, vol. 9, no. 11, pp. 1920–1929, August 1992.

[6]  M. Yao and K.J. Parker, "Modified approach to the construction of a blue noise mask," *Journal of Electronic Imaging*, vol. 3, no. 1, pp. 92–97, January 1994.

[7]  J. Allebach and Q. Lin, "FM screen design using the DBS algorithm," in *Proceedings of the IEEE International Conference on Image Processing*, Lausanne, Switzerland, September 1996, pp. 549–552.

[8]  B.E. Cooper, T.A. Knight, and S.T. Love, "Method for halftoning using stochastic dithering with minimum density variance," U.S. Patent 5 696 602, December 1997.

[9]  K.E. Spaulding, R.L. Miller, and J. Schildkraut, "Methods for generating blue-noise dither matrices for digital halftoning," *Journal of Electronic Imaging*, vol. 6, no. 2, pp. 208–230, April 1997.

[10] D.L. Lau, G.R. Arce, and N.C. Gallagher, "Green-noise digital halftoning," *Proceedings of the IEEE*, vol. 86, no. 12, pp. 2424–2444, December 1998.

[11] R.W. Floyd and L. Steinberg, "An adaptive algorithm for spatial gray-scale," *Proceedings of the Society of Information Display*, vol. 17, no. 2, pp. 75–78, 1976.

[12] R.A. Ulichney, "Dithering with blue noise," *Proceedings of the IEEE*, vol. 76, no. 1, pp. 56–79, January 1988.

[13] R.L. Stevenson and G.R. Arce, "Binary display of hexagonally sampled continuous-tone images," *Journal of the Optical Society of America*, vol. 2, no. 7, pp. 1009–1013, July 1985.

[14] M. Analoui and J.P. Allebach, "Model based halftoning using direct binary search," *Proceedings of SPIE*, vol. 1666, pp. 96–108, August 1992.

[15] V.F.Q. Du and M. Gunzburger, "Centroidal Voronoi tessellations: Applications and algorithms," *SIAM Review*, vol. 41, no. 4, pp. 637–676, April 1999.

[16] O. Deussen, S. Hiller, C.V. Overveld, and T. Strothotte, "Floating points: A method for computing stipple drawings," *Computer Graphics Forum*, vol. 19, no. 3, pp. 41–50, March 2000.

[17] A. Secord, "Weighted Voronoi stippling," in *Proceedings of the 2nd International Symposium on Non-Photorealistic Animation and Rendering*, Annecy, France, June 2002, pp. 37–43.

[18] B.W. Kolpatzik and J.E. Thornton, "Image rendering system and method for generating stochastic threshold arrays for use therewith," U.S. Patent 5 745 660, April 1998.

[19] D. Kacker and J.P. Allebach, "Aperiodic micro-screen design using DBS and training," *Proceedings of SPIE*, vol. 3300, pp. 386–397, January 1998.

[20] Z. Baharav and D. Shaked, "Watermarking of dither halftone images," *Proceedings of SPIE*, vol. 3657, pp. 307–316, January 1999.

[21] S. Kim and J.P. Allebach, "High quality, low complexity halftoning with good compressibility," in *Proceedings of the IEEE International Conference on Image Processing*, Rochester, NY, USA, September 2002, vol. 1, pp. 453–456.

[22] D.L. Lau and T. Smith, "Model-based error-diffusion for high-fidelity lenticular screening," *Optics Express*, vol. 14, no. 8, pp. 3214–3224, April 2006.

[23] D.L. Lau and R. Ulichney, "Blue-noise halftoning for hexagonal grids," *IEEE Transactions on Image Processing*, vol. 15, no. 5, pp. 1270–1284, May 2005.

[24] K.S.A. Okabe, B. Boots, and S.N. Chiu, *Spatial Tessellations: Concepts and Applications of Voronoi Diagrams*. Chichester, UK: John Wiley & Sons Ltd, 2nd edition, 2000.

[25] D.L. Lau and G.R. Arce, *Modern Digital Halftoning*. New York, USA: Marcel Dekker, 2001.

[26] R. Eschbach and K.T. Knox, "Error-diffusion algorithm with edge enhancement," *Journal of the Optical Society of America*, vol. 8, no. 12, pp. 1844–1850, December 1991.

[27] B.W. Kolpatzik and C.A. Bouman, "Optimized error diffusion for image display," *Journal of Electronic Imaging*, vol. 1, no. 3, pp. 277–292, July 1992.

[28] T. Zeggle and O. Bryngdahl, "Halftoning with error-diffusion on an image-adaptive raster," *Journal of Electronic Imaging*, vol. 3, no. 3, pp. 288–294, July 1994.

[29] T.D. Kite, B.L. Evans, A.C. Bovik, and T.L. Sculley, "Digital halftoning as 2-D delta-sigma modulation," in *Proceedings of the IEEE International Conference on Image Processing*, Santa Barbara, CA, USA, October 1997, vol. 1, pp. 799–802.

[30] M.S. Fu and O.C. Au, "Data hiding watermarking for halftone images," *IEEE Transactions on Image Processing*, vol. 11, no. 4, pp. 477–484, April 2002.

[31] Z. Zhou, G.R. Arce, and G.D. Crescenzo, "Halftone visual cryptography," *IEEE Transactions on Image Processing*, vol. 15, no. 8, pp. 2441–2453, August 2003.

# 11

## Perceptual Color Descriptors

**Serkan Kiranyaz, Murat Birinci, and Moncef Gabbouj**

## 11.1 Introduction

Color in general plays a crucial role in today's world. Color alone can influence the way humans think, act, and even react. It can irritate or soothe eyes, create ill-tempers, or result in loss of appetite. Whenever used in a right form, color can save on energy; otherwise, it can cause detrimental effects. Particularly, the color composition of an image can turn out to be a powerful cue for the purpose of content-based image retrieval (CBIR), if extracted in a perceptually oriented way and kept semantically intact. Furthermore, the color structure in a visual scenery is robust to noise, image degradations, changes in size, resolution, and orientation. Eventually, most of the existing CBIR systems use various color descriptors in order to retrieve relevant images (or visual multimedia material); however, their retrieval performance is usually limited especially on large databases due to lack of discrimination power of such color descriptors. One of the main reasons for this is because most of such systems are designed based on some heuristics or naive rules that are not formed with respect to what humans or more specifically the human visual system (HVS) finds relevant in terms of color similarity. The word relevance is described as "the ability (as of an information retrieval system) to retrieve material that satisfies the needs of the user." Therefore, it is of decisive importance that human color perception is respected while modeling and describing any color composition of an image. In other words, if and only when a particular color descriptor is designed based entirely on HVS and human color perception rules, further discrimination power and hence certain improvements in the retrieval performance can be achieved.

Accordingly, the study of human color perception and similarity measurement in the color domain become crucial and there is a wealth of research performed in this field. For example, Reference [1] focused on the utilization of color categorization (termed as focal colors) for CBIR purposes and introduced a new color matching method, which takes human cognitive capabilities into account. It has exploited the fact that humans tend to think and perceive colors only in eleven basic categories. Reference [2] performed a series of psychophysical experiments, analyzing how humans perceive and measure similarity in the domain of color patterns. These experiments resulted in five perceptual criteria (so-called basic color vocabulary), which are important for comparing the color patterns, as well as a set of rules (so-called basic color grammar), which are governing the use of these criteria in similarity judgment. One observation worth mentioning here is that the human eye cannot perceive a large number of colors at the same time, nor is it able to distinguish similar colors well. At the coarsest level of judgment, the HVS judges similarity primarily using dominant colors, that is, few prominent colors in the scenery. Henceforth, the two rules are particularly related for modeling the similarity metrics of the human color perception. The first one indicates that the two color patterns that have similar dominant colors (DCs) are perceived as similar. The second rule states that two multicolored patterns are perceived as similar if they possess the same (dominant) color distributions regardless of their content, directionality, placement, or repetitions of a structural element.

It is known that humans focus on a few dominant colors and their spatial distributions while judging the color similarity between images. Humans have the ability to extract such a global color view out of a visual scenery, irrespective of its form, be it a digital image or a natural three-dimensional view. However, accomplishing this is not straightforward while dealing with digital images for CBIR purpose. In a standard 24-bit representation, there is a span of 16 million colors, which can be assigned on thousands of individual pixels. Such a high-resolution representation might be required for current digital image technologies; however, it is not too convenient for the purpose of describing color composition or performing a similarity measurement based on the aforementioned perception rules. Nevertheless, it is obvious that humans can neither see individual pixels, nor perceive even a tiny fracture of such a massive amount of color levels, and thus it is crucial to perform certain steps in order to extract the true perceivable elements (the true DCs and their global distributions). In other words the imperceivable elements, called here as outliers, which do not have significant contribution or weight over the present color structure, in both color and spatial domain, should be suppressed or removed. Recall that according to two color perception rules presented in Reference [3], two images that are perceived as similar in terms of color composition have similar DC properties; however, the color properties of their outliers might be entirely different and hence this can affect (degrade, bias, or shift) any similarity measurement if not handled accordingly. For example, popular perceptual audio coding schemes, such as MP3 and AAC [4], maximize the coding efficiency by removing outlying sound elements, that humans cannot hear, in both time and frequency domains and thus more bits can be spent for the dominant sound elements. In a similar fashion, the outliers both in color and spatial domain should be removed for description efficiency. Henceforth, this chapter mainly focuses on a systematic approach to extract such a perceptual color descriptor and efficient similarity metrics to achieve the highest discrimination power possible for color-based retrieval in general-purpose image databases.

Furthermore, one can also note that human color perception is strongly dependent on the variations in their adjacency, thus the spatial distribution of the prominent colors should also be inspected in a similar manner. The color correlogram [5], [6], is one of the most promising approaches in the current state of art, which is a table where the $k$th entry for the color histogram bin pair $(i, j)$ specifies the probability of finding a pixel of color bin $j$ at a distance $k$ within a maximum range $d$ from a pixel of color bin $i$ in an image $I$ with dimensions $W \times H$ quantized with $m$ colors. Accordingly, Reference [7] conducted a comprehensive performance evaluation among several global/spatial color descriptors for CBIR and reported that the correlogram achieves the best retrieval performance among many others, such as color histograms, color coherence vector (CCV) [8], color moments, and so on. However, it comes along with many infeasibility problems, such as its massive memory requirement and computational complexity. This chapter shall first present a detailed feasibility analysis for the correlogram in the next section. Apart from the aforementioned feasibility problems, using the probability alone makes the descriptor insensitive to the dominance of a color or its area (weight) in the image. This might be a desirable property in finding the similar images, such as images that are simply zoomed [6], and hence the color areas significantly vary while the distance probabilities do not. On the other hand, it also causes severe mismatches especially in large databases, since the probability of the pairwise color distances might be the same or close independent of their *area* and hence regardless of their dominance (whether they are DCs or outliers). Therefore, in this chapter, stemming from the earlier discussions, a perceptual approach will be presented in order to overcome the shortcomings of the correlogram, namely, its lack of perceptuality and infeasibility. This chapter shall present a color descriptor that utilizes a dynamic color clustering and extracts the DCs present in an image in order to bring in the nonexistent perceptuality to the correlogram. Global color properties are thus obtained from these clusters. Dominant colors are then back-projected onto the image to capture the spatial color distribution via a DC-based correlogram. Finally, a penalty-trio model combines both global properties extracted from the DCs and their spatial similarities obtained through the correlogram, forming a perceptual distance metric.

The perceptual correlogram, although an efficient solution for the lack of color perceptual model and computational infeasibility, cannot yet properly address the outlier problem. In order to remove the spatial outliers and to secure the global (perceptual) color properties, one alternative is to apply nonlinear filters (for example, median or bilateral [9]). However, there would be no guaranty that such a filter will remove all or the majority of the outliers and yet several filter parameters are needed to be set appropriately for an acceptable performance, which is not straightforward to do so, especially for large databases. Instead, a top-down approach is adopted both in extracting DCs and modeling their global spatial distribution. This approach is in fact phased from the well-known Gestalt rule of perception [10] formulated as "humans see the whole before its parts." Therefore, the method strives to extract what is the next global element both in color and spatial domain, which are nothing but the DCs and their spatial distribution within the image. In order to achieve such a global spatial representation within an image, starting from the entire image, quad-tree decomposition is applied to the current (parent) block only if it cannot host the majority of a particular DC; otherwise, it is kept intact (non-decomposed) representing a single, homogeneous DC presence in it. So, this approach tries to capture the whole before going through its parts, and whenever the whole body can be perceived with a single DC, it is kept as is. Hence outliers can be suppressed from the spatial distribution and furthermore, the resultant (blockwise) partitioned scheme can be efficiently used for a global modeling and due description of the spatial distribution. Finally, a penalty-trio model uses both global and spatial color properties, and performs an efficient similarity metric. After the image is (quad-tree) decomposed, this global spatial distribution is then represented via inter-proximity statistics of the DCs, both in scalar and directional modes. These modes of spatial color distribution (SCD) can both describe the distribution of a particular DC with itself (auto-SCD) and with other DCs (inter-SCDs). Integrating both techniques as a feature extraction (FeX) module into MUVIS framework [11] allows for testing the mutual performance in the context of multimedia indexing and retrieval.

The rest of this chapter is organized as follows. Section 11.2 surveys the related studies in the area of color-based CBIR, stressing particularly their limitations and drawbacks under the light of the earlier discussion on human color perception. Section 11.3 presents the perceptual correlogram and compares it with the original correlogram in sample image databases. Section 11.4 introduces a generic overview of the spatial color descriptor (SCD) along with the extraction, formation of the feature vector, and calculation of the similarity distances. Finally, this chapter concludes with Section 11.5.

## 11.2   Related Work

There is a wealth of research done and still going on in developing content-based multimedia indexing and retrieval systems, such as MUVIS [11], QBIC [12], PicHunter [13], Photobook [14], VisualSEEk [15], Virage [16], Image-Rover [17], and VideoQ [18]. In such frameworks, database primitives are mapped into some high-dimensional feature domain, which may consist of several types of descriptors, such as visual, aural, and so on. From the latitude of low-level descriptors, careful selection of some sets to be used for a particular application may capture the semantics of the database items in a content-based multimedia retrieval (CBMR) system. Although color is used in many areas, such as object and scene recognition [19], this chapter shall restrict the focus on CBIR domain, which employs only color as the descriptor for image retrieval.

### 11.2.1 Global Color Descriptors

In one of the earlier studies on color descriptors, Reference [20] used the color of every corresponding pixel in two images for comparison and the number of corresponding pixels having the same color to determine the similarity between them. Recall the HVS fact mentioned earlier about the inability of humans to see individual pixels or to perceive large amount of color levels. Hence, this approach does not provide robust solutions, that is, slight changes in camera position, orientation, noise, or lightning conditions may cause significant degradations in the similarity computation. Reference [21] proposed the first color histogram, which solves this sensitivity problem. Namely, color histograms are extracted and histogram intersection method is utilized for comparing two images. Since this method is quite simple to implement and gives reasonable results especially in small- to medium-sized databases, several other histogram-based approaches emerged [11], [12], [13], [14], [17], [22], [23], [24], [25], [26], [27]. The MPEG-7 color structure descriptor (CSD) [28] is also based on color histogram; however, it provides a more accurate color description by identifying localized color distributions of each color. Unlike the conventional color histograms, CSD is extracted by accumulating from a $8 \times 8$ structuring window. The image is scanned and CSD counts the number of occurrences of a particular color within the structuring window. Additional information, including an efficient representation of color histograms based on Karhunen-Loeve transform (KLT), can be found in Reference [29].

The primary feature of such histogram-based color descriptors (be it in the RGB, CIE-Lab, CIE-Luv, or HSV color space) is that they cluster the pixels into fixed color bins, which are quantizing the entire color space using a predefined color palette. This twofold approach, clustering all the pixels having similar color and reducing the color levels from millions to usually thousands or even hundreds via quantization, is the main reason behind the limited success that the color histograms achieved since both operations are indeed the small steps through obtaining the perceivable elements (the true DCs and their global distributions); yet their performance is still quite limited and usually degrades drastically in large databases due to several reasons. First and the foremost, they apply static-quantization where the color palette boundaries are determined empirically or via some heuristics; none of these are based on the perception rules. If, for example, the number of bins are set too high (fine quantization), then similar color pairs will end up in different bins. This will eventually cause erroneous similarity computations whenever using any of the naive metrics, such as $L_1$ and $L_2$, or using the histogram intersection method [21]. On the other hand, if the number of bins is set too low (coarse quantization), then there is an imminent danger of completely different colors falling into the same bin and this will obviously degrade the similarity computation and reduce the discrimination power. No matter how the quantization level (number of bins) is set, pixels with such similar colors happens to be opposite sides of the quantization boundary, separating two consecutive bins will be clustered into different bins and this is an inevitable source of error in all histogram-based methods.

The color quadratic distance [12] proposed in the context of QBIC system provides a solution to this problem by fusing the color bin distances into the total similarity metric. Writing the two color histograms $X$ and $Y$, each with $N$ bins, as pairs of color bins and weights gives $X = \left\{ \left(c_1, w_1^X\right), \left(c_2, w_2^X\right), ..., \left(c_N, w_N^X\right) \right\}$ and $Y = \left\{ \left(c_1, w_1^Y\right), \left(c_2, w_2^Y\right), ..., \left(c_N, w_N^Y\right) \right\}$. Then, the quadratic distance between $X$ and $Y$ is expressed as follows:

$$D_Q(X,Y)^2 = (X-Y)^T A(X-Y)$$

$$= \sum_{i}^{N} \sum_{j}^{N} (w_i^X - w_i^Y)(w_j^X - w_j^Y)a_{ij}, \qquad (11.1)$$

where $A = [a_{ij}]$ is the matrix of color similarities between the bins $c_i$ and $c_j$. This formulation allows the comparison of different histogram bins with some inter-similarity between them; however, it underestimates distances because it tends to accentuate the color similarity [30]. Furthermore, Reference [31] showed that the quadratic distance formulation has serious limitations, as it does not match the human color perception well enough and may result in incorrect ranks between regions with similar salient color distributions. Hence, it gives even worse results than the naive $L_p$ metrics in some particular cases.

Besides the aforementioned clustering drawbacks and the resultant erroneous similarity computation, color histograms have computational deficiencies due to the hundreds or even thousands of redundant bins created for each image in a database, although ordinary images usually contain few DCs (less than eight). Moreover, they cannot be perceived by the HVS [3] according to the second color perception rule mentioned earlier. Therefore, color histograms do not only create a major computational deficiency in terms of storage, memory limit, and computation time due to spending hundreds or thousands of bins for several present DCs, but their similarity computations will be biased by the *outliers* hosted within those redundant bins. Recall that two images with similar color composition will have similar DC properties; however, there is no such requirement for the outliers as they can be entirely different. Hence, including color outliers into similarity computations may cause misinterpreting two similar images as dissimilar or vice versa, and usually reduces the discrimination power of histogram-based descriptors, which eventually makes them unreliable especially in larger databases.

In order to solve the problems of static quantization in color histograms, various DC descriptors [3], [27], [28], [32], [33], [34] have been developed using dynamic quantization with respect to image color content. DCs, if extracted properly according to the aforementioned color perception rules, can indeed represent the prominent colors in any image. They have a global representation, which is compact and accurate, and they are also computationally efficient. Here, a top-down DC extraction scheme is used; this scheme is similar to the one in Reference [33], which is entirely designed with respect to HVS color perception rules. For instance, HVS is more sensitive to the changes in smooth regions than in detailed regions. Thus, colors can be quantized more coarsely in the detailed regions, while smooth regions have more importance. To exploit this fact, a smoothness weight $w(p)$ is assigned to each pixel $p$ based on the variance in a local window. Afterward, the *general Lloyd algorithm* (GLA), also referred to as *Linde-Buzo-Gray* and equivalent to the well-known *K-means* clustering method [35], is used for color quantization.

### 11.2.2 Spatial Color Descriptors

Although the true DCs, which are extracted using such perceptually oriented scheme with the proper metric, can address the aforementioned problems of color histogram, global color properties (DCs and their coverage areas) alone are not enough for characterizing and

**FIGURE 11.1**

Different color compositions of red (light gray), blue (dark gray), and white with the same proportions.

describing the real color composition of an image since they all lack the crucial information of spatial relationship among the colors. In other words, describing what and how much color is used will not be sufficient without specifying where and how the (perceivable) color components (DCs) are distributed within the visual scenery. For example, all the patterns shown in Figure 11.1 have the same color proportions (be it described via DCs or color histograms), but different spatial distributions, and thus cannot be perceived as the same. Especially in large image databases, this is the main source of erroneous retrievals, which makes accidental matches between images with similar global color properties but different in the color distribution.

There are several approaches to address such drawbacks. Segmentation-based methods may be an alternative; however, they are not feasible since in most cases automatic segmentation is an ill-posed problem and therefore, it is not reliable and robust for applications on large databases. For example, in Reference [36], DCs are associated with the segmented regions; however, the method is limited for use on a small size *National Flags* database where segmentation is trivial. Some studies relied on the local positions of color blocks to characterize the spatial distributions. For instance, Reference [22] divides the image into nine equal subimages and represented each of them by a color histogram. Similarly, Reference [30] splits the image into five regions (an oval central region and four corners) and tries to combine color similarity from each region while attributing more weight to the central region. In Reference [37], the image is split into $16 \times 16$ blocks, and each block is represented by a unique dominant color. Due to the fixed partitioning, such methods become strictly domain-dependant solutions. Reference [38] enhances the idea of using a statistically derived quad-tree decomposition to obtain homogeneous blocks, but again the matching blocks (in the same position) are compared to determine the SCD similarity. Basically, in such approaches, the local position of a certain color in an image cannot really describe the true SCD due to several reasons. First, the image dimensions, resolution, and their aspect ratio can vary significantly. Thus, an object with a certain size can fall (perhaps partially) into different blocks in different locations. Furthermore, such a scheme is not rotation and translation invariant.

Reference [8] presented the color coherence vector (CCV) technique, which partitions the histogram bins based on the spatial coherence of the pixels. A given pixel is coherent if its color is similar color to a colored region; otherwise, it is considered as incoherent. For each color $c_i$, let $\alpha(c_i)$ and $\beta(c_i)$ be the number of coherent and incoherent pixels. The pair $(\alpha(c_i), \beta(c_i))$ is called a coherence pair for the $i$th color. Thus, the coherent vector for an image $I$ can be defined as

$$CCV(I) = \{(\alpha(c_1), \beta(c_1)), (\alpha(c_2), \beta(c_2)), ..., (\alpha(c_N), \beta(c_N))\}. \quad (11.2)$$

A nice property of this method is the classification of the outlying (that is, incoherent) color pixels in spatial domain from the prominent (that is, coherent) ones. A better retrieval performance than that of the traditional histogram-based methods was reported. Yet apart from the aforementioned drawbacks of histogram-based methods with respect to individual pixels, classifying color pixels alone, without any metric or characterization for the SCD, will not describe the real color composition of an image.

Another variation of this approach is characterizing adjacent color pairs through their boundaries. Using a color matching technique to model color boundaries, two images are expected to be similar if they have similar sets of color pairs [39]. Reference [30] used the boundary histograms to describe the length of the color boundaries. Another color adjacency-based descriptor can be found in Reference [40]. Such heuristic approaches of using color adjacency information might be more intuitive than the ones using fixed blocks, since they at least used relative features instead of static ones. Yet the approach is likely to suffer from changes in background color or relative translations of the objects in an image. The former case implies a strong dissimilarity, although only the background color is changed while the rest of the object(s) or color elements stay intact. In the latter case, there is no change in the adjacent colors; however, the inter-proximities of the color elements (hence the entire color composition) are changing and hence a certain dissimilarity should occur. Therefore, the true characterization of SCD lies in the inter-proximities or relative distances of color elements with respect to each other. In other words, characterizing inter- or self-color proximities, such as the relative distances of the DCs, shall be a reliable and discriminative cue about the color composition. This property is invariant to translations, rotations and variations in image properties (dimensions, aspect ratio, and resolution), and hence will be the basis of the SCD presented in this chapter.

### 11.2.3 Color Correlogram

One of the most promising approaches among all SCD descriptors is the color correlogram [5], [6], which is a table where the $k$th entry for the color histogram bin pair $(i,j)$ specifies the probability of finding a pixel of color bin $j$ at a distance $k$ from a pixel of color bin $i$ in the image. A similar technique, the color edge co-occurrence histogram, has been used for color object detection in Reference [41].

Let $I$ be an $W \times H$ image quantized with $m$ colors $(c_1, \ldots, c_i, \ldots, c_m)$ using the RGB color histogram. For a pixel $p = (x,y) \in I$, let $I(p)$ denote its color value and let $I < c_i > \equiv \{p | I(p) = c_i\}$. Thus, the color histogram value $h(c_i, I)$ of a quantized color $c_i$ can be defined as follows:

$$h(c_i, I) = W H \Pr(p \in I < c_i >). \tag{11.3}$$

Accordingly, for the quantized color pair and a pixel distance, the color correlogram can be expressed as

$$\gamma_{c_i,c_j}^{(k)} = \Pr_{p_1 \in I < c_i >, p_2 \in I} \left( p_2 \in I < c_j > \big| \ |p_1 - p_2| = k \right), \tag{11.4}$$

where $c_i, c_j \in \{c_1, \ldots, c_m\}$, $k \in \{1, \ldots, d\}$ and $|p_1 - p_2|$ is the distance between pixels $p_1$ and $p_2$ in $L_\infty$ norm. The feature vector size of the correlogram is $O(m^2 d)$, as opposed to the auto-correlogram [5], which captures only the spatial correlation between the same colors and thus reduces the feature vector size to $O(md)$ bytes. A variant of the correlogram based on the HSV color domain can be found in Reference [42].

In the spatial domain and at the pixel level, the correlogram can characterize and thus describe the relative distances of distinct colors between each other; therefore, such a description can indeed reveal a high-resolution model of SCD. Accordingly, References [7] and [43] conducted comprehensive performance evaluations among several global/spatial color descriptors for CBIR and reported that the (auto-)correlogram achieves the best retrieval performance among the others, such as color histograms, CCV, and color moments. In another work [44], Markov stationary features (MSFs) were proposed (this method is an extension of the color auto-correlogram) and compared with the auto-correlogram and other MSF-based CBIR features, such as color histograms, CCVs, texture, and edge. Among all color descriptors, the auto-correlogram extended by MSF performs the best, although only slightly better than the auto-correlogram. Another extension is the wavelet correlograms [45]; it performs slightly better than the correlogram and surpasses other color descriptors, such as color histograms and scalable color descriptor. Reference [46] proposed another approach, called Gabor wavelet correlogram, for image indexing and retrieval and further improved the retrieval performance.

This chapter shall make comparative evaluations of the proposed technique against the color correlogram whenever applicable, since it suffers from a serious computational complexity and a massive memory requirement problems. Nowadays, digital image technology offers several Megapixel (Mpel) image resolutions. For a conservative assumption, consider a small size database with only 1000 images, each image with only 1 Mpel resolution. Without any loss of generality, it is here assumed that $W = H = 1000$. In such image dimensions, a reasonable setting for $d$ would be $100 < d < 500$, corresponding to $10\% - 50\%$ image dimension range. Any $d$ setting less than 100 pixels would be too small for characterizing the true SCD of the image, probably describing only a thin layer of adjacent colors (that is, colors that can be found within a small range). Assume the lowest range setting $d = 100$ (yet a correlogram working over only a 10% range of the image dimension is hardly a spatial color descriptor). Even with such minimal settings, the naive algorithm will require $O(10^{10})$ computations, including divisions, multiplications, and additions. Even with fast computers, this will require several hours of computations per image and infeasible time is required to index even the smallest databases. In order to achieve a feasible computational complexity for the naive algorithm, the range has to be reduced drastically (that is, $d \sim 10$), and the images should be decimated by three to five times in each dimension. Such a solution unfortunately changes (decimates) the color composition of the scheme and with such limited range, the true SCD cannot anymore be characterized.

The other alternative is to use the fast algorithm. A typical quantization for RGB color histogram can be eight partitions in each color dimension, resulting in $8 \times 8 \times 8 = 512$ bins. The fast algorithm will speed up the process around 25 times while requiring a massive memory space (more than 400 GB per image); and this time neither decimation nor drastic reduction on the range will make it feasible. Practically speaking, one can hardly make it work only for thumbnail size images and only when $d < 10$ and much coarser quantization (for example, using a $4 \times 4 \times 4$ RGB histogram) is used. Furthermore, a massive storage requirement is another serious bottleneck of the correlogram. Note that for the minimal range ($d = 100$) and typical quantization settings (that is, $8 \times 8 \times 8$ RGB partitions), the amount of space required for the feature vector storage of a single image is above 400 MB. This allows the correlogram barely applicable only for small size databases, that is, for 1000 image database the storage space required is above 400 GB. To make it work, the range value has to be reduced drastically along with using a much coarser quantization ($4 \times 4 \times 4$ bins or less), resulting in similar problems as discussed for color histograms. The only alternative is to use the auto-correlogram instead of the correlogram, as proposed in Reference [5]. However, without characterizing spatial distribution of distinct colors with respect to each other, the performance of the color descriptor may be degraded.

Apart from all such feasibility problems, the correlogram may exhibit several limitations and drawbacks. The first and the foremost is its pixel-based structure, which characterizes the color proximities at a pixel level. Such a high-resolution description not only makes it too complicated and infeasible to perform, it also becomes meaningless with respect to HVS color perception rules, simply because individual pixels do not mean much for the human eye. As an example, consider a correlogram description such as "the probability of finding a red pixel within a 43 pixel proximity of a blue pixel is 0.25;" so what difference does it make to have this probability in 44 or 42 pixels proximity for the human perception? Another similar image might have the same probability, but in 42 pixels proximity, which makes it indifferent or even identical for the human eye; however, a significant dissimilarity will occur via correlogram's naive dissimilarity computations. Furthermore, since the correlogram is a pixel-level descriptor working over RGB color histogram, the outliers, both in color and spatial domains, have an imminent effect over the computational complexity and the retrieval performance of the descriptor. Hundreds of color outliers hosted in the histogram, even though not visible to the human eye, will cause computational problems in terms of memory, storage, and speed; thus making the correlogram inapplicable in many cases. Yet the real problem lies in the degradation of the description accuracy caused by the *outliers*; this is the case of bias or shift that affects the true/perceivable probabilities or inter-color proximities. Finally, using the probability alone makes the descriptor insensitive to the dominance of a color or its area (weight) in the image. This is basically due to the normalization by $h(c_i, I)$ denoting the amount (weight or area) of color; and such an important perceptual cue is omitted in the correlogram's description. An example of such a descriptor deficiency can be seen in a query of the sample image shown in Figure 11.2. To this end, these properties make the correlogram more of a colored *texture* descriptor rather than a

(a)          (b)

**FIGURE 11.2 (See color insert.)**

Top six ranks of correlogram retrieval in two image databases with (a) 10000 images and (b) 20000 images. Top-left is the query image.

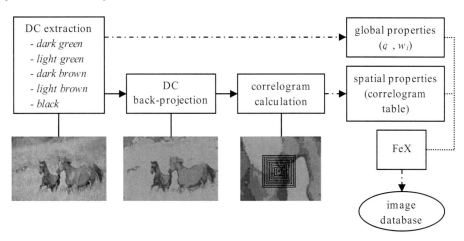

**FIGURE 11.3**

Overview of the perceptual correlogram descriptor.

*color* descriptor, since its pixel-level, area-insensitive, co-occurrence description is quite similar to texture descriptors based on co-occurrence statistics (for example, gray-level co-occurrence matrix) only with a major difference of describing color co-occurrences instead of gray-level (intensity) values.

## 11.3 Perceptual Correlogram

In order to obtain the global color properties, DCs are extracted in a way human the visual system perceives them and then back-projected onto the image to extract their spatial relations using the DC-based correlogram. The overview of the feature extraction (indexing) process is illustrated in Figure 11.3. In the retrieval phase, a penalty-trio model is utilized that penalizes both global and spatial dissimilarities in a joint scheme.

### 11.3.1 Formation of the Color Descriptor

The adopted DC extraction algorithm is similar to one in Reference [33], where the method is entirely designed with respect to human color perception rules and configurable with few thresholds: $T_S$ (color similarity), $T_A$ (minimum area), $\varepsilon_D$ (minimum distortion), and $N_{DC}^{max}$ (maximum number of DCs).

As the first step, the true number of DCs present in the image (that is, $1 \leq N_{DC} \leq N_{DC}^{max}$) are extracted in the CIE-Luv color domain. Let $C_i$ represent the $i$th DC class (cluster) with the following members: the color value (centroid) $c_i$ and the weight (unit normalized area) $w_i$. Due to the DC thresholds set beforehand, $w_i > T_A$ and $|c_i - c_j| > T_S$ for $1 \leq i \leq N_{DC}$ and $1 \leq j \leq N_{DC}$. The extracted DCs are then back-projected to the image for further analysis, namely, extraction of the SCD. Spatial distribution of the extracted DCs is represented as a correlogram table as described in Reference [5], except that the colors utilized are the ones coming from the DC extraction; hence, $m = N_{DC}$. Thus, for every DC pair on the back-projected image, the probability of finding a pixel with DC $c_i$ at distance of $k$ pixels from a pixel with DC $c_j$ is calculated. The extracted correlogram table and the global features $(c_i, w_i)$ are stored in the database's feature vector.

### 11.3.2 Penalty-Trio Model for Perceptual Correlogram

In a retrieval operation in an image database, a particular feature of the query image $Q$ is used for similarity measurement with the same feature of a database image $I$. Repeating this process for all images in the database $D$ and ranking the images according to their similarity distances yield the retrieval result. In the retrieval phase, while judging the similarity of two image's global and spatial color properties, two DC lists together with their correlogram tables are to be compared. The more dissimilar colors are present, the larger their similarity distance should be. In order to do so, DCs in both images are subjected to a one-to-one matching. This model penalizes for the mismatching colors together with the global and spatial dissimilarities between the matching through $P_\phi$, which denotes the amount of different (mismatching) DCs, and $P_G$ and $P_{Corr}$, which denote, respectively, the differences of the matching DCs in terms of the global differences and correlogram differences.

The overall penalty (color dissimilarity) between query image $Q$ and a database image $I$ over all color properties can be expressed as

$$P_\Sigma(Q,I) = P_\phi(Q,I) + (\alpha P_G(Q,I) + (1-\alpha)P_{Corr}(Q,I)), \qquad (11.5)$$

where $P_\Sigma \leq 1$ is the (unit) normalized total penalty, which corresponds to (total) color similarity distance, and $0 < \alpha < 1$ is the weighting factor between global and spatial color properties of the matching DCs. Thus, the model computes a complete distance measure from *all* color properties. Note that all global color descriptors mentioned in Section 11.2.1 use only the first two (penalty) terms while discarding $P_{SCD}$ entirely. The correlogram, on the other hand, works only over $P_{SCD}$ without considering any global properties. Therefore, the SCD penalty-trio model fuses both approaches to compute a complete distance measure from all color properties. Since color (DC) matching is a key factor in the underlying application, a two-level color partitioning is proposed here. The first level partitions the group of color elements, which are too close for the human eye to distinguish, using a minimum (color) threshold $T_C^{min}$. Recall from the earlier discussion that such close color elements are clustered into DC classes, that is, $|c_i - c_j| \leq T_S$ for $\forall c_j \in C_i$, and using the same analogy $T_C^{min}$ can conveniently be set as $T_S$. Another threshold $T_C^{max}$ is empirically set for second-level partitioning, above which no color similarity can be perceived. Finally, for a given two DCs with the inter-color distance falling between the two levels, that is, $T_C^{min} < |c_i - c_j| < T_C^{max}$, there exists a certain level of (color) similarity, but not too close to be perceived as identical.

Define such colors, which show some similarity, as matching, and let $T_C^{max}$ be used to partition the mismatching colors from the matching ones. In order to get a better understanding of the penalty terms, consider a set of matching colors ($S^M$) and a set of mismatching colors ($S^\phi$) selected among the colors of a database $I(C^I)$ and query $Q(C^Q)$ images, thus $S^M + S^Q = C^I + C^Q$. Here $S^\phi$ consists of the colors that holds $|c_i - c_j| > T_S$ for $\forall c_i, c_j \in S^\phi$, and the subset $S^M$ holds the rest of the (matching) colors. Accordingly, $P_\phi$ can be defined directly as

$$P_\phi(Q,I) = \frac{\sum (w_i | c_i \in S^\phi)}{2} \leq 1. \tag{11.6}$$

Consider the case where $S^M = \{\phi\}$, which means there are no matching colors between images $Q$ and $I$. Thus $P_\Sigma = P_\phi = 1$, which makes sense since *colorwise* there is nothing similar between two images. Conversely, if $S^\phi = \{\phi\} \Rightarrow P_\phi = 0$, the dissimilarity will *only* emerge from the global and spatial distribution of colors. These distributions are denoted here as $P_G$ and $P_{Corr}$, respectively, and can be expressed as follows:

$$P_G(Q,I) = \beta \sum_{i=1}^{N_M} \left| w_i^Q - w_i^I \right| + (1-\beta) \frac{1}{T_S N_M} \sqrt{\sum_{i=1}^{N_M} (c_i^Q - c_i^I)^2} \leq 1, \tag{11.7}$$

$$P_{Corr}(Q,I) = \sum_{i,j \in S^M} \sum_{k=1}^{d} \left\{ \begin{array}{ll} 0 & \text{if } \gamma_{c_i,c_j}^{(k)}(Q) = \gamma_{c_i,c_j}^{(k)}(I) = 0 \\ \frac{\left| \gamma_{c_i,c_j}^{(k)}(Q) - \gamma_{c_i,c_j}^{(k)}(I) \right|}{\gamma_{c_i,c_j}^{(k)}(Q) + \gamma_{c_i,c_j}^{(k)}(I)} & \text{otherwise} \end{array} \right\}, \tag{11.8}$$

where $\gamma_{c_i,c_j}^{(k)}$ is the probability of finding DC $c_i$ at a distance $k$ from DC $c_j$, and $0 < \beta < 1$ in $P_G$ provides the adjustment between color area (weight) difference indicated by the first term and DC (centroid) difference of the matching colors indicated by the second term. Note that the distance function used for comparing two correlogram tables ($P_{Corr}$) differs from the formulation proposed in Reference [5], which avoids divisions by zero by adding the value of one to the denominator. However, correlogram values are the probabilities that hold $0 < \gamma_{c_i,c_j}^{(k)} < 1$, thus such an addition becomes relatively significant and introduces a bias to distances; hence, it is strictly avoided in proposed distance calculations. As a result, the weighted combination of $P_G$ and $P_{Corr}$ represents the amount of dissimilarity that occurs in the color properties; through the weight $\alpha$ one color property can be favored over another. With the combination of $P_\phi$, the penalty trio models a complete similarity distance metric between two color compositions.

### 11.3.3 Retrieval Performance on Synthetic Images

The perceptual correlogram and the two traditional descriptors, that is, correlogram and MPEG-7 dominant color descriptor (DCD), are implemented as a feature extraction (FeX) module within MUVIS framework [11] in order to test and perform performance comparisons in the context of CBIR. All descriptors are tested using two separate databases on a personal computer with 2 GB of RAM and Pentium 4 central processing unit (CPU) operating at 3.2 GHz.

First database has 1088 synthetic images that are formed of arbitrary shaped and colored regions. In this way, the color matching accuracy can visually be tested and the penalty terms $P_\phi$, $P_G$, and $P_{Corr}$ can individually be analyzed via changing the shape and/or size of any color regions. Furthermore, the robustness against resolution, translation, and rotation changes can be compared against the traditional correlogram. First, twelve retrievals of two queries from the synthetic database are shown in Figure 11.4. In Q1, correlogram's area insensitivity, due to its probabilistic nature, can clearly be observed. It can retrieve the first relevant image only at the eight rank, whereas the perceptual correlogram retrieves

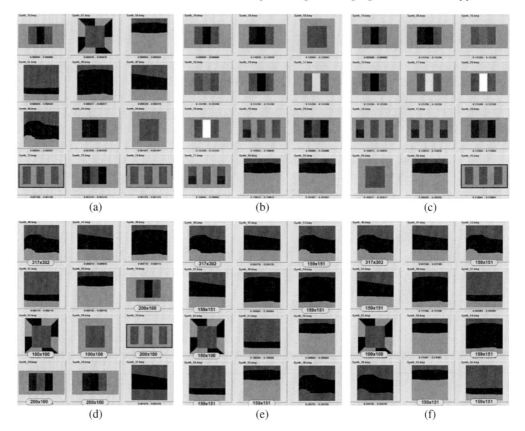

(a)                  (b)                  (c)

(d)                  (e)                  (f)

**FIGURE 11.4**

Two queries Q1 (a–c) and Q2 (d–f) in the synthetic database with typical and perceptual correlograms with different parameters: (a,d) typical correlogram, (b,e) perceptual correlogram for $d = 10$, and (c,f) perceptual correlogram for $d = 40$. Top-left is the query image.

seven similar color compositions on the first page using $d = 10$ and still fails to bring them to higher ranks due to the limited range value ($d = 10$). For $d = 40$, all relevant images sharing with similar color compositions can be retrieved in the highest ranks. In the same figure, Q2 demonstrates the resolution dependency of the correlogram, which fails to retrieve the image with identical color structure but different resolution in the first page. Even though the penalty-trio model aids to overcome the resolution dependence problem, note that this descriptor still utilizes the correlogram as the spatial descriptor and hence the perceptual correlogram manages to retrieve it only at the third rank and despite severe resolution differences, it can further retrieve most of the relevant images on the first page.

### 11.3.4   Retrieval Performance on Natural Image Databases

The second database has 10000 medium resolution ($384 \times 256$ pixels) images from diverse contents selected from the Corel image collection. The recommended distance metrics are implemented for each FeX module, that is the quadratic distance for MPEG-7 DCD and $L_1$ norm for the correlogram. Due to its memory requirements, the correlogram could

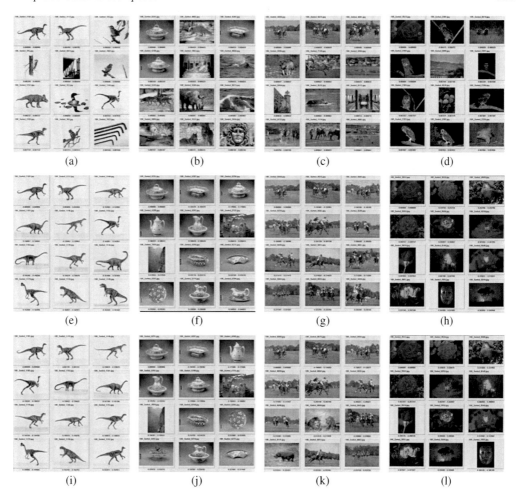

**FIGURE 11.5 (See color insert.)**

Four queries in *Corel 10K* database: (a–d) correlogram for $d = 10$, (e–h) perceptual correlogram for $d = 10$, and (i–l) perceptual correlogram for $d = 40$. Top-left is the query image.

be implemented only with $m = 27$ colors ($3 \times 3 \times 3$ bins RGB histogram) and $d = 10$, since further increase in either setting is infeasible in terms of computational complexity. In order to measure the retrieval performance, an unbiased and limited formulation of normalized modified retrieval rank $NMRR(q)$ is used here, which is defined in MPEG-7 as the retrieval performance criteria per query $q$. It combines both of the traditional hit-miss counters; *Precision – Recall*, and further takes the ranking information into account as follows:

$$AVR(q) = \frac{1}{N(q)} \sum_{k=1}^{N(q)} R(k) \quad \text{with} \quad W = 2N(q), \tag{11.9}$$

$$NMRR(q) = \frac{2AVR(q) - N(q) - 1}{2W - N(q) + 1} \leq 1, \tag{11.10}$$

$$ANMRR = \frac{1}{Q} \sum_{q=1}^{Q} NMRR(q) \leq 1, \tag{11.11}$$

**TABLE 11.1**

ANMRR results for *Corel 10K* database.

| Method | Score |
|---|---|
| MPEG7 DCD | 0.495 |
| correlogram for $d = 10$ | 0.422 |
| perceptual correlogram for $d = 10$ | 0.391 |
| perceptual correlogram for $d = 40$ | 0.401 |

where $N(q)$ is the minimum number of relevant images (based on ground-truth) in a set of $Q$ retrieval experiments, $R(k)$ is the rank of the $k$th relevant retrieval within a window of $W$ retrievals, which are taken into consideration during per query $q$. If there are less than $N(q)$ relevant retrievals among $W$, then a rank of $W + 1$ is assigned for the remaining ones. The term $AVR(q)$ is the average rank obtained from the query $q$. Since each query item is selected within the database, the first retrieval will always be the item queried and this obviously yields a biased $NMRR(q)$ calculation; therefore, it is excluded from ranking. Hence, the first relevant retrieval $R(1)$ is ranked by counting the number of irrelevant images *a priori* and if all $N(q)$ retrievals are relevant, then $NMRR(q) = 0$; the best retrieval performance is thus achieved. On the other hand, if none of relevant items can be retrieved among $W$, then $NMRR(q) = 1$ as the worst case. Therefore, the lower $NMRR(q)$, the better (more relevant) the retrieval is for the query $q$. Keeping the number of query-by-example experiments sufficiently high, the average $NMRR$ and $ANMRR$ can thus be used as the retrieval performance criteria.

In total, 104 query-by-example experiments were performed for each descriptor, with some retrieval results shown in Figure 11.5. The average scores are listed in Table 11.1.

The perceptual correlogram extracts both global and spatial color properties throughout the whole image, thus erroneous queries such as in Figure 11.2 can be avoided where global color properties vary significantly (for example, see Figure 11.6).

**FIGURE 11.6 (See color insert.)**

Retrieval result of the perceptual correlogram for the query in Figure 11.2.

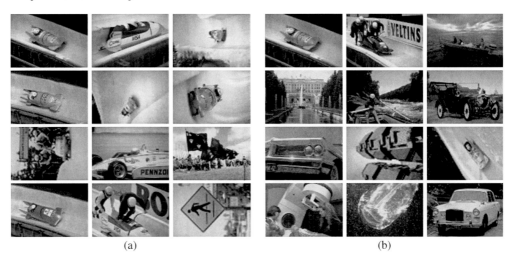

(a)                                              (b)

**FIGURE 11.7 (See color insert.)**

A special case where the correlogram works better than the perceptual correlogram descriptor: (a) correlogram and (b) perceptual correlogram.

As discussed earlier, in some particular cases, the traditional correlogram may yield a better retrieval performance than the perceptual correlogram. Figure 11.7 shows a typical example, where the images retrieved using the correlogram are the zoomed versions of the query image or they are all taken from different viewpoints. Apart from such special query examples, the perceptual correlogram achieved better retrieval performance on the majority of the queries performed. Yet the most important point is the feasibility issue of both descriptors. Note that the utilization of the traditional correlogram is only possible because image resolution is fairly low ($384 \times 256$ pixels), database contains only 10000 images, and the range value $d$ is quite limited ($d = 10$). Even with such settings, the correlogram required 556 MB disk space and 850 MB memory (fast algorithm) during the feature extraction phase, whereas the perceptual correlogram takes 53 MB disk space and 250 MB memory for $N_{DC}^{max} = 8$ and $d = 10$, thus reducing the storage space by approximately 90% and the computational memory (fast algorithm) by 70%. Henceforth, only the perceptual correlogram descriptor can be used if a further increase occurs in any of the settings.

## 11.4 Spatial Color Descriptor

Under the light of the earlier discussion, an efficient spatial color descriptor (SCD) is designed to address the drawbacks and problems of the color descriptors, particularly the color correlogram. In order to achieve this goal, various perception rules are followed by extracting and describing the global and spatial color properties in a way as perceived by the HVS. Namely, *outliers* in color and spatial domains are suppressed or eliminated by adopting a top-down approach during feature extraction. The SCD is formed by a proper combination of global and spatial color features. During the retrieval phase, the similarity between two images is computed using a penalty-trio model, which penalizes the individual differences in global and spatial color properties. In the following subsections, both indexing (feature extraction) and retrieval schemes are discussed in detail.

### 11.4.1 Formation of the Spatial Color Descriptor

As previously discussed, the DCs represent the prominent colors in an image while the imperceivable color components (outliers) are discarded. As a result, they have a global representation, which is compact and accurate, and they represent few dominant colors that are present and perceivable in an image. For a color cluster $C_i$, its centroid $c_i$ is given by

$$c_i = \frac{\sum w(p)x(p)}{\sum w(p)}, \quad \text{for } x(p) \in C_i, \tag{11.12}$$

and the initial clusters are determined by using a weighted distortion measure, defined as

$$D_i = \sum w(p) \|x(p) - c_i\|^2, \quad \text{for } x(p) \in C_i. \tag{11.13}$$

These definitions are used to determine which clusters to split until either $N_{DC}^{\max}$ denoting a maximum number of clusters (DCs) is achieved or $\varepsilon_D$ denoting a maximum allowed distortion criteria is met. Hence, pixels with smaller weights (detailed sections) are assigned to fewer clusters so that the number of color clusters in the detailed regions, where the likelihood of outliers' presence is high, is suppressed. As the final step, an agglomerative clustering (AC) is performed on the cluster centroids to further merge similar color clusters so that there is only one cluster (DC) hosting all similar color components in the image. A similarity threshold $T_S$ is assigned to the maximum color distance possible between two similar colors in a certain color domain (for instance, CIE-Luv, CIE-Lab). Another merging criterion is the color area, that is, any cluster should have a minimum amount of coverage area $T_A$ so as to be assigned as a DC; otherwise, it will be merged with the closest color cluster since it is just an outlier. Another important issue is the choice of the color space since a proper color clustering scheme for DC extraction tightly relies on the metric. Therefore, a perceptually uniform color space should be used and the most common ones are CIE-Luv and CIE-Lab, which are designed such that perceived color distances are qual for the Euclidean ($L_2$) distance in these spaces. The HSV space, although it represents an intuitive color domain, suffers from discontinuities and the RGB color space is not perceptually uniform. To this end, among CIE-Luv and CIE-Lab, the former one is selected here since it yields a lower transformation cost from native RGB space. For CIE-Luv, a typical value for $T_S$ is between 10 and 20, whereas $T_A$ is between 2 to 5% [28], and $\varepsilon_D < 0.05$. Based on the earlier remarks, $N_{DC}^{\max}$ can be conveniently set to 8. As shown in Figure 11.8, the employed DC extraction method is similar to the one in Reference [33], which is entirely designed with respect to HVS color perception rules and configurable with few thresholds, $T_S$ (color similarity), $T_A$ (minimum area), $\varepsilon_D$ (minimum distortion), and $N_{DC}^{\max}$ (maximum number of DCs). As the first step, the true number of DCs present in the image (that is, $1 \leq N_{DC} \leq N_{DC}^{\max}$) is extracted in CIE-Luv color domain and back-projected to the image for further analysis involving extraction of the spatial properties (SCD) of DCs. Let $C_i$ represents the $i$th DC class (cluster) with the following members: $c_i$ is the color value (centroid), $w_i$ is the weight (unit normalized area), and $\sigma_i$ is the standard deviation obtained from the distribution of (real) colors clustered by $C_i$. Due to the DC thresholds set beforehand, $w_i > T_A$ and $|c_i - c_j| > T_S$ for $1 \leq i \leq N_{DC}$ and $1 \leq j \leq N_{DC}$.

During the back-projection phase, the DC with the closest centroid value to a particular pixel color will be assigned to that pixel. As a natural consequence of this process, spatial *outliers*, that is, isolated pixel(s), which are not populated enough to be perceivable, can emerge (see Figure 11.8) and should thus be eliminated. Due to the perceptual approach based on the Gestalt rule that states "humans see the whole before its parts," a top-down approach, such as quad-tree decomposition, can process the whole first, meaning the largest blocks possible which can be described (and perceived) by a single DC, before going into its parts. Due to its top-down structure, the SCD scheme does not suffer from the aforementioned problems of some pixel-based approaches.

Two parameters are used to configure the quad-tree (QT): $T_W$, which is the minimum weight (dominance) within the current block required from a DC not to go down for further partition and $D_{QT}^{max}$, which is the depth limit indicating the maximum amount of partition (decomposition) allowed. Note that with the proper setting of $T_W$ and $D_{QT}^{max}$, QT decomposition can be carried out to reach the pixel level; however, such an extreme partitioning should not be permitted to avoid the aforementioned problems of pixel-level analysis. Using a similar analogy, $T_W$ can be set in accordance with $T_A$, that is, $T_W \cong 1 - T_A$. Therefore, for the typical $T_A$ setting (between 2 and 5%), $T_W$ can be conveniently set as $T_W \geq 95\%$. Since $D_{QT}^{max}$ determines when to stop the partitioning abruptly, it should not be set too low so that it does not cause inhomogeneous (mixed) blocks. On the other hand, extensive experimental results suggest that $D_{QT}^{max} > 6$ is not required even for the most complex scenes since the results are almost identical to the one with $D_{QT}^{max} = 6$. Therefore, the typical range is $4 \leq D_{QT}^{max} \leq 6$. Let $B^p$ corresponds to the $p$th partition of the block $B$, where $p = 0$ is the entire block and $1 \leq p \leq 4$ represents the $p$th *quadrant* of the block. The four *quadrants* can be obtained simply by applying equal partitioning to the parent block or via any other partitioning scheme, which can be optimized to yield most homogenous blocks possible. For simplicity the former case is used here, and accordingly a generic QT algorithm can be expressed as depicted in Algorithm 11.1.

**ALGORITHM 11.1**  QuadTree(parent, depth).

1. Let $w_{max}$ be the weight of the DC which has the minimum coverage in *parent* block.

2. If $w_{max} > T_w$, then return.

3. Let $B_0 = $ parent.

4. For $\forall p \in [1, 2, 3, 4]$ do QuadTree($B^p$, depth).

5. Return.

The QT decomposition of a (back-projected) image $I$ can then be initiated by calling QuadTree($I, 0$) and once the process is over, each QT block carries the following data: its depth $D \leq D_{QT}^{max}$ where the partitioning is stopped, its location in the image, the major DC with the highest weight in the block (that is, $w_{max} > T_W$), and perhaps some other DCs

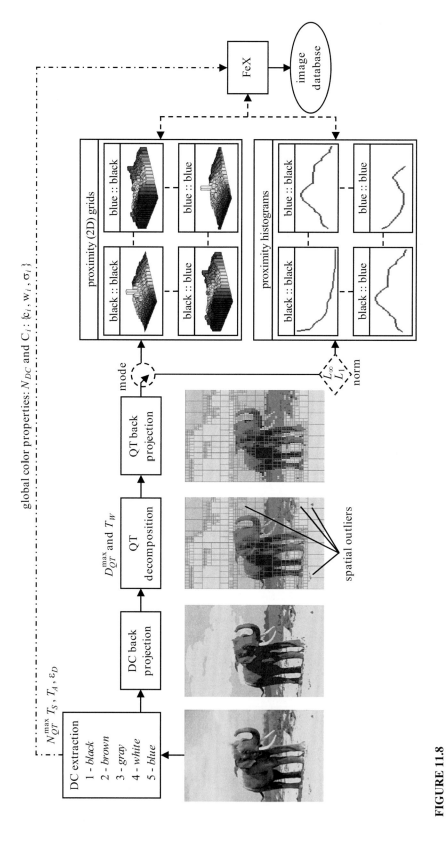

**FIGURE 11.8**
Overview of the SCD formation.

which are eventually some spatial outliers. In order to remove those spatial outliers, a QT back-projection of the major DC into its host block is sufficient. Figure 11.8 illustrates the removal of some spatial outliers via QT back-projection on a sample image. The final scheme, where outliers in both color and spatial domains are removed and the (major) DCs are assigned (back-projected) to their blocks, can be conveniently used for further (SCD) analysis to extract spatial color features. Note that QT blocks can vary in size depending on the depth, yet even the smallest (highest depth) block is large enough to be perceivable, and carry a homogenous DC. So instead of performing pixel-level analysis, such as in the correlogram, the uniform grid of blocks in the highest depth ($D = D_{QT}^{\max}$) can be used for characterizing the global SCD and extracting the spatial features in an efficient way. As shown in Figure 11.8, one of the two modes, which perform two different approaches to extract spatial color features, can be used. The first is the scalar mode, over which inter-DC proximity histograms are computed within the full image range. These histograms indicate the amount of a particular DC that can be found from a certain distance of another DC; however, this is a scalar measure which lacks the direction information. For example, such a measure can state "17% of red is eight units (blocks) away from blue," which does not provide any directional information. Therefore, the second mode is designed to represent inter-occurrence of one DC with respect to another over a two-dimensional (proximity) grid, from which both distance and direction information can be obtained. Note that inter-color distances are crucial for characterizing the SCD of an image; however, the direction information may or may not be useful depending on the content. For example, the direction information in "17% of red is eight units (blocks) right of blue" is important for describing a national flag and hence the content, but "one black and one white horse are running together on a green field" is sufficient to describe the content without any need to know the exact directional order of black, white, and green. The following subsections will first detail both modes and then evaluate their computational and retrieval performances individually.

### 11.4.2 Proximity Histograms-Based Description

Once the QT back-projection of major DCs into their host blocks are completed, all QT blocks hosting a single (major) DC with a certain depth ($D \leq D_{QT}^{\max}$) are further partitioned into the blocks in highest depth ($D = D_{QT}^{\max}$) to achieve a proximity histogram in the highest blockwise resolution. Therefore, in such a uniform block-grid, the image $I$ will have $N \times N$ blocks for $N = 2^{D_{QT}^{\max}}$, each of which hosts a single DC. Accordingly the problem of computing inter-DC proximities turns out to be block distances and hence the block indices in each dimension (that is, $\forall x, y \in [1, N]$) can directly be used for distance (proximity) calculation. Since the number of blocks does not change with respect to image dimension(s), resolution invariance is achieved (for example, the same image in different resolutions will have identical proximity histograms/grids as opposed to significantly varying correlograms due to its pixel-based computations). As shown in Figure 11.8, one can use either $L_1$ or $L_\infty$ norms for block-distance calculations. Let $b_1 = (x_1, y_1)$ and $b_2 = (x_2, y_2)$ be two blocks, their distance for the $L_1$ norm can be defined as $\|b_1 - b_2\| = |x_1 - x_2| + |y_1 - y_2|$, and for the $L_\infty$ norm as $\|b_1 - b_2\| = \max(|x_1 - x_2|, |y_1 - y_2|)$, respectively. Using the block indices in both norms, the block distances become integer numbers and note that for a full range histogram, the maximum (distance) range will be $[1, L]$, where $L$ is $N - 1$ in $L_\infty$ and $2N - 2$

$k=1$

| 3 | 5 | 5 | 5 | 5 | 5 | 5 | 3 |
|---|---|---|---|---|---|---|---|
| 5 | 8 | 8 | 8 | 8 | 8 | 8 | 5 |
| 5 | 8 | 8 | 8 | 8 | 8 | 8 | 5 |
| 5 | 8 | 8 | 8 | 8 | 8 | 8 | 5 |
| 5 | 8 | 8 | 8 | 8 | 8 | 8 | 5 |
| 5 | 8 | 8 | 8 | 8 | 8 | 8 | 5 |
| 5 | 8 | 8 | 8 | 8 | 8 | 8 | 5 |
| 3 | 5 | 5 | 5 | 5 | 5 | 5 | 3 |

$k=2$

| 5 | 6 | 9 | 9 | 9 | 9 | 6 | 5 |
|---|---|---|---|---|---|---|---|
| 6 | 7 | 11 | 11 | 11 | 11 | 7 | 6 |
| 9 | 11 | 16 | 16 | 16 | 16 | 11 | 9 |
| 9 | 11 | 16 | 16 | 16 | 16 | 11 | 9 |
| 9 | 11 | 16 | 16 | 16 | 16 | 11 | 9 |
| 9 | 11 | 16 | 16 | 16 | 16 | 11 | 9 |
| 6 | 7 | 11 | 11 | 11 | 11 | 7 | 6 |
| 5 | 6 | 9 | 9 | 9 | 9 | 6 | 5 |

$k=7$

| 15 | 8 | 8 | 8 | 8 | 8 | 8 | 15 |
|---|---|---|---|---|---|---|---|
| 8 | 0 | 0 | 0 | 0 | 0 | 0 | 8 |
| 8 | 0 | 0 | 0 | 0 | 0 | 0 | 8 |
| 8 | 0 | 0 | 0 | 0 | 0 | 0 | 8 |
| 8 | 0 | 0 | 0 | 0 | 0 | 0 | 8 |
| 8 | 0 | 0 | 0 | 0 | 0 | 0 | 8 |
| 8 | 0 | 0 | 0 | 0 | 0 | 0 | 8 |
| 15 | 8 | 8 | 8 | 8 | 8 | 8 | 15 |

$k=1$

| 2 | 3 | 3 | 3 | 3 | 3 | 3 | 2 |
|---|---|---|---|---|---|---|---|
| 3 | 4 | 4 | 4 | 4 | 4 | 4 | 3 |
| 3 | 4 | 4 | 4 | 4 | 4 | 4 | 3 |
| 3 | 4 | 4 | 4 | 4 | 4 | 4 | 3 |
| 3 | 4 | 4 | 4 | 4 | 4 | 4 | 3 |
| 3 | 4 | 4 | 4 | 4 | 4 | 4 | 3 |
| 3 | 4 | 4 | 4 | 4 | 4 | 4 | 3 |
| 2 | 3 | 3 | 3 | 3 | 3 | 3 | 2 |

$k=2$

| 3 | 4 | 5 | 5 | 5 | 5 | 4 | 3 |
|---|---|---|---|---|---|---|---|
| 4 | 6 | 7 | 7 | 7 | 7 | 6 | 4 |
| 5 | 7 | 8 | 8 | 8 | 8 | 7 | 5 |
| 5 | 7 | 8 | 8 | 8 | 8 | 7 | 5 |
| 5 | 7 | 8 | 8 | 8 | 8 | 7 | 5 |
| 5 | 7 | 8 | 8 | 8 | 8 | 7 | 5 |
| 4 | 6 | 7 | 7 | 7 | 7 | 6 | 4 |
| 3 | 4 | 5 | 5 | 5 | 5 | 4 | 3 |

$k=14$

| 1 | 0 | 0 | 0 | 0 | 0 | 0 | 1 |
|---|---|---|---|---|---|---|---|
| 0 | 0 | 0 | 0 | 0 | 0 | 0 | 0 |
| 0 | 0 | 0 | 0 | 0 | 0 | 0 | 0 |
| 0 | 0 | 0 | 0 | 0 | 0 | 0 | 0 |
| 0 | 0 | 0 | 0 | 0 | 0 | 0 | 0 |
| 0 | 0 | 0 | 0 | 0 | 0 | 0 | 0 |
| 0 | 0 | 0 | 0 | 0 | 0 | 0 | 0 |
| 1 | 0 | 0 | 0 | 0 | 0 | 0 | 1 |

**FIGURE 11.9**

$N(b_i, k)$ templates in $8 \times 8$ block grid ($D_{QT}^{max} = 3$) for three range values in $L_\infty$ (top) and $L_1$ (bottom) norms.

in $L_1$ norms, respectively. A blockwise proximity histogram for a DC pair $c_i$ and $c_j$ stores in its $k$th bin the number of blocks hosting $c_j$ (that is, $\forall b_j | I(b_j) = c_j$, equivalent to the amount of color $c_j$ in $I$) from all blocks hosting $c_i$ (that is, $\forall b_i | I(b_i) = c_i$, equivalent to amount of color $c_i$ in $I$) at a distance $k$. Such a histogram clearly indicates how close or far two DCs and their spatial distribution with respect to each other are. Yet the histogram bins should be normalized by the total number of blocks, which can be found $k$ blocks away from the source block $b_i$ hosting the DC $c_i$, because this number will significantly vary with respect to the distance ($k$), the position of source block ($b_i$), and the norm ($L_1$ or $L_\infty$) used. Therefore, the $k$th bin $\Phi_{c_i}^{c_j}(k)$ of the normalized proximity histogram between the DC pair $c_i$ and $c_j$ can be expressed as follows:

$$\Phi_{c_i}^{c_j}(k) = \sum_{b_i} \sum_{b_j} \Delta(b_i, b_j, k), \tag{11.14}$$

where

$$\Delta(b_i, b_j, k) = \begin{cases} N(b_i, k)^{-1} & \text{if } b_i \in I(c_i), b_j \in I(c_j), \|b_i - b_j\| = k, \\ 0 & \text{otherwise.} \end{cases} \tag{11.15}$$

The normalization factor $N(b_i, k)$, denoted as the total number of neighbor blocks in distance $k$, is independent from the DC distribution and hence it is only computed once and used for all images in the database. Figure 11.9 shows $N(b_i, k)$ templates computed for all blocks ($\forall b_i \in I$), both norms, and some range values. In this figure, $N$ is set to eight ($D_{QT}^{max} = 3$) for illustration purposes. Note that normalization cannot be applied for blocks with $N(b_i, k) = 0$, since $k$ is out of image boundaries and hence $\Phi_{c_i}^{c_j}(k) = 0$ for $\forall c_i$.

Once the $N(b_i, k)$ templates are formed, computing the normalized proximity histogram takes $O(N^4)$. Note that this is basically independent from the image dimensions $W$ and $H$, and it is also a full-range computation (that is, $k \in [1, L]$), which may not be necessary

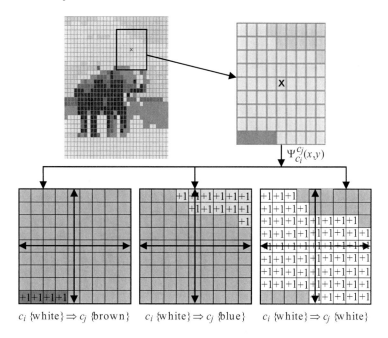

**FIGURE 11.10**

The process of proximity grid formation for the block (X) for L=4.

in general (say, half image range may be quite sufficient since above this range most of the central blocks will have either out-of-boundary case, where $\Phi_{c_i}^{c_j}(k) = 0$ for $\forall c_i$, or only few blocks are within the range, which is too low for obtaining useful statistics). A typical setting $D_{QT}^{\max} = 5$, which implies $N = 32$ and $N_{DC}^{\max} = 8$, requires $O(10^6)$ computations, which is 10000 times less compared to the correlogram with a minimal range setting (10% of image dimension range). In fact, the real speed enhancement is much more since the computations in the correlogram involve several additions, multiplications, and divisions (worst of all) for probability computations, whereas only additions are sufficient for computing $\Phi_{c_i}^{c_j}(k)$ as long as $N(b_i, k)^{-1}$ is initially computed and stored as the template. The memory requirements for the full-range computation are $O(N^2 L)$ for storing $N(b_i, k)^{-1}$ and plus $O(N_{DC}^2 L)$ for computing $\Phi_{c_i}^{c_j}(k)$, respectively. The memory space required for the typical settings given earlier will thus be 500 Kb, which is a significant reduction compared to that of the correlogram. The typical storage space required per database image is less than 17 Kb with $L_\infty$, and less than 33 Kb with $L_1$ norm, which is eventually 50 times smaller than the auto-correlogram's requirement $O(md)$ with minimal $m$ and $d$ settings.

### 11.4.3 Two-Dimensional Proximity Grids-Based Description

An alternative approach for characterizing the inter-DC distribution uses the respective proximities and also their inter-occurrences accumulated over a two-dimensional proximity grid. The process starts from the same configuration outlined earlier. Let the image $I$ have $N \times N$ blocks, each of which hosts a single DC. The two-dimensional proximity grid, $\Psi_{c_i}^{c_j}(x, y)$, is formed by cumulating the co-occurrence of blocks hosting $c_j$ (that is, $\forall b_j | I(b_j) = c_j$) in a certain vicinity of the blocks hosting $c_i$ (that is, $\forall b_i | I(b_i) = c_i$) over a

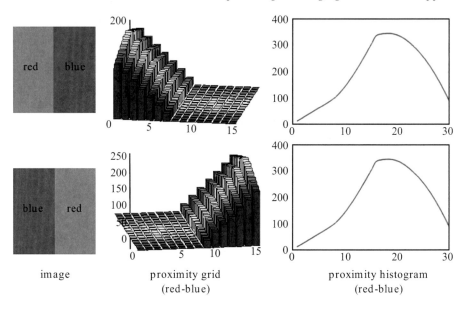

image      proximity grid
(red-blue)      proximity histogram
(red-blue)

**FIGURE 11.11**

Proximity grid compared against histogram for a sample red-blue color pair.

two-dimensional proximity grid. In other words, by fixing the block $b_i$ (hosting $c_i$) in the center bin of the grid ($x = y = 0$), the particular bin corresponding to the relative position of block $b_j$ (hosting $c_j$) is incremented by one and this process is repeated for all blocks hosting $c_j$ in a certain vicinity of $b_i$. Then, the process is repeated for the next block (hosting $c_i$) until the entire image blocks are scanned for the color pair ($c_i, c_j$). As a result, the final grid bins represent the inter-occurrences of the $c_j$ blocks with respect to the ones hosting color $c_i$ within a certain range $L$ (that is, $\forall x, y \in [-L, L]$, with $L \leq N - 1$). Although $L$ can be set as $N - 1$ for a full-range representation, this is a highly redundant setting since $L \geq N/2$ cannot be fit exactly for any block without exceeding the image (block) boundaries. Therefore, $L < N/2$ would be a reasonable choice for $L$.

The computation of $\Psi_{c_i}^{c_j}(x, y)$ can be performed in a single pass through all image blocks. Let $b_i(x_i, y_i)$ be the next block hosting the DC $c_i$. Fixing $b_i$ in the center (that is, $\Psi_{c_i}^{c_j}(0,0)$), all image blocks within the range $L$ from $b_i$ (that is, $\forall b_j = (x_i + x, y_i + y) \in I \mid \forall x, y \in [-L, L]$) are scanned and the corresponding (proximity) grid bin $\Psi_{c_i}^{c_j}(x, y)$ for a color $c_j$ in a block $b_j = (x_i + x, y_i + y) \in I$ is incremented by one. This process is illustrated on a sample image shown in Figure 11.10. During the raster-scan of uniform blocks, the block with *white* DC updates only three proximity grids (*white* to *white*, *brown*, and *blue*), since those DCs can only be found within the range of $\pm L$. For illustration purposes, $D_{QT}^{\max} = 5$ which implies $N = 32$, and $L = 4$. As a result, such a proximity grid characterizes both inter-DC proximities and the relative spatial position (inter-DC direction) between two DCs. This is straightforward to see in the sample images in Figure 11.11, where the proximity grid distinguishes the relative direction of a DC pair (red-blue), while the proximity histogram cannot do it due to its scalar metric. Note that $\Psi_{c_i}^{c_j}(0,0) = 0$ for $i \neq j$, and $\Psi_{c_i}^{c_j}(0,0)$ indicates the total number of blocks hosting $c_i$. Since this is not a SCD property, rather a local DC property showing a noisy approximation of $w_i$ (weight of $c_i$), it can be conveniently excluded from the feature vector and the remaining $(2L+1)^2 - 1$ grid bins are (unit) normalized by the total number of blocks $N^2$ to form the final descriptor $\overline{\Psi}_{c_i}^{C_j}(x,y)$, where $\overline{\Psi}_{c_i}^{C_j}(x,y) \leq 1$ for $\forall x,y \in [-L,L]$.

The proximity grid computation takes $O(N^2 L^2)$. Similar to the proximity histogram, this is also independent from original image dimensions $W$ and $H$, and for a full-range process $(L = N/2)$, the same number of computations equal to $O(N^4)$ is obtained. However, instead of regular addition operations required for the proximity histogram, or multiplications and divisions for the correlogram, the proximity grid computation requires only incrementing operations. Thus, for a typical grid dimension range, for example, $N/8 < L < N/4$, computing the proximity grid takes the shortest time. The memory space requirement is in $O(N_{DC}^2 L^2)$ and for a full-range process $(L = N/2)$ with the typical settings $D_{QT}^{max} = 5$ implying $N = 32$, and $N_{DC}^{max} = 8$, the memory required per database image will be 256 Kb. This is still smaller than $O(md)$ by the auto-correlogram even with the minimal $m$ and $d$ settings, and it is equivalent to half of the memory required for the proximity histogram. Since $\Psi_{c_i}^{C_j}(x,y) = \Psi_{c_j}^{c_i}(-x,-y)$, which indicates the symmetry with respect to origin, the storage (disk) space requirement is even less, namely, $O(N_{DC}^2 L^2)$. However, it requires eight times more space than the proximity histogram. This is the cost of computing full-range proximity grid, and therefore, it is recommended to employ the typical grid dimension range (for example, $N/8 < L < N/4$) to reduce this cost to an acceptable level.

### 11.4.4 Penalty-Trio Model for Spatial Color Descriptors

As previously discussed, the penalty-trio model used therein provides a balanced similarity metric that encapsulates three crucial global and spatial color properties. Therefore, a similar penalty-trio model is adopted for computing the similarity distance between two SCDs. As shown in Figure 11.8, the SCDs of $Q$ and $I$ contain both global and spatial color properties. Let $C_i^Q$ and $C_j^I$ represent the $i$th and $j$th $(i \leq N_{DC}^Q, j \leq N_{DC}^I)$ DC classes, where $N_{DC}^Q$ and $N_{DC}^I$ are the number of DCs in $Q$ and $I$, respectively. Along with these global properties, the SCD descriptors of $Q$ and $I$ contain either proximity histogram $\Phi_{c_i}^{c_j}(k)$ or grid $\Psi_{c_i}^{c_j}(x,y)$ depending on the SCD mode. Henceforth, for the similarity distance computation over the SCD, both global and spatial color properties are used within a similar penalty-trio model used earlier, which basically penalizes the mismatches between $Q$ and $I$. The three mismatches considered here are $P_\phi$ denoting the amount of different DCs, and the differences of the matching DCs in global color properties expressed via $P_G$ and SCD properties expressed via $P_{SCD}$. Thus, the penalty-trio over all color properties can be expressed as

$$P_\Sigma(Q,I) = P_\phi(Q,I) + (\alpha P_G(Q,I) + (1-\alpha)P_{SCD}(Q,I)), \tag{11.16}$$

where $P_\Sigma \leq 1$ is the normalized total penalty, which corresponds to the total color similarity distance, and $0 < \alpha < 1$ is the weighting factor between global and spatial color properties.

As mentioned earlier, one can form two sets, that is, matching $(S^M)$ and mismatching $(S^\phi)$ DC classes from $C^Q$ and $C^I$. This is accomplished by assigning each DC $c_i \in C_i$ in one set, which cannot match any DC $c_j \in C_j$ in the other (implying $|c_i - c_j| > T_C^{max}$ for all $i$ and $j$) into $S^\Phi$ and the rest (with at least one match) into $S^M$. Note that $S^M + S^\Phi = C^Q + C^I$, and using the DCs in $S^\Phi$, the term $P_\Phi$ can directly be expressed as given in Equation 11.6.

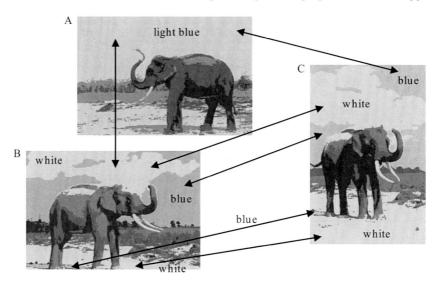

**FIGURE 11.12**

One-to-one matching of DC pairs among three images.

Recall that the dissimilarity (penalty $P_\phi$) increases proportionally with the total amount (weight) of mismatching DCs. In one extreme case, where there are no colors matching, $S^M = \{\phi\}$ implying $P_\Sigma = P_\phi = 1$ means that the two images have no similar (matching) colors. In another extreme case, where all DCs are matching, so $S^\phi = \{\phi\}$ implying $P_\phi = 0$, color similarity will only emerge from global $P_G$ and spatial $P_S CD$ color properties of the (matching) DCs. Typically, $P_\phi$ contributes a certain color distance as a natural consequence of mismatching colors between $Q$ and $I$, yet the rest of the distance will result from the cumulated difference of color matching. This is, however, not straightforward to compute since one DC in $Q$ can match one or more DCs in $I$ (or vice versa). One solution is to apply color quadratic distance [12] to fuse DC distances into the total similarity metric. However, besides its serious drawbacks mentioned earlier, this formulation can be applied only to distance calculation from *global* DC properties and hence cannot address the problem of fusing SCD distances (from proximity grid or histogram of each individual DC pair). Another alternative is to enforce a one-to-one DC matching, that is, one DC alone in $Q$ can match a single DC in $I$ by choosing the best match and discarding the other matches. This, as well, induces serious errors due to the fact that DC extraction is a dynamic clustering algorithm in color domain and due to variations in color composition of the scenery or its predetermined parameters (thresholds), it can result in over-clustering or under-clustering. Therefore, similar color compositions can be clustered into different number of DCs and enforcing one-to-one matching may miss part of matching DCs from both global and spatial similarity computation.

A typical example of such a consequence can be seen in Figure 11.12 which shows three images with highly similar content, that is, "an elephant under cloudy sky." In two images (B and C), the cloud and sky are distinguished during DC extraction with separate blue and white DCs; however, in image A, only one DC (light blue) is extracted with the same parameters. Consequently, there is no (one-to-one) matching problem between B and C and such a matching will naturally reflect similar global and spatial DC properties. However, between A and B or C, if the single DC (light blue) is matched only with one DC (white or blue), this will obviously yield an erroneous result on both global and spatial similarity computations since neither DC (white or blue) properties such as weight, distribution, and proximities to other DCs are similar to the one in A (light-blue).

As a result, before computing $P_G$ and $P_{SCD}$, the DC sets in $Q$ (or $I$), which are in a close vicinity of a single DC in $I$ (or $Q$) should be first fused into a single DC. For instance, in Figure 11.12, the DC *light-blue* in image A is close to both *white* and *blue* in image B (and C); therefore, both colors in B should be fused into a new DC (perhaps a similar light-blue color) and then $P_G$ and $P_{SCD}$ can be computed accurately between A and B. In order to accomplish this, $T_C^{min} = T_S$ is used for matching the close DCs and a twofold matching process is performed via function TargetMatch, which first verifies and then fuses some DCs in the target set $T$, if required by any DC in the source set $S$. Let $S_Q^M \subset S^M$ and $S_I^M \subset S^M$ be the sets of matching DCs for $Q$ and $I$, respectively. Since any DC in any set can request fusing two or more DCs in the other set, the function is called twice, that is, first TargetMatch($S_Q^M, S_I^M$) and then TargetMatch($S_I^M, S_Q^M$). Accordingly, TargetMatch can be expressed as depicted in Algorithm 11.2.

**ALGORITHM 11.2** TargetMatch($S,T$).

1. For $\forall c_i \in S$, let $L_i^M$ be the matching DC list for $c_i$.
2. For $\forall c_j \in T$, if $|c_i - c_j| \leq T_S$, then $c_j \to L_i^M$.
3. If $|L_i^M| \geq 2$ then:
   - Let $L_i^N = T - L_i^M$ be the non-matching list.
   - $C_x = FuseDCs(L_i^M, L_i^N)$
   - Update $T = L_i^N + C_x$.
4. Return.

**ALGORITHM 11.3** FuseDCs($L_i^M, L_i^N$).

1. Create $C_x : \{c_x, w_w, \sigma_x\}$ by fusing $\forall c_j \in L_i^M$.
   - $w_x = \sum\limits_{C_j \in L_i^M} w_j$
   - $c_x = \dfrac{\sum\limits_{C_j \in L_i^M} w_j c_j}{\sum\limits_{C_j \in L_i^M} w_j}$ and $\sigma_x = \dfrac{\sum\limits_{C_j \in L_i^M} w_j \sigma_j}{\sum\limits_{C_j \in L_i^M} w_j}$

2. Form both $X_{c_x}^{c_x}$ and $(X_{c_x}^{c_j} - X_{c_j}^{c_x})|\forall c_j \in L_i^N$.
   - $X_{c_x}^{c_x} = \sum\limits_{C_j \in L_i^M} \sum\limits_{C_k \in L_i^M} X_{c_j}^{c_k}$
   - $X_{c_j}^{c_x} = \sum\limits_{C_k \in L_i^M} X_{c_j}^{c_k} | \forall c_j \in L_i^N$
   - Compute $X_{c_j}^{c_x}$ from $X_{c_j}^{c_x}, \forall c_j \in L_i^N$.
3. Return $C_x$.

The function FuseDCs depicted in Algorithm 11.3 fuses all DCs in the list $L_i^M$, reforms the SCD descriptors of all (updated) DC pairs $\Phi_{c_i}^{c_j}(k)$ or $\Psi_{c_i}^{c_j}(k)$, and finally returns a new (fused) DC $C_X$. Then, the target set $T$ is updated accordingly. Let $X_{c_i}^{c_j}$ be the SCD operator (that is, $\Phi_{c_i}^{c_j}$ or $\Psi_{c_i}^{c_j}$ depending on the SCD mode as shown in Figure 11.8) and $X_{c_i}^{c_1} + X_{c_i}^{c_2}$ be defined as

$$X_{c_i}^{c_1} + X_{c_i}^{c_2} = \begin{cases} \Phi_{c_i}^{c_1}(k) + \Phi_{c_i}^{c_2}(k) & \text{for } \forall k \in [1, L] \\ \Psi_{c_i}^{c_1}(x,y) + \Psi_{c_i}^{c_2}(x,y) & \text{for } \forall x,y \in [-L, L] \end{cases} \quad (11.17)$$

Furthermore, let $\oplus$ be the fusing operator over DC classes. It is simple to show that $X_{c_i}^{c_1 \oplus c_2} = X_{c_i}^{c_1} + X_{c_i}^{c_2}$ for $c_{1,2} \neq c_i$. Once the DCs in $L_i^M$ are fused, they are removed along with their SCD descriptors while keeping the DCs (and their internal SCD descriptors) in $L_i^N$ intact. The new (fused) DC $C_X$ (along with its SCD descriptors) is inserted into the target set $T$. Recall from the earlier remarks on SCD descriptor properties that $\Phi_{c_i}^{c_j}(k) = \Phi_{c_j}^{c_i}(k)$ and $\Psi_{c_i}^{c_j}(x,y) = \Psi_{c_j}^{c_i}(-x,-y)$; therefore, once $X_{c_j}^{c_x}$ for $\forall c_j \in L_i^N$ are formed, it is straightforward to compute $X_{c_x}^{c_j}$ for $\forall c_j \in L_i^N$. After the consecutive calls of TargetMatch function, all DC sets in each set, which are close (matching) to a particular DC in the other set, are fused. Thus, one-to-one matching can be conveniently performed by selecting the best matching pair in both sets. As a result, the number of DCs in both (updated) sets $S_Q^M$ and $S_I^M$ becomes equal, that is, $|S_Q^M| = |S_I^M| = N_M$). Assume without loss of generality that $i$th DC class in set $C_i^Q : \{c_i^Q, w_i^Q, \sigma_i^Q\} \in S_Q^M$ matches the $i$th DC in set $C_i^I : \{c_i^I, w_i^I, \sigma_i^I\} \in S_I^M$, via sorting one set with respect to the other. So the penalties for global and SCD properties can be expressed as follows:

$$P_G(Q,I) = \beta \sum_{i=1}^{N_M} \left| w_i^Q - w_i^I \right| + (1-\beta) \frac{1}{T_C^{\max} N_M} \sqrt{\sum_{i=1}^{N_M} \left( c_i^Q - c_i^I \right)^2} \leq 1 \quad (11.18)$$

$$P_{SCD}(Q,I) = \begin{cases} \dfrac{1}{N_M^2(2L+1)^2} \sum_{i=1}^{N_M} \sum_{j=1}^{N_M} \sum_{x,y=-L}^{L} \Delta \left( \overline{\Psi}_{c_i^Q}^{c_j^Q}(x,y) - \overline{\Psi}_{c_i^I}^{c_j^I}(x,y) \right) \leq 1 \\[3mm] \dfrac{1}{N_M^2 L} \sum_{i=1}^{N_M} \sum_{j=1}^{N_M} \sum_{k=1}^{L} \Delta \left( \dfrac{\Phi_{c_i^Q}^{c_j^Q}(k)}{\max(w_i^Q,w_j^Q)} - \dfrac{\Phi_{c_i^I}^{c_j^I}(k)}{\max(w_i^I,w_j^I)} \right) \leq 1 \end{cases} \quad (11.19)$$

where

$$\Delta(x-y) = \begin{cases} 0 & \text{if } x = y = 0 \\ |x-y|/(x+y) & \text{otherwise} \end{cases} \quad (11.20)$$

and $0 < \beta < 1$, similar to $\alpha$, is the weighting factor between the two global color properties that are DC weights and centroids. The term $\Delta$ denotes the normalized difference operator which emphasizes the difference from zero to nonzero pairs. This is a common consequence when the DC pairs' area is relatively small whereas their SCDs are quite different. It also suppresses the bias from similar SCDs of two DCs with large weights. Note that computing $P_{SCD}$ should be independent from the effect of DC weights since this is already taken into consideration within $P_G$ computation. As a result, the combination of $P_G$ and $P_{SCD}$ represents the amount of dissimilarity present in all color properties and the unit normalization allows the combination in a configurable way with weights $\alpha$ and $\beta$, which can favor one color property to another. With the combination of $P_\phi$, which represents the natural color dissimilarity due to mismatching, the penalty trio models a complete similarity distance between two color compositions.

### 11.4.5 Experimental Results

Simulations are performed to evaluate the SCD efficiency subjectively and to compare retrieval performances, in terms of the query-by-example results, within image databases indexed by SCD and competing correlogram and MPEG-7 DCD [28], [31] FeX modules. Comparisons with other color descriptors, such as color histograms, CCV, and color moments, are not included since the study in Reference [7] clearly demonstrates the correlogram's superiority over other descriptors. The experiments performed in this section are based on four sample databases described below.

- *Corel 1K* image database: There are total of 1000 medium resolution ($384 \times 256$ pixels) images from 10 classes with diverse contents, such as wildlife, city, buses, horses, mountains, beach, food, African natives, and so on.

- *Corel 10K* image database: Same database used in the previous section.

- *Corel 20K* image database: There are 20000 images from Corel database bearing 200 distinct classes, each of which contains 100 images with a similar content.

- Synthetic image database: Same synthetic database used in the previous section.

The classes in Corel databases are extracted by the *ground-truth*, considering the content similarity and not the color distribution similarity. For instance, a red car and blue car are still in the same *Cars* class, although their colors do not match at all. Accordingly, color-based retrievals are also evaluated using the same ground-truth methodology, that is, considering a retrieval as relevant only if its content matches with the query. Note that all sample databases containing images with mediocre resolutions had to be selected; otherwise, it is not feasible to apply the correlogram method due to its computational complexity, especially for *Corel 10K* and *Corel 20K* databases. Finally, the performance evaluation is presented for the synthetic database in order to demonstrate the true description power of the SCD whenever color alone entirely characterizes the content of the image. Moreover, the robustness of the SCD is also evaluated against the changes of resolution, aspect ratio, color variations, translation, and so on. All experiments are carried out on a Pentium 5 computer with 1.8 GHz CPU and 1024 MB memory. If not stated otherwise, the following parameters are used for all the experiments performed throughout this section: $N_{DC}^{\max} = 6$, $T_A = 2\%$, and $T_S = 15$ for DC extraction; $T_W = 96\%$ and $D_{QT}^{\max} = 6$ for QT decomposition; and $T_C^{\min} = 45$, $T_C^{\min} = T_S$, and $\alpha = \beta = 0.5$ for the penalty-trio model. For the auto-correlogram, RGB color histogram quantization is set as $8 \times 8 \times 8$ (that is, $m = 512$ colors) with $d = 20$ for *Corel 1K* and $4 \times 4 \times 4$ (that is, $m = 64$ colors) with $d = 10$ for *Corel 10K* and *Corel 20K*. For the correlogram, $4 \times 4 \times 4$ bins are used for *Corel 1K* and $3 \times 3 \times 3$ bins for *Corel 10K* with $d = 10$. Note that the auto-correlogram was used only for *Corel 20K* due to its infeasible memory requirement for this database size. The same DC extraction parameters were used for MPEG-7 DCD and SCD. A MUVIS application, called DbsEditor, dynamically uses the respective FeX modules for feature extraction to index sample

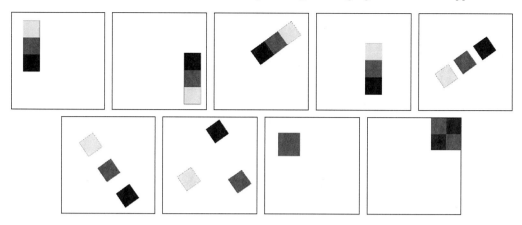

**FIGURE 11.13**

Query of a three-color object (top-left) in the synthetic database.

databases with the aforementioned parameters. Afterward, MBrowser application is used to perform similarity-based retrievals via query-by-example operations. A query image is chosen among the database items to be the example and a particular FeX module, such as MPEG-7 DCD, is selected to retrieve and rank the similar images in terms of color using only the respective features and an appropriate distance metric implemented within the FeX module. The recommended distance metrics are implemented for each FeX module, that is, the quadratic distance for MPEG-7 DCD and the $L_1$ norm for the correlogram.

### 11.4.5.1 Retrieval Performance on Synthetic Images

Recall that the images in the synthetic database contain colored regions in geometric and arbitrary shapes within which uniform samples from the entire color space are represented. In this way, the color matching accuracy can be visually evaluated and the first two penalty terms $P_\phi$ and $P_G$ can be individually tested. Furthermore, the same or matching colors form different color compositions by varying their region's shape, size, and/or inter-region proximities. This allows for testing both individual and mutual penalty terms $P_G$ and $P_{SCD}$. Finally, the penalty-trio's cumulative accuracy and robustness against variations of resolution, translation, and rotation can also be tested and compared against the correlogram.

Figure 11.13 presents a snapshot of the query of an image with three-color squares on a white background. The color descriptor is used with proximity histogram as the SCD descriptor and the retrieval results are ranked from left to right and top to bottom and the similarity distances are given on the bottom of the images. Among the first six retrievals, the same amount of identical colors are used and hence $P_\phi = P_G = 0$, which allows for testing the accuracy of $P_{SCD}$ alone. The first three retrievals have insignificant similarity distances, which demonstrates the robustness of $P_{SCD}$ against the variations of rotation and translation. The fourth, fifth, and sixth ranks present cases where spatial proximity between the three colors starts to differentiate, and hence SCD descriptor reflects the proximity differences successfully. For the seventh ($P_\phi \neq 0$ ) and the eight ($P_G \neq 0$) ranks, $P_\Sigma$ starts to build up significantly, since the color composition changes drastically due to emerging and missing color components.

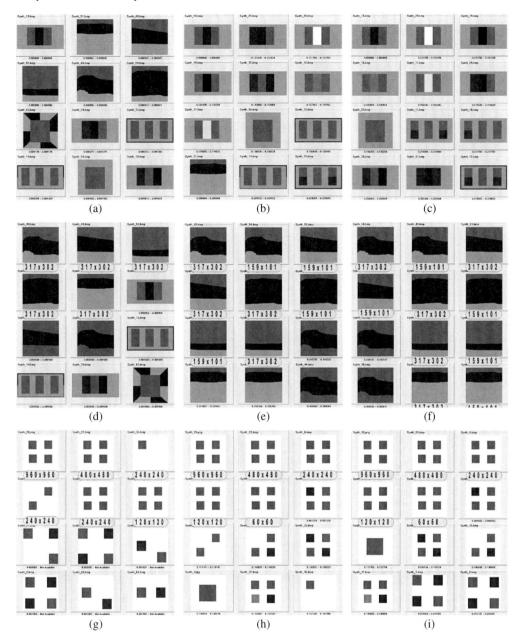

**FIGURE 11.14**

Three queries in the synthetic database processed using (a,d,g) correlogram, (b,e,h) SCD with proximity histogram, and (c,f,i) SCD with proximity grid. Some dimensions are tagged in boxes below pictures. Top-left is the query image.

Figure 11.14 shows three queries in the synthetic database with different color compositions and resolutions. In the case of Figures 11.14a to 11.14c, both proximity histogram and proximity grid successfully retrieve images with similar color compositions, whereas the correlogram cannot do it due to its invariance to weight (area) and limited range. The

**TABLE 11.2**
Similarity distances and ranks of A and B in Figure 11.12 when C is queried in *Corel 1K* database.

| Approach | $P_\Sigma$ | | Rank | |
|---|---|---|---|---|
| | A | B | A | B |
| with DC fusing | 0.176 | 0.156 | 3 | 1 |
| without DC fusing | 0.585 | 0.205 | 258 | 1 |

area invariance effect can be easily seen in the second and particularly the third ranks, where entirely different red and green weights occur. The same comments can be made in the case of Figures 11.14d to 11.14f for the fifth and all ranks above the seventh. Moreover in this case, it is obvious that the correlogram cannot retrieve the image with identical color composition among the first eleven ranks due to its resolution (pixel-based) sensitivity. Note further that the SCD with both proximity histogram and grid first retrieves the color compositions, where all colors are perfectly matching ($P_\phi = 0$) with the weights in a close vicinity ($P_G \neq 0$), and then balances between mismatching colors and weight differences of the matching ones. The case of Figures 11.14g to 11.14i is particularly shown here to emphasize the effect of image resolution over the correlogram and the SCD. The query of the largest image among the others with dimensions in five different resolutions logarithmically scaled from 60 to 960, but the same color composition (four red squares over white background), resulted in accurate ranking for the SCD. On the other hand, the correlogram retrieves accurately only one while the other two are shifted to lower ranks and the one (with $60 \times 60$ dimension) is missed within the first twelve ranks.

### 11.4.5.2 Retrieval Performance on Natural Images

In this section, three sample databases (*Corel 1K*, *Corel 10K*, and *Corel 20K*) are indexed using each FeX module and each individual (sub)feature is used for retrieval. As presented in Table 11.2, the first retrieval experiment is performed to demonstrate the effect of DC fusing over the retrieval accuracy. Similar results of several retrieval experiments suggest that DC fusing becomes the key factor for the success of the SCD. Therefore, DC fusing is applied for the rest of the experiments presented in this section.

**TABLE 11.3**
ANMRR scores of the SCD and the competing descriptors for databases *Corel 1K* (34 queries), *Corel 10K* (176 queries), and *Corel 20K* (22 queries).

| Descriptor | Database | | |
|---|---|---|---|
| | *Corel 1K* | *Corel 10K* | *Corel 20K* |
| MPEG-7 DCD | 0.180 | 0.458 | 0.461 |
| auto-correlogram | 0.222 | 0.381 | 0.444 |
| correlogram | 0.195 | 0.357 | N/A |
| SCD (proximity histogram) | 0.154 | 0.263 | 0.357 |
| SCD (proximity grid) | 0.162 | 0.291 | 0.390 |

**FIGURE 11.15 (See color insert.)**

Four typical queries using three descriptors in *Corel 10K* database: (a–d) dominant color, (e–h) correlogram, and (i–l) proposed method. Top-left is the query image.

Table 11.3 presents ANMRR results and the query dataset size of each of the three Corel databases. The query dataset is prepared *a priori* by regarding a certain degree of color content coherency. This means that the content similarity can mostly be perceived by color similarity. However, unlike previously presented synthetic images, a unique one-to-one correspondence between content and color similarities can never be guaranteed in natural images due to the presence of other visual cues, such as texture and shape. Nevertheless, according to the presented ANMRR scores, SCD with either mode achieves superior retrieval performance than the competing methods (correlogram, auto-correlogram, and MPEG-7 DCD combined with the quadratic distance computation) for all Corel databases. Moreover, it was observed that in the majority of the queries (between 58 and 78%), the SCD outperforms the (auto-)correlogram, whereas the figure is even higher (76–92%) with the MPEG-7 DCD. Finally, for shorter descriptor size with proximity histograms, a norm was used since comparative retrieval results promise no significant gain of using $L_1$ (for example, ANMRR score of SCD with proximity histogram using $L_1$ is 0.254 for *Corel 10K*).

**FIGURE 11.16 (See color insert.)**

Four typical queries using three descriptors in *Corel 20K* database: (a–d) dominant color, (e–h) auto-correlogram, and (i–l) proposed method. Top-left is the query image.

For visual evaluation, four retrieval results are presented for both *Corel 10K* and *Corel 20K* databases using all three descriptors. For the queries as shown in Figure 11.15, the proximity grid in SCD was used against the correlogram and MPEG-7 DCD. In the first, second, and fourth queries, one can easily notice the erroneous retrievals of the correlogram due to its color area insensitivity (for instance, compare the amount of red, white, and black colors between the query and the fifth ranked image in the first query). As already discussed, in such large databases the co-occurrence probabilities can (accidentally) match images with significantly different color proportions. Particularly in the first and fourth queries, erroneous retrievals of MPEG-7 DCD occur due to the lack of SCD description, which also makes accidental matches between dissimilar images with close color proportions (for instance, in the first query, the amount of white, red, and black colors is quite close between the query and sixth, seventh, and eight ranks; however, their SCDs are not).

For the queries shown in Figure 11.16, the proximity histogram in SCD was used against the auto-correlogram and the MPEG-7 DCD. Similar conclusions as before can be drawn

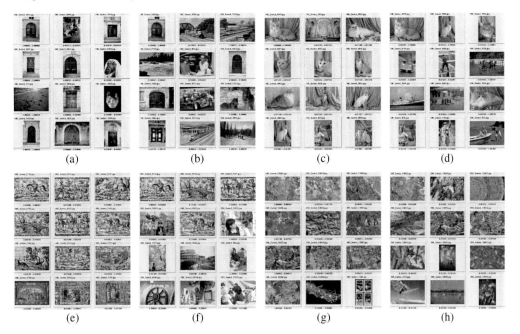

**FIGURE 11.17 (See color insert.)**

Two queries in (a,b,e,f) *Corel 10K* and (c,d,g,h) *Corel 20K* databases where (auto-)correlogram performs better than SCD: (a,e) correlogram, (b,f) proposed method, (c,g) auto-correlogram, and (d,h) proposed method. Top-left is the query image.

for the retrieval results. Furthermore, it should be noted that the amount of erroneous retrievals is increased particularly in the second and third queries, since the database size is doubled and hence accidental matches occur more often than before. However, in both databases, the (auto-)correlogram may occasionally perform better than SCD; the case of the queries shown in Figure 11.17, where significant (color) textures are present in all query images. This is indeed in accordance with the earlier remark stating that the correlogram is a colored texture descriptor and hence it can outperform any color descriptor whenever a textural structure is dominant.

## 11.5 Conclusion

This chapter presented two perceptual and spatial color descriptors. The first one, the perceptual correlogram is primarily designed to overcome the drawbacks of the color correlogram and brings a solution to its infeasibility problems. Utilization of dynamic color quantization instead of a static color space quantization not only achieves a more perceptual description of the color content, but also significantly decreases the number of colors. By doing so, the size of the correlogram table is reduced to feasible limits and since the memory limitations are also suppressed, implementation with higher-range values has become possible. However, it should be still noted that such a pixel-based proximity measure is highly resolution dependent, and if large image resolutions are considered, increasing the range value would become a restrictive factor in terms of memory and speed. Since a dynamic color quantization is applied, the obtained DCs do not have fixed centroids as in static quantization. Therefore, during the retrieval phase, such colors were compared using a one-to-one color matching scheme, which is preferred due to its accurate and fast computation. Yet much complex color matching methods are also possible, such as one-to-many matching, that involves comparison and/or combination of many global and spatial features. Apart from efficient and perceptual description of the color content, using spatial differences only, particularly in terms of probabilities, leaves a color descriptor incomplete. Retrieval experiments on the synthetic databases also demonstrate that global properties are indispensable in judging the color content similarity. These experiments further reveal that the penalty-trio model effectively combines spatial and global features into a similarity distance metric. Although the main aim of this descriptor was to solve the infeasibility problems of the traditional correlogram, the induced perceptuality and addition of global properties further increased the discrimination power. Experimental results, including the ANMRR scores, also demonstrate such superiority. However, it should be noted that color alone does not fully describe the entire image content and it can be associated with it only in a certain extent.

The spatial color descriptor (SCD) presented next in this chapter characterizes the perceptual properties of the color composition in a visual scenery in order to maximize the description power. In other words, the so-called outliers, which are the imperceivable color elements, are discarded for description efficiency using a top-down approach while extracting global and spatial color properties. In this way, severe problems and limitations of traditional pixel-based methods are effectively avoided and in spatial domain only the perceived (visible) color components can be truly extracted using QT decomposition. In order to reveal the true SCD properties, the proximity histogram and the proximity grid, representing the inter-proximity statistics in scalar and directional modes, are presented. During the retrieval phase, one-to-many DC matching is performed in order to apply the penalty-trio model over matching (and possibly fused) DC sets. This greatly reduces the faulty mismatches and erroneous similarity distance computations. The penalty-trio model computes the normalized differences in both spatial and global color properties and combines all so as to yield a complete comparison between two color compositions. Experimental results approve the superiority of the SCD over the competing methods in terms of discrimination power and retrieval performance especially over large databases. The SCD has a major advantage of being applicable to any database size and image resolution. Thus, it does not suffer from the infeasibility problems and severe limitations of the correlogram. Finally, it achieves a significant performance gain in terms of ANMRR scores. However, this gain remained below higher performance expectations, particularly when compared with the correlogram due to two reasons. First and foremost, the correlogram has the advantage of describing texture in color images thanks to its pixel-level analysis via co-occurrence probabilities. Yet the major reason is that the color similarity alone does not really imply the content-similarity. This significantly degrades the retrieval performance of SCD in several experiments. For instance, when an image with gray horse on a green field and a blue sky is queried, all retrievals with a gray elephant and similar background are counted as irrelevant (since they do not belong to horse class), although the color distribution is quite similar.

Many other irrelevant retrievals with similar color properties can be seen in the figures presented above. In summary, color properties correlate with the true content only in a certain extent and cannot be used as the single cue to characterize the entire content [47].

## Acknowledgment

This chapter is adapted from Reference [48] with the permission from Elsevier.

## References

[1] E. van den Broek, P. Kisters, and L. Vuurpijl, "The utilization of human color categorization for content-based image retrieval," *Proceedings of SPIE*, vol. 5292, pp. 351–362, January 2004.

[2] A. Mojsilovic, J. Kovacevic, J. Hu, R. Safranek, and K. Ganapathy, "Matching and retrieval based on the vocabulary and grammar of color patterns," *IEEE Transactions on Image Processing*, vol. 9, no. 1, pp. 38–54, January 2000.

[3] A. Mojsilovic, J. Hu, and E. Soljanin, "Extraction of perceptually important colors and similarity measurement for image matching, retrieval, and analysis," *IEEE Transactions on Image Processing*, vol. 11, no. 11, pp. 1238–1248, November 2002.

[4] K.H. Brandenburg, "MP3 and AAC explained," in *Proceedings of the AES 17th International Conference on High-Quality Audio Coding*, Florence, Italy, September 1999.

[5] J. Huang, S. Kumar, M. Mitra, W. Zhu, and R. Zabih, "Image indexing using color correlograms," in *Proceedings of the IEEE Conference on Computer Vision and Pattern Recognition* San Juan, Puerto Rico, June 1997, pp. 762–768.

[6] I.Kunttu, L. Lepisto, J. Rauhamaa, and A. Visa, "Image correlogram in image database indexing and retrieval," in *Proceedings of the 4th European Workshop on Image Analysis for Multimedia Interactive Services*, London, UK, April 2003, pp. 88–91.

[7] W. Ma and H. Zhang, "Benchmarking of image features for content-based retrieval," in *Proceedings of the Thirty-Second Asilomar Conference on Signals, Systems, and Computers*, Pacific Grove, CA, USA, November 1998, vol. 1, pp. 253–256.

[8] G. Pass, R. Zabih, and J. Miller, "Comparing images using color coherence vectors," in *Proceedings of the Fourth ACM International Conference on Multimedia*, Boston, MA, USA, November 1996, pp. 65–73.

[9] C. Tomasi and R. Manduchi, "Bilateral filtering for gray and color images," in *Proceedings of the International Conference on Computer Vision*, Bombay, India, January 1998, pp. 839–846.

[10] M. Wertheimer, "Laws of organization in perceptual forms," in *A Source Book of Gestalt Psychology*, London, UK: Routledge & Kegan Paul, 1938, pp. 71–88.

[11] "Muvis," Available online, http://muvis.cs.tut.fi.

[12] J. Hafner, H. Sawhney, W. Equitz, M. Flickner, and W. Niblack, "Efficient color histogram indexing for quadratic form distance functions," *IEEE Transactions on Pattern Analysis and Machine Intelligence*, vol. 17, no. 7, pp. 729–736, July 1995.

[13] I. Cox, M. Miller, S. Omohundro, and P. Yianilos, "Pichunter: Bayesian relevance feedback for image retrieval," in *Proceedings of the IEEE International Conference on Pattern Recognition*, Vienna, Austria, August 1996, vol. 3, pp. 361–369.

[14] A. Pentland, R. Picard, and S. Sclaroff, "Photobook: Tools for content-based manipulation of image databases," *Proceedings of SPIE*, vol. 2185, pp. 34–47, April 1994.

[15] J. Smith and S. Chang, "Visualseek: A fully automated content-based image query system," in *Proceedings of the Fourth ACM International Conference on Multimedia*, Boston, MA, USA, November 1996, pp. 87–98.

[16] "Virage," Available online, http://www.virage.com.

[17] S. Sclaroff, L. Taycher, and M.L. Cascia, "Imagerover: A content-based image browser for the World Wide Web," in *Proceedings of the IEEE Workshop on Content-Based Access of Image and Video Libraries*, San Juan, Puerto Rico, June 1997, pp. 2–9.

[18] S. Chang, W. Chen, J. Meng, H. Sundaram, and D. Zhong, "Videoq: An automated content based video search system using visual cues," in *Proceedings of the Fifth ACM International Conference on Multimedia*, Seattle, WA, USA, November 1997, pp. 313–324.

[19] K. van de Sande, T. Gevers, and C. Snoek, "Evaluation of color descriptors for object and scene recognition," *IEEE Transactions on Pattern Analysis and Machine Intelligence*, vol. 32, no. 9, pp. 1582–1596, September 2010.

[20] T. Kato, T. Kurita, and H. Shimogaki, "Intelligent visual interaction with image database systems – Toward the multimedia personal interface," *Journal of Information Processing*, vol. 14, no. 2, pp. 134–143, April 1991.

[21] M. Swain and D. Ballard, "Color indexing," *International Journal of Computer Vision*, vol. 7, no. 1, pp. 11–32, November 1991.

[22] Y. Gong, C.H. Chuan, and G. Xiaoyi, "Image indexing and retrieval based on color histograms," *Multimedia Tools and Applications*, vol. 2, no. 2, pp. 133–156, March 1996.

[23] V. Ogle and M. Stonebraker, "Chabot: Retrieval from a relational database of images," *Computer*, vol. 28, no. 9, pp. 40–48, September 1995.

[24] J. Smith and S. Chang, "Single color extraction and image query," in *Proceedings of the IEEE International Conference on Image Processing*, Washington, DC, USA, October 1995, vol. 3, pp. 528–531.

[25] A. Utenpattanant and O. Chitsobhuk, "Image retrieval using Haar color descriptor incorporating with pruning techniques," in *Proceedings of the 9th International Conference on Advanced Communication Technology*, Phoenix Park, Korea, February 2007, vol. 2, pp. 1123–1126.

[26] S. Wang, L.T. Chia, and R. Deepu, "Image retrieval using dominant color descriptor," in *Proceedings of the International Conference on Imaging Science, Systems and Technology*, Las Vegas, NV, USA, June 2003, vol. 1, pp. 107–110.

[27] K.M. Wong, L.M. Po, and K.W. Cheung, "A compact and efficient color descriptor for image retrieval," in *Proceedings of the IEEE International Conference on Multimedia and Expo*, Beijing, China, July 2007, pp. 611–614.

[28] B. Manjunath, J.R. Ohm, V. Vasudevan, and A. Yamada, "Color and texture descriptors," *IEEE Transactions on Circuits and Systems for Video Technology*, vol. 11, no. 6, pp. 703–715, June 2001.

[29] L. Tran and R. Lenz, "Compact colour descriptors for colour-based image retrieval," *Signal Processing*, vol. 85, no. 2, pp. 233–246, February 2005.

[30] M. Stricker and M. Orengo, "Similarity of color images," *Proceedings of SPIE*, vol. 2420, pp. 381–392, February 1995.

[31] L.M. Po and K.M. Wong, "A new palette histogram similarity measure for MPEG-7 dominant color descriptor," in *Proceedings of the International Conference on Image Processing*, Singapore, October 2004, vol. 3, pp. 1533–1536.

[32] G. Babu, B. Mehtre, and M. Kankanhalli, "Color indexing for efficient image retrieval," *Multimedia Tools and Applications*, vol. 1, no. 4, pp. 327–348, November 1995.

[33] Y. Deng, C. Kenney, M. Moore, and B. Manjunath, "Peer group filtering and perceptual color image quantization," in *Proceedings of the 1999 IEEE International Symposium on Circuits and Systems*, Orlando, FL, USA, June 1999, vol. 4, pp. 21–24.

[34] J. Fauqueur and N. Boujemaa, "Region-based image retrieval: Fast coarse segmentation and fine color description," *Journal of Visual Languages & Computing*, vol. 15, no. 1, pp. 69–95, February 2004.

[35] J. MacQueen, "Some methods for classification and analysis of multivariate observations," in *Proceedings of the Fifth Berkeley Symposium on Mathematical Statistics and Probability*, Berkeley, CA, USA, 1967, vol. 1, pp. 281–297.

[36] M. Sudhamani and C. Venugopal, "Grouping and indexing color features for efficient image retrieval," *International Journal of Applied Mathematics and Computer Sciences*, vol. 4, no. 3, pp. 150–155, May 2007.

[37] I. Valova and B. Rachev, "Retrieval by color features in image databases," in *Proceedings of the European Conference on Advances in Databases and Information Systems*, Budapest, Hungary, September 2004.

[38] B. Ooi, K.L. Tan, T. Chua, and W. Hsu, "Fast image retrieval using color-spatial information," *VLDB Journal*, vol. 7, no. 2, pp. 115–128, May 1998.

[39] A. Nagasaka and Y. Tanaka, "Automatic video indexing and full-video search for object appearances," in *Proceedings of the Second Working Conference on Visual Database Systems*, Budapest, Hungary, October 1991, pp. 113–127.

[40] H. Lee, H. Lee, and Y. Ha, "Spatial color descriptor for image retrieval and video segmentation," *IEEE Transactions on Multimedia*, vol. 5, no. 3, pp. 358–367, September 2003.

[41] J. Luo and D. Crandall, "Color object detection using spatial-color joint probability functions," *IEEE Transactions on Image Processing*, vol. 15, no. 6, pp. 1443–1453, June 2006.

[42] T. Ojala, M. Rautiainen, E. Matinmikko, and M. Aittola, "Semantic image retrieval with HSV correlograms," in *Proceedings of the 12th Scandinavian Conference on Image Analysis*, Bergen, Norway, June 2001, pp. 621–627.

[43] Y. Chun, N. Kim, and I. Jang, "Content-based image retrieval using multiresolution color and texture features," *IEEE Transactions on Multimedia*, vol. 10, no. 6, pp. 1073–1084, October 2008.

[44] J. Li, W. Wu, T. Wang, and Y. Zhang, "One step beyond histograms: Image representation using markov stationary features," in *Proceedings of the IEEE International Conference on Computer Vision and Pattern Recognition*, Anchorage, AK, USA, June 2008, pp. 1–8.

[45] S.J. Lee, Y.H. Lee, H. Ahn, and S.B. Rhee, "Color image descriptor using wavelet correlogram," in *Proceedings of the the 23rd International Technology Conference on Circuits/Systems, Computers and Communications*, Shimonoseki, Japan, July 2008, pp. 1613–1616.

[46] H. Moghaddam and M. Saadatmand-Tarzjan, "Gabor wavelet correlogram algorithm for image indexing and retrieval," in *Proceedings of the IEEE International Conference on Pattern Recognition*, Hong Kong, August 2006, vol. 2, pp. 925–928.

[47] B. Rogowitz, T. Frese, J. Smith, C. Bouman, and E. Kalin, "Perceptual image similarity experiments," *Proceedings of SPIE*, vol. 3299, pp. 576–590, January 1998.

[48] S. Kiranyaz, M. Birinci, and M. Gabbouj, "Perceptual color descriptor based on spatial distribution: A topdown approach," *Image and Vision Computing*, vol. 28, no. 8, pp. 1309–1326, August 2010.

# 12

## Concept-Based Multimedia Processing Using Semantic and Contextual Knowledge

**Evaggelos Spyrou, Phivos Mylonas, and Stefanos Kollias**

## 12.1 Introduction

Most of today's content-based multimedia analysis and retrieval systems tend to follow a low-level approach when tackling both content analysis and retrieval tasks, thus falling short of benefits uprising from higher-level interpretation and knowledge. The role of additional information in the sense of semantics, context, and implicit or explicit knowledge is gaining focus on the task of bridging the semantic and conceptual gap that exists between humans and computers, in order to further facilitate human-computer interaction and scene content understanding. This chapter focuses on modeling and exploiting contextual knowledge toward efficient multimedia content understanding. As discussed below, this type of information acts as a simulation of the human visual perception scheme, by taking into account all contextual information relative to the visual content of a scene [1]. As a result, the notion of context, provided that it will be properly modeled and justified, may be used to improve the performance of knowledge-assisted analysis, semantic indexing, and retrieval of multimedia content.

When tackling the well-known problems of semi-automated high-level concept detection or scene classification, the researcher faces a challenging and broad research area. In order to achieve better semantic results during any multimedia content analysis phase, the influence of additional contextual information may be of great help, because although the well-known semantic gap [2] has been acknowledged for a long time, current multimedia analysis approaches are still divided into two rather discrete categories as low-level multimedia analysis methods and tools (for example, Reference [3]) and high-level semantic annotation methods and tools (for example, References [4] and [5]). Semantic knowledge technologies, like ontologies [6] and folksonomies [7], are only lately being successfully incorporated within multimedia analysis and retrieval frameworks, especially when using them for creation, manipulation, and postprocessing of multimedia metadata.

Still, one of the most interesting problems in multimedia content analysis is detection of high-level concepts within multimedia documents. Recognizing the need for such an analysis, many research works set focus on low-level feature extraction to efficiently describe various audiovisual characteristics of a multimedia document. However, the semantic gap often characterizes the differences between descriptions of a multimedia object by different representations and the linking from the low-level to the high-level features. Moreover, the semantics of each object depend on the context it is regarded within. For multimedia applications this means that any formal representation of real-world analysis and processing tasks requires the translation of high-level concepts and relations, for instance, in terms of valuable knowledge, into the elementary and extensively evaluated characteristics of low-level analysis, such as visual descriptions and low-level visual features.

An important step for narrowing this gap is to automate the process of semantic feature extraction and annotation of multimedia content objects, by enhancing image and video classification with semantic characteristics. The main idea introduced herein relies on the integrated handling of concepts evident within multimedia content. Recent advances in the research field of knowledge-assisted multimedia analysis, along with the emerge of new content and metadata representations, have driven more and more researchers looking beyond solely low-level features (such as color, texture, and shape) in pursuit of more effective high-level multimedia representation and analysis methods. Current and previous multimedia research efforts are starting to focus on the combination of both low-level descriptors computed automatically from raw multimedia content and semantics focusing in extracting high-level features.

In the following, Section 12.2 describes the motivation in utilizing the notion of visual context in concept detection and scene classification. Section 12.3 presents the notions of bag-of-words image analysis techniques and visual context, and surveys the relevant state-of-the art methods. Section 12.4 deals with a novel proposition of an enhanced visual conceptualization of relative knowledge, as well as the instantiation of an image's region types. Section 12.5 presents three different types of context knowledge formalization, together with the proposed contextual adaptation in terms of the visual context algorithm and its optimization steps, according to the utilized knowledge. Some self-explanatory examples are presented in Section 12.6, whereas Section 12.7 lists experimental results derived from the *beach* domain. Finally, this chapter concludes with Section 12.8.

## 12.2 Motivation and Overview

Visual context forms a rather classical approach to context, tackling it from the scope of environmental or physical parameters that are evident in multimedia applications. The discussed context representation supports audiovisual information (for example, lighting conditions, environmental information) and is separately handled by visual context models. Research objectives in the field include visual context analysis, that is, to take into account the extracted/recognized concepts during content analysis in order to find the specific context, express it in a structural description form, and use it for improving or continuing the content analysis, indexing, and searching procedures, as well as personalization aspects. The following text refers to the term *visual context*, by interpreting it as *all information related to the visual scene content of a still image or video sequence that may be useful during its analysis phase*.

Since there is no globally applicable aspect of context in the multimedia analysis chain, it is very important to establish a working representation for context, in order to benefit from and contribute to the proposed enhanced multimedia analysis. The problems to be addressed include how to represent and determine context, how to use it, and how to define and model corresponding analysis features to take advantage of it. Additionally, efficient ways to utilize the new content and context representations must be investigated, in order to optimize the results of content-based analysis. In general, the lack of contextual information significantly hinders optimal analysis performance [8] and, along with similarities in low-level features of various object types, results in a significant number of misinterpretations. Taken into account the current state-of-the-art, both in terms of works dealing with content classification and regional visual dictionaries, as well as context modeling techniques, this work aims at a hybrid unification of them, in order to achieve optimized content analysis results and strengthen its high-level and low-level correlation.

According to the previous statements, visual context is strongly related to two main problems of image analysis; that is, *scene classification* and *high-level concept detection/recognition*. Scene classification forms a *top-down* approach, where low-level visual features are typically employed to globally analyze the scene content and classify it in one of a number of predefined categories, such as indoor/outdoor, city/landscape, and so on. Concept detection/recognition is a *bottom-up* approach that focuses on local analysis to detect and recognize specific objects in limited regions of an image, without explicit knowledge of the surrounding context (for example, recognize a building or a tree). The above two major fields of image analysis actually comprise a chicken-and-egg problem. For instance, detection of a *building* in the middle of an image might imply a picture of a *city* with a high probability, whereas pre-classification of the picture as *city* would favor the recognition of a *building* versus a *tree*.

However, a significant number of misclassifications usually occur because of the similarities in low-level color and texture characteristics of various object types and the lack of contextual information, which is a major limitation of individual object detectors. Toward the solution to the latter problem, an interesting approach is the one presented in Reference [9]. A spatial context-aware object-detection system is proposed, initially combining the output of individual object detectors in order to produce a composite belief vector for the objects potentially present in an image. Subsequently, spatial context constraints, in the form of probability density functions obtained by learning, are used to reduce misclassification by constraining the beliefs to conform to the spatial context models. Unfortunately, such an approach alone is not considered sufficient, as it does not utilize the significant amount of available additional knowledge in the form of semantic relations.

So far, none of the existing methods and techniques utilizes the herein proposed contextual modeling in any form. This tends to be the main drawback of these individual object detectors, since they only examine isolated strips of pure object materials, without taking into consideration the context of the scene or individual objects themselves. This is very important and also extremely challenging even for human observers. The notion of visual context is able to aid in the direction of natural object detection methodologies, simulating the human approach to similar problems. For instance, many object materials can have the same appearance in terms of color and texture, while the same object may have different appearances under different imaging conditions, such as lighting and magnification. However, one important trait of humans is that they examine all the objects in the scene before making a final decision on the identity of individual objects. The use of visual context in the visual analysis process is the one that provides the necessary added value and forms the key for such a solid unambiguous recognition process and will be extensively presented and exploited in the following.

More specifically, this chapter presents an integrated approach, offering unified and unsupervised manipulation of multimedia content. It acts complementary to the current state-of-the-art, as it tackles both aforementioned challenges. Focusing on semantic analysis of multimedia, it contributes toward bridging the gap between the semantic and raw nature of multimedia content. It tackles one of the most interesting problems in multimedia content analysis, namely, detection of high-level concepts within multimedia documents, based on the semantics of each object, in terms of its visual context information. The latter is based on semantic relationships that are inherent within the visual part of the content. The proposed approach proves also that the use of such information enhances the results obtained from traditional knowledge-assisted image analysis techniques, based on both *visual* and *contextual* information.

## 12.3 Image Analysis Based on Regions and Context

If focused solely on the visual part of analysis, it is rather true that high-level concept detection remains still a challenging and unsolved problem. Its most interesting aspects are first the low-level feature extraction, aiming to capture and describe the visual content of images or regions, and last the way that these features will be assigned to high-level concepts. This chapter deals with the latter part of the analysis process, and aims to create image descriptions from image regions, using standardized visual features, that is, the MPEG-7 (Moving Picture Experts Group) descriptors.

The most common approach in detection and recognition tasks begins with the extraction of a low-level description of the visual content of concepts. Then, for each concept, a detector is trained based on one or more examples. This is typically done using various machine learning techniques, such as neural networks, support vector machines (SVMs), and fuzzy systems.

In order to train the concept detectors, it is important to use or create a specific dataset, appropriately annotated either globally or locally. For a globally annotated image, one only gets the knowledge of the existence of certain concepts within it. For a locally annotated image, one also knows the exact location of concepts within it. However, despite the continuous growth of audiovisual content, the available locally annotated image collections remain few. This is not surprising, since such an annotation process is a difficult and tedious task. Two of the most important locally annotated collections are LabelMe [10], a collaboratively annotated collection for a very large number of concepts, and the PASCAL [11] collection. On the other hand, a global annotation of a given image is a much easier and less time-consuming task. There exist many such datasets, among which one should note the collaborative annotation of the LSCOM workshop [12], which focused on sequences from news videos of the TRECVID 2005 collection, and a similar attempt presented in Reference [13], focusing on cultural videos from the TRECVID 2007 collection. It should be noted here that in many popular social networks, such as Flickr,[1] many thousands of photos are globally annotated, while research efforts toward the annotation of such datasets are still increasing [14], [15], [16].

It is now clear that image analysis techniques that focus on the extraction of high-level concepts from globally annotated content are of greater importance and may have a broader field of applications. Thus, the algorithms that are presented in this chapter aim toward this exact problem, that is, how to extract and manipulate efficiently the visual content of globally annotated images in order to detect the presence of high-level concepts, without specifying their exact location.

---

[1] http://www.flickr.com

### 12.3.1   The Bag-of-Words Model

A very popular model that combines the aforementioned aspects of visual analysis is the *bag-of-words* model. Visual descriptions are extracted locally, from groups of image pixels. Using an appropriate visual dictionary, these descriptions are quantized to the corresponding visual words. An image is then described by a set of visual words, without considering any spatial or semantic relations. Finally, an appropriate detector is trained for each concept.

In order to develop a bag-of-words based technique, the first step to consider is how to select image parts, from whom visual descriptors should be extracted. Early approaches used a grid, dividing images to blocks. References [17] and [18] used square blocks of equal size. These techniques were very fast, but lacked in terms of the semantic interpretation of each block. To overcome this, Reference [19] used random sampling and a variable block size. Since this disintegration proved not to be very robust, later techniques were based on the extraction of points of interest. References [20] and [21] extracted features from the neighborhood of Harris affine points. Reference [18] selected points detected by the difference-of-Gaussian and extracted multi-resolution features. All these approaches aimed to selected invariant points under scale and some geometrical transforms and are very effective in the case of object detection. In parallel, Reference [22] applied a segmentation algorithm, in order to split an image to regions, based on their color and texture properties, with many advantages in material or scene detection.

The next step to consider is the extraction of the visual descriptions. In the case of grid-selected regions, color and texture descriptors are extracted. When regions are selected by a segmentation process, shape descriptors may also be extracted, if applicable to the targeted concepts. The MPEG-7 standard [23] contains many visual descriptors that are broadly used. Finally, in the point-of-interest-based regions, appropriate and popular features are those generated by the scale-invariant feature transform (SIFT) [24] and speeded-up robust feature (SURF) [25] methods and their various variations.

The success or failure of each technique is highly related to the creation of an appropriate codebook, based on which an image region is assigned to a visual word. Most of the techniques use typical clustering algorithms, such as the traditional $K$-means [26], or a hierarchical variation, such as the one proposed in Reference [27]. The selection of $K$ is usually selected empirically or by a trial-and-error process on a validation set. However, certain techniques, such as the minimum description length (MDL) approach, are often applied in order to determine the appropriate dictionary size [28]. Typical sizes of dictionaries vary from a few tenths for grid approaches to many thousands of visual words for point-of-interest approaches.

The last step is to select an appropriate bag-of-words representation, with which the detectors will be trained. A number techniques have been proposed; many of them have been inspired by text categorization. A brief overview of the most important techniques is presented below.

Reference [29] created a texton library and trained texton histogram-based detectors. Similarly, Reference [30] constructed a codebook of prototype regions with specific geometric and photometric properties, namely, three-dimensional textons. Reference [31] used mean-shift clustering to split images to regions based on color features. Each image was then represented by a binary vector indicating the presence of absence of visual words. In another work [32] pixel-based concept detectors were used in order to achieve semantic image clustering. Visual features were extracted from the resulting regions and scenes were detected using a codebook. Reference [20] replaced visual words with points of interest (bag-of-keypoints) and used $K$-means clustering to construct the visual dictionary. Both SVM and naive Bayesian techniques were then applied for concept detection. Reference [33] used semantic descriptions instead of low-level descriptors. In Reference [34], the authors used the bag-of-words model to a contour-based object recognition problem. They constructed the visual dictionary based on curve parts. Another work [35] divided images to subregions and calculated local histograms within these subregions. Then, the bag-of-words model was used. Reference [36] proposed the use of *keyblocks*, which is the equivalent of keywords in the field of images. A codebook, that contained those keyblocks in various resolutions, was built. Reference [18] investigated various methods of splitting images to parts and used a Bayesian hierarchical model.

During the past few years, there have been a few notable attempts to enhance the bag-of-words model with spatial relations. Reference [37] used adaptive correlation histograms and presented a model robust to geometric transformations. Reference [38] suggested a hierarchical probabilistic model, based on a set of parts, each describing the expected position of objects. Reference [39] suggested a hierarchical bag-of-words model of spatial and spatiotemporal features of video sequences, whereas another work [40] suggested a segmentation algorithm as an extension of object detection techniques with the aid of a visual dictionary.

As the bag-of-words model is inspired by text processing techniques, it is not surprising that it has been enhanced by popular methods in this field. For example, the latent semantic analysis (LSA) approach [41] aims at exploiting the latent relations among the visual words. References [42] and [43] extended the bag-of-words model using LSA and probabilistic LSA (pLSA). Reference [21] modeled images as a mixture of concepts and also applied pLSA, in a fully unsupervised model.

### 12.3.2   The Role of Context

*Scene context* is probably the simplest way to model context in an image analysis problem. According to Reference [44], scene context may be defined as a combination of objects/concepts which are correlated under human perception and share the property to complement each other. Thus, in order to exploit scene context, one has first to detect present concepts and optionally their location within the image and then use contextual information to infer the scene the image depicts. For example, in a concept detection problem, if the concept *sea* is detected with high confidence, one could also expect the image to contain *sky* and depict a *beach* scene. Reference [45] used scene context to detect events. Reference [46] applied the expectation-maximization algorithm to the low-level features of image parts that depict concepts. Reference [47] investigated the use of Bayesian networks, Boltzmann machines, Markov random fields, and random fields to model the contextual relations between concepts. Reference [48] used scene context as an extra source in global descriptions and faced scene classification and object detection as a unified problem.

*Spatial context* aims at modeling the spatial relations among concepts. This step is used to enhance context models, after modeling scene context. It may be defined as a combination of objects/concepts, which, apart from their co-occurrences, occupy spatial positions within an image that are not independent. For example, to further extend the previous scene context example, if an image is detected to depict a *beach* scene and for a region it cannot be determined with high confidence whether it depicts *sky* or *sea*, two concepts that share similar visual features, its location should be considered. Thus, if it is located on the top of all other regions it should depict *sky*. Reference [49] used quantitative spatial and photometric relations among regions in a scene classification problem. Reference [50] used probabilistic models to model spatial relations among materials; this effort was further enhanced in Reference [9] by modeling spatial relations among concepts after a learning process. Reference [51] proposed an expectation-maximization model, which after a learning process low-level features to visual words that describe concepts. Reference [52] used a grid to split images to blocks and encoded both relations among low-level features and concepts and spatial relations among concepts. References [53] and [54] also used a grid and a two-dimensional hidden Markov model to form spatial relations among concepts. Another work [55] used graphs and statistics and showed that spatial relation-driven models increase precision. In Reference [56], the authors combined knowledge about scene layout and a generative model using material detectors and further improved their model in Reference [57].

*Temporal context* considers the relations of an image with other images from the same collection, taken with a small time difference, independently of their visual content. For example, if an image depicts a *beach* scene with a confidence, concepts present in images taken by the same user with a small time difference should also be in the context of *beach*. Reference [58] exploited the idea that the visual content of a photo is correlated to the one of photos taken with a small time interval. Reference [59] used many successive images from different viewpoints and exploited their temporal context to solve a three-dimensional photo retrieval problem by employing a Bayesian model. Reference [60] used temporal context in event detection in text documents. References [61] and [62] combined temporal context and photo location, whereas Reference [63] constructed hierarchical event models based on the visual content and temporal context of photo collections.

*Metadata context* involves the relations among the metadata that are available in digital images, such as the camera settings with which the image was taken. These metadata are embedded in image files according to the EXIF (Exchangeable Image File) standard [64]. For example, for an *indoor/outdoor* scene classification problem, the knowledge of the focus distance can let one assume the depicted scene; for example, a large distance usually indicates an *outdoor* scene. Reference [65] used metadata of camera settings combined with low-level features to automatically annotate photos. Reference [66] used a boosting algorithm and metadata in a scene classification problem. Reference [67] combined metadata with color features and *face* and *natural place* detectors. References [68], [69], and [70] showed that metadata information can significantly assist in the problems of *indoor/outdoor* classification and *sunset* detection, using Bayesian networks. Finally, Reference [71] proposed a system that combines metadata and visual features in order to automatically construct photo collections.

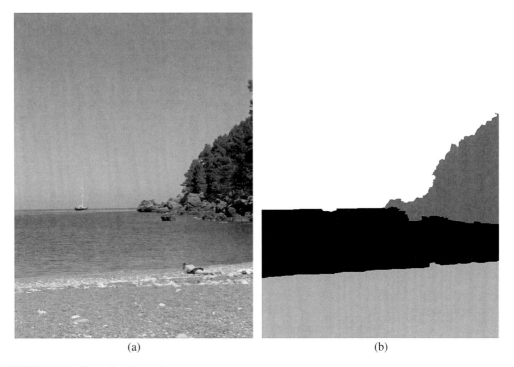

(a)                              (b)

**FIGURE 12.1 (See color insert.)**

An input image and its coarse segmentation.

## 12.4    Image Description and High-Level Concept Detection Using a Region Thesaurus

The following presents the proposed approach to tackle the problems of image description and high-level concept detection from a different and at the same time innovative aspect, that is, based on a region thesaurus containing a set of region types [72]. This research effort was expanded and further strengthened in References [73], [74], [75] and [76] by exploiting visual context in the process and achieving promising research results. The main focus remains to provide an ad hoc ontological knowledge representation containing both high-level features (that is, high-level concepts) and low-level features and exploit them toward efficient multimedia analysis.

Generally, the visual features extracted from an image or video can be divided into two major categories. The first contains typical *low-level* visual features that may provide a qualitative or quantitative description of the visual properties. Often these features are standardized in the form of a *visual descriptor*. The second category contains *high-level* features that describe the visual content of an image in terms of its semantics. One fundamental difference between these categories is that low-level features may be calculated directly from an image or video, while high-level features cannot be directly extracted, but are often determined by exploiting the low-level features. A human observer can easily recognize high-level features, even in situations when it could be rather difficult to provide their qualitative description and almost impossible to provide a quantitative one.

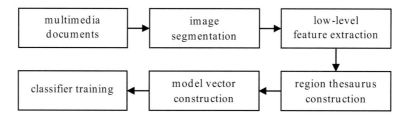

**FIGURE 12.2**

Offline part of the high-level concept detection algorithm.

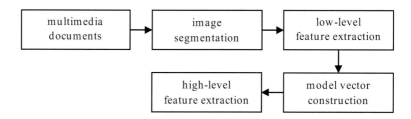

**FIGURE 12.3**

Online part of the high-level concept detection algorithm.

In this sense, this chapter tries to enhance the notion of a visual context knowledge model with *mid*-level concepts. Those concepts are referred to as region types, for reasons clarified in Section 12.4.2. Such concepts may provide an in-between description, which can be described semantically, but does not express neither a high-level nor a low-level feature. Thus, this work will focus on a unified multimedia representation by combining low-level and high-level information in an efficient manner and attach it to the context model by defining certain relations. To better understand the notion of region types, Figure 12.1 presents a visual example. In this example, a human could easily describe the visual content of the image either in a high-level manner (that is, the image contains *sky*, *sea*, *sand*, and *vegetation*) or in a lower level, but higher than a low-level description (that is, an *azure* region, a *blue* region, a *green* region, and a *gray* region). Although a quantitative description cannot be provided, each image can be intuitively and even efficiently described by a set of such features, that is, the region types. Therefore, it is of crucial importance to encode the set of region types in an effective manner that can efficiently describe almost every image in a given domain. To achieve this, a *region thesaurus* needs to be constructed. The next sections briefly describe the extraction of the low-level features and the construction of the region thesaurus. Figure 12.2 presents the offline part of the overall methodology for the high-level concept detection process that leads to a trained set of classifiers, while Figure 12.3 presents the online part that leads to the extraction of high-level features.

## 12.4.1   Low-Level Feature Extraction

To represent the color and texture features of a given image, this chapter follows an approach of extracting visual descriptors *locally*, that is, from image regions. First, color segmentation is performed using a multi-resolution implementation of the well-known RSST method [77], tuned to produce a coarse segmentation. Note that this under-segmentation both facilitates image description and prevents the problem from being too complex. After extracting image regions, low-level visual descriptors from the MPEG-7 standard [23] are extracted to capture a description of their visual content. Here, the extraction process focuses on color and texture features using appropriate MPEG-7 color and texture descriptors, since the employed high-level concepts belong to the categories of *materials* and *scenes*. To perform such feature extraction, a descriptor extraction tool [78], which is fully compatible with the MPEG-7 eXperimentation Model (XM) [79], is used.

More specifically, four color and two texture descriptors are selected: the *dominant color descriptor* (DCD), the *color layout descriptor* (CLD), the *scalable color descriptor* (SCD), the *color structure descriptor* (CSD), the *homogeneous texture descriptor* (HTD), and the *edge histogram descriptor* (EHD). To obtain a single region description from all the extracted region descriptions, features are merged after their extraction [80] into a feature vector. The feature vector $f_i$ that corresponds to a region $r_i \in R$, for $R$ denoting the set of all regions, is defined as follows:

$$f_i = f(r_i) = [DCD(r_i), CLD(r_i), SCD(r_i), CSD(r_i), HTD(r_i), EHD(r_i)]. \tag{12.1}$$

## 12.4.2 Construction of a Region Thesaurus

After extracting color and texture features, the next step aims to bridge these low-level features to the high-level concepts aimed at detection. To achieve this, first a region thesaurus will be constructed to assist with quantizing regions and forming an intermediate image description. This description will contain all the necessary information to connect one image with every region type of the dictionary. In this way, a fixed-size image description can be achieved, tackling the problem that the number of segmented regions is not fixed. Moreover, this description will prove again useful when contextual relations will be exploited, as described in Section 12.5.

Given the entire training set of images and their extracted regions, one can easily observe that regions belonging to similar semantic concepts also have similar low-level descriptions, and that images containing the same high-level concepts consist of similar regions. This gives a hint to exploit region similarity, as region co-existences often characterize the concepts that exist within an image [72].

The first step is the selection of region types that will form the region thesaurus. Based on the aforementioned observations, the proposed method starts from an arbitrary large number of segmented regions and applies a *hierarchical clustering* algorithm [81], adjusted for the problem at hand and with the clustering level empirically selected. After the clustering process, each cluster may or may not represent a high-level feature and each high-level feature may be contained in one or more clusters. This means that the concept *sand* can have many instances differing, for example, in color or texture. Moreover, in a cluster that may contain instances from a semantic entity (for example, *sea*), these instances could be mixed up with parts from another visually similar concept (for example, *sky*). Here, a single region is selected to represent each cluster, that is, the region type.

Finally, a region thesaurus $T$ can be formally described as a set of $N_T$ visual words $t_i$:

$$T = \left\{ t_i, \quad i = 1, 2, \ldots, N_T \right\}, \quad t_i \subset R, \tag{12.2}$$

$$\bigcup_i t = R, \quad i = 1, 2, \ldots, N_T, \bigcap_{i,j} t = \emptyset, \quad i \neq j. \tag{12.3}$$

Generally, a thesaurus combines a list of every term in a given domain of knowledge and a set of related terms for each term in the list, which are the synonyms of the current term. In the proposed approach, the constructed region thesaurus contains all the region types that are encountered in the training set. Each region type is represented by its feature vector that contains all the extracted low-level information. As it is obvious, a low-level descriptor does not carry any semantic information. It only constitutes a formal representation of the extracted visual features of the region. On the other hand, a high-level concept carries only semantic information. It is now clear that a region type lies in-between those features. It contains the necessary information to formally describe the color and texture features, but can also be described with a *lower* description than the high-level concepts. Namely, one can describe a region type as *a green region with a coarse texture*.

### 12.4.3 Construction of Model Vectors

This section presents the algorithm, which is used here to describe each image with the aid of the region thesaurus. First, it must be noted that the MPEG-7 standard does not specify strict distance measures. It only suggests some, so as to allow for other measures to be used and test their efficiency. As depicted in the experiments presented in Reference [72], the use of the Euclidean distance provides a simple yet effective way to fuse all extracted low-level information, leading also to satisfactory results. Then, the distance $d(r_1, r_2)$ between two regions $r_1$ and $r_2$ defined in $\mathscr{R}$ is calculated by the Euclidean distance of their feature vectors $f_1$ and $f_2$ as follows:

$$d(r_1, r_2) = d(f_1, f_2) = \sqrt{\sum_{i=1}^{n} (f_1^i - f_2^i)^2}. \tag{12.4}$$

Having calculated the distance of each image region to all the words of the constructed thesaurus, the model vector that semantically describes the visual content of the image is formed by keeping the smaller distance for each intermediate concept (region type), thus containing all the necessary information to associate an image with the whole set of the region thesaurus. In particular, the model vector $m_p$ describing image $p$ is given by

$$m_p = [m_p(1), m_p(2), \ldots, m_p(j), \ldots, m_p(N_T)], \ i = 1, 2, \ldots, N_K, \tag{12.5}$$

where

$$m_p(j) = \min_{r \in R(p)} \{d(f(t_j), f(r))\}, \ i = 1, 2, \ldots, N_T; \ j = 1, 2, \ldots, N_T, \tag{12.6}$$

and $R(p)$ denotes the set of all regions of image $p$.

In order to better understand the above process, Figure 12.4 presents an indicative example, where an image is segmented in regions and a region thesaurus is formed by six region types. On the left, this figure presents the distances of each region type from the sky region; the distances of each image region from region type 5 are presented on the right. The model vector is constructed by the smallest distances for each region type. In this case and considering region type $t_5$, the minimum distance is equal to 0.1. The model vector for the specific image, given the region thesaurus, is defined as

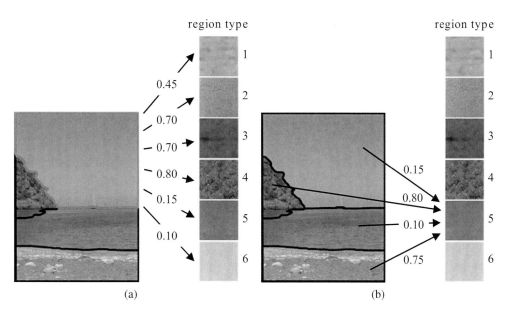

**FIGURE 12.4 (See color insert.)**

Distances between regions and region types: (a) distances between an image region and all region types; (b) distances between all regions and a specific region type.

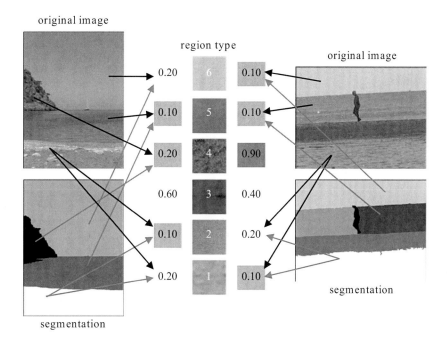

**FIGURE 12.5 (See color insert.)**

Construction of model vectors for two images and a visual thesaurus of six region types; lowest values of model vectors are highlighted (light gray) to note which region types of the thesaurus are most contained in each image, whereas a high value (dark gray) indicates a high distance between the corresponding region type and the image.

$$m = [m(1), m(2), \ldots, m(5), m(6)],\qquad\qquad(12.7)$$

where $m(5)$ will be equal to 0.1. Taking into consideration all distances between all four image regions and all six region types (that is, a total of 24 distances), the corresponding model vector is formed. Figure 12.4 presents the model vectors for two images, using the region thesaurus shown in Figure 12.5.

### 12.4.4  High-Level Feature Detection

After extracting model vectors from all images in the (annotated) training set, an SVM-based detector is trained separately for each high-level concept. A model vector $m_i$, describing a keyframe in terms of the region thesaurus, is fed to the detectors. The output of the network is the confidence that the image in question contains the specific concept, this is done for all concepts. It is important to note that the detectors are trained based on annotation per image and not per region. The same stands for their output, thus providing the confidence that the specific concept exists *somewhere* within the keyframe in question.

## 12.5  Visual Context Optimization

In order to fully exploit the notion of visual context and combine it with the aforementioned bag-of-words technique, a threefold approach is introduced next. The proposed methodology could be divided into the following three sections, according to the effect of visual context regarding concepts and region types:

- a scene context approach that aims to refine initial high-level concept detection results by exploiting solely the contextual relations between high-level concepts;

- an approach that aims to refine the input of trained high-level concept detectors based on the contextual relations between region types of the given training set; and

- a unified approach that utilizes contextual relations among high-level concepts and region types.

It should be emphasized here that this research effort focuses on the integrated approach of the subject, which offers a unified and unsupervised management of multimedia content. It is proved that the use of enhanced intermediate information can improve the results of traditional, knowledge-assisted image analysis, based on both *visual* and *contextual* information.

### 12.5.1  Scene Context

The proposed approach differentiates itself from most of the related research works, because it deals with a global interpretation of the image and the concepts that are present in it. In other words, high-level concepts either exist or do not exist within the entire image under consideration and not within a specific region of interest (for example, the image might contain concept *water*, but there is no information regarding its spatial location). Now, in order to further adapt the results of low-level and descriptor-based multimedia analysis, utilizing the notion of region types, a scene context method, founded on an enhanced high-level contextual ontology, is introduced. The proposed visual context application optimizes the high-level concept detection results (in terms of the classifiers' output) that were obtained based on the detailed methodology described in the previous sections.

### 12.5.1.1 A Scene Context Knowledge Model

The high-level concept ontology proposed herein is described as a set of concepts and semantic relations between concepts within a given universe. This set is introduced in order to efficiently express the real-world relations that exist between the concepts of the domain at hand. In general, one may decompose such an ontology $O_c$ into two parts:

- Set $C = \{c_i\}$, for $i = 1, 2, \ldots, n$, of all semantic concepts in the domain of interest.

- Set $R_c = \{R_{c,ij}\}, i, j = 1, 2, \ldots, n$ of all semantic relations among concepts. Note that $R_{c,ij} = r_{c,ij}^{(k)}$ contains the $K$ relations that can be defined among concepts $c_i$ and $c_j$. Moreover, for a given relation $r_{c,ij}^{(k)}$, its inverse $\bar{r}_{c,ij}^{(k)}$ can be defined.

More formally:

$$O_c = \{C, R_c\}, \quad R_{c,ij} : C \times C \to \{0, 1\}. \tag{12.8}$$

However, modeling of a domain using $O_c$ is inappropriate, since it does not model relations among concepts as fuzzy as in real-world domains. Therefore, the aforementioned model is expanded to produce a fuzzified version of the scene context ontology, formally denoted as follows:

$$\mathcal{O}_c = \{C, \mathcal{R}_c\}, \tag{12.9}$$

where $\mathcal{R}_c = \{\mathcal{R}_{c,ij}\}$ is the set of fuzzy relations among concepts, with $\mathcal{R}_{c,ij} = r_{c,ij}^{(k)}$ and $\bar{r}_{c,ij}^{(k)}$ denoting again the corresponding inverse relation. Now, the following can be written:

$$r_{c,ij} : C \times C \to [0, 1]. \tag{12.10}$$

Since for two concepts the existence of more than one relations is possible, a combination of all relations among $c_i$ and $c_j$ is defined as

$$\mathcal{U}_{c,ij} = \bigcup_k [r_{c,ij}^{(k)}]^p, \ i, j = 1, 2, \ldots, N, \ k = 1, 2, \ldots, K. \tag{12.11}$$

The final combination of the MPEG-7 originating relations forms a resource description framework (RDF) graph and constitutes the abstract contextual knowledge model to be used (Figure 12.6). The value of $p$ is determined by the semantics of each relation $r_{c,ij}$ used in the construction of $\mathcal{U}_{c,ij}$. More specifically:

- $p = 1$, if the semantics of $r_{c,ij}$ imply that it should be considered as is;

- $p = -1$, if the semantics of $r_{c,ij}$ imply that its inverse $\bar{r}_{c,ij}$ should be considered; and

- $p = 0$, if the semantics of $r_{c,ij}$ do not allow its participation in the construction of the combined relation $\mathcal{U}_{c,ij}$.

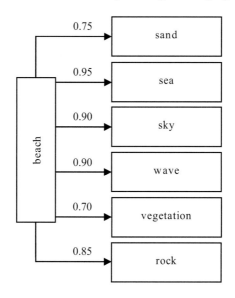

**FIGURE 12.6**

A fragment of the *Beach* domain ontology depicting the relations between concept *Beach* (the root element) and seven high-level concepts.

As indicated in Reference [82], any kind of semantic relation may be represented by such an ontology, however, herein it is restricted to a fuzzified ad hoc context ontology. The latter is introduced in order to optimally express the real-world relationships that exist between each domain's participating concepts. In order for this ontology to be highly descriptive, it must contain a representative number of distinct and even diverse relations among concepts, so as to scatter information among them and thus describe their context in a rather meaningful way. Moreover, the utilized relations need to be meaningfully combined, so as to provide a view of the knowledge that suffices for context definition and estimation. Since modeling of real-life information is usually governed by uncertainty and ambiguity, it is believed that these relations must incorporate fuzziness in their definition. Therefore, the proposed method extends a subset (Table 12.1) of the MPEG-7 semantic relations [83] that are suitable for image analysis and specified, in this case, by a domain expert. It should be noted at this point that since the proposed semantic relations are redefined in a way to represent fuzziness, a degree of confidence is associated to each of them. To further un-

**TABLE 12.1**

Fuzzy scene context semantic relations between concepts.

| Name | Inverse | Symbol | Meaning |
| --- | --- | --- | --- |
| Specialization | Generalization | $Sp(a,b)$ | $b$ is a specialization in the meaning of $a$ |
| Part | PartOf | $P(a,b)$ | $b$ is a part of $a$ |
| Example | ExampleOf | $Ex(a,b)$ | $b$ is an example of $a$ |
| Instrument | InstrumentOf | $Ins(a,b)$ | $b$ is an instrument of or is employed by $a$ |
| Location | LocationOf | $Loc(a,b)$ | $b$ is the location of $a$ |
| Patient | PatientOf | $Pat(a,b)$ | $b$ is affected by or undergoes the action of $a$ |
| Property | PropertyOf | $Pr(a,b)$ | $b$ is a property of $a$ |

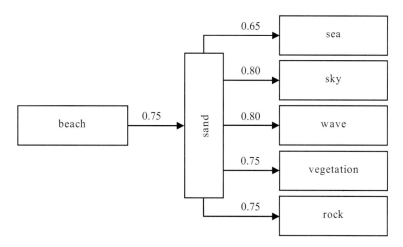

**FIGURE 12.7**

A fragment of the *Beach* domain ontology depicting the relations between concept *sand* (the root element) and six high-level concepts.

derstand the meaning of these semantic relations, some indicative examples are presented below. For example, a *bear* is a *Specialization* of an *animal*, whereas *tree* is a *part* of *forest*. Moreover, *clay* is an *Example* of a *material* and a *wheel* is the *Instrument* of a *car*. *Beach* might be the *Location* of an *umbrella*, *gun* is a *Patient* of the action of *soldier*, and *wavy* is a *Property* of a *sea*.

The graph of the proposed model contains nodes (that is, domain concepts) and edges (that is, an appropriate combination[2] of contextual fuzzy relations between concepts). The degree of confidence of each edge represents fuzziness in the model. Non-existing edges imply non-existing relations, meaning that relations with zero confidence values are omitted. An existing edge between a given pair of concepts is produced based on the set of contextual fuzzy relations that are meaningful for the particular pair. For instance, the edge between concepts *rock* and *sand* is produced by the combination of relations *Location* and *Patient*, whereas the edge between *water* and *sea* utilizes *Specialization*, *PartOf*, *Example*, *Instrument*, *Location*, and *Patient*, in order to be constructed. Since each concept has a different probability to appear in the scene, a flat context model would not have been sufficient in this case. On the contrary, concepts are related to each other, implying that the graph relations used are in fact transitive. The degree of confidence is implemented using the RDF reification technique [84].

**12.5.1.2   Scene Context Optimization**

Once the contextual knowledge structure is finalized and the corresponding representation is implemented, a variation of the context-based confidence value readjustment algorithm [8] is applied to the output of the neural network-based classifier. The proposed contextualization approach empowers a postprocessing step on top of the initial set of re-

---

[2]The combination of different contextual fuzzy relations toward the generation of a practically exploitable knowledge view is conducted by utilizing fuzzy algebraic operations in general and the default *t*-norm in particular.

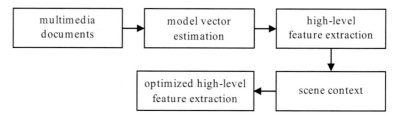

**FIGURE 12.8**

Contextual influence on high-level concepts.

gion types extracted. It provides an optimized re-estimation of the initial concepts' degrees of confidence for each region type and updates each model vector. In the process, it utilizes the high-level contextual knowledge from the constructed contextual ontology. The corresponding methodology is presented in Figure 12.8.

The degree of membership for each concept is estimated from direct and indirect relationships of the concept with other concepts using a meaningful compatibility indicator or distance metric. Again, depending on the nature of the domains provided in the domain ontology, the best indicator could be selected using the *max* or the *min* operator, respectively. Algorithm 12.1 depicts the general structure of the degree of membership re-evaluation procedure. The number of the iterations is defined empirically and usually three to five iterations are enough. The final output of the algorithm are the updated degrees of confidence for the presence of all concepts of the given domain within image $p$.

**ALGORITHM 12.1**  Procedure for evaluating the degree of membership.

1. The considered domain imposes the use of a domain similarity (or dissimilarity) measure $w_s \in [0,1]$.
2. For each image $p$, a fuzzy set $L_p$ with the degrees of confidence $\mu_p(c_i)$ is defined for all concepts $c_i$ of the domain.
3. For each $c_i$ in the fuzzy set $L_p$ with a degree of membership $\mu_p(c_i)$, the particular contextual information is obtained in the form of the set $\mathcal{U}_{c,i} = \{\mathcal{U}_{c,ij} : c_i, c_j \in C, \forall i \neq j\}$.
4. The new degree of membership $\mu'_p(c_i)$ is calculated by taking into account each domain's similarity measure. In the case of multiple concept relations in the ontology, when relating concept $c_i$ to more concepts apart the *root* concept (Figure 12.6), an intermediate aggregation step should be applied for the estimation of $\mu'_p(c_i)$ by considering the *context relevance* notion, $cr_i = \max_j\{\mathcal{U}_{c,ij}\}$, $j = 1, 2, \ldots, c_k$, defined in Reference [8]. The calculation of $\mu_p(c_i)$, which is the degree of confidence for $c_i$ at the $l$-th iteration, is expressed with the recursive formula

$$\mu_p^l(c_i) = \mu_p^{l-1}(c_i) - w_s(\mu_p^{l-1}(c_i) - cr_i).$$

Equivalently, for an arbitrary iteration $l$:

$$\mu_p^l(c_i) = (1 - w_s)^l \cdot \mu_p^0(c_i) + (1 - (1 - w_s)^l) \cdot cr_i,$$

where $\mu_p^0(c_i)$ represents the initial degree of membership for concept $c_i$.

## 12.5.2 Region Type Context

This section presents a different approach on scene context. The context optimization will have an effect on region types rather than concepts. It is relatively easy to prove that the utilization of this context information will improve the results of traditional image analysis. Indeed, initial analysis results are enhanced through the utilization of semantic knowledge, in terms of region-independent *region types* and semantic relations between them. In general, this information may be described by an intermediate description, which can be semantically described, but does not express the high-level concepts.

### 12.5.2.1 A Region Type Knowledge Model

The proposed methodology, that is presented in this section, follows precisely the steps of the scene context optimization from Section 12.5.1. Thus, an appropriate fuzzified ontology will be first defined in order to model in an appropriate way the real-world relations among the region types. In this case, the crisp ontology $O_T$ may be described as set $T$ of $m$ region types and a set $R_{T,ij}$ of semantic relations among them. More specifically, let:

1. $T = \{t_i\}$, for $i = 1, 2, \ldots, m$, be the set of all region types of the visual thesaurus used in the problem at hand, and

2. $R_T = \{R_{T,ij}\}$, for $i, j = 1, 2, \ldots, m$, be the set of the semantic relations among region types. Set $R_{T,ij}^{(k)}$, for $k = 1, 2, \ldots, K'$, includes $K'$ relations that can be defined between region types $t_i$ and $t_j$. Moreover, for a given relation $r_{T,ij}^{(k)}$, its inverse $\bar{r}_{T,ji}^{(k)}$ can be defined.

Thus, for a given ontology $O_T$ and sets $T$ and $R_{T,ij}$, the following can be written:

$$O_T = \{C, R_T\}, \tag{12.12}$$

$$R_{T,ij} : T \times T \to \{0, 1\}. \tag{12.13}$$

Now, the ontology modeling should be redefined to include fuzziness. A fuzzified version $\mathscr{O}_T$ of $O_T$ is defined as

$$\mathscr{O}_T = \{T, \mathscr{R}_T\}, \tag{12.14}$$

where $\mathscr{R}_T = \{\mathscr{R}_{T,ij}\}$ denotes the set of fuzzy semantic relations. Set $\mathscr{R}_{T,ij} = r_{T,ij}^{(k)}$, for $k = 1, 2, \ldots, K'$, includes all $K'$ relations that can be defined between two region types $t_i$ and $t_j$. Moreover, for each relation $r_{T,ij}^{(k)}$, its inverse $\bar{r}_{T,ji}^{(k)}$ can be defined. Finally, since relations are fuzzy, the following can be written:

$$r_{T,ij} : T \times T \to [0, 1]. \tag{12.15}$$

Since it is possible that more than one relation may be valid simultaneously between two region types, a combination of relations can be defined as

$$\mathscr{U}_{T,ij} = \bigcup_k [r_{T,ij}^{(k)}]^p, \ i, j = 1, 2, \ldots, N; \ k = 1, 2, \ldots, K'. \tag{12.16}$$

The value of $p$ is once again determined by the semantics of each relation $r_{ij}$ used to construct $\mathscr{U}_{,ij}$. More specifically:

- $p = 1$, if the semantics of $r_{T,ij}$ imply that it should be considered as is;

- $p = -1$, if the semantics of $r_{T,ij}$ imply the use of its inverse $\bar{r}_{T,ji}$; and

- $p = 0$, if the semantics of $r_{T,ij}$ do not allow its participation in the construction of the combined relation $\mathscr{U}_{T,ij}$.

### 12.5.2.2 Relations between Region Types

Once again, semantic relations defined by the MPEG-7 standard [83] are chosen and redefined to include fuzziness. The relations that may be applicable between region types are summarized in Table 12.2.

In this case, these relations may be calculated after a statistical analysis in an appropriate training set, that is, the one used to form the region thesaurus. To make their semantics and the calculations clear, a few indicative examples are presented below.

- *Similar* denotes that a region type is similar to another region type, under a certain degree of confidence. To calculate this degree, their low-level features should be compared using an appropriate similarity function.

- *Accompanier* denotes the degree to which two region types co-occur in an image. It is calculated as the percentage of the images in the training set that contain both region types to the images that contain either of them.

- *PartOf* denotes that a region type is part of another. This relation is defined by an expert, when this knowledge derives from observations to the visual thesaurus construction.

- *Combination* denotes that two region types are combined to form another region type. This is a special case where the inverse relation cannot be defined.

It becomes obvious that modeling region type context with an ontology leads again to the construction of an RDF graph (Figure 12.9). Its nodes correspond to region types and its edges to their combined relations. RDF reification [84] is used again here to estimate the corresponding degrees of confidence. This way RDF triplets are formed, for instance, *blue partOf green*, with a degree of confidence equal to 0.85. This triplet does not imply that a *blue* region type will *always* be part of a *green* region type.

The region types of the ontology are those of the region thesaurus that have been constructed for the visual analysis. The final ontology relations are formed after calculations among these regions.

**TABLE 12.2**

Contextual relations between region types.

| Relation | Inverse | Symbol | Meaning |
|---|---|---|---|
| Similar | Similar | $Sim(a,b)$ | region type $a$ is similar to region type $b$ |
| Accompanier | Accompanier | $Acc(a,b)$ | region type $a$ is accompanier of region type $b$ |
| Part | PartOf | $P(a,b)$ | region type $a$ is part of region type $b$ |
| Combination | – | $Comb(a,b)$ | combines two or more region types |

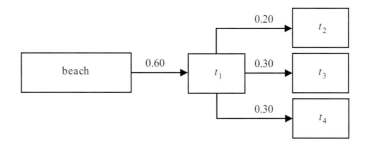

**FIGURE 12.9**

A fragment of the *Beach* region type knowledge model.

### 12.5.2.3 Region Type Context Optimization

After the construction of the model vector of an image $p$, an appropriately modified version of Algorithm 12.1 is applied. The algorithm now aims at refining the model vector by modifying its constituents, each corresponding to the degree of confidence for a region type. This is a preprocessing step that outputs an improved version of the model vector after considering the context of its region types. This leads to an increase in the detectors' precision, as the refined model vectors are closer to those used for their training.

To make this clear, a simplistic example is considered below. The *sea* detector in the *Beach* domain may have correlated this concept with the existence of a *blue*, a *light blue*, and a *brown* region, which corresponds to a typical *Beach* image that depicts *sea*, *sky*, and *sand*, respectively. If an image is presented where the region type that corresponds to *sea* is *green* (as in Figure 12.10) while the others remain as described before, the *sea* detector is certain to produce either a wrong result or a correct result with a small confidence, which will decrease overall precision.

Algorithm 12.2, which is a modified version of the previous algorithm, aims at this exact problem. In the given example, the model vector will be altered in a way that the confidence of the existence of a *blue* region type is increased, while the one of a *green* region type is decreased.

(a)                               (b)

**FIGURE 12.10  (See color insert.)**

An example from the *Beach* domain, where the region types of an image are different than a typical *Beach* image.

**ALGORITHM 12.2**   Improved procedure for evaluating the degree of membership.

1. The model vector for the image in question is calculated using the procedure presented in Section 12.4.3.

2. The considered domain imposes the use of a domain similarity (or dissimilarity) measure $w_T \in [0,1]$.

3. For each image $p$, the fuzzy set $L_T$ with the degrees of confidence $\mu_p(t_i)$ is defined for all region types $t_i$, with $i = 1, 2, \ldots, k'$ of the visual thesaurus.

4. For each region type $t_i$ in the fuzzy set $L$ with a degree of confidence $\mu_p(T_i)$, the particular contextual information is obtained in the form of the set $\mathcal{U}_{T,i} = \{\mathcal{U}_{T,ij} : t_i, t_j \in T, \forall i \neq j\}$.

5. The new degrees of confidence $\mu'_p(T_i)$ are calculated by taking into account the current domain's similarity measure. In the case of multiple concept relations, when $t_i$ is related with one or more types, apart from the *root* of the ontology, an intermediate aggregation step should be applied in order to calculate $\mu'_p(T_i)$ using the context relevance notion $cr_i$ defined in Reference [8] as $cr_i = \max_j\{\mathcal{U}_{T,ij}\}$, for $j = 1, 2, \ldots, c_k$. The calculation of $\mu'_p(t_i)$, which is the degree of confidence for $t_i$ at the $l$-th iteration, is expressed with the recursive formula

$$\mu_p^l(t_i) = \mu_p^{l-1}(t_i) - w_t(\mu_p^{l-1}(t_i) - cr_i).$$

Equivalently, for an arbitrary iteration $l$:

$$\mu_p^l(t_i) = (1 - w_t)^l \cdot \mu_p^0(t_i) + (1 - (1 - w_t)^l) \cdot cr_i,$$

where $\mu_p^0(c_i)$ represents the initial degree of confidence for $t_i$.

Figure 12.11 depicts a flowchart that describes the region type context. The number of the iterations is defined empirically and usually three to six iterations are enough also in this case. The final output of the algorithm is a refined model vector that is fed to the concept detectors instead of the one that is calculated by the visual features.

### 12.5.3   Unified Context

This section further advances the proposed conceptualization; it introduces a novel knowledge representation approach in the form of an extended mixed context model [74]. The classical notion of a contextual ontology is enhanced with *mid*-level concepts, that is, the region types and relations among different types of entities. These provide an intermediate description, which may be semantically described, but they do not express a high-level

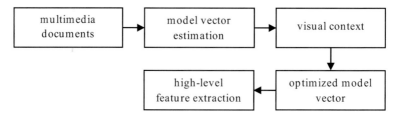

**FIGURE 12.11**

Contextual influence on region types.

nor a low-level concept. As a result, the focus here is on an integrated multimedia representation, combining efficiently low-level and high-level information and the description of a typical context model by defining new and expanding older relations. Within this section, both high-level concepts and region types will be simply referred to as *entities*.

### 12.5.3.1 A Unified Knowledge Model

This section describes a mixed fuzzified ontology that aims to model real-world relations among all entities present in images, that is, high-level concepts and region types. An ontology $O$ that will model the unified context of a given domain contains:

- $C = \{c_i\}$, for $i = 1, 2, \ldots, n$, which is the set of all high-level concepts of a given domain;

- $T = \{t_i\}$, for $i = 1, 2, \ldots, m$, which is the set of all region types of the visual thesaurus used in the analysis process; and

- $R = R_{ij}$, for $i, j = 1, 2, \ldots, n+m$, which is the set of all semantic relations among two entities $x_i$ and $x_j$. The set $R_{ij} = r_{ij}^{(k)}$, for $k = 1, 2, \ldots, K + K'$, includes at most $K + K'$ relations among $x_i$ and $x_j$.

Thus, for a unified context ontology $O$ and the aforementioned sets $C$, $R$, and $T$, the following can be written:

$$O = \{C, T, R_{ij}\}, \tag{12.17}$$

$$r_{ij}^{(k)} : (C \cup T) \times (C \cup T) \rightarrow \{0, 1\}, \quad i, j = 1, 2, \ldots m+n, \quad i \neq j. \tag{12.18}$$

As can be observed in Equation 12.18, unified context includes relations among concepts and region types. To model these relations as they exist in real-world problems, a fuzzified ontology $\mathscr{O}$ should be defined as follows:

$$\mathscr{O} = \{C, T, \mathscr{R}\}, \tag{12.19}$$

where $\mathscr{R}$ contains fuzzified relations among entities. As in the crisp ontology, $\mathscr{R} = \mathscr{R}_{ij}$. Set $\mathscr{R}_{ij} = r_{ij}^{(k)}$, for $k = 1, 2, \ldots, K + K'$, includes $K + K'$ among two entities $x_i$ and $x_j$. For a given relation $r_{ij}^{(k)}$, its inverse $\bar{r}_{ji}^{(k)}$ can be defined. Finally, $r_{ij}$ can be formally expressed as

$$r_{ij} : (C \cup T) \times (C \cup T) \rightarrow [0, 1]. \tag{12.20}$$

Since there often exist more than one relation among two entities, their combination is defined as

$$\mathscr{U}_{ij} = \bigcup_k [r_{ij}^{(k)}]^p, \quad i, j = 1, 2, \ldots, N; \; k = 1, 2, \ldots, K'. \tag{12.21}$$

This way allows constructing the model to be used in the analysis step. The value of $p$ is defined by the semantics of each relation $r_{ij}$. More specifically:

- $p = 1$, if the semantics of $r_{ij}$ imply that it should be considered as is;

- $p = -1$, if the semantics of $r_{ij}$ imply the use of its inverse $\bar{r}_{ji}$; and

- $p = 0$, if the semantics of $r_{ij}$ do not allow its participation in the construction of the combined relation $\mathscr{U}_{ij}$.

**TABLE 12.3**
Semantic relations used in unified context.

| Relation | Inverse | Symbol | Meaning | $C \times C$ | $T \times T$ | $C \times T$ |
|---|---|---|---|---|---|---|
| Similar | Similar | $Sim(a,b)$ | similarity between $a$ and $b$ | – | • | – |
| Accompanier | Accompanier | $Acc(a,b)$ | co-occurrence between $a$ and $b$ | • | • | • |
| Part | PartOf | $P(a,b)$ | $a$ is part of $b$ | • | • | • |
| Component | ComponentOf | $Comp(a,b)$ | $a$ is a component of $b$ | • | • | • |
| Specialization | Generalization | $Sp(a,b)$ | $b$ specializes the meaning of $a$ | • | – | – |
| Example | ExampleOf | $Ex(a,b)$ | $b$ is an example of $a$ | • | – | – |
| Location | LocationOf | $Loc(a,b)$ | $b$ is a location of $a$ | • | – | – |
| Property | PropertyOf | $Pr(a,b)$ | $b$ is a property of $a$ | – | • | • |

### 12.5.3.2 Relations between Entities

The relations between entities, as defined in Sections 12.5.1 and 12.5.2, are summarized in Table 12.1. Note that each entity may be related to another with more than one relation. However, it should be made clear that not all of the relations are appropriate for any two given entities. For instance, *Similar* cannot be defined between two concepts or a concept and a region type, whereas *sea* cannot be *Similar* to *sand* or to a *brown* region type. All applicable pairs of entities for each relation are summarized in Table 12.3.

The appropriate degrees of confidence of the semantic relations are either defined by an expert or calculated as described in Sections 12.5.1 and 12.5.2. For example:

- *Similar* may be defined only between two region types, with their visual similarity denoting the degree of confidence.

- *Accompanier* denotes the co-occurrence of any two entities in the same image. It should be noted here that a region type that co-occurs with a high degree of confidence with a concept does not necessarily depicts this concept. This degree is calculated statistically.

- *PartOf* is defined for any two given entities, when one is part of the other. For example, in case of two concepts, *sea* is *PartOf Beach*. In case of two region types, a *green and textured* region type is *PartOf* a *green* region type. Finally, in case of a concept and a region type, a *green* region type is *PartOf* a *tree*.

- *Component* is defined for any two given entities. For example, in case of two concepts, *tree* is a *Component* of *forest*, and in case of two region types, a *dark green* region type is a *Component* of a *green* region type. Finally, in case of a concept and a region type, an *orange* region type is a *Component* of a *sunset*. It should be noted here that there is also the case that a concept may be *Component* of a region type, due to undersegmentation. However, this case is not considered in this approach, since local annotation per image are unavailable.

- *Specialization* may be defined between two concepts, as defined in Section 12.5.1.

- *Example* may be defined between two concepts, as defined in Section 12.5.1.

- *Location* may be defined between two concepts, as defined in Section 12.5.1.

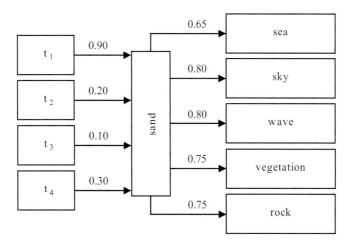

**FIGURE 12.12**

A fragment of unified context ontology that includes relations among *sand* and all other entities.

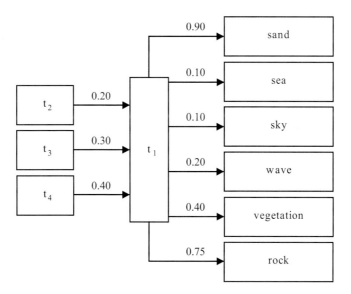

**FIGURE 12.13**

A fragment of unified context ontology that includes relations among region type $t_1$ and all other entities.

- *Property* may be defined between a concept and a region type, or between two concepts. In the first case, a *green* region type is a *Property* of *vegetation*. In the latter case, *wave* is a *Property* of *sea*.

The aforementioned relations model the unified context among all concepts and region types of a given domain. Between any two given entities $x_i$ and $x_j$ a single relation $\mathcal{U}_{ij}$ is formed and the occurring ontology $\mathcal{O}$ forms again an RDF graph, using RDF reification [85] to describe the degree of confidence for each edge. It is noted again that an edge between two entities is formed based on the set of valid relations for this pair. For example, the edge between *rock* and *sand* is formed by *Location* and *Accompanier*, while the edge between *water* and *sea* is formed by *Specialization, PartOf, Example,* and *Location.* Similarly, a *green* and a *blue* region types are related by the combination of *Similar, Accompanier,* and *Component,* while a *blue* region type and *sea* are related by the combination of *Accompanier, PartOf,* and *Component.*

Figures 12.12 and 12.13 present two fragments of the graph built for the *Beach* domain. More specifically, Figure 12.12 depicts relations among *sea* and other entities, while Figure 12.13 depicts relations between region type $T_4$ and other entities.

### 12.5.3.3 Unified Context Optimization

Algorithm 12.3 is used for the optimization in the case of the unified context. This algorithm is a mixture of the ones presented in Sections 12.5.1 and 12.5.2. The target now is to refine model vectors and detector results in an iterative way.

**ALGORITHM 12.3**   Optimized iterative procedure for evaluating the degree of membership in the case of the unified context.

1. The considered domain imposes the use of a domain similarity (or dissimilarity) measure $w_m \in [0,1]$.

2. For each image $p$, the fuzzy set $L_p$ with the degrees of confidence $\mu_p(c_i)$ is defined for all concepts $c_i$, with $i = 1, 2, \ldots, k$ of the given domain.

3. For each image $p$, the fuzzy set $L_T$ with the degrees of confidence $\mu_p(t_i)$ is defined for all region types $t_i$, with $i = 1, 2, \ldots, k'$ of the visual thesaurus.

4. For each concept $c_i$ in $L_p$ with a degree of confidence $\mu_p(c_i)$, the particular contextual information is obtained in the form of the set $\mathscr{U}_{c,i} = \{\mathscr{U}_{c,ij} : c_i, c_j \in C, \forall i \neq j\}$.

5. For each region type $t_i$ in $L$ with a degree of confidence $\mu_p(T_i)$, the particular contextual information is obtained in the form of the set $\mathscr{U}_{T,i} = \{\mathscr{U}_{T,ij} : t_i, t_j \in T, \forall i \neq j\}$.

6. The new degrees of confidence $\mu'_p(c_i)$ and $\mu'_p(T_i)$ are calculated by taking into account the similarity measure of the given domain. In the case of multiple concept relations, when $x_i$ is related with one or more types, apart from the root of the ontology, an intermediate aggregation step should be applied in order to calculate $\mu'_p(c_i)$ and $\mu'_p(T_i)$ by applying the context relevance notion [8], that is, $cr_i = \max_j\{\mathscr{U}_{ij}\}$, for $j = 1, 2, \ldots, c_k$. The calculation of $\mu'_p(c_i)$ and $\mu'_p(t_i)$, which correspond to the degrees of confidence for the presence of $c_i$ and $t_i$ at the $l$-th iteration of the algorithm, is expressed with the recursive formula

$$\mu_p^l(x_i) = \mu_p^{l-1}(x_i) - w_m(\mu_p^{l-1}(x_i) - cr_i). \tag{12.22}$$

Equivalently, for the $l$-th iteration:

$$\mu_p^l(x_i) = (1 - w_m)^l \cdot \mu_p^0(x_i) + (1 - (1 - w_m)^l) \cdot cr_i,$$

where $\mu_p^0(x_i)$ denotes the initial degree of confidence for $x_i$.

The number of the iterations is defined empirically and usually three to six iterations are enough also in this case. A flowchart that describes the influence of the unified context is depicted in Figure 12.14.

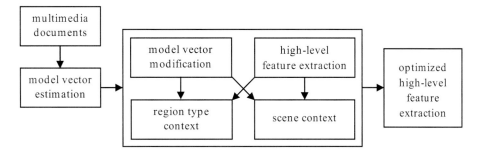

**FIGURE 12.14**

Unified contextual influence.

## 12.6 Context Optimization Examples

In order to clarify the influence of the context optimization on image analysis, this section presents simplistic examples for all three above cases.

### 12.6.1 Scene Context Example

Focused on scene context optimization, a simple example is presented to illustrate the way in which this optimization influences on the initial degrees of confidence. Based on the context ontology, whose fragments are depicted in Figures 12.6 and 12.7, and using appropriately trained detectors, Table 12.4 depicts degrees of confidence for all concepts before and after context optimization for the images shown in Figure 12.15. It can be observed that concepts detected with high confidence are considered to appear on images with a higher confidence. The opposite may be observed for concepts initially detected with low confidence.

### 12.6.2 Region Type Context Example

Next, a simple example is presented to illustrate region type context influence. In this case, context acts as a preprocessing step. For the image shown in Figure 12.10 and the

**TABLE 12.4**

Degrees of confidence before and after scene context optimization for the images shown in Figure 12.15.

| Concept | Figure 12.15a | | Figure 12.15b | | Figure 12.15c | |
|---|---|---|---|---|---|---|
| | Before | After | Before | After | Before | After |
| sea | 0.77 | 0.85 | 0.65 | 0.75 | 0.62 | 0.72 |
| water | 0.63 | 0.70 | 0.60 | 0.69 | 0.58 | 0.67 |
| vegetation | 0.35 | 0.43 | 0.35 | 0.40 | 0.62 | 0.72 |
| sky | 0.45 | 0.57 | 0.55 | 0.60 | 0.53 | 0.61 |
| sand | 0.69 | 0.75 | 0.45 | 0.56 | 0.52 | 0.60 |
| rock | 0.25 | 0.35 | 0.63 | 0.68 | 0.65 | 0.75 |
| wave | 0.00 | 0.00 | 0.25 | 0.34 | 0.20 | 0.27 |

**FIGURE 12.15 (See color insert.)**

Three examples from the *Beach* domain. Initial images and their segmentation maps.

simplistic ontology, a fragment of which is depicted in Figure 12.9, the model vector calculated from the visual features is as follows:

$$\mathbf{MV}_{before} = \begin{bmatrix} 0.723 \ 0.220 \ 0.753 \ 0.364 \end{bmatrix}. \tag{12.23}$$

As it can be observed from the example image, it depicts *sky* and *sea* and intuitively one would expect that it should contain region types similar to those of the region thesaurus. However, in this case, *sea* is significantly different, perhaps more similar to a *rock*. After region type optimization, the model vector takes the following form:

$$\mathbf{MV}_{after} = \begin{bmatrix} 0.778 \ 0.452 \ 0.800 \ 0.338 \end{bmatrix}. \tag{12.24}$$

Thus, the degree of confidence for the region type that corresponds to *sea* (2nd constituent) is increased while the one that corresponds to *rock* (4th constituent) is decreased.

### 12.6.3 Unified Context Example

Finally, in the unified context case, the ontology with fragments depicted in Figures 12.12 and 12.13 is used. The proposed algorithm is applied to the image shown in Figure 12.10. Model vector $T$ is initially set as

$$\mathbf{T} = \{T_i\} = \begin{bmatrix} 0.89 \ 0.62 \ 0.21 \ 0.68 \ 0.670.31 \end{bmatrix}, \tag{12.25}$$

while the degrees of confidence $c_i$ are

$$\mathbf{C} = \{c_i\} = \begin{bmatrix} 0.32 \ 0.91 \ 0.12 \ 0.87 \ 0.35 \end{bmatrix}. \tag{12.26}$$

As can be seen, the input image depicts *sea*, *sky*, and *wave*. However, the initial confidence for *sea* was low because no similar instances of *sea* were part of the training set. After unified context optimization, the improved value of the model vector $T'$ is

$$\mathbf{T}' = \{T_i'\} = \begin{bmatrix} 0.89 & 0.62 & 0.21 & 0.68 & 0.67 & 0.31 \end{bmatrix}, \tag{12.27}$$

and the degrees of confidence for all concepts are

$$\mathbf{C}' = \{c_i'\} = \begin{bmatrix} 0.62 & 0.95 & 0.18 & 0.90 & 0.29 \end{bmatrix}. \tag{12.28}$$

In brief, it should be emphasized that the unified context algorithm exploited the following information that was stored in the unified ontology:

- This was a *Beach* image, thus using the appropriate ontology.

- *Sky* was initially detected with a high confidence.

- *Wave* was initially detected with a high confidence.

- Image contains a *blue* region type.

- Image contains a *white* region type.

- *sky* and *wave* are related with a high degree with *sea*.

- *blue* and *white* region type are related with a high degree with *sea*.

Thus, the model vector and the degrees of confidence were modified in a way that:

- The confidence for the *blue* region type was increased.

- The confidence for the other region types remained invariable.

- The confidence for *sea* was increased.

- The confidence for the other concepts remained invariable.

## 12.7 Experimental Results

The following presents an indicative selection of experimental results. It includes results from the application of the proposed visual context utilization methodology, as presented in Sections 12.5.1.2 to 12.5.3.3. The utilized expert knowledge is rather ad hoc; however, this is not considered to be a liability, nor part of the discussed context model and is aligned to the current application datasets. More specifically, the evaluation focuses on both utilizing parts of the well-known Corel and TRECVID datasets and compares the efficiency of relevant state-of-the-art techniques.

**FIGURE 12.16** (See color insert.)

Indicative Corel images.

Initially, a set of experimental results are presented from the application of the proposed visual context approaches on a dataset containing 750 images, 40 region types, and 6 high-level concepts (*sea*, *vegetation*, *sky*, *sand*, *rock*, and *wave*). The number of the region types was selected based on the size of the region thesaurus and was verified using the minimum description length (MDL) method [28]. The amount and type of utilized concepts is imposed by the problem/dataset at hand; the employed dataset was a subset of the well-known Corel image collection [86], an indicative sample of which is presented in Figure 12.16. For the non-contextual detection of high-level concepts, the methodology described in Section 12.4 was applied. Overall, 525 images were used to train six individual SVM-based concept classifiers and 225 images were used as the test set.

Some additional results are also presented from the application of the proposed unified contextualization approach on a second dataset, consisting of 4000 images from the TRECVID collection [87], 100 region types and 7 high-level concepts (*vegetation*, *road*, *fire*, *sky*, *snow*, *sand*, *water*). The number of region types for this dataset was again decided based on experiments on the size of the region thesaurus and verified by using the minimum description length methodology introduced in Reference [28]. Figure 12.17 shows a characteristic sample of this dataset. In total, 250 of those images were used to train 7 individual SVM classifiers, whereas other 997 images were used for testing.

To evaluate the proposed approaches, they are compared to similar techniques used in previous research work. The results of all approaches on both Corel and TRECVID datasets are summarized in Tables 12.5 and 12.6. Note that *Region Types (RT)* refers to the results based only on the detection scheme presented in Section 12.4, without exploiting any contextual knowledge. Results from the application of contextual approaches correspond to *RT+Scene Context (SC)*, *RT+RT Context (RTC)*, and *RT+Unified Context* (UC).

To further evaluate the last proposed approach, two other techniques are implemented. The first technique, known as *relative LSA* (RLSA) [42], adds directly structural constraints

**FIGURE 12.17** (See color insert.)

Indicative TRECVID images.

to the visual words of the thesaurus. The fundamental difference between the traditional LSA and RSLA is that every possible unordered pair of clusters is this time considered as a visual word. In this way, a visual thesaurus with too many words (that is, pairs of clusters) is created. Nevertheless, the low-level features extracted from each region are simpler than the MPEG-7 low-level features used here. More specifically, a 64-bin histogram expressed in the hue-saturation-value (HSV) color space is used to capture the color features of 24 Gabor filters whose energies capture the texture features. The number of the words that form the visual thesaurus is determined empirically. The second implemented technique [88] starts with the extraction of *local interest points* (LIPs). The local interest points, often denoted as *salient*, tend to have significantly different properties compared to all other pixels in their neighborhood. To extract these points, a method called *difference of Gaussians* is applied. From each LIP, a SIFT descriptor is extracted from an elliptic region. A visual thesaurus is generated by an offline quantization of LIPs. Then, using this thesaurus, each image is described as a vector of visual keywords. Finally, for each high-level concept, a classifier is trained. It should be noted that the proposed method is compared against the above techniques and methodologies mainly because they try to face the same problem with more or less the same motivation as the presented work. The first one tries to exploit the co-occurrence of region types and to incorporate structural knowledge when building a visual thesaurus, while the the other one defines the LIPs as the regions of interest and extracts therein appropriate low-level descriptors. Moreover, both works have been successfully applied to the TRECVID dataset. Finally, as it is obvious from the interpretation of Tables 12.5 and 12.6, the proposed contextual unified approach outperforms in principle all compared approaches in terms of the achieved precision, whereas in some cases it lacks in terms of the recall criterion.

**TABLE 12.5**

Comparative precision $P$, recall $R$, and $F$-measure scores per concept for six different concept detection methodologies applied on the Corel dataset.

| Concepts | RT | | | SC | | | RTC | | | UC | | | LIPs [88] | | | RLSA [42] | | |
|---|---|---|---|---|---|---|---|---|---|---|---|---|---|---|---|---|---|---|
| | $P$ | $R$ | $F$ | $P$ | $R$ | $F$ | $P$ | $R$ | $F$ | $P$ | $R$ | $F$ | $P$ | $R$ | $F$ | $P$ | $R$ | $F$ |
| road | 0.22 | 0.40 | 0.28 | 0.25 | 0.37 | 0.30 | 0.39 | 0.32 | 0.35 | 0.43 | 0.35 | 0.39 | 0.34 | 0.37 | 0.35 | 0.42 | 0.35 | 0.38 |
| sand | 0.38 | 0.50 | 0.43 | 0.40 | 0.46 | 0.43 | 0.50 | 0.41 | 0.45 | 0.55 | 0.44 | 0.49 | 0.47 | 0.46 | 0.46 | 0.52 | 0.45 | 0.48 |
| sea | 0.72 | 0.85 | 0.78 | 0.71 | 0.81 | 0.76 | 0.81 | 0.78 | 0.79 | 0.89 | 0.80 | 0.84 | 0.77 | 0.83 | 0.80 | 0.80 | 0.82 | 0.81 |
| sky | 0.81 | 0.88 | 0.84 | 0.79 | 0.80 | 0.79 | 0.77 | 0.81 | 0.79 | 0.88 | 0.82 | 0.85 | 0.86 | 0.85 | 0.85 | 0.88 | 0.83 | 0.85 |
| snow | 0.48 | 0.68 | 0.56 | 0.51 | 0.62 | 0.56 | 0.62 | 0.57 | 0.59 | 0.72 | 0.57 | 0.64 | 0.58 | 0.61 | 0.59 | 0.64 | 0.57 | 0.60 |
| vegetation | 0.67 | 0.81 | 0.73 | 0.67 | 0.76 | 0.71 | 0.67 | 0.71 | 0.69 | 0.81 | 0.74 | 0.77 | 0.73 | 0.76 | 0.74 | 0.76 | 0.73 | 0.74 |
| Total: | 0.55 | 0.69 | 0.61 | 0.56 | 0.64 | 0.59 | 0.63 | 0.60 | 0.61 | 0.71 | 0.62 | 0.57 | 0.62 | 0.65 | 0.54 | 0.67 | 0.63 | 0.55 |

**TABLE 12.6**

Comparative precision $P$, recall $R$, and $F$-measure scores per concept for six different concept detection methodologies applied on the TRECVID dataset.

| Concepts | RT | | | SC | | | RTC | | | UC | | | LIPs [88] | | | RLSA [42] | | |
|---|---|---|---|---|---|---|---|---|---|---|---|---|---|---|---|---|---|---|
| | $P$ | $R$ | $F$ | $P$ | $R$ | $F$ | $P$ | $R$ | $F$ | $P$ | $R$ | $F$ | $P$ | $R$ | $F$ | $P$ | $R$ | $F$ |
| vegetation | 0.50 | 0.64 | 0.56 | 0.50 | 0.60 | 0.55 | 0.68 | 0.49 | 0.57 | 0.78 | 0.45 | 0.57 | 0.50 | 0.59 | 0.54 | 0.52 | 0.55 | 0.54 |
| road | 0.22 | 0.31 | 0.26 | 0.25 | 0.30 | 0.27 | 0.41 | 0.27 | 0.33 | 0.43 | 0.24 | 0.31 | 0.30 | 0.30 | 0.30 | 0.37 | 0.27 | 0.31 |
| sand | 0.83 | 0.82 | 0.81 | 0.87 | 0.80 | 0.83 | 0.93 | 0.69 | 0.79 | 1.00 | 0.76 | 0.86 | 0.93 | 0.76 | 0.84 | 0.94 | 0.71 | 0.81 |
| water | 0.60 | 0.67 | 0.63 | 0.57 | 0.68 | 0.62 | 0.70 | 0.58 | 0.63 | 0.81 | 0.57 | 0.67 | 0.60 | 0.66 | 0.63 | 0.61 | 0.64 | 0.63 |
| sky | 0.60 | 0.79 | 0.68 | 0.62 | 0.77 | 0.69 | 0.74 | 0.68 | 0.71 | 0.90 | 0.57 | 0.70 | 0.59 | 0.79 | 0.67 | 0.60 | 0.76 | 0.67 |
| snow | 0.43 | 0.50 | 0.46 | 0.40 | 0.44 | 0.42 | 0.51 | 0.38 | 0.44 | 0.57 | 0.37 | 0.44 | 0.50 | 0.44 | 0.47 | 0.56 | 0.40 | 0.47 |
| fire | 0.30 | 0.47 | 0.37 | 0.22 | 0.44 | 0.29 | 0.46 | 0.37 | 0.41 | 0.55 | 0.36 | 0.43 | 0.38 | 0.45 | 0.41 | 0.45 | 0.43 | 0.44 |
| Total: | 0.50 | 0.60 | 0.55 | 0.49 | 0.58 | 0.54 | 0.63 | 0.50 | 0.56 | 0.72 | 0.47 | 0.57 | 0.54 | 0.57 | 0.55 | 0.58 | 0.54 | 0.55 |

Experimental results presented above show that existing relationships between concepts improve the precision of results, not only for the well trained, but also for weak SVM classifiers. The proposed unified context algorithm uses and exploits these relations and provides an expanded view of the research problem, which is based on a set of meaningful semantic relations. The interpretation of presented experimental results depicts that the proposed contextualization approach will favor rather certain degrees of confidence for the detection of a concept that exists within an image. On the contrary, it will also discourage rather uncertain or misleading degrees. It will strengthen the concepts' differences, but it will treat smoothly almost certain concepts' confidence values. Finally, based on the constructed knowledge, the algorithm is able to disambiguate cases of similar concepts or concepts being difficult to be detected from the simple low-level analysis steps.

## 12.8   Conclusion

Research effort summarized in this chapter clearly indicates that high-level concepts can be efficiently detected when an image is represented by a model vector with the aid of a visual thesaurus and visual context. The role of the latter is crucial and significantly aids the image analysis process. The core contributions of this work include, among others, the implementation of a novel threefold visual context interpretation utilizing a fuzzy, ontology-based representation of knowledge. Experimental results presented in this chapter indicate significant high-level concept detection optimization over the entire datasets. Although the improvement is not considered to be impressive, it is believed that the proposed approach successfully incorporates the underlying contextual knowledge and further exploits visual context in the multimedia analysis value chain. Moreover, minor enhancements of the implemented contextual model, for example, in terms of additional spatial, temporal, or semantic relationships exploitation, would further boost its performance.

## Acknowledgment

The work presented in this book chapter was partially supported by the European Commission under contract FP7-215453 WeKnowIt.

## References

[1]  I. Biederman, R. Mezzanotte, and J. Rabinowitz, "Scene perception: Detecting and judging objects undergoing relational violations," *Cognitive Psychology*, vol. 14, no. 2, pp. 143–177, April 1982.

[2] A. Smeulders, M. Worring, S. Santini, A. Gupta, and R. Jain, "Content-based image retrieval at the end of the early years," *IEEE Transactions on Pattern Analysis and Machine Intelligence*, vol. 22, no. 12, pp. 1349–1380, December 2000.

[3] K. Rapantzikos, Y. Avrithis, and S. Kollias, "On the use of spatiotemporal visual attention for video classification," in *Proceedings of the International Workshop on Very Low Bitrate Video Coding*, Sardinia, Italy, September 2005.

[4] G. Tsechpenakis, G. Akrivas, G. Andreou, G. Stamou, and S. Kollias, "Knowledge-assisted video analysis and object detection," in *Proceedings of the European Symposium on Intelligent Technologies, Hybrid Systems and their Implementation on Smart Adaptive Systems*, Algarve, Portugal, September 2002, pp. 497–504.

[5] A. Benitez and S. Chang, "Image classification using multimedia knowledge networks," in *Proceedings of the IEEE International Conference on Image Processing*, Barcelona, Spain, September 2003, pp. 613–616.

[6] S. Staab and R. Studer, *Handbook on Ontologies*. New York, USA: Springer Verlag, 2004.

[7] A. Mathes, "Folksonomies-cooperative classification and communication through shared metadata," *Computer Mediated Communication*, vol. 47, pp. 1–28, December 2004.

[8] P. Mylonas, T. Athanasiadis, and Y. Avrithis, "Image analysis using domain knowledge and visual context," in *Proceedings of the 13th International Conference on Systems, Signals and Image Processing*, Budapest, Hungary, September 2006, pp. 483–486.

[9] J. Luo, A. Singhal, and W. Zhu, "Natural object detection in outdoor scenes based on probabilistic spatial context models," in *Proceedings of the IEEE International Conference on Multimedia and Expo*, Baltimore, USA, July 2003, pp. 457–460.

[10] B. Russell, A. Torralba, K. Murphy, and W. Freeman, "LabelMe: A database and Web-based tool for image annotation," *International Journal of Computer Vision*, vol. 77, no. 1, pp. 157–173, May 2008.

[11] M. Everingham, L. Van Gool, C.K.I. Williams, J. Winn, and A. Zisserman, "The PASCAL visual object classes challenge 2008 results." *International Journal of Computer Vision*, vol. 88, no. 2, June 2010, pp. 303–338.

[12] L. Kennedy, A. Hauptmann, M. Naphade, A. Smith, and S. Chang, "LSCOM lexicon definitions and annotations version 1.0," in *Proceedings of the DTO Challenge Workshop on Large Scale Concept Ontology for Multimedia*, ADVENT Technical Report 217-2006-3, New York, USA, March 2006.

[13] S. Ayache and G. Quenot, "TRECVID 2007 collaborative annotation using active learning," in *Proceedings of the TRECVID Workshop*, Gaithersburg, MD, USA, November 2007.

[14] C. Cusano, "Region-based annotation of digital photographs," in *Proceedings of the Computational Color Imaging Workshop*, Milan, Italy, April 2011, pp. 47–59.

[15] N. Aslam, J. Loo, and M. Loomes, "Adding semantics to the reliable object annotated image databases," *Procedia Computer Science*, vol. 3, pp. 414–419, February 2011.

[16] M. Wang and X. Hua, "Active learning in multimedia annotation and retrieval: A survey," *ACM Transactions on Intelligent Systems and Technology*, vol. 2, no. 2, pp. 1–21, February 2011.

[17] J. Vogel and B. Schiele, "A semantic typicality measure for natural scene categorization," in *Proceedings of DAGM Pattern Recognition Symposium*, vol. 3175, September 2004, pp. 195–203.

[18] L. Fei-Fei and P. Perona, "A Bayesian hierarchical model for learning natural scene categories," in *Proceedings of the IEEE Conference on Computer Vision and Pattern Recognition*, San Diego, CA, USA, June 2005, pp. 524–531.

[19] S. Ullman and M. Vidal-Naquet, "Visual features of intermediate complexity and their use in classification," *Nature Neuroscience*, vol. 5, no. 7, pp. 682–687, June 2002.

[20] G. Csurka, C. Dance, L. Fan, J. Willamowski, and C. Bray, "Visual categorization with bags of keypoints," in *Proceedings of the International Workshop on Statistical Learning in Computer Vision*, Prague, Czech Republic, May 2004.

[21] J. Sivic, B. Russell, A. Efros, A. Zisserman, and W. Freeman, "Discovering objects and their location in images," in *Proceedings of the IEEE International Conference on Computer Vision*, Beijing, China, October 2005, pp. 370–377.

[22] K. Barnard, P. Duygulu, D. Forsyth, N. de Freitas, D. Blei, and M. Jordan, "Matching words and pictures," *The Journal of Machine Learning Research*, vol. 3, pp. 1107–1135, February 2003.

[23] S. Chang, T. Sikora, and A. Purl, "Overview of the MPEG-7 standard," *IEEE Transactions on Circuits and Systems for Video Technology*, vol. 11, no. 6, pp. 688–695, June 2001.

[24] D. Lowe, "Object recognition from local scale-invariant features," in *Proceedings of the 7th IEEE International Conference on Computer Vision*, Corfu, Greece, September 1999, pp. 1150–1157.

[25] H. Bay, T. Tuytelaars, and L. Van Gool, "SURF: Speeded up robust features," *Lecture Notes in Computer Science*, vol. 3951, pp.404–417, February 2006.

[26] J. Hartigan and M. Wong, "A k-means clustering algorithm," *Applied Statistics*, vol. 28, no. 1, pp. 100–108, 1979.

[27] D. Nister and H. Stewenius, "Scalable recognition with a vocabulary tree," in *Proceedings of the IEEE Conference on Computer Vision and Pattern Recognition*, New York, USA, June 2006, pp. 2161–2168.

[28] S. Kim and I. Kweon, "Simultaneous classification and visual word selection using entropy-based minimum description," in *Proceedings of the International Conference on Pattern Recognition*, Hong Kong, August 2006, pp. 650–653.

[29] O. Cula and K. Dana, "Compact representation of bidirectional texture functions," in *Proceedings of the IEEE Conference on Computer Vision and Pattern Recognition*, Kauai, HI, USA, December 2001, pp. 1041–1047.

[30] T. Leung and J. Malik, "Representing and recognizing the visual appearance of materials using three-dimensional textons," *International Journal of Computer Vision*, vol. 43, no. 1, pp. 29–44, June 2001.

[31] B. Le Saux and G. Amato, "Image recognition for digital libraries," in *Proceedings of the ACM SIGMM International Workshop on Multimedia Information Retrieval*, New York, USA, October 2004, pp. 91–98.

[32] D. Gokalp and S. Aksoy, "Scene classification using bag-of-regions representations," in *Proceedings of the IEEE Conference on Computer Vision and Pattern Recognition*, Minneapolis, MN, USA, June 2007, pp. 1–8.

[33] J. Smith, M. Naphade, and A. Natsev, "Multimedia semantic indexing using model vectors," in *Proceedings of the International Conference on Multimedia and Expo*, Baltimore, MD, USA, July 2003, pp. 445–448.

[34] A. Opelt, A. Pinz, and A. Zisserman, "Incremental learning of object detectors using a visual shape alphabet," in *Proceedings of the IEEE Conference on Computer Vision and Pattern Recognition*, New York, USA, June 2006, pp. 3–10.

[35] S. Lazebnik, C. Schmid, and J. Ponce, "Beyond bags of features: Spatial pyramid matching for recognizing natural scene categories," in *Proceedings of the IEEE Conference on Computer Vision and Pattern Recognition*, New York, USA, June 2006, pp. 2169–2178.

[36] L. Zhu and A. Zhang, "Theory of keyblock-based image retrieval," *ACM Transactions on Information Systems*, vol. 20, no. 2, pp. 224–257, April 2002.

[37] S. Savarese, J. Winn, and A. Criminisi, "Discriminative object class models of appearance and shape by correlatons," in *Proceedings of the IEEE Conference on Computer Vision and Pattern Recognition*, New York, USA, June 2006, pp. 2033–2040.

[38] E. Sudderth, A. Torralba, W. Freeman, and A. Willsky, "Learning hierarchical models of scenes, objects, and parts," in *Proceedings of the IEEE International Conference on Computer Vision*, Beijing, China, October 2005, pp. 1331–1338.

[39] J. Niebles and L. Fei-Fei, "A hierarchical model of shape and appearance for human action classification," in *Proceedings of the IEEE Conference on Computer Vision and Pattern Recognition*, Minneapolis, MN, USA, June 2007, pp. 1–8.

[40] B. Leibe and B. Schiele, "Interleaved object categorization and segmentation," in *Proceedings of the British Machine Vision Conference*, Norwich, UK, September 2003, pp. 759–768.

[41] S. Deerwester, S. Dumais, G. Furnas, T. Landauer, and R. Harshman, "Indexing by latent semantic analysis," *Journal of the American Society for Information Science*, vol. 41, no. 6, pp. 391–407, September 1990.

[42] F. Souvannavong, B. Merialdo, and B. Huet, "Region-based video content indexing and retrieval," in *International Workshop on Content-Based Multimedia Indexing*, Riga, Latvia, June 2005.

[43] A. Bosch, A. Zisserman, and X. Munoz, "Scene classification using a hybrid generative/discriminative approach," *IEEE Transactions on Pattern Analysis and Machine Intelligence*, vol. 30, no. 4, pp. 712–727, April 2008.

[44] A. Torralba, A. Oliva, M. Castelhano, and J. Henderson, "Contextual guidance of eye movements and attention in real-world scenes: The role of global features in object search," *Psychological Review*, vol. 113, no. 4, pp. 766–786, October 2006.

[45] M. Naphade and J. Smith, "A hybrid framework for detecting the semantics of concepts and context," *Lecture Notes in Computer Science*, vol. 2728, pp. 196–205, Springer, 2003.

[46] J. Fan, Y. Gao, and H. Luo, "Multi-level annotation of natural scenes using dominant image components and semantic concepts," in *Proceedings of the ACM International Conference on Multimedia*, New York, USA, ACM 2004, pp. 540–547.

[47] R. Yan, M. Chen, and A. Hauptmann, "Mining relationship between video concepts using probabilistic graphical model," in *Proceedings of the IEEE International Conference on Multimedia and Expo*, Toronto, ON, Canada, pp. 301–304.

[48] K. Murphy, A. Torralba, and W. Freeman, "Using the forest to see the trees: A graphical model relating features, objects and scenes," *Advances in Neural Information Processing Systems*, vol. 16, 2003.

[49] P. Lipson, E. Grimson, and P. Sinha, "Configuration based scene classification and image indexing," in *Proceedings of the IEEE Conference on Computer Vision and Pattern Recognition*, San Juan, Puerto Rico, June 1997, pp. 1007–1013.

[50] A. Singhal, J. Luo, and W. Zhu, "Probabilistic spatial context models for scene content understanding," in *Proceedings of the IEEE Conference on Computer Vision and Pattern Recognition*, Madison, WI, USA, June 2003, pp. 235–241.

[51] P. Carbonetto, N. de Freitas, and K. Barnard, "A statistical model for general contextual object recognition," *Lecture Notes in Computer Science*, vol. 3021, pp. 350–362, May 2004.

[52] W. Li and M. Sun, "Semi-supervised learning for image annotation based on conditional random fields," *Lecture Notes in Computer Science*, vol. 4071, pp. 463–472, July 2006.

[53] J. Li, A. Najmi, and R. Gray, "Image classification by a two-dimensional hidden Markov model," *IEEE Transactions on Signal Processing*, vol. 48, no. 2, pp. 517–533, February 2000.

[54] J. Jiten, B. Merialdo, and B. Huet, "Semantic feature extraction with multidimensional hidden Markov model," *Proceedings of SPIE*, vol. 6073, pp. 211–221, January 2006.

[55] J. Yuan, J. Li, and B. Zhang, "Exploiting spatial context constraints for automatic image region annotation," in *Proceedings of the ACM International Conference on Multimedia*, Ausburg, Germany, September 2007, pp. 595–604.

[56] M. Boutell, C. Brown, and J. Luo, "Learning spatial configuration models using modified Dirichlet priors," in *Proceedings of the ICML Workshop on Statistical Relational Learning and Its Connections to Other Fields*, Banff, AB, Canada, July 2004, pp. 29–34.

[57] M. Boutell, J. Luo, and C. Brown, "Improved semantic region labeling based on scene context," in *Proceedings of the IEEE International Conference on Multimedia and Expo*, Amsterdam, the Netherlands, July 2005, pp. 980–983.

[58] M. Boutell, J. Luo, and C. Brown, "A generalized temporal context model for classifying image collections," *Multimedia Systems*, vol. 11, no. 1, pp. 82–92, November 2005.

[59] L. Paletta, M. Prantl, and A. Pinz, "Learning temporal context in active object recognition using Bayesian analysis," in *Proceedings of the International Conference on Pattern Recognition*, Barcelona, Spain, September 2000, pp. 695–699.

[60] D. Moldovan, C. Clark, and S. Harabagiu, "Temporal context representation and reasoning," in *Proceedings of the International Joint Conference on Artificial Intelligence*, Edinburgh, Scotland, August 2005, pp. 1099–1104.

[61] N. O'Hare, C. Gurrin, H. Lee, N. Murphy, A. Smeaton, and G. Jones, "My digital photos: Where and when?," in *Proceedings of the 13th Annual ACM International Conference on Multimedia*, Singapore, November 2005, pp. 261–262.

[62] J. Pauty, P. Couderc, and M. Banâtre, "Using context to navigate through a photo collection," in *Proceedings of the ACM International Conference on Human Computer Interaction with Mobile Devices and Services*, Salzburg, Austria, September 2005, pp. 145–152.

[63] P. Mulhem and J. Lim, "Home photo retrieval: Time matters," *Lecture Notes in Computer Science*, vol. 2728, pp. 321–330, July 2003.

[64] Japan Electronics and Information Technology Industries Association, "Exchangeable image file format for digital still cameras: Exif Version 2.2," Technical report, JEITA CP-3451, April 2002.

[65] P. Sinha and R. Jain, "Classification and annotation of digital photos using optical context data," in *Proceedings of the International Conference on Content-Based Image and Video Retrieval*, Niagara Falls, ON, Canada, July 2008, pp. 309–318.

[66] X. Liu, L. Zhang, M. Li, H. Zhang, and D. Wang, "Boosting image classification with LDA-based feature combination for digital photograph management," *Pattern Recognition*, vol. 38, no. 6, pp. 887–901, June 2005.

[67] M. Tuffield, S. Harris, D. Dupplaw, A. Chakravarthy, C. Brewster, N. Gibbins, K. O Hara, F. Ciravegna, D. Sleeman, and N. Shadbolt, "Image annotation with photocopain," in *Proceedings of the Semantic Web Annotation of Multimedia Workshop at the World Wide Web Conference*, Edinburgh, Scotland, May 2006.

[68] M. Boutell and J. Luo, "Bayesian fusion of camera metadata cues in semantic scene classification," in *Proceedings of the IEEE Conference on Computer Vision and Pattern Recognition*, Washington, DC, USA, July 2004, vol. II, pp. 623–630.

[69] M. Boutell and J. Luo, "Beyond pixels: Exploiting camera metadata for photo classification," *Pattern Recognition*, vol. 38, no. 6, pp. 935–946, June 2005.

[70] M. Boutell and J. Luo, "Photo classification by integrating image content and camera meta-data," in *Proceedings of the International Conference on Pattern Recognition*, Cambridge, UK, August 2004, vol. 4, pp. 901–904.

[71] S. Boll, P. Sandhaus, A. Scherp, and S. Thieme, "MetaXa – context- and content-driven meta-data enhancement for personal photo books," in *Proceedings of the International Multi-Media Modeling Conference*, Singapore, January 2007, pp. 332–343.

[72] E. Spyrou and Y. Avrithis, "A region thesaurus approach for high-level concept detection in the natural disaster domain," in *Proceedings of the International Conference on Semantic and Digital Media Technologies*, Genova, Italy, December 2007, pp. 74–77.

[73] P. Mylonas, E. Spyrou, and Y. Avrithis, "Enriching a context ontology with mid-level features for semantic multimedia analysis," in *Proceedings of the 1st Workshop on Multimedia Annotation and Retrieval Enabled by Shared Ontologies*, Genova, Italy, December 2007, pp. 16–30.

[74] P. Mylonas, E. Spyrou, and Y. Avrithis, "High-level concept detection based on mid-level semantic information and contextual adaptation," in *Proceedings of the 2nd International Workshop on Semantic Media Adaptation and Personalization*, London, UK, December 2007, pp. 193–198.

[75] E. Spyrou, P. Mylonas, and Y. Avrithis, "Semantic multimedia analysis based on region types and visual context," in *Proceedings of the Artificial Intelligence and Innovations: From Theory to Applications*, Athens, Greece, September 2007, pp. 389–398.

[76] E. Spyrou, G. Tolias, P. Mylonas, and Y. Avrithis, "A semantic multimedia analysis approach utilizing a region thesaurus and LSA," in *Proceedings of the Ninth International Workshop on Image Analysis for Multimedia Interactive Services*, Klagenfurt, Austria, May 2008, pp. 8–11.

[77] Y. Avrithis, A. Doulamis, N. Doulamis, and S. Kollias, "A stochastic framework for optimal key frame extraction from MPEG video databases," *Computer Vision and Image Understanding*, vol. 75, no. 1, pp. 3–24, July 1999.

[78] G. Tolias, *VDE: Visual descriptor extraction*, 2008. Available online, http://image.ntua.gr/smag/tools/vde.

[79] A. Yamada, M. Pickering, S. Jeannin, L. Cieplinski, J. Ohm, and M. Kim, "MPEG-7 visual part of experimentation model version 9.0 ISO/IEC JTC1/SC29/WG11/N3914," International Organisation for Standardisation ISO, 2001, pp. 1–83.

[80] E. Spyrou, H. Le Borgne, T. Mailis, E. Cooke, Y. Avrithis, and N. O'Connor, "Fusing MPEG-7 visual descriptors for image classification," *Proceedings of the Artificial Neural Networks: Formal Models and Their Applications-ICANN*, Warsaw, Poland, September 2005, pp. 847–852.

[81] P. Mylonas, M. Wallace, and S. Kollias, "Using k-nearest neighbor and feature selection as an improvement to hierarchical clustering," *Lecture Notes in Artificial Intelligence*, vol. 3025, pp. 191–200, Springer 2004.

[82] P. Mylonas and Y. Avrithis, "Context modeling for multimedia analysis," in *Proceedings of the International and Interdisciplinary Conference on Modeling and Using Context*, Paris, France, July 2005.

[83] A. Benitez, D. Zhong, S. Chang, and J. Smith, "MPEG-7 MDS content description tools and applications," *Lecture Notes in Computer Science*, vol. 2124, pp. 41–52, September 2001.

[84] W3C, "RDF Reification," 2004. Available online, http://www.w3.org/TR/rdf-schema/#ch\_reificationvocab.

[85] D. Beckett and B. McBride, "RDF/XML syntax specification (revised)," W3C Recommendation, vol. 10, 2004.

[86] J. Wang, J. Li, and G. Wiederhold, "SIMPLIcity: Semantics-sensitive integrated matching for picture libraries," *IEEE Transactions on Pattern Analysis and Machine Intelligence*, vol. 23, no. 9, pp. 947–963, September 2001.

[87] A.F. Smeaton, P. Over, and W. Kraaij, "Evaluation campaigns and trecvid," in *Proceedings of the ACM International Workshop on Multimedia Information Retrieval*, Santa Barbara, CA, USA, October 2006, pp. 321–330.

[88] Y. Jiang, W. Zhao, and C. Ngo, "Exploring semantic concept using local invariant features," in *Proceedings of the Asia-Pacific Workshop on Visual Information Processing*, Beijing, China, November 2006.

# 13

# *Perceptually Driven Video Shot Characterization*

**Gaurav Harit and Santanu Chaudhury**

## 13.1 Introduction

Analyzing a video shot requires identifying the meaningful components that influence the semantics of the scene through their behavioral and perceptual attributes. This chapter describes an unsupervised approach to identify meaningful components in the scene using perceptual grouping and perceptual prominence principles. The perceptual grouping approach is unique in the sense that it makes use of an organizational model that encapsulates the criteria that govern the grouping process. The proposed grouping criteria rely on spatiotemporal consistency exhibited by emergent clusters of grouping primitives. The principle of perceptual prominence models the cognitive saliency of the subjects. Prominence is interpreted by formulating an appropriate model based on attributes which commonly influence human judgment about prominent subjects. The video shot is categorized based on the observations pertaining to the mise-en-scène aspects of the depicted content.

Perceptual organization is the process of identifying inherent organizations in visual primitives. Simpler primitives can be organized to obtain higher-level meaningful structures. In order to group simpler primitives one needs to formulate a grouping model that specifies the grouping criteria that influence the formation of meaningful (perceptual) clusters. For example, a model of a human being can be specified as a spatial organization of parts, such as torso, head, legs, hands, and so on. Being able to detect the parts and their requisite spatial organization reinforces the belief that the human being is present in the scene. For a general video the grouping primitives need to be simple enough to be extracted from all types of objects. Further, the organization model should make use of general grouping criteria and be flexible enough to incorporate grouping evidences from an object category-specific grouping model, if available. One also needs unsupervised algorithmic procedures that can make use of the grouping model to identify salient clusters.

The principle of perceptual prominence [1] is motivated from film grammar that states that scene understanding depends on the characteristics of the scene components. Human attention gets focused on the most over-riding or prominent components in the scene. A film-maker or a painter uses the arrangement of the mise-en-scène [2] to direct visual attention across the visualization space (a scene in space and time). Perceptual prominence captures the cognitive aspects of attention, which deals with perceiving the perceptible (visual) attributes in the context of prior knowledge or expectation from the scene, and is used to identify the perceived prominent subjects. The semantics of a scene are closely linked with the way humans interpret the scene and thereby establish the visual attributes and behavioral patterns of the prominent subjects. The perceptual attribute specifications are parametrized into different prominent models that correspond to different interpretations of prominence. The behavioral aspects of prominent subjects and the mise-en-scène features are used to generate a taxonomy of interpretation of the video scene.

Many classical perceptual organization studies, which have addressed organizing and grouping primitives in images, have been recently extended to organizing primitives in the spatiotemporal domain of video. Given a pair of frames, the organization principle of smoothness (or continuity) has been used for computing optical flow, organizing moving points for planar motion [3], and estimating the dense motion field in a tensor voting framework [4]. Organization of motion fields to detect patterns, such as saddles, sinks, and sources, has been used to study qualitative behaviors of objects [5], [6]. Perceptual organization has been applied to primitives that exist over extended sequence of frames in different applications like grouping feature trajectories [7], tracking structures [8], and object extraction [9].

Attempts to quantify the saliency of organizational structures have made use of different properties like transformational invariance [10], structural information [11], local characteristics [12], and probability of occurrence of the structure [13]. There has been work on modeling pre-attentive vision [14] and computation of visual saliency [15]. Reference [16], has used perceptual features of the groupings and computed qualitatively significant perceptual groupings by repeated applications of rules. The "interestingness" of a structure is evaluated by considering its size, contrast, extent, type of similarity, and the number of groupings the structure is associated with.

This chapter presents a perceptual grouping scheme for the spatiotemporal domain of video. The proposed grouping model is motivated from the Gestalt law of common fate that states that the entities having a coherent behavior in time are likely to be parts of the same organization. The researchers in the Gestalt school of psychophysics [17], [18], [19] have identified certain relationships, such as continuation of boundaries, proximity, adjacency [20], enclosures, and similarity of shape, size, intensity, and directionality, which play the key role in the formation of "good" structures in the spatial domain. These associations have been used to identify salient groupings of visual patterns (primitive entities) in a two-dimensional visual domain [21]. However, in the spatiotemporal domain, the spatially salient grouping must show coherence (common fate) in the temporal domain. Therefore, this chapter also presents formulations for association measures which characterize temporal coherence between primitives to group them into salient spatiotemporal clusters, termed as perceptual clusters. The foreground perceptual clusters identified by the proposed perceptual grouping algorithm correspond to the subjects in the scene.

Semantic information from a video is extracted here in the form of a collection of concepts interpreted by analyzing the behavior and properties of the identified meaningful components. In order to do so, video shots are categorized into two broad categories:

1. *Subject-centric scenes*, which have one or few prominent subjects and are more suitable for object-centric descriptions that facilitate an object-oriented search [22], [23].

2. *Frame-centric scenes*, which have many subjects, but none of them can be attributed a high prominence. Hence, the overall activity depicted in the scene is the subject of interest. Such scenes can be described in terms of features and semantic constructs that do not directly pertain to the objects but characterize the gross expressiveness of the medium at the frame level or the shot level, for example, features like color histogram, dynamicity [24], tempo [25], dialogue [26], etc.

Note that the annotations for a given video scene would be more efficacious when they align with the viewer's perception of the content depicted in the scene. Therefore, perceptual grouping and prominence principles are employed here toward the end goal of automatic semantic annotation of videos into general categories. Content specific semantic metadata can facilitate effective utilization of video information in various applications, such as transcoding, browsing, navigation, and querying.

This chapter is organized as follows. Section 13.2 gives an overview of proposed clustering methodology to extract space-time blob tracks, which are the visual primitives for grouping. Extraction of meaningful components in a video is done by applying perceptual organization principles on visual primitives. Section 13.3 describes the basic ideas behind spatiotemporal grouping and outlines the formulations of association measures that play a key role in formation of salient groupings. Section 13.4 describes the proposed spatiotemporal grouping model which is then used in a perceptual grouping algorithm to identify the foreground subjects in the video shot. Section 13.5 reports results obtained by applying the perceptual grouping algorithm to video scenes. Section 13.6 provides a general framework to evaluate perceptual prominence of the identified subjects by making use of perceptual attributes such as life span and movement, and mise-en-scène features like placement in the scene, and color contrast. The attribute specifications are parametrized into different *prominence models* that signify different interpretations of prominence. A taxonomy of interpretation of video scenes based on the behavior of prominent subjects is also generated. Video shot interpretation into generic categories follows from the analysis of prominent subjects in the scene. Section 13.7 uses domain knowledge in the form of learned object-part models to perform object recognition through a grouping process. This chapter concludes with Section 13.8.

## 13.2  Space-Time Blob Tracks

The grouping primitives are identified here as space-time regions, homogeneous in color using unsupervised clustering techniques, and modeled as tracks of two-dimensional blobs along time. Video data are multidimensional; they can be represented using two spatial coordinates $(x, y)$, three color components (for example, RGB or LUV), and one time dimension. This makes a heterogeneous feature vector because color, space, and time have different semantics. Clustering such a six-dimensional feature space can result in undesirable smoothing across features of different semantics. Pixels having similar color but actually lying on different objects may become part of the same cluster. This leads to a higher variance along certain dimensions, which may be semantically unacceptable because the cluster is encompassing patches that belong to different objects. To ensure that the clusters model patches that are homogeneous in color and belong to the same object, one can apply clustering in stages and consider a subspace of semantically homogeneous features at each stage. Each cluster is divided into further subclusters at the subsequent level, resulting in a hierarchical clustering tree called the *decoupled semantics clustering tree* (DSCT) [27]. The clustering methodology and the feature subspace can be chosen as different for each level. The choice is guided by the distribution of feature points in the feature space used for partitioning the cluster.

The root node (level 0) of the clustering tree represents all the feature points, that is, pixels, in the video shot. At level 1 of the clustering tree, the entire set of pixels is partitioned along time to form video stacks or subclusters, each comprising ten consecutive video frames. Making such video stacks or sets of consecutive frames reduces the amount of data handled by the clustering algorithm at the next level. At level 2, pixels in the video stack are decomposed into regions homogeneous in color. This is done by applying hierarchical mean shift [28] to obtain clusters in the color subspace, such as the LUV color space. Hierarchical mean shift results in a dendrogram of color modes. A set of color modes are chosen as the color model. Each color cluster comprises the set of pixels that had converged to the mode during the mean shift process. At level 3, the pixels in the color cluster are partitioned along time by projecting them onto the individual frames. The clusters at this level comprise pixels having similar color and belonging to the same frame. Pixels in clusters at level 3 are modeled as two-dimensional Gaussians (blobs) using Gaussian mixture model clustering. These blobs are then tracked across successive frames to obtain blob tracks that model space-time regions. These space-time blobs are the primitive patterns to be grouped to identify the perceptual clusters, as discussed in the next section.

## 13.3 Perceptual Grouping in Spatiotemporal Domain

The perceptual units that serve as input to the proposed perceptual grouping algorithm are the spatiotemporal blob tracks that model regions homogeneous in color. The goal is to identify meaningful groupings of these blob tracks to form perceptual clusters.

Consider a set of patterns $S = \{p_1, p_2, ... p_n\}$ and a set $c_k$ such that $c_k \subseteq S$. Let $G(c_k)$ denote a spatial grouping of the patterns in $c_k$. The set of attributes for all the member patterns in $c_k$ at time $t$ are denoted as $c_k^t$ and the attributes of the complete spatial grouping considered as a single entity are denoted by $G^t(c_k)$. A set $c_k \in S$ is a perceptual cluster in the spatiotemporal domain if $\{c_k^t, G^t(c_k)\}$ obeys a spatiotemporal grouping model that defines a salient temporally coherent organization. This model specifies a set of temporal behavior attributes for the cluster $c_k$ and an operational method that uses these attributes to compute a grouping saliency measure for $c_k$. Temporal behavior attributes, for example, related to visual characteristics or contextual behavior, can be formulated for the overall grouping and the individual patterns. The proposed definition requires that in addition to the overall grouping $G^t(c_k)$, the individual patterns in the grouping must show temporal consistency in terms of their attributes $c_k^t$, which could be, for example, visual characteristics or contextual behavior. Such a formalism allows enforcement of specific visual and temporal behavior characteristics on some or all patterns of a grouping.

Following the proposed definition of a perceptual cluster, there are three possible types of attributes that can be used to characterize spatiotemporal saliency:

- *Properties of the patterns*: The individual patterns can be characterized using visual properties like appearance or shape. Let $p_i^t$ denote the attribute values of pattern $p_i$ at time $t$ and let $\ddot{p}_i$ be the attributes that characterize the temporal behavior of $p_i^t$. Specific knowledge about the temporal change in appearance or deformation of a meaningful grouping allows constraining the values for $\ddot{p}_i$, for a grouping to be computationally valid. A computationally valid grouping should correspond to a meaningful perceptual cluster.

- *Associations amongst the patterns*: The inter-pattern associations characterize the structural (that is, grouping) compatibilities for the patterns. Let $E^t(p_i, c_j)$ denote the attribute values for the association of pattern $p_i$ with cluster $c_j$ at time $t$. The attributes characterizing the temporal behavior of $E^t(p_i, c_j)$ are denoted as $\ddot{E}(p_i, c_j)$.

- *Structural characteristics of the overall cluster*: These attributes are derived for the cluster *as a whole* (not viewing it as a set of patterns). Let $G^t(c_j)$ denote the structural characteristics of the grouping $c_j$ at time $t$. The term $\ddot{G}(c_j)$ denotes the temporal behavior attributes of the structural characteristics $G^t(c_j)$.

Depending on the convenience of modeling and the available computer vision tools for gathering observations, the saliency of a cluster can be formulated based on any or all of the aforementioned attributes. A spatiotemporal grouping model defines a set of temporal behavior attributes $\ddot{p}_i$, $\ddot{E}(p_i, c_j)$, and $\ddot{G}(c_j)$, and an operational method that uses $\ddot{p}_i$, $\ddot{E}(p_i, c_j)$, and $\ddot{G}(c_j)$ for all $p_i \in c_j$ to compute a grouping saliency measure for the cluster $c_j$. A cluster that evaluates a high value of saliency measure is taken as meaningful and also as computationally valid. The proposed formulation for temporal consistency makes use of temporal behavior attributes that characterize the temporal coherence of a grouping. The remaining part of this section presents the formulations for the attributes from Reference [1], which are used here to characterize the spatiotemporal saliency of a cluster. All measures are formulated to produce a value between zero and one. Higher values depict stronger associations.

### 13.3.1 Motion Similarity Association

A real-world object generally comprises several parts, each of which shares an adjacency to at least one of the other parts. When a real-world object undergoes translation, all its parts appear to exhibit a similar motion over an extended period of time. The motion similarity association between two blob tracks can be formulated using the steps depicted in Algorithm 13.1a, where the resulting measure takes into account only the translational motion of the blobs.

> **ALGORITHM 13.1** Motion similarity association between two blob tracks.
>
> 1. Identify the smooth sections of the trajectories of the blobs A and B.
> 2. Consider one such smooth section of $f$ frames; the measured displacements of the two blobs are $\vec{d}_A$ and $\vec{d}_B$. The motion difference between the two blobs, computed as pixels per frame, is formulated as $|\vec{d}_A - \vec{d}_B|/f$.
> 3. Average the motion difference values for all the smooth sections in the trajectories of the two blobs.
> 4. Divide the computed motion difference value by ten and clamp the result to one in order keep the computed value of motion difference in the range $[0, 1]$.
> 5. Subtract the motion difference value from one to give the motion similarity measure.

### 13.3.2 Adjacency Association

The parts of a real-world object are connected to each other. The adjacency association is formulated such that it exhibits strongly for the blob patterns that belong to a scene object. A strong adjacency link should involve a good amount of overlap between adjacent patterns [1]. However, some adjacent patterns may have a small overlap, thus leading to a weak link. In the proposed algorithm, a representative blob (one with the longest life span) in the cluster is selected and labeled as the generator pattern of the cluster. A given pattern is considered to have a qualified adjacency with a cluster if it has a direct or indirect connectivity with the generator pattern of the cluster. The strength of the adjacency association between a pattern and a cluster depends on the strengths of all the links involved in the direct or indirect connectivity. Presence of weaker links have a debilitating effect on the adjacency association. Assimilating the temporal information, the adjacency association measure can be formulated as the ratio of the number of frames in which the pattern has a qualified spatial adjacency with the cluster, to the total life span (in terms of number of frames) of the cluster.

### 13.3.3   Cluster Bias Association

The cluster bias for a pattern signifies the affinity of the cluster toward the pattern. A pattern may be important for a cluster if it could facilitate the adjacency of other patterns with the cluster. Consider a pattern $p_i$ whose removal from a cluster $c$ would cause a set of patterns $q$ (including $p_i$) to get removed from the cluster. The bias applied by a cluster $c$ toward the pattern $p_i$ is formulated as:

$$\text{Cluster bias association measure} = \frac{\sum_{\forall q_k \in q} \text{duration\_of\_disconnection}(q_k)}{\sum_{\forall p_i \in c} \text{life span}(p_i)} \qquad (13.1)$$

The duration of disconnection and the life span are measured in number of frames.

### 13.3.4   Self-Bias Association

The self-bias association signifies a pattern's affinity toward a cluster. Self-bias is incorporated to facilitate an appropriate grouping for patterns that remain mostly occluded in the cluster but show up for a small time span. The purpose is to facilitate grouping such a pattern to an adjacent cluster, even if the life span of the pattern is small compared to the life span of the cluster. A pattern will have a self-bias to a cluster if it shares an adjacency for a duration that is a large fraction of the pattern's life span. The measure is formulated as the ratio of the duration of qualified adjacency of the pattern with the cluster to the life span of the pattern.

### 13.3.5   Configuration Stability Association

The configuration stability association measure favors a grouping of a pattern with a cluster only if the relative configuration of the pattern with respect to the other patterns in the cluster remains stable. To assess the configuration stability of a given pattern with a cluster, the generator of the cluster is considered here to be a reference pattern, and the placement (orientation and position) of the centroid of each of the member patterns relative to the generator is computed. Any change observed in the relative orientation of a given pattern with respect to the other member patterns penalizes its association strength with the cluster.

The measure is formulated [1] by computing the average of the relative change in the orientation of the pattern $p_i$ with respect to the other patterns $p_j$ (except for the generator). The computed value is normalized by dividing it by 180, and then subtracting the result from one.

## 13.4   Spatiotemporal Grouping Model: Computing the Cluster Saliency

Given the observed values of inter-pattern associations for a putative cluster, it is necessary to develop a computational framework that can assimilate all these grouping evidences and give the grouping saliency of the cluster. A belief network (Figure 13.1) is used here as the evidence integration framework [1]. Some salient points about this methodology for computing the grouping saliency for a putative cluster $c_j$ are mentioned below:

1. The node S shown in the belief network takes on two states [*groupingIsSalient*, *groupingIsNotSalient*]. The probability that a cluster $c_j$ is salient is denoted as $P(c_j)$. The grouping saliency of a cluster $c_j$ is formulated as $\mathscr{S}_{c_j} = P(c_j)$.

2. The nodes A, B, C, D, and E correspond to the grouping evidences arising due to different inter-pattern associations and grouping characteristics of the emergent cluster $c_j$. Each of these nodes also takes on two states [*groupingIsSalient*, *groupingIsNotSalient*], which concern with the saliency arising from the grouping evidence contributed by the respective node. The grouping probability computed at either of nodes A, B, C, D, and E as denoted by $P_g(c_j)$, where $g$ denotes the particular node.

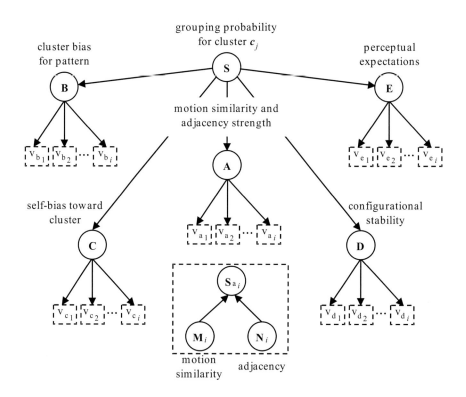

**FIGURE 13.1**

Belief network for spatiotemporal grouping [1]. The nodes shown in dotted rectangles are the virtual evidences.

3. Each pattern (except the generator) in a putative grouping contributes an evidence for the saliency of the grouping. An evidence is computed by evaluating a specified association of a pattern with the *rest of the grouping*. In the proposed belief network, such an evidence instantiates a virtual evidence node which contributes probability values in favor and against the saliency of a grouping. Figure 13.1 shows the virtual evidence $Va_i$ contributed by the $i$th pattern when assessing the grouping criterion of node A. Likewise, other virtual evidences contribute for their respective nodes.

4. Association measures described in the previous section are used for computing evidences at the leaf nodes $Va_i$, $Vb_i$, $Vc_i$, and so on, for the different grouping criteria. Every pattern (except the generator) in a cluster $c_j$ provides one evidence for each type of association. If a cluster $c_j$ has a pattern that exhibits a weak association (say, of type $A$) with the cluster, then it would contribute an evidence against the grouping, leading to a low probability $P_a(c_j)$ and hence a low $P(c_j)$, which implies an invalid cluster. Specifically, a cluster is considered to be invalid if $P(c_j) < 0.5$, which would imply a higher probability for the negative state *groupingIsNotSalient*.

5. The *a priori* probabilities for the proposition states in node S are set as $\{0.5, 0.5\}$. The conditional probability values for the nodes A, B, C, D, and E are set as $P(x|S) = 0.95$, $P(\bar{x}|S) = 0.05$, $P(x|\bar{S}) = 0.05$, and $P(\bar{x}|\bar{S}) = 0.95$, where $x$ is the positive state of any of the nodes A, B, C, D, and E.

6. For the virtual evidence node ($Vb_i$, $Vc_i$, or $Vd_i$), the probability of the parent states $x$ and $\bar{x}$ is calculated as $P(v|x) = \{associationmeasure\}$ and $P(v|\bar{x}) = 1 - P(v|x)$, where $x$ and $\bar{x}$ are the two states for each of the nodes A, B, C, and D. All the association measures are computed in the range $[0, 1]$.

7. The association strength given the combined observations of motion similarity and adjacency is computed using a three-node belief network shown in Figure 13.1b, where the node $Sa_i$ has two states [*groupingIsSalient*, *groupingIsNotSalient*]. The nodes $M_i$ and $N_i$ represent virtual evidences corresponding to motion similarity and adjacency in the belief network of Figure 13.1b. The conditional probability values for $P(Sa_i|m_i, n_i)$ are set as:

   - 0.5, when the adjacency measure is small and motion similarity is large (implying a weak evidence in favor of the grouping),
   - 0.1, when motion similarity is small (implying a strong evidence against the grouping), and
   - 0.9, when both the adjacency measure and the motion similarity measure are large (implying a strong evidence in favor of the grouping).

8. The node E of the network incorporates any kind of evidence which relates to the visual or the overall structural characteristics of a valid grouping. This kind of evidence is not generic and essentially requires the use of domain knowledge. Such evidences play an important role in the grouping process when the evidences due to other associations do not exhibit strongly enough to identify the valid clusters.

### 13.4.1 Perceptual Grouping Algorithm

Given a spatiotemporal grouping model, an algorithm [1], which makes use of the grouping model and finds the perceptual cluster in the scene, can now be designed. The grouping saliency for the entire scene can be formulated as:

$$\mathscr{S}_{scene} = \sum_{\forall c_j \in C} \mathscr{S}_{c_j} \tag{13.2}$$

where $C$ is the set of all perceptual clusters identified in the scene.

The spatiotemporal grouping problem is to identify the set $C$ of perceptual clusters in the scene so as to maximize $\mathscr{S}_{scene}$ such that the final set of clusters have:

$$P(c_i) \geq 0.5, \quad P(c_i \cup c_j) < 0.5, \quad \text{and} \quad c_i \cap c_j = \phi, \forall c_i, c_j \in C, \ i \neq j \tag{13.3}$$

The first condition in Equation 13.3 mandates that each cluster in the set of groupings be valid. The second condition constrains the set of valid groupings so that no two clusters can be further grouped together to form a valid cluster as per the specified grouping model. The third condition specifies that a pattern can be a member of only one cluster.

This formulates the perceptual grouping problem as an optimization problem. A naive perceptual grouping algorithm would explore all possible sets of clusters and output the one that has the maximum scene's grouping saliency and all the clusters valid as per conditions in Equation 13.3. However, these conditions restrict the set of groupings which need to be explored. Therefore, Algorithm 13.2 outlines a procedure which maximizes $\mathscr{S}_{scene}$ to a local maximum and obtains a set of clusters satisfying conditions in Equation 13.3. The algorithm is graphically depicted in Figure 13.2.

---

**ALGORITHM 13.2**  Perceptual grouping algorithm (PGA).

Input: Set of patterns $P = \{p_1, p_2, ..., p_n\}$.
Output: Set of clusters $C$ which satisfy conditions in Equation 13.3.

1. Initialize the set of clusters $C = \{\}$. Hence, $\mathscr{S}_{scene} = 0$. Instantiate a queue, *clusterQ*, to be used for storing the clusters. Initially, none of the patterns have any cluster label.

2. If all the patterns have cluster labels and the *clusterQ* is empty, then exit. Otherwise, instantiate a new cluster if there are patterns in set $C$ that do not have any cluster label. From among the unlabeled patterns, pick up the one that has the maximum life span. This pattern is taken as the generator pattern for the new cluster $c_{new}$.

3. (a) Compute associations of the generator pattern of $c_{new}$ with the unlabeled patterns. An unlabeled pattern $p_i$ is added to the cluster $c_{new}$ if $P(p_i \cup c_{new}) > 0.5$. Repeat this step till no more unlabeled patterns get added to $c_{new}$. Any unlabeled patterns that exist now are the ones that do not form a valid grouping with any of the existing clusters.

   (b) Compute associations of the generator pattern of $c_{new}$ with the labeled patterns of other clusters. If there are patterns, say $p_k$ (of another cluster, say, $c_j$) which form a valid grouping with $c_{new}$, that is, $P(p_k \cup c_{new}) > 0.5$, then put $c_{new}$ and $c_j$ into the *clusterQ* if they are not already there inside it.

4. If the *clusterQ* is empty, go to step 2. Otherwise, take the cluster at the front of the *clusterQ*. Let this cluster be $c_f$.

5. Pick a pattern, say, $p_i$, of the cluster $c_f$. The pattern $p_i$ is considered to be a part of, one by one, every cluster in the set $C$. The association measures between $p_i$ and a cluster $c_j$ need to be recomputed if $c_j$ has been updated. For each cluster $c_j$ in $C$, compute the grouping probability for the putative grouping $\{p_i \cup c_j\}$. If the grouping $\{p_i \cup c_j\}$ is valid (that is, $P(p_i \cup c_j) > 0.5$), note the change in the scene grouping saliency as $\delta_S = P(p_i \cup c_j) - P(c_j)$. The pattern $p_i$ is labeled with the cluster (say, $c_m$) for which $\delta_S$ computes to the highest value. If $c_m$ is not $c_f$, then put $c_m$ into the *clusterQ* if it is not already inside it. Repeat step 4 for all the patterns in $c_f$.

6. If any pattern in $c_f$ changes its label as a result of step 5, then go to step 4. If none of the patterns changes its label, remove the front element of the *clusterQ* and go to step 4.

The iterative process terminates when the pattern labels have stabilized. It attempts to hypothesize new clusters and organizes them such that $\mathscr{S}_{scene}$ reaches a local maximum. Convergence to a local maximum is guaranteed since $\triangle \mathscr{S}_{scene} \geq 0$ is ensured at every step of the algorithm. A grouping formed out of a single pattern always leads to an increase in $\mathscr{S}_{scene}$. A pattern would change its cluster label only when the increase in saliency of the cluster to which it is joined is larger compared to the decrease in saliency of the cluster from which it is defected. Thus, instantiation of clusters as well as relabeling of patterns always leads to an increase in the $\mathscr{S}_{scene}$. The conditions in Equation 13.3 are enforced in step 2, where instantiation of clusters is done only when the *clusterQ* is empty. This implies that all existing clusters are stable at that stage and merging any two clusters would lead to an invalid grouping (since otherwise the two clusters would not have emerged as separate). The proposed algorithm performs a local optimization (similar to *k*-means), since at every step the set of groupings being explored depends on the set of groupings in the previous step. A simple heuristic is used to label a cluster as a foreground or a background. A cluster which touches two or more frame borders (top, bottom, left, right) for a reasonable proportion (over 80%) of its life span is marked as a background cluster. Most of the videos have the background as a single cluster which touches all four frame borders.

Association measures, which rely on the evaluation of adjacencies between patterns and clusters, are dependent on the groupings that emerge as the algorithm proceeds. Two patterns which do not share boundary pixels can be taken as connected only if they can be linked together by a contiguous path of intermediate adjacent patterns, all of which belong to the same cluster. Thus, the adjacency strength of two indirect neighbor patterns depends on the cluster labels of the intermediate patterns. The dependence of adjacency on cluster labels implies that the existing adjacency associations for the member patterns in a cluster are examined every time when any member pattern is defected from the cluster. Similarly, a new member joining a cluster may lead to an indirect adjacency of other patterns (which carry a label of some other cluster) to this cluster. To summarize, the pattern-to-cluster adjacencies need to be re-evaluated only for those clusters whose compositions get affected as a result of reorganization. With the revised adjacencies, the association measures, such as self-bias, cluster bias, and configuration stability, need to be recomputed. This is not the case of motion similarity since it does not depend on adjacency.

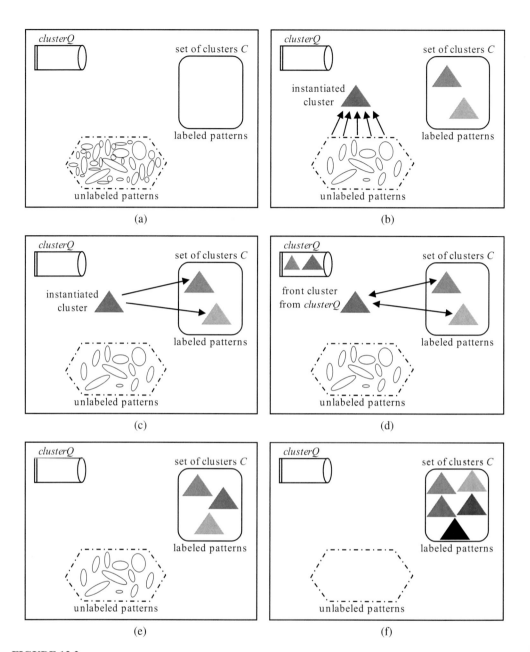

**FIGURE 13.2**

Illustration of the steps performed in Algorithm 13.1. (a) Initial stage, where the set of clusters C is empty. (b) Step 2, where a cluster is instantiated by using the pattern with longest life span as the generator. The new cluster computes associations with the unlabeled patterns in step 3a of PGA. (c) Step 3b, where the new cluster computes associations with labeled patterns. (d) Step 5, where re-organization of patterns for the clusters that have been put in the *clusterQ* takes place. (e) *ClusterQ* is empty at the end of the re-organization step. This state leads to step 6 followed by step 4 and step 2. Further instantiation of new clusters and reorganization takes place till no more unlabeled patterns are left. (f) Termination stage.

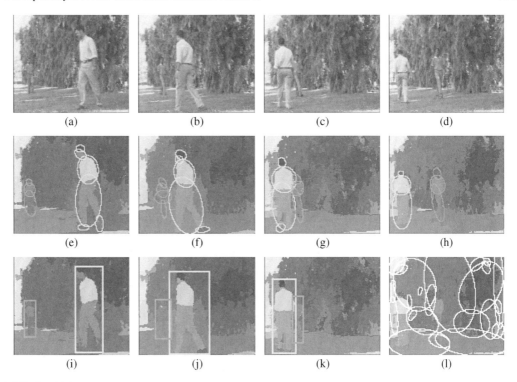

**FIGURE 13.3**

PGA performance demonstration: (a,e,i) frame 4, (b,f,j) frame 25, (c,g,k) frame 50, and (d,h,l) frame 83. (a–d) Some original frames from a sequence in which two persons are walking in the background of trees. The total length of the sequence is 356 frames. (e–h) Color segmented (clustered) frames produced by the DSCT algorithm. (i–k) The PGA groups foreground clusters, which are marked using rectangles of distinct colors. Blobs that constitute a foreground cluster are highlighted in (e–h) with the same color as that of the bounding rectangle. (l) All other blobs in the scene get grouped into the background cluster. Total number of blob tracks processed is 32 of which 11 were grouped as foreground clusters. Grouping of the background cluster took four iterations of step 5 of PGA, while the two foreground clusters took two iterations each.

## 13.5 Evaluation of Cluster Identification Using Perceptual Grouping

The output of the perceptual grouping algorithm (PGA) is verified manually and is considered to be correct if the output clusters correspond to meaningful foreground subjects, and no actual foreground region is incorrectly merged with the background or any other foreground cluster. PGA groups the background as a single perceptual cluster; meaningful components in the background are not searched here. Figure 13.3 shows a sequence in which the background has dynamic textures (trees). The color clustering step in the decoupled semantics clustering tree algorithm smooths out the textured background and models all homogeneous color regions as space-time blob tracks. The blobs belonging to the background are more or less static, while those that are part of the two foreground subjects show distinct motion. The PGA identifies three clusters, two foreground clusters and one background cluster, which comprises all the remaining blobs. The decoupled se-

**FIGURE 13.4 (See color insert.)**

PGA performance demonstration on a sequence from *Titanic* movie: (a,e,i) frame 77, (b,f,j) frame 103, (c,g,k) frame 142, and (d,h,l) frame 175. (a–d) Original frames, (e–h) PGA-identified foreground blobs on the DSCT color-clustered frame, and (i–l) perceptual clusters marked with the same color as the blobs belonging to the distinct foreground clusters. Note that the person wearing the red jacket, initially seen on the right of the person wearing the yellow jacket, comes to the left starting from frame 110.

mantics clustering tree algorithm can handle dynamic textures without the need of explicit modeling in contrast to Reference [29].

Figure 13.4 shows a sequence in which people walk along a passage in a ship. The camera is not static and there are some frames for which the distinct subjects get adjacent to each other and they even change their orientation with respect to each other. As can be seen, the PGA identifies the foreground subjects correctly. Some more representative results of PGA are shown in Figure 13.5. The detailed results and explanations can be found in Reference [1].

Overall, it can be said that the PGA successfully identifies foreground subjects for the class of scenes where the subjects show distinct characteristics compared to the background. The use of node E plays a crucial role in the grouping process when associations like motion similarity and adjacency do not exhibit strongly enough. The output of the PGA crucially depends on what qualifies to be a valid grouping. The factors governing a valid grouping are the observed Gestalt associations, the evidences formulated corresponding to the associations, and the Bayesian network parameters. The PGA reasonably outputs the chunk of pixels that belong to the foreground subjects even in the presence of some inconsistencies in the blob tracks.

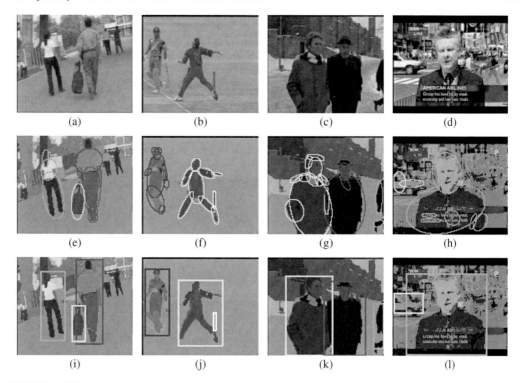

**FIGURE 13.5 (See color insert.)**

PGA performance demonstration on various scenes: (a–d) original frames in four different scenes, (e–h) identified foreground blobs, and (i–l) perceptual clusters.

Having identified the meaningful objects in the scene foreground, the next problem toward the task of video understanding is to interpret the scene. An interpretation establishes the *context* in which humans perceive the information about the scene content. The context, among other things, depends significantly on the prominent components in the scene. These ideas and a computational model for perceptual prominence are discussed below.

## 13.6  Perceptual Prominence

The following describes the proposed principle of perceptual prominence for the subjects in the scene. Prominence refers to cognitive interest in the subjects. The semantics of the scene is closely associated with the prominent subjects, that is, the subjects that deserve attention. Perceptual prominence is different from visual saliency [15], which refers to the identification of salient locations in the scene by the psychophysical sensory mechanisms *before* any cognitive interpretation is carried out.

Context plays an important role in influencing the prominence measure for scene components (the perceptual clusters). Though a perceptual cluster may be prominent as a standalone cluster, the presence of other clusters may establish a context in which it may not be prominent. Hence, it may happen that within a context and an interpretation, one, few, or none of the clusters come out as prominent. The proposed approach to modeling the context

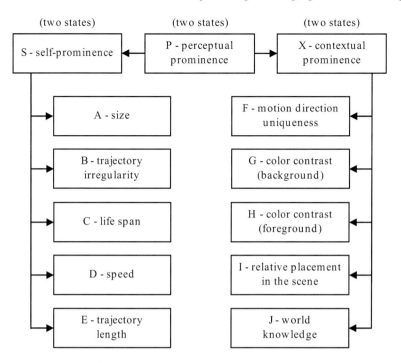

**FIGURE 13.6**

Belief network [1] for computing the prominence measure for a perceptual cluster $c_i$.

is to use perceptual attributes that characterize the context. A contextual attribute captures the distinguishing properties of a cluster with respect to the surrounding background blobs or other foreground clusters present in the scene.

Prominence can arise in several ways. In many scenes, a subject occupying a position that is in imbalance with the arrangement of the scene components turns out to be prominent. A marked contrast in the appearance of a subject relative to the background or other foreground subjects also directs visual attention, making the subject perceptually prominent. Different interpretations of prominence can be encoded into different prominence models. A specific prominence model denotes a set of perceptual attributes and the way the computed values for the attributes influence the prominence measure for a subject. The choice of a prominence model depends on the type and interpretation of the scene. For example, in a subject-oriented scene, the leading subject generally lasts for a long duration in the shot period and hence is prominent. In surveillance scenes, the subjects appearing for a short span of time or subjects moving in a direction different from other subjects are considered as prominent (deserving attention). Thus, it is possible to formulate one prominence model that identifies subjects showing zigzag motion as prominent, and another model that identifies large-sized subjects with a long life span as prominent, and so on.

The proposed methodology for computing the prominence measure makes use of a belief network, shown in Figure 13.6, which models the prominence of a cluster as a propositional node P. The prominence measure is formulated as the probability computed for the proposition that the cluster is prominent. Each of the nodes P, S, and X takes on two discrete states {*isProminent, isNotProminent*} for the cluster. The evidences are computed based on a characterization of the perceptual attributes over the entire life span of the subject. All the attributes do not need to be contextual. If none of the perceptual attributes take the context into account, the prominence measure thus computed from the non-contextual (that is, subject specific) attributes is called as the *self-prominence* measure. In the belief network there are separate nodes for computing measures (that is, probabilities) for self-prominence and contextual prominence. The virtual evidence nodes A through J in the belief network characterize the observation of the perceptual attributes over the entire life span of the subject, and hence are termed as the observation nodes. Instantiation of evidence nodes leads to belief propagation [30] in the Bayesian network and each node updates the posterior probability for its states. The marginalized probability for the positive state of node P is taken as the prominence value. The proposed framework for computing the prominence can be used to model different interpretations of prominence by using an appropriate subset of relevant observation nodes. The computed attribute value is mapped to its contribution toward the prominence by using an evidence function.

### 13.6.1 Evaluation of Perceptual Prominence

For the sequence shown in Figure 13.7, the proposed grouping algorithm groups the moving subjects (since they have relative motion with respect to the background, they can be grouped without any world knowledge) and the subjects who are stationary (sitting) by making use of the face detector and a skin detector (with vertical oval-shaped blobs signifying a face) and grouping the face regions with the body region below it. Table 13.1 lists the normalized perceptual attributes for four subjects (who are eventually prominent in the scene), along with the computed probability of prominence. A subject with a frontal face gets a higher contribution (at node E) toward its prominence (because a frontal face is generally in focus), compared to non-frontal view faces. The prominence computed for subjects sp1 and sp2 (see Figure 13.7) is high, primarily because of their large life span and size. The motion of the moving subjects, and their higher color contrast compared to the other subjects give them a higher prominence.

A scene can be interpreted in terms of the behavioral analysis of prominent subjects and the mise-en-scène concepts. Cognitive analysis forms the link between the extracted meaningful components and the perceived interpretation of the content. The next section describes the proposed scene interpretation framework.

### 13.6.2 Scene Categorization Using Perceptual Prominence

Perceptual prominence helps in identifying the meaningful subjects which influence many of the cognitive abstractions from the scene. A scene is characterized here using two broad categories:

- *Frame-centric scene* – A cognitive understanding of a frame-centric scene requires shifting the focus of attention to different spots on the entire frame. A frame-centric scene generally comprises multiple prominent subjects that may be engaged in a group-specific activity or independent activities.

- *Subject-centric scene* – In a subject-centric scene the focus of attention is mainly on the activity of one or two subjects. A subject-centric scene comprises fewer prominent subjects that normally exist for most of the duration of the shot.

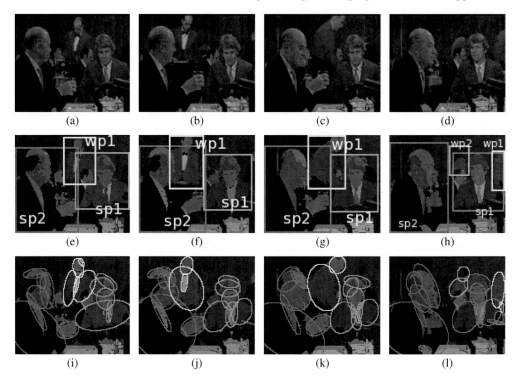

|  |  |  |  |
|:---:|:---:|:---:|:---:|
| (a) | (b) | (c) | (d) |
| (e) | (f) | (g) | (h) |
| (i) | (j) | (k) | (l) |

**FIGURE 13.7**

Perceptual prominence computation for a sequence in which two persons are sitting in a restaurant: (a,e,i) frame 25, (b,f,j) frame 37, (c,g,k) frame 61, and (d,h,l) frame 231. Note other persons sitting in the background and sporadically moving persons, such as the waiter. The foreground subjects are named as sp1 and sp2, for the persons sitting on the right and the left side, respectively; and wp1 and wp2 for the waiter wearing a red uniform and the moving person wearing a dark green coat, respectively.

**TABLE 13.1**

Perceptual attributes, as depicted in Figure 13.6, computed (and normalized) for four foreground subjects in the sequence shown in Figure 13.7. The normalized perceptual attributes are shown only for prominent subject (with prominence measure $> 0.5$). The probabilities computed for self-prominence (S), contextual prominence (X), and the overall prominence (P) are also shown. The prominence computed for subjects sp1 and sp2 is high because of their longer life span and a larger size. The subject sp1 has a frontal face that contributes a higher prominence compared to sp2, which has a side-view face. The prominence for the moving subjects wp1 and wp2 is mainly contributed by their motion and higher color contrast. ©2007 IEEE.

| Subject | Self-Prominence | | | | | Contextual Prominence | | | | | Prominence | | |
|---|---|---|---|---|---|---|---|---|---|---|---|---|---|
| | A | B | C | D | E | F | G | H | I | J | S | X | P |
| sp1 | 0.34 | 0.10 | 1.00 | 0.09 | 0.05 | 0.50 | 0.38 | 0.41 | 0.70 | 0.90 | 0.92 | 0.97 | 0.94 |
| sp2 | 0.62 | 0.10 | 1.00 | 0.10 | 0.10 | 0.50 | 0.38 | 0.44 | 0.60 | 0.70 | 0.96 | 0.87 | 0.92 |
| wp1 | 0.23 | 0.60 | 0.30 | 0.64 | 0.79 | 0.50 | 0.39 | 0.80 | 0.50 | 0.70 | 0.80 | 0.86 | 0.83 |
| wp2 | 0.47 | 0.20 | 0.19 | 0.52 | 0.86 | 0.50 | 0.64 | 0.65 | 0.50 | 0.50 | 0.81 | 0.59 | 0.70 |

**FIGURE 13.8 (See color insert.)**

Scene categorization: (a–d) subject-centric scenes, as they all have one or two prominent subjects, which exist throughout the duration of the shot, and (e–h) frame-centric scenes because the subjects in these scenes last only for a small duration in the shot. Prominence values for the subjects taken from left to right: (a) {0.96,0.1}, (b) {0.3, 0.8, 0.95}, (c) {0.8, 0.8, 0.18}, (d) {0.61, 0.66}, and (e–h) values smaller than 0.5 according to the proposed prominence model which includes life span as an attribute.

Examples of subject-centric and frame-centric scenes are given in Figure 13.8. Each of the subject-centric and frame-centric scene category is divided into three subcategories based on the kind of activity exhibited by the subjects:

- *Independent activity* scenes, where a subject follows an independent movement pattern without any influence from the other subjects. The subjects are likely to move in a uniform direction.

- *Involved activity* scenes, where a subject normally moves in response to the movement of the other subjects and are more likely to exhibit an irregular motion compared to those in an independent activity scene. Moreover, the subjects may move in different directions.

- *No-activity* scenes, where the subjects do not show any significant activity. There is little or no motion.

A Bayesian network is a convenient way to model the uncertain relationships between observations and concepts. In the proposed belief network (Figure 13.9) for characterizing the scenes, the nodes G, S, F, $V_s$, $D_s$, $O_s$, $V_f$, $D_f$, and $O_f$ are the scene type nodes, and each has two discrete states that correspond to the positive and negative propositions of whether a given scene belongs to the specified type. The nodes shown in dotted rectangles are observation nodes, each contributes to the belief of the scene type node to which it is connected. The node $N_{subjects}$ corresponds to the total number of subjects identified in the scene. Three types of prominence models are used, which could identify subjects showing a prominent zigzag motion (subscript zz), or a prominent uniform motion (subscript u) or a prominent no motion (subscript nm). Observations related to these prominent subjects are used in virtual evidence nodes $P_{zz}^1$, $P_u^1$, $P_{nm}^1$, $P_{zz}^2$, $P_u^2$, or $P_{nm}^2$ to provide evidences for the relevant scene types.

### 13.6.3  Evaluation of Scene Categorization

Different scenes are characterized by making use of key mise-en-scène aspects and subject behaviors specific to the scene types. For illustration, the subject behavior observations and scene categorization results are discussed below for a few example scenes:

- *Cricket*: A cricket scene is frame centric, with the subjects generally showing an independent or an involved activity. For a typical scene in which a batsman plays a ball, the camera is moved fast to follow the ball, and hence the subjects exhibit an apparent fast motion in a uniform direction. The subjects show adherence to a prominence model highlighting high-speed motion in a uniform direction.

- *Tennis*: A tennis scene is subject-centric and exhibits an involved activity of the subjects. The subjects show adherence to a prominence model highlighting zigzag motion behavior.

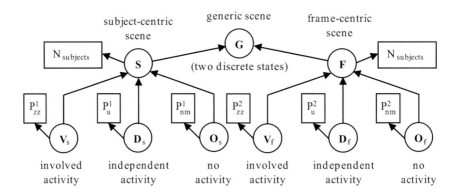

**FIGURE 13.9**

The belief network for the proposed scene model [1]. ©2007 IEEE.

**TABLE 13.2**

Scene interpretation on various test videos. Scene 1 evaluated a higher probability for frame centric (0.911) compared to subject centric (0.161). Within frame centric, the subcategory that got the highest probability is involved activity. Hence, this scene is categorized as frame centric with involved activity. Likewise the final categorization of other examples is listed in the table. ©2007 IEEE

| Scene Example | Subject Centric (SC) | | | | Frame Centric (FC) | | | | Final Scene Categorization |
|---|---|---|---|---|---|---|---|---|---|
| | Subj. Cent. | Invol. Act. | Indep. Act. | No Act. | Frame Cent. | Invol. Act. | Indep. Act. | No Act. | |
| Scene 1 (football) | 0.161 | 0.285 | 0.631 | 0.224 | 0.911 | 0.952 | 0.822 | 0.539 | FC, invol. act. |
| Scene 2 (person walking) | 0.808 | 0.394 | 0.970 | 0.394 | 0.353 | 0.411 | 0.539 | 0.411 | SC, indep. act. |
| Scene 3 (tennis) | 0.884 | 0.955 | 0.815 | 0.383 | 0.391 | 0.534 | 0.537 | 0.410 | SC, invol. act. |
| Scene 4 (cricket) | 0.097 | 0.237 | 0.291 | 0.237 | 0.868 | 0.548 | 0.961 | 0.548 | FC, indep. act. |
| Scene 5 (crowd walking) | 0.164 | 0.225 | 0.284 | 0.628 | 0.877 | 0.546 | 0.960 | 0.607 | FC, indep. act. |
| Scene 6 (news reader) | 0.655 | 0.350 | 0.350 | 0.966 | 0.364 | 0.411 | 0.457 | 0.527 | SC, no activity |
| Scene 7 (crowd sitting) | 0.102 | 0.236 | 0.236 | 0.321 | 0.877 | 0.546 | 0.593 | 0.969 | FC, no activity |

**TABLE 13.3**

Evaluation of subject- and frame-centric scene categorization on a test set of 215 video shots.

| | Subject Centric | Frame Centric |
|---|---|---|
| Subject Centric | 100% | 0% |
| Frame Centric | 21% | 79% |

**TABLE 13.4**

Evaluation of activity-based scene categorization on a test set of 215 video shots.

| | Independent Activity | Involved Activity | No Activity |
|---|---|---|---|
| Independent Activity | 92% | 8% | 0% |
| Involved Activity | 19% | 77% | 4% |
| No Activity | 0% | 0% | 100% |

- *Football*: A football scene is frame centric with involved or independent activity of the subjects. The subjects show adherence to a prominence model highlighting zigzag motion or a uniform direction motion.

- *News report*: A news report has many subject-centric scenes, for example, the scenes where a reporter (a prominent subject) may stand in front of a still background or one having moving objects such as traffic. The foreground subject adheres to a prominence model highlighting subjects with a longer life span and showing no motion.

- *Crowd scenes*: Crowd scenes are frame centric, and generally exhibit independent activity of the subjects. The subjects may be walking in a uniform direction, or may be just sitting.

A scene is classified as subject centric or frame centric depending on which of the nodes (S or F) computes a higher belief. A subject-centric scene is further classified as belonging to any of the subcategories that (propositional node) compute the highest belief (among $V_s$, $D_s$, and $O_s$); and likewise for a frame-centric scene. The classification accuracy for a given scene category is computed as the ratio of the number of shots that were correctly classified as belonging to that category to the number of shots that actually belonged to that category. Table 13.2 lists the results of scene interpretation on various example videos. The classification performance is summarized in Tables 13.3 and 13.4. The performance figure is lower for frame-centric scenes; that was mainly because of missed subject detection by the PGA, since generally, such scenes have subjects of small sizes. Further, the involved activity scenes also have a lower classification accuracy and are more likely to get misclassified as independent activity scenes. This happens mostly when the subject life span is small or the shot duration is small, so that the subject motion appears more or less uniform and in similar directions. As it can be seen, the various scene examples have been correctly interpreted into the scene hierarchy. Extensibility of the proposed interpretation framework to incorporate other classes of scenes requires formulating appropriate observations (evidences) for characterizing the scenes.

## 13.7 Perceptual Grouping with Learned Object Models

This section explores the use of learned object-model knowledge with the perceptual organization process. A framework, which not only performs detection of foreground objects but also recognizes the object category, is proposed. The advantages of the add-on grouping evidences, as contributed by the object models for a more robust perceptual organization in the spatiotemporal domain, are demonstrated. The object is modeled as a pictorial structure [31], [32] since this kind of representation involves modeling an object as a composition of parts. It offers a probabilistic model, which allows computing the likelihood of the part or the object as a whole in the process of grouping. This in turn provides a scheme for object recognition through a grouping process in a video sequence. Using perceptual grouping in spatiotemporal domain avoids the need of bootstrapping, that is, having a correct recognition of the object in the first frame. Further, it gathers grouping evidences from all the frames of a video, and is thus more reliable in detecting specific (with known models) objects as well as general objects.

Section 13.7.1 describes the proposed object model formulated as pictorial structures. The perceptual grouping algorithm that uses the domain knowledge is discussed in Section 13.7.4. The experimental results are reported in Section 13.7.5.

### 13.7.1 Object Model as a Pictorial Structure

Consider an object with parts $v_1, v_2, ..., v_n$, where each part $v_i$ has a configuration-label specified as $l_i$ on a common coordinate frame. This label is formulated as $l_i = (x_i, y_i, \theta_i, \sigma_i)$, where $(x_i, y_i)$ denotes the location, $\theta_i$ is the orientation, and $\sigma_i$ is the scale at which the object part occurs in the image plane. An instance of the pictorial structure model for an object can be defined using its part-configuration parameters $\mathbf{l} = \{l_1, l_2, ..., l_n\}$ and the part-appearance parameters $\mathbf{a} = \{a_1, a_2, .., a_n\}$, where the subscript indicates the part index. The Bayesian formulation is adopted for pictorial structures as given in Reference [32]. Let $n_P$ be the set of parts that together comprise the object model. Let $D_i$ denote the image region of the $i$th part. The posterior distribution from the model parameters can be defined as follows:

$$P(\mathbf{a}, \mathbf{l}|D) \propto \prod_{i=1}^{i=n_P} \frac{P(D_i|a_i)}{P(D_i|a_{bg})} \exp\left(-\sum_{j\neq i} \Psi(l_i, l_j)\right) \tag{13.4}$$

where $\Psi(l_i, l_j)$ is the potential between parts $v_i$ and $v_j$, and $a_i$ and $a_{bg}$ are appearance parameters for part $v_i$ and background, respectively. The part labels which maximize the posterior probability are taken as the best fit for the pictorial structure model.

### 13.7.2 Learning the Object Model

An object model is learned in terms of the configuration of the body parts and the appearance of each of the body parts. Considering analogy with a Markov random field model, each object part is considered to be a site. Configuration compatibility of the two parts is formulated as the potential between them. To compute a continuous measure of conformance of an observed configuration with the expected (true) configuration, the relative configuration of the two parts is modeled using parameters $h_{ij} = \{length, orientation\}$ of a hypothetical line joining the centroid of the parts $v_i$ and $v_j$. Then, the distribution of the two parameters is modeled as a mixture of Gaussians.

The potential $\Psi(l_i, l_j)$ between two parts is defined as

$$\Psi(l_i, l_j) = \sum_{x=0}^{C} w_x \frac{\left|\Sigma_{x_{ij}}^{-1}\right|^{1/2}}{2\pi} \exp\left(-\frac{1}{2}(h_{ij} - \mu_{x_{ij}})^t \Sigma_{x_{ij}}^{-1}(h_{ij} - \mu_{x_{ij}})\right) \tag{13.5}$$

where $C$ is the number of Gaussians, $w_x$ is the weight given to the $x$th Gaussian modeled with covariance matrix $\Sigma_{x_{ij}}$ and mean $\mu_{x_{ij}}$. While learning the potential, it was found that the Gaussian mixture model learning algorithm gives the maximum log-likelihood score (sum of log-likelihoods over all the training samples) for a mixture of two Gaussians.

### 13.7.3 Formulation of Appearance Parameters for Object Parts

An object part may have fine distinguishing image features and may also contain important information in lower frequencies. It is desirable that the visual attributes provide a representation of the appearance that is jointly localized in space, frequency, and orientation. For this purpose, a wavelet transform is used to decompose the image into various subbands localized in orientation and frequency. The subbands represent different frequencies in horizontal, vertical, and diagonal orientations, and in multiple resolutions. The coefficients within a subband are spatially localized as well. Here, a three-level wavelet decomposition using Haar filter is performed. Level 1 describes the lowest octave of frequency, whereas each subsequent level represents the next higher octave of frequencies.

Algorithm 13.3 summarizes the steps in learning the object-part detector using the appearance attributes formulated on wavelet coefficients. Note that the attribute histograms need to be learned on a fairly large labeled dataset in order to be representative of the object and non-object categories.

**ALGORITHM 13.3**   Learning of the object-part detector using the appearance attributes formulated on wavelet coefficients.

1. Decompose the known object region (rectangular) into subwindows.

2. Within each subwindow $(w_x, w_y)$, model the joint statistics of a set of quantized wavelet coefficients using a histogram. Several such histograms are computed, each capturing the joint distribution of a different set of quantized wavelet coefficients. Here, 17 such histograms are obtained. The coefficient sets are formulated as given in Reference [33].

3. A histogram is a nonparametric statistical model for the wavelet coefficients evaluated on the subwindow. Hence, it can be used to compute the likelihood for an object being present or not being present. The detector makes its decision using the likelihood ratio test:

$$\frac{\prod\limits_{w_x, w_y} \prod\limits_{k=1}^{17} P_k \left(\text{histogram}_k(w_x, w_y), w_x, w_y\right) \mid \text{object}}{\prod\limits_{w_x, w_y} \prod\limits_{k=1}^{17} P_k \left(\text{histogram}_k(w_x, w_y), w_x, w_y\right) \mid \text{non-object}} > \lambda$$

where the outer product is taken over all the subwindows $(w_x, w_y)$ in the image region of the object. The threshold $\lambda$ is taken as $P(\text{non} - \text{object})/P(\text{object})$, as proposed in Reference [33].

### 13.7.4 Spatiotemporal Grouping Model with Object Model Knowledge

The spatiotemporal grouping framework relies on making use of evidences formulated on the observed associations amongst the patterns. In Figure 13.1, the evidence contributed by a pattern signifies its association strength with the *rest of the grouping*. The virtual evidence nodes $v_{e1}, v_{e2}, ..., v_{en}$ contribute evidences corresponding to the conformance of a putative grouping of blob patterns to a known object model. A virtual evidence $v_{ei}$ computes the grouping probability of a pattern $p_i$ with the rest of the patterns in the grouping, given the knowledge of the object model. A pattern $p_i$ may exhibit only a partial overlap with an object. Assuming an object $O_m$ comprising of $v_1^m, v_2^m, ..., v_n^m$ parts, it is necessary to find whether $p_i$ overlaps with any of the object parts. Algorithm 13.4 summarizes the steps of this procedure.

**ALGORITHM 13.4** Searching for overlaps between pattern $p_i$ and object parts.

1. Define a search window (around the pattern) that is larger than the size of the pattern.

2. At all locations $(w_x, w_y)$ within that window do a search for each as yet undetected object part at three different scales. An object part is taken to be detected correctly if its posterior (Equation 13.4) is above a threshold. Note that when using Equation 13.4, the product index $i$ is taken over the $n_D$ parts that have been already detected up to this step.

3. The overlap of $p_i$ with object part(s) provides a measure of the degree to which the pattern *belongs* to the object. If the fractional overlap (in number of pixels) of $p_i$ with the object is found to be more than 30% of the area of $p_i$, the pattern is said to have a *qualified* overlap with the object $O_m$.

4. Step 3 is repeated for all the frames $F_{p_i}$ for which the pattern (blob track) $p_i$ exists. The association strength of the pattern $p_i$ with the object $O_m$ is formulated as $\left(\sum_{\forall f \in F_{p_i}} \alpha_{p_i}^f\right)/F_{p_i}$, where $\alpha_{p_i}^f = 1$ if $p_i$ has a qualified part match in the frame $f$, and $\alpha_{p_i}^f = 0$ otherwise. This association measure gets computed in the range $[0, 1]$ and is taken as the probabilistic evidence contributed at nodes $v_{e1}, v_{e2}, ..., v_{en}$, in favor of the putative cluster.

#### 13.7.4.1 Grouping-Cum Recognition

Having formulated the grouping evidence for a pattern using the object-model knowledge, the following lists the steps of the grouping-cum recognition algorithm. Using object model knowledge requires an extra procedure (step 3 in Algorithm 13.5) to be added to the grouping algorithm discussed in Section 13.4.1. The node E is used to provide the object model-based grouping evidences when the cluster $c_j$ has been associated with a hypothesized object-class label. The object class label is chosen depending on the object part to which the generator pattern matches. If the generator pattern does not match with any object part, then it would initiate a general grouping without any object class label. Given a set of object models $\mathscr{O} = \{O_1, O_2, ..., O_s\}$ and a set of patterns $P = \{p_1, p_2, ..., p_n\}$, the proposed algorithm outputs the set of perceptual clusters $C = \{c_1, c_2, ..., c_r\}$.

**ALGORITHM 13.5** Perceptual grouping-cum recognition.

Input: Set of object models $\mathscr{O} = \{O_1, O_2, ..., O_s\}$ and a set of patterns $P = \{p_1, p_2, ..., p_n\}$.
Output: Perceptual clusters with a labeled object class (if any).

1. Initialize the set of clusters $C = \{\}$. Hence, $\mathscr{S}_{scene} = 0$. Instantiate a queue, *clusterQ*, to be used for storing the clusters. Initially none of the patterns have any cluster label.

2. If all the patterns have cluster labels and the *clusterQ* is empty, then exit. Otherwise, instantiate a new cluster if there are patterns in set $C$ that do not have any cluster label. From among the unlabeled patterns, pick up the one that has the maximum life span. This pattern is taken as the generator pattern for the new cluster $c_{new}$.

3. The generator may or may not be a part of any of the objects in set $\mathcal{O}$. Define a search region around the generator and look for possible parts of any of the objects, as depicted in Figure 13.10. The object part(s) which compute (by virtue of their appearance) the *a posteriori* probability in Equation 13.4 to be greater than a threshold are considered as *present* in the search region. The generator pattern is labeled as belonging to the object with which it shares the maximum overlap (in terms of number of pixels). The correctness of the object label is further verified by checking its consistency in five other frames in which the generator blob exists. Once an object label (say, $O_m$) has been associated with the generator, the node E is used to provide a probability of a pattern being a part of the object $O_m$. If the generator pattern cannot be associated with any object label from set $\mathcal{O}$, the node E is not used in the grouping model. The cluster formed by such a generator corresponds to a general grouping without any specific object tag. Moreover, such a grouping would use only the general associations for grouping.

4. (a) Compute associations of the generator pattern of $c_{new}$ with the unlabeled patterns. An unlabeled pattern $p_i$ is added to the cluster $c_{new}$ if $P(p_i \cup c_{new}) > 0.5$. Repeat this step till no more unlabeled patterns get added to $c_{new}$. Any unlabeled patterns that exist now are the ones that do not form a valid grouping with any of the existing clusters.

   (b) Compute associations of the generator pattern of $c_{new}$ with the labeled patterns of other clusters. If there are patterns, say, $p_k$ (of another cluster, say, $c_j$) that form a valid grouping with $c_{new}$, that is, $P(p_k \cup c_{new}) > 0.5$, then put $c_{new}$ and $c_j$ into the *clusterQ* if they are not already there inside it.

5. If the *clusterQ* is empty, go to step 2. Otherwise, take the cluster at the front of the *clusterQ*. Let this cluster be $c_f$.

6. Pick a pattern, say, $p_i$, of the cluster $c_f$. The pattern $p_i$ is considered to be a part of, one by one, every cluster in the set $C$. The association measures between $p_i$ and a cluster $c_j$ need to be recomputed if $c_j$ has been updated. For each cluster $c_j$ in $C$, compute the grouping probability for the putative grouping $\{p_i \cup c_j\}$. If the grouping $\{p_i \cup c_j\}$ is valid (that is, $P(p_i \cup c_j) > 0.5$), the change in the scene grouping saliency is noted as $\delta_S = P(p_i \cup c_j) - P(c_j)$. The pattern $p_i$ is labeled with the cluster (say, $c_m$) for which $\delta_S$ computes to the highest value. If $c_m$ is not $c_f$, then put $c_m$ into the *clusterQ*, if it is not already inside it. Repeat step 4 for all the patterns in $c_f$.

7. If any pattern in $c_f$ changes its label as a result of step 5, then go to step 4. If none of the patterns changes its label, remove the front element of the *clusterQ* and go to step 4.

Apart from step 3 (which gives an object label to the generator), the algorithm follows the steps in the original PGA. The iterative process terminates when the pattern labels have stabilized. Each cluster has a label of the object class to which it belongs.

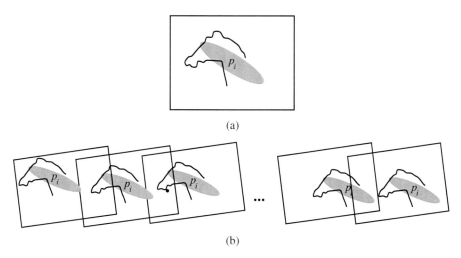

(a)

(b)

**FIGURE 13.10**

Illustration of the steps performed in Algorithm 13.2: (a) a bounding box defined around a pattern $p_i$, (b) the consistency of the overlap of the object part with the pattern $p_i$ is established for a sequence of frames.

### 13.7.5 Evaluation of Perceptual Grouping with Object Models

The models were trained for two object categories: cows and horses. For the purpose of learning the appearance attribute histograms for the object parts and the non-object regions, the appearance features were computed for each object part for about 650 images, sampled from cow/horse videos. A large training set is needed since the skin texture/color has substantial variations within certain animal categories. The features were also learned for the mirror images of the object regions. To minimize the classification error, different weights were used for the different attributes [33]. The Gaussian mixture model distribution of configuration for each pair of object parts was also learned. The appearance attribute histograms for non-object regions were learned from about 1200 random images that did not have the object.

The experiment has helped to establish two advantages offered by the perceptual grouping process through an effective use of object model knowledge. First, once an object part has been reliably detected (with high probability), other (computationally costly) evidences, such as adjacency and cluster-bias, do not need to be gathered for all the frames since an object model is a very strong grouping evidence that would allow grouping of only the blobs that overlap with some part of the same object. Second, there are scenes in which the foreground subjects do not show a distinct motion with respect to the background. For such a case, the general grouping criteria would group all the blobs in the scene to output a single cluster having both the background and the foreground. Such failure cases are shown to be successfully handled by the use of object models that provide a stronger grouping evidence to compensate for the (non-exhibited) general grouping evidences. Searching an object part around a blob pattern is a costly operation; however, it needs to be done only once, and not in every iteration of the grouping algorithm. Moreover, it can be computed on a few frames (around five) and not on all the frames of the blob track, since a consistent overlap of the blob with an object part on a few frames can be taken as an indicator that

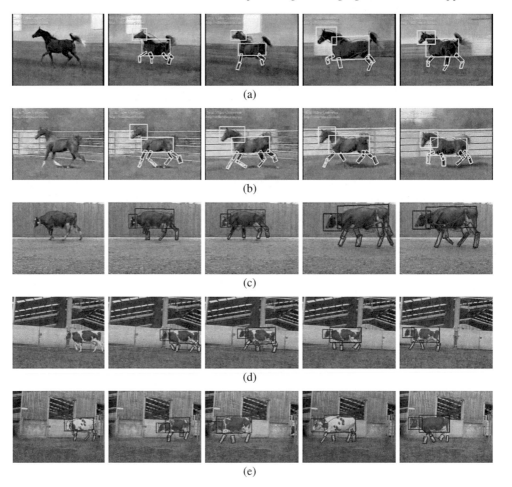

**FIGURE 13.11 (See color insert.)**

Results obtained for various horse and cow videos. Probability at node E: (a) 0.814, (b) 0.795, (c) 0.910, (d) 0.935, and (e) 0.771, 0.876, 0.876, 0.800, and 0.911 for different videos with the object model cow.

the blob belongs to the object region. It was observed that certain object parts get missed by the detectors when a large object motion causes blur in the object appearance, since the appearance gets confused with the background. For such cases, the generic grouping evidences play the major role in the grouping process.

Figure 13.11 presents the results of grouping with learned object models. Namely, Figures 13.11a and 13.11b show the result of applying the grouping algorithm to test videos of a horse galloping in a barn. The camera is panning and the background remains almost constant. Thus, blob patterns belonging to both the object and the background have almost no motion. Hence, simple generic grouping (using motion similarity, adjacency, and so on), when applied to such a video, fails to successfully cluster the patterns into meaningful organizations and clusters all patterns into one large cluster. This situation is resolved in the proposed model-guided perceptual grouping scheme, where the evidence from the object model knowledge is used to correctly groups the object region blobs into a separate cluster. In the presented figures, the recognized object parts are shown as rectangles overlaying the blobs belonging to the horse region.

Figures 13.11c to 13.11e show results on various test videos with cows of very different textures and appearances. The algorithm is successfully able to cluster all the blob patterns belonging to the cows and detect and recognize parts around the patterns (shown as rectangles).

The proposed algorithm has demonstrated the benefits of using object models as domain knowledge to assist, in both computational speed-up and detection accuracy, in the perceptual organization process applied to videos. The results have been presented on two object classes; and as future work, the goal will be to demonstrate use of object models for correctly grouping multiple foreground-objects, which could be sharing similar motion and adjacency in the video frames.

## 13.8 Conclusion

This chapter presented unsupervised techniques to identify objects that show distinct perceptual characteristics with respect to the background. The decoupled semantics clustering tree methodology is an unsupervised clustering tool for analyzing video data. It employs a decoupled clustering scheme applied to feature subspaces with different semantics for the six-dimensional heterogeneous feature space of video to model the homogeneous color regions in a video scene as temporal tracks of two-dimensional blobs.

The space-time blob tracks (simpler primitives or visual patterns) obtained by the decoupled semantics clustering tree methodology are subsequently grouped together to compose the meaningful objects. A perceptual grouping scheme for the spatiotemporal domain of video was proposed to complete this task. The grouping scheme relies on a specified grouping model. This model makes use of inter-pattern association measures motivated from the Gestalt law of common fate, which states that the entities having a coherent behavior in time are likely to be parts of the same organization. Temporal coherence is established with the premise that time is not an added dimension to the data to be grouped, but it is an added dimension to the grouping such that the grouping should maintain coherence in time. This chapter presented formulations for novel association measures that characterize the temporal coherence between primitives in order to group them into salient spatiotemporal clusters, termed as perceptual clusters. The foreground perceptual clusters correspond to the subjects in the scene. The grouping saliency was defined as a measure of goodness of a grouping, and quantified as a probabilistic measure computed using a Bayesian network model. The perceptual grouping problem was formulated as an optimization problem. The perceptual grouping algorithm (PGA) maximizes the overall scene saliency to a local maximum and obtains a set of perceptual clusters. The PGA successfully identifies foreground subjects for the class of scenes where the subjects show distinct characteristics compared to the background. As part of future work it would be interesting to explore if such grouping models could be learned. This would mean learning the structure of the grouping belief network, the optimal set of association attributes to be used, and their relative influence in the computation of grouping saliency. This chapter also demonstrated the use of learned object models as domain knowledge to assist in grouping as well as recognition of the object class instances in a video shot. Use of object models was also shown to assist in computational speed-up and improved detection accuracy in the perceptual organization process.

The problem of general scene characterization was explored by making use of different interpretation models to detect and analyze the meaningful components in the scene. A computable model for characterizing a video scene into two broad categories was designed. The two categories are subject-centric scenes, which have one or few prominent subjects, and frame-centric scenes, in which none of the objects can be attributed a high prominence and hence the overall activity depicted in a sequence of frames is the subject of interest. In addition, a novel concept of perceptual prominence for the identified subjects was proposed. Perceptual prominence has been motivated from film grammar that states that human understanding of the scene depends on the characteristics of scene components focused on. It is an extension of the concept of saliency to the cognitive level. Further, it depends on chosen interpretation of prominence, as different subjects may turn out to be prominent when using different interpretations of prominence. Prominence is interpreted by formulating an appropriate model based on attributes that commonly influence human judgment about meaningful subjects. Here, empirical prominence models for formulating perceptual evidences for various types of scenes were used. The attribute specifications were parametrized into different prominence models that signify different interpretations of prominence. A set of perceptual and contextual attributes that influence the prominence were described and a methodology that uses the perceptual attributes and computes a value signifying the prominence was developed. Moreover, a taxonomy of interpretation of video scenes was generated; each of the subject-centric and frame-centric scene classes was divided into three subcategories (independent activity scenes, involved activity scenes, and no-activity scenes), based on the kind of activity exhibited by the subjects. Presented experimental results established the effectiveness of the proposed perceptual attribute-based prominence formulation and showed that such attributes indeed have a direct relationship to many of the cognitive abstractions from a video scene. The scene interpretation framework can be extended to incorporate other classes of scenes by formulating appropriate observations (evidences) for the more specific scene classes.

## Acknowledgment

Figure 13.9 and Tables 13.1 and 13.2 are reprinted from Reference [1] with the permission of IEEE.

## References

[1] G. Harit and S. Chaudhury, "Video shot characterization using principles of perceptual prominence and perceptual grouping in spatio-temporal domain," *IEEE Transactions on Circuits and Systems for Video Technology*, vol. 17, no. 12, pp. 1728–1741, December 2007.

[2] D. Bordwell and K. Thompson, *FILM ART An Introduction*. New York, USA: McGraw-Hill Higher Education, 2001.

[3] D.D. Hoffman and B.E. Flinchbaugh, "The interpretation of biological motion," *Biological Cybernetics*, vol. 42, no. 3, pp. 195–204, 1982.

[4] M. Nicolescu and G. Medioni, "Perceptual grouping from motion cues using tensor voting in 4-D," in *Proceedings of the 7th European Conference on Computer Vision*, Copenhagen, Denmark, May 2002, vol. 3, pp. 303–308.

[5] A. Verri and T. Poggio, "Against quantitative optic flow," in *Proceedings of the First International Conference on Computer Vision*, London, UK, June 1987, pp. 171–180.

[6] R. Nelson and J. Aloimonos, "Towards qualitative vision: Using flow field divergence for obstacle avoidance in visual navigation," in *Proceedings of the Second International Conference on Computer Vision*, Tarpon Springs, FL, USA, December 1988, pp. 188–196.

[7] M. Shah, K. Rangarajan, and P.S. Tsai, "Generation and segmentation of motion trajectories," in *Proceedings of 11th International Conference on Pattern Recognition*, The Hague, The Netherlands, September 1992, pp. 74–77.

[8] S. Sarkar, "Tracking 2D structures using perceptual organization principles," in *Proceedings of the International Symposium on Computer Vision*, Coral Gables, FL, USA, November 1995, pp. 283–288.

[9] S. Sarkar, D. Majchrzak, and K. Korimilli, "Perceptual organization based computational model for robust segmentation of moving objects," *Computer Vision and Image Understanding*, vol. 86, no. 3, pp. 141–170, June 2002.

[10] S.E. Palmer, "The psychology of perceptual organization: A transformational approach," in *Human and Machine Vision*, J. Beck, B. Hope, and A. Rosenfeld (eds.), New York, USA: Academic Press, 1983, pp. 269–339.

[11] E.L.J. Leeuwenberg, "Quantification of certain visual pattern properties: Salience, transparency, similarity," in *Formal Theories of Visual Perception*, E.L.J. Leeuwenberg and J.F.J.M. Buffart (eds.), New York, USA: Wiley, 1978, pp. 277–298.

[12] S. Sarkar and K.L. Boyer, *Computing Perceptual Organization in Computer Vision*. River Edge, NJ, USA: World Scientific, 1994.

[13] D.G. Lowe, *Perceptual Organization and Visual Recognition*. Boston, MA, USA: Kluwer Academic Publishers, 1985.

[14] A. Shashua and S. Ullman, "Structural saliency: The detection of globally salient structures using locally connected network," in *Proceedings of the Second International Conference on Computer Vision*, Tarpon Springs, FL, USA, December 1988, pp. 321–327.

[15] L. Itti, *Models of Bottom-Up and Top-Down Visual Attention*. PhD thesis, California Institute of Technology, Pasadena, CA, USA, 2000.

[16] D.T. Lawton and C.C. McConnell, "Perceputal organization using interestingness," in *Proceedings of the Workshop on Spatial Reasoning and Multi-Sensor Fusion*, St. Charles, IL, USA, October 1987, pp. 405–419.

[17] M. Wertheimer, "Laws of organization in perceptual forms," in *A Sourcebook of Gestalt Psychology*, Harcourt, Brace and World, 1938, pp. 71–88.

[18] K. Koffka, *Principles of Gestalt Psychology*. New York, USA: Harcourt, 1935.

[19] W. Kohler, *Gestalt Psychology*. New York, USA: Liveright Publishing Corporation, 1929.

[20] I. Rock and S. Palmer, "The legacy of Gestalt psychology," *Scientific American*, vol. 263, no. 6, pp. 84–90, December 1990.

[21] S. Sarkar and K.L. Boyer, "Perceptual organization in computer vision: A review and a proposal for a classificatory structure," *IEEE Transactions on Systems, Man, and Cybernetics*, vol. 23, no. 2, pp. 382–399, March 1993.

[22] S.F. Chang, W. Chen, H.J. Meng, H. Sundaram, and D. Zhong, "A fully automated content-based video search engine supporting spatiotemporal queries," *IEEE Transactions on Circuits and Systems for Video Technology*, vol. 8, no. 5, pp. 602–615, September 1998.

[23] J. Sivic and A. Zisserman, "Video Google: A text retrieval approach to object matching in videos," in *Proceedings of International Conference on Computer Vision*, Nice, France, October 2003, vol. 2, pp. 1470–1477.

[24] N. Vasconcelos and A. Lippman, "Toward semantically meaningful feature spaces for the characterization of video content," in *Proceedings of the IEEE International Conference on Image Processing*, Washington, DC, USA, October 1997, vol. 1, pp. 25–29.

[25] B. Adams, C. Dorai, and S. Venkatesh, "Toward automatic extraction of expressive elements from motion pictures: Tempo," *IEEE Transactions on Multimedia*, vol. 4, no. 4, pp. 472–481, December 2002.

[26] H. Sundaram and S.F. Chang, "Computable scenes and structures in films," *IEEE Transactions on Multimedia*, vol. 4, no. 4, pp. 482–491, December 2002.

[27] G. Harit and S. Chaudhury, "Clustering in video data: Dealing with heterogeneous semantics of features," *Pattern Recognition*, vol. 39, pp. 789–811, May 2006.

[28] D. DeMenthon and R. Megret, "Spatio-temporal segmentation of video by hierarchical mean shift analysis," Technical report, LAMP-TR-090, University of Maryland, College Park, MD USA, July 2002.

[29] Y. Sheikh and M. Shah, "Bayesian modeling of dynamic scenes for object detection," *IEEE Transactions on Pattern Analysis and Machine Intelligence*, vol. 27, no. 11, pp. 1778–1792, November 2005.

[30] R.E. Neapolitan, *Probabilistic Reasoning in Expert Systems: Theory and Algorithms*. New York, USA: John Wiley and Sons, 1990.

[31] P. Felzenszwalb and D. Huttenlocher, "Efficient belief propagation for early vision," in *Proceedings of the IEEE International Conference on Computer Vision and Pattern Recognition*, Washington, DC, USA, June 2004, pp. 261–268.

[32] M.P. Kumar, P.H.S. Torr, and A. Zisserman, "Extending pictorial structures for object recognition," in *Proceedings of the British Machine Vision Conference*, London, UK, September 2004, vol. 2, pp. 789–798.

[33] H. Schneiderman and T. Kanade, "A statistical method for 3D object detection applied to faces and cars," in *Proceedings of the IEEE International Conference on Computer Vision and Pattern Recognition*, Hilton Head, SC, USA, June 2000, vol. 1, pp. 746–751.

# 14

# *Perceptual Encryption of Digital Images and Videos*

**Shujun Li**

## 14.1 Introduction

Digital images and videos are so ubiquitous nowadays that it would be difficult to live without them in the heavily digitized and networked world. Various digital devices, such as digital cameras, camcorders, scanners, digital televisions, personal computers, mobile phones, and printers, rely on digital image and video processing and coding technologies. Despite the extreme usefulness of digital images and videos, there are more and more concerns about security and privacy rising from a heavy dependence on digital images and videos: how to protect sensitive information recorded as digital images and videos from unauthorized access and misuse? Different people may have completely different reasons for having such a security concern. For instance, digital content providers are worrying about pirate copies of their products, end users want to keep their private images and videos in their mobile phones and their online web spaces safe from unwanted hands, and governmental/military/diplomatic bodies need to carefully protect classified documents (many of which are recorded in the form of digital image and video). All of these demands require content protection and access control, which can be fulfilled by applying cryptography to multimedia data, that is, by encrypting digital images and videos and by defining who can access the encrypted contents.

A unique feature of digital images and videos is their bulky size. Despite the increasing power of computers and image/video coding standards, the demand for high-resolution and high-quality multimedia contents, such as high-definition television (HDTV) and stereoscopic television, keeps asking for more storage space of digital multimedia data. For example, in the era of video compact disc (VCD), a typical movie was of size around 700 MB. Today, a high-definition movie on a Blu-ray disc can require more than 20 GB space. A natural consequence of the extensive size of multimedia data is that full encryption of digital images and videos may not be necessary or economical in some applications. Because of the entertaining nature of many commercial multimedia content, selectively encrypting parts of the whole content to downgrade the visual quality to a predefined degree becomes a desirable solution in applications like pay-per-view television (TV) services. This leads to a new concept called *perceptual encryption* (or *transparent encryption*), whose goal is not to conceal all perceptual information but only a specific amount of it. Here, the leakage of partial perceptual information is intended because the low-quality contents can be used to attract potential buyers and to allow further postprocessing (such as digital watermarking) of encrypted multimedia contents by an agent without access to the decryption key.

While perceptual encryption has to be implemented via *selective encryption*, the former has higher technical requirements than the latter, thus leading to much more technical challenges. There are mainly three issues making perceptual encryption more complicated. First, while selective encryption can be achieved, for example, by randomly encrypting one bit out of $n$ bits of the image/video bitstream, perceptual encryption cannot be done this way because the bitstream has a strict syntax format that has to be respected so that the partial unencrypted perceptual information can be decoded. This calls for joint multimedia encryption-encoding, which is not trivial because most image and video coding standards are not designed to be encryption-friendly and recent research revealed that there exists trade-offs among different requirements of compression and encryption. Second, perceptual encryption naturally asks for a definition of perceptual information, which has been a long-standing open problem in the field because of a very limited understanding of the human visual system. Last, dividing multimedia data into two parts implies that there are two ways of attacking the system: breaking the encrypted part directly and recovering the encrypted part from unencrypted part. The second approach is possible in practice because existing compression techniques cannot completely remove correlative information among different parts of digital images and videos. Very recent research has demonstrated that attacks based on such correlative information are far more effective than originally expected. The attacks also challenge experts to re-think how perceptual information should be encoded more effectively and how encryption should be combined with compression to ensure a desired level of security.

This chapter gives an overview of perceptual encryption of digital images and videos, paying special attention to very recent research on advanced attacks on perceptual encryption schemes for images and videos coded using the discrete cosine transform (which covers most mainstream image and video coding standards like JPEG and MPEG-1/2/4). Namely, Section 14.2 introduces some fundamental backgrounds of multimedia coding and cryptography. Section 14.3 surveys multimedia encryption to prepare some basic facts about joint multimedia encryption-encoding (JMEE) as the foundation of perceptual encryption. Section 14.4 focuses on different aspects of perceptual encryption, including a brief survey of various encryption schemes and classifications of different attacks on these schemes. Section 14.5 introduces a special form of attacks, error-concealment attack (ECA), which represents the most latest and interesting advances in the field. Finally, this chapter concludes with Section 14.6 by pointing some future research trends.

## 14.2 Fundamentals of Multimedia Coding and Cryptography

### 14.2.1 Image and Video Coding

Multimedia technology has been rapidly developed in the past several decades. Multimedia coding – compression and binary representation of multimedia data – plays a key role in multimedia applications [1]. Among different forms of multimedia data, digital images and videos are widely used. A digital image is a two-dimensional matrix of *pixels* whose values represent the brightness or the color of discrete points on the image plane. Depending on the nature of an image, a pixel value can be either a scalar, which is the case of binary and grayscale images, or a triplet representing some color, which is the case of red-green-blue (RGB) color images, for instance. A pixel value can also contain more than three color components, which is the case for multispectral images, but this chapter will not discuss such cases. However, note that all facts about RGB images discussed in this chapter can be generalized to multispectral images because they are largely independent of the dimension of pixel values. Similarly, a digital video is a three-dimensional matrix of pixels where the additional dimension is time. In other words, a digital video is simply a collection of digital images with a sequential order in temporal domain. Each image contained in a digital video

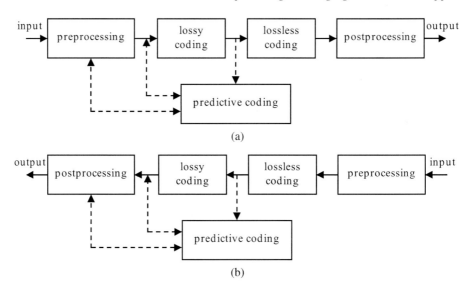

**FIGURE 14.1**

The general structure of an image or video codec, where the dotted parts represent optional components: (a) encoder, (b) decoder.

is called a *frame*, and the number of video frames per second is called the *frame rate* of the video. In some video coding systems, a video frame is divided into a *top field* composed of odd rows and a *bottom field* composed of even rows; the two fields are coded separately, which is called *interlaced video coding*. This chapter does not differentiate video frames from fields to simplify discussion.

Whenever talking about coding, both encoding and decoding are always considered here. The encoding process (or an encoder) transforms an image or a video into a bitstream consuming less space, and the decoding process (or an decoder) recovers the encoded bitstream back to the original uncompressed image/video. An encoder and a decoder are often called a *codec* collectively.

The encoding and decoding processes are basically symmetric in structure, but they can have completely different computational complexities. Normally, the decoder is much less computationally costly than the encoder. The asymmetry is mainly due to different goals of the two processes: encoding tries to find the compression transform with the best overall encoding performance (compression efficiency as one major but not the only factor), but decoding merely translates what has been encoded in a bitstream into perceptual information following the transform selected by the encoding process. In some cases, however, there is a different scenario: the encoding process should be computationally light but the decoding process can be more computationally expensive. A typical application is wireless multimedia sensor networks (WMSNs) [2], in which encoding is done by resource-constrained sensor nodes (often powered by battery) and decoding by computationally more powerful devices like multicore computers. A special multimedia coding architecture called distributed coding has been developed to handle such asymmetric coding [3].

Although there are many different image and video coding schemes, almost all of them follow the basic structure shown in Figure 14.1. Because of the similarity between the encoder and the decoder, their functionality is briefly explained on the example of an encoder.

In the first step, the input image or video passes a preprocessing step, which can include several different processes, such as color space conversion, signal partitioning, and pre-filtering (for example, subsampling and noise removal). The main goal of this step is to prepare data for future steps and remove some redundant information that does not need to be processed by later steps. For instance, encoding of color images and videos normally involves a conversion from RGB space to another color space, such as YCbCr and YUV, to separate the luminance and chrominance components so that the lower sensitivity of the human visual system to chrominance components can be exploited to reduce information needed for encoding. The reduction of chrominance information is done by subsampling the chrominance components, meaning to sample them with a lower (normally halved) sampling rate than the luminance component. While this immediately leads to a 2:1 compression ratio, the visual quality of the image/video reconstructed from the halved data remains almost as good as that of the original image/video. Partitioning is another widely use preprocessing technique, which is often useful for speeding up the encoding process.

The second step is lossy coding. It aims at removing psycho-visually redundant information, thus creating an image that is perceptually nearly indistinguishable from the original. The lossy coding part is usually realized by converting the data from spatial domain to frequency domain via a transform, such as the discrete cosine transform (DCT) [4] and the discrete wavelet transform (DWT) [5], and then quantizing the resultant transform coefficients to be integers. The main purpose of using a transform is to decorrelate the original image/video so that less correlative information exists in the new transform domain than in spatial domain. Performing a transform on the whole image/video is computationally expensive, so the common practice is to first partition the image and each video frame into $n \times n$ blocks and apply the transform locally on each block. The lossy coding step is optional because in some applications, such as medical imaging and forensics, it may not be desired to lose any information (even when it is invisible to human eyes).

The lossy coding step is followed by lossless coding, which aims at encoding the data in a more compact form so that less bits are needed to represent the same amount of information. Since no information loss is involved in lossless coding, all information can later be exactly recovered by the decoder from the encoded bitstream. Lossless coding relies on information theory [6], which has introduced a concept called *information entropy* to measure the amount of information contained in a random source of messages and offered a number of effective ways of representing messages more effectively while still losslessly. All lossless coding methods based on information entropy are collectively called entropy coding. The two commonly used entropy coding algorithms are Huffman coding and arithmetic coding; both algorithms can be made context adaptive to improve their encoding performance. Besides entropy coding, some other lossless coding algorithms have also been developed. The two most important ones are dictionary coding and run-length coding. Dictionary coding compresses a message by building a static or dynamic dictionary of strings and replacing an input message by a sequence of dictionary entries. Run-length coding represents a message as a sequence of (run,level) pairs, which can effectively compress the message if the distribution of "run" has a long tail. Both methods have their roots in information theory, but normally they are not classified as entropy coding. Many lossless coding algorithms, including Huffman coding and arithmetic coding, belong to *variable-length coding* (VLC) because they encode an input symbol into a variable number of bits.

The last step of the encoding process is postprocessing. It mainly aims at collecting the output from the previous step and assembling the final bitstream. In some applications, the final bitstream may contain several quality layers that should be encoded in such a way that separated decoding is possible. When there are multiple streams involved, for example, audio and video streams of a TV program or the left- and right-eye video streams of a stereoscopic video, a container is needed to hold the multiple streams and some mechanism is needed to guarantee the synchronization among them. Different from previous steps, this last step does not involve any compression technique.

In addition to the above four steps, there is another optional step called *predictive coding*. It can be done together with the first two steps, which explains its multiple connections with other parts of the encoding process. As its name implies, the goal of predictive coding is to predict not-yet-encoded part from already-encoded data to reduce the amount of information for further encoding. If the prediction error (also called residual data) has a more skewed distribution than the original data, which is the normal case, information theory guarantees that the compression efficiency will be higher. For instance, for image and video coding based on DCT transform, the first transform coefficient (called DC coefficient) of each block can be predicted from the previous block's DC coefficient with a fairly high accuracy. This prediction can also happen in temporal domain, where the motion of objects in the input video is estimated and the obtained *motion vectors* are encoded as part of the final bitstream. Predictive coding is particularly important for video coding because consecutive frames are often highly correlated in both spatial and temporal domains.

Different image and video coding schemes can be developed by combining different techniques for different steps of the whole encoding/decoding process. Various image and video coding standards have been established, among which JPEG and MPEG-1/2/4 are the most successful standards. The development of image and video standards on one hand takes advantage of existing technologies, but on the other hands also promotes new research by offering a platform for testing different kinds of new technologies. As a matter of fact, the latest image and video coding standards, such as JPEG2000 [7], JPEG-XR [8], MPEG-4 AVC/H.264 [9], and SMPTE VC-1 [10], represent the state of the art of the field.

Generally speaking, a multimedia coding standard defines a syntax format of the encoded bitstream and a method of decoding an encoded bitstream to reconstruct the original multimedia data (that is, the input of the encoder) with an acceptable perceptual quality. Multimedia coding standards are designed to offer balanced solutions to the following goals simultaneously: good perceptual quality, high compression efficiency, and low computational complexity of the codec. In addition, some other factors are often taken into account to enhance the applicability of the designed standard, which include flexibility to support diverse applications, backward compatibility with old standards, scalability, resilience against transmission errors, and so forth. Because of the limited space, existing image and video coding standards are not discussed in detail; this chapter aims to describe perceptual encryption schemes working with those standards in a more general sense.

### 14.2.2 Modern Cryptography and Cryptanalysis

Modern cryptography is the science of securing digital systems, which is often considered as a subfield of computer science although it does have a strong link to information

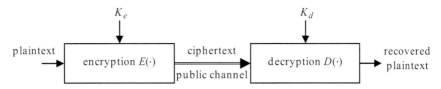

**FIGURE 14.2**

The encryption and decryption procedures of an encipher/decipher pair.

theory (thus communications in general) and mathematics (especially number theory, finite field, and combinatorics) [11], [12]. Traditionally, cryptography is mainly about data encryption, but nowadays more complicated cryptographic systems, such as cryptographic hash functions, digital signatures, and security protocols, have been developed. This chapter is mainly focused on data encryption. The nature of modern cryptography implies that cryptanalysis, the study on how to break existing cryptographic systems, is extremely important and plays a key role in the development of any secure systems. In some sense, it can be said that modern cryptography is built on top of cryptanalysis; or that cryptology is composed of two subfields: cryptography and cryptanalysis, one is about designing secure systems and the other about breaking them. This subsection introduces only some very basic concepts and terms of modern cryptography and tries to avoid discussing any ad hoc designs and attacks; as these topics are widely discussed in References [11] and [12].

An encryption system is also called an *encipher*, and a decryption system is called a *decipher*. When there is no need to differentiate an encipher and a decipher or the same system can be used for both encryption and decryption, such a system can be simply called a *cipher* or a *cryptosystem*.[1] The plain message taken as the input of an encipher (and the output of a decipher) is called a *plaintext*. Similarly, the encrypted message as the output of an encipher (and the input of a decipher) is called a *ciphertext*. Denoting the plaintext and the ciphertext by $P$ and $C$, respectively, the encryption procedure of an encipher can be described by

$$C = E_{K_e}(P), \tag{14.1}$$

where $K_e$ is the *encryption key* and $E(\cdot)$ is the encryption function implemented by the encipher. Similarly, the decryption procedure of a decipher can be described by

$$P = D_{K_d}(C), \tag{14.2}$$

where $K_d$ is the *decryption key* and $D(\cdot)$ is the decryption function implemented by the decipher. If an encipher and a decipher are concatenated by feeding the ciphertext $C$ to the decipher, the following is obtained:

$$P = D_{K_d}(E_{K_e}(P)). \tag{14.3}$$

Basically, the decryption function $D(\cdot)$ can be considered as the inverse of the encryption function $E(\cdot)$. Figure 14.2 illustrates how an encipher and a decipher can work together to allow secure transmission of data over a public insecure channel.

---

[1]The term *cryptosystem* actually can be used to represent more cryptographic systems beyond ciphers, but in many application scenarios it is a synonym of *cipher*.

Note that if the encryption and decryption keys should be made public it is possible to have two different kinds of ciphers. When both keys have to be kept secret, the cipher is called a *private-key* cipher or a *symmetric* cipher. The second name comes from the fact that for most private-key ciphers the encryption key and the decryption key are identical, or the decryption key can be derived from the encryption key so that they can be considered as identical. The main problem with private-key ciphers is the key management issue. Since the keys have to be kept secret, they have to be transmitted to both the sender and the receiver in advance via a secret channel. For $n$ users in a community, communications between any of these users will require secure transmission of $n(n-1)/2$ keys, which can become difficult or impossible to manage when $n$ becomes large. This problem can be solved if *only* the decryption key needs to be kept secret but the encryption key can be made public, thus leading to the so-called *public-key* cipher or *asymmetric* cipher. Apparently, for an $n$-user community only $n$ encryption keys need to be published to support secure communications between any two users. Normally, public-key ciphers are much slower than private-key ciphers due to their special properties, but they can be combined in such a way that a public-key cipher is used to implement the key management process of a private-key cipher. This chapter is more concerned about private-key ciphers because encryption of multimedia data is always performed by a private-key cipher.

Private-key ciphers can be further divided according to their encryption structure into *block ciphers* and *stream ciphers*. Block ciphers encrypt the plaintext block by block in the same manner, meaning that the same plaintext block will lead to the same ciphertext block independent of the position of the block in the whole plaintext/ciphertext. In contrast, stream ciphers encrypt the plaintext based on a pseudo-random sequence (called a *keystream*) generated under the control of the encryption key, thus allowing producing different ciphertext blocks for two identical plaintext blocks at different positions. Over the years, many different block ciphers and stream ciphers have been proposed, among which Advanced Encryption Standard (AES) [13] was the most widely used block cipher and (Alleged) Rivest Cipher 4 (RC4) the most widely used stream cipher [14] at the time this chapter was written.

While block ciphers are memoryless compared with stream ciphers, they can be made to behave in a way similar to a stream cipher by running it under different *modes of operation*. There are many different modes of operations, among which the naive way of running a block cipher is called the electronic code book (ECB) mode. By introducing ciphertext feedback, that is, revising the encryption procedure as follows:

$$C_i = E(P_i \oplus C_{i-1}), \tag{14.4}$$

the so-called cipher block chaining (CBC) mode is obtained, where $C_0$ is a given initial vector (IV) and $\oplus$ denotes XOR operation. There are also other modes that can transform a block cipher into a stream cipher, thus making block ciphers a foundation of building stream ciphers. This can be done, for example, by repeatedly encrypting an initial vector, a counter, or a combination of both. One of the reasons why more advanced modes of operation are needed is a security problem with the ECB mode, as shown in Figure 14.3.

Following a well-known principle proposed in References [15] and [16], the security of a cipher should rely *solely* on the decryption key $K_d$, but not about secrecy of the algorithm itself. This is widely accepted in modern cryptography because an obscured algorithm can

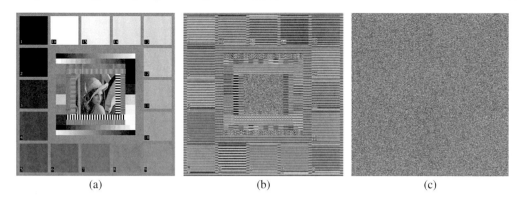

(a)    (b)    (c)

**FIGURE 14.3**

The encryption results of an image by AES running in ECB and CBC modes: (a) plaintext, (b) ciphertext of ECB mode, and (c) ciphertext of CBC mode.

be easily reverse engineered by crackers or exposed by insiders. Keeping this principle in mind, an upper bound of the security of a cipher is defined by the size of the key space, that is, the number of all possible keys. This is because the simplest attack on a cipher is to exhaustively try every possible key and see which one can give a meaningful output. Such a naive attack is called *brute-force attack* in cryptanalysis. Considering the computational power of today's computers and distributed computing, the key space is required to be as large as $2^{128}$ or even $2^{256}$. In addition to brute force attack, there are also attacks defined according to the capability of the attack to access/choose plaintexts and/or ciphertexts:

- *ciphertext-only attack*: the attacker can access the ciphertexts only;

- *known-plaintext attack*: the attacker has access to some plaintexts and the corresponding ciphertexts;

- *chosen-plaintext attack*: the attacker can choose plaintexts and get the corresponding ciphertexts; and

- *chosen-ciphertext attack*: the attacker can choose ciphertexts and get the corresponding plaintexts.

The ciphertext-only attack represents the most common attacking scenario because the ciphertext is supposed to be transmitted through a public channel (otherwise, data encryption is not needed at all). The known-plaintext attack is often possible in practice because many files have fixed headers and the contents of encrypted messages may be guessed in some cases (or simply exposed by the sender or receiver intentionally or unintentionally[2]). Chosen-plaintext and chosen-ciphertext attacks become possible when the attacker gets temporary access to the encipher and the decipher (but not the key), respectively. Note that the goal of a ciphertext-only attack is usually the plaintext, but the other attacks normally target the decryption key.

---

[2] For instance, a receiver may decide to reveal a secret message once he/she believes that the value of the message itself has expired. But the exposure of this message may be useful for an attacker to accumulate information for breaking future messages encrypted with the same key.

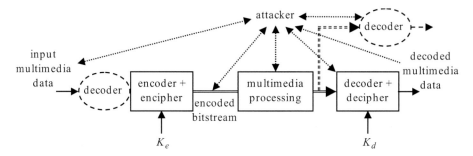

**FIGURE 14.4**

The threat model of a typical multimedia encryption system.

Besides the above attacks, there are also many other attacks defined in different contexts. For instance, when a cipher is implemented in a real product, the energy and/or the time consumed on encrypting different plaintexts with different encryption keys may differ. Such differences may be captured by an eavesdropper to derive useful information about the plaintext and the key, if one has access to the environment of the encryption machine. Such implementation-oriented attacks are called *side channel attacks* and they have been proved very effective in breaking many real implementations of theoretically secure ciphers.

## 14.3   Multimedia Encryption in General

As pointed out in the introductory section, depending heavily on multimedia creates the need for securing the data for both security and privacy purposes. As a consequence, multimedia encryption has attracted attention from all parties on the multimedia production and consumption chain, including content creators, producers, distributors, and consumers. Some standardization efforts have been made to facilitate incorporation of security tools (including ciphers) into multimedia codecs to build digital rights management (DRM) system working with image and video coding standards like JPEG2000 and MPEG-2/4 [17], [18], [19]. Contents and copy protection schemes have also been deployed in digital versatile discs (DVDs), advanced video coding high definition (AVCHD) media, and Blu-ray discs for thwarting unauthorized access to encrypted multimedia contents [20]. This section discusses the threat model and main technique challenges related to multimedia encryption in general.

### 14.3.1   Threat Model

A typical multimedia encryption system consists of an encoder-encipher (equipped with an encryption key $K_e$) and a decoder-decipher (equipped with a decryption key $K_d$), as shown in Figure 14.4. The input data may be already compressed in some applications, in such cases a decoder (or a partial decoder) is needed to decode the data first before encoding and encryption. A decoder without access to the decryption key $K_d$ may try to decode the encoded-encrypted multimedia data. It is also possible that some multimedia processing

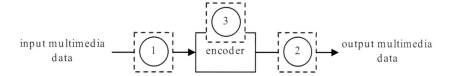

**FIGURE 14.5**

Three possible approaches to multimedia encryption at the encoder side.

units are inserted between the encoder-encipher and the decoder-decipher to perform some additional manipulation on encoded-encrypted multimedia data, such as watermark embedding and detection, re-encryption, filtering, bitrate re-control, re-packetization, transcoding, and so on. In most cases, these in-between multimedia processing units have no access to the encryption key nor to the decryption key. They may even be encryption-agnostic, that is, unaware whether the bitstream has been encrypted. On the other hand, the encoder-encipher often has no clue which multimedia processing operations may happen *after* the encoded-encrypted data are sent out to the transmission channel.

The threat model of such a multimedia encryption system is as follows. An attacker may appear at any point after the encoder-encipher and may have access to a compliant decoder, but no access to the decryption key. The attacker is assumed to know all the details of the encryption algorithm and of the key management process. The attacker may have access to a database of multimedia data of the same type as the plaintext. The main goal of an attacker will be to break the decryption key, or to completely/partially recover perceptual information in the encrypted multimedia data. In other words, both ciphertext-only attacks, known/chosen-plaintext attacks, and chosen-ciphertext attacks are possible threats. Side channel attacks may also be possible if the attacker has access to the ambience of the encoder-encipher or the decoder-decipher.

### 14.3.2 Three Approaches to Multimedia Encryption

According to where encryption is added into a multimedia encoder, there are three possible approaches to realize multimedia encryption, as shown in Figure 14.5: *encryption before encoding, encryption after encoding*, and *joint encryption-encoding*. The third approach also covers the case when an encoded multimedia file/bitstream is (partially) decoded and a joint process of encryption and (partial) re-encoding is performed to generate the encrypted multimedia file/bitstream. Note that the (partial) decoding and re-encoding processes may not involve decompression and/or re-compression. One typical example is sign bit encryption, in which only partial decompression is involved in the decoder at the encipher side, since the encrypted sign bits can be directly put back into their original places in the input (already encoded) multimedia data.

The first two approaches are much simpler due to their independence of the encoding process. Unfortunately, they suffer from some nontrivial drawbacks that cannot be easily overcome without introducing dependence on the encoding process. The following briefly discusses the three approaches of multimedia encryption and explains why the third approach, that is, joint encryption-encoding, is of greater interest.

### 14.3.2.1 Encryption before Encoding

This approach has been followed in some multimedia encryption algorithms designed to encrypt raw multimedia data. However, it hardly works with any existing multimedia coding systems because encryption generally leads to a random-like output that cannot be further compressed effectively. There do exist some special algorithms supporting compression after encryption. One solution based on distributed source coding was reported in Reference [21], taking the encryption key as useful side information available at the decoder side. For this solution, decoding is generally impossible without the knowledge of the key, so decryption and decoding have to be carried out simultaneously, which is not desired in applications requiring format compliance. In addition, this specific scheme puts some requirements on the encryption algorithm used such that not all available ciphers can be chosen and deployed.

### 14.3.2.2 Encryption after Encoding

This is the most direct and simplest approach, and is often called *naive encryption* in the literature [22]. Compared with the previous approach, this approach has no negative influence on the compression process and can run (much) faster since the data encrypted have been efficiently compressed by the encoder. The main drawback of naive encryption is its incapability of maintaining *format compliance* of compressed multimedia data, which is a natural result of its independence of the encoding process. Here, format compliance is defined as follows: the encoded-encrypted data can still be (probably partially) decoded without access to the key. Although format compliance is not always very important in all applications, it is indeed an essential requirement of many multimedia applications such as the following:

- *in-between multimedia processing of encrypted multimedia data without access to the key*;

- *perceptual encryption*: encryption of selective perceptual information, which requires that the unencrypted part of the multimedia data can still be decoded without access to the key;

- *scalable encryption*: different resolutions and quality levels of the same multimedia bitstream have different encryption configurations; and

- *region-of-interest encryption*: encryption is selectively performed on regions of interest (maybe dynamically detected objects) instead of the whole multimedia data.

The following example demonstrates why format compliance is useful in practice. Imagine a web site hosting millions of compressed video files uploaded by a large number of registered users. One user, Alice, wants some of her video files publicly accessible to all registered users, but with a visual quality determined by a threshold $q$. Interested users have to pay a fee to Alice for the high-resolution videos. The video hosting site automatically embeds a watermark into each video file to track each specific user for the anti-piracy purposes. This digital watermarking process has to be done without access to the decryption key, because Alice does not fully trust the server. Apparently, blindly applying a cipher like

AES to an encoded video file cannot help Alice. Instead, the encipher has to be aware of the syntax format of the video file, in order to know what syntax elements should be encrypted. That is, Alice needs joint encryption-encoding – the third approach to multimedia encryption.

It deserves mentioning that homomorphic encryption [23] may be used to achieve format compliance for some applications. However, the performance of homomorphic encryption is still very limited because a really practical encryption scheme should be homomorphic to all possible signal processing and compression operations, which include not only addition and multiplication, but also nonlinear operations like non-uniform quantization. Although designing an encryption scheme homomorphic to a single operation like addition or multiplication is easy, it remains an open problem to design a practical encryption scheme homomorphic to various operations simultaneously [24], [25].

### 14.3.2.3   Joint Encryption and Encoding

While joint encryption-encoding is the only approach to keep compression efficiency and format compliance simultaneously, it actually has more advantages over the other two approaches. Essentially, joint encryption-encoding is a problem of how to apply a cipher to multimedia data so that some application-oriented requirements can still be fulfilled. As part of the encoding process, the encipher can adapt itself to achieve an optimal/acceptable solution to the target application. When there are multiple performance factors, which is the normal case in practice, jointly performing encryption and encoding is often helpful to achieve a better trade-off among all the performance factors.

Since the 1990s, a large number of joint multimedia encryption-encoding (JMEE) algorithms have been proposed. The basic encryption techniques [26] include secret permutations of syntax elements at different levels, fixed-length codewords encryption, variable-length codewords index encryption, secret entropy coding, "virtual full encryption" working with context-adaptive coding, header encryption, and so forth. Three main aspects of joint encryption-encoding are what cipher should be used, what syntax elements should be encrypted, and how the selected syntax elements should be encrypted. Most JMEE algorithms are independent of selection of the underlying cipher, although some can only work with stream ciphers or newly designed ciphers tailored for multimedia encryption.

For a typical JMEE algorithm, several performance factors need to be considered simultaneously, which mainly include security, format compliance, compression efficiency, computational complexity, energy efficiency, size preservation, and parallelism. The need for format compliance has been clarified in the previous subsection and the security requirement is obvious, so the following discusses other factors.

- *Compression efficiency* becomes a concern when the encryption algorithm changes the behavior of the encoder, which often decreases the compression efficiency to some extent. Note that compression efficiency is always calculated and compared with respect to a level of perceptual quality.

- *Computational complexity* determines the processing speed and the throughput of the JMEE system. Since encryption will always add more computations on top of the multimedia encoding process, it is beneficial to minimize the additional complexity induced by encryption.

- *Energy efficiency* is one of the most crucial factors for resource-constrained devices like wireless multimedia sensor networks (WMSNs) [2]. It has a close link to computational complexity, but cannot be replaced by the latter because different computing components (for example, CPU and the wireless module) often have different energy consumption rates. This implies that the fastest solution is not necessarily the most energy-efficient one.

- *Size preservation* can become a desirable feature, when the multimedia data have already been encoded. In this case, one may want that (ideally) each syntax element keeps its original size. Such a feature is useful to avoid bitrate re-control, and to support some interesting functionalities, such as on-the-fly encryption/decryption and concurrent encryption/decryption of different parts of a single file. Size preservation is a non-trivial issue for multimedia encryption due to the use of variable-length coding (VLC) in most multimedia coding systems.

- *Parallelism* is possible for some selective encryption schemes of encoded multimedia data. For instance, sign bit encryption can run in parallel for different random access units (for example, frames or slices). The structure of a specific multimedia coding system may allow different kinds of parallelism at different levels, and the added encryption may enhance or compromise the parallelism.

As a whole, an ideal JMEE system should have the following features:

- high security, meaning that no perceptual information leaked from the encrypted part;

- low computational complexity, which corresponds to fast encryption/decryption and high throughput;

- high energy efficiency or low energy consumption;

- strict format compliance, meaning that every syntax element can be decoded without access to the key;

- strict size preservation, that is, every syntax element preserves its original size;

- high parallelism or high throughput on multicore and distributed platforms; and

- other desirable features, such as supporting perceptual encryption, working with any cipher, and so forth.

Many multimedia encryption schemes are designed following the idea of *selective encryption*. Although many modern ciphers (like AES and RC4) are fast enough to handle multimedia data, it is always nice to avoid handling the whole bulky data. Even if the reduction in computational complexity is considered insignificant, the saved energy may still be meaningful for power-constrained devices, such as sensor nodes. For instance, a 10% reduction in computational complexity is not too much, but the saved energy will extend the lifetime of a power-constrained device from 9 days to 10 days. Selective encryption is specifically important when the input plaintext is compressed. In this case, there is no full encoding process, but only a partial decoding process followed by encryption and probably by a partial re-encoding process. This partial decoding and encoding process can be significantly simpler than the whole coding process, and thus selective encryption can help reduce the total computational complexity and energy consumption dramatically. For instance, for sign bit encryption of MPEG videos, the partial decoding process does not involve motion compensation, inverse DCT and some other time-consuming steps. In addition, the number of sign bits is significantly smaller than the total number of bits. Therefore, selective encryption will be much more efficient than naive full encryption in terms of computational complexity and energy efficiency.

Since multiple performance factors have to be considered, it is a challenging task to design a multimedia encryption scheme fulfilling all the desired features. As a matter of fact, recent research has shown trade-offs among different performance factors for all the available multimedia encryption methods. Some of these trade-offs are listed below.

- Encryption of variable-length codewords or their indices can maintain format compliance, but normally cannot preserve the size. Leaving variable-length codewords unencrypted can preserve the size, but may not be able to offer a sufficient level of security.

- Secret permutations can be easily implemented to ensure format compliance and size preservation, but it is insecure against known/chosen-plaintext attacks [27] and even ciphertext-only attacks [22].

- FLC encryption can maintain format compliance and size preservation easily, but it is unable to provide a very high security level against error-concealment attacks (ECA) [28].

- VLC index encryption can ensure format compliance and security when used properly, but it has an uncontrollable influence on compression efficiency and cannot maintain size preservation. By constraining the index encryption in a small number of VLCs, the influence on compression efficiency will be less problematic, but the security will be compromised to some extent [29].

- Secret entropy coding is unable to keep format compliance and size preservation. In addition, most secret entropy coders are either insecure against chosen-plaintext attacks [30], [31], [32] or unable to provide a better performance than simple naive encryption (that is, encryption after compression) [32].

- "Virtual full encryption" and header encryption cannot ensure format compliance for some multimedia coding standards, mainly due to the dependence of the encrypted syntax elements on the context.

- Most format-compliant selective encryption schemes cannot conceal all perceptual information, thus they cannot be used for some applications requiring a high level of perceptual security [33], [34], [35], [36], [37], [38].

- High parallelism can be achieved by some size-preserving JMEE schemes, which are unable to offer a high level of perceptual security because some syntax elements are left unencrypted.

It has been known that most of the trade-offs have their roots in the underlying multimedia coding process: independent coding of different syntax elements, scattering of visual information over different syntax elements, visual information contained in bit sizes of syntax elements, embedded error-concealment mechanisms, and so on. In case when the underlying multimedia coding standard cannot be changed, the problem of multimedia encryption becomes how to achieve a better trade-off among all the performance factors for the target application.

## 14.4 Perceptual Multimedia Encryption

As mentioned above, perceptual encryption is one of the key reasons for joint encryption-encoding. As a major branch of multimedia encryption, perceptual encryption can be useful in many applications, such as pay-per-view and video on demand (VoD) services. When a multimedia application involves multiple parties and not all of them are trusted, perceptual encryption is also useful to allow those untrusted parties to perform some postprocessing on encrypted multimedia data. For instance, in the example discussed in the previous section, digital watermarking may be allowed by a third party without access to the decryption key, where the watermark can be embedded into the unencrypted part without influencing the encrypted part but still remain detectable at the user end.

The concept of perceptual encryption was introduced in Reference [39], where the authors used the term "transparent encryption" in the context of digital TV broadcasting. Some other researchers later used the term "perceptual encryption" because "transparent encryption" may be confused with another term "format-compliant encryption" (meaning encryption is transparent to encoding and vice versa). This chapter adopts the term "perceptual encryption." In the literature, there are different perceptual encryption proposals for audio, speech, image, and video. This chapter focuses on perceptual encryption for digital images and videos only.

### 14.4.1 Definition and Performance Evaluation

Perceptual encryption is selective multimedia encryption where only a specific amount of perceptual information is encrypted by encryption. Equivalently, perceptual encryption can be seen as intended quality degradation by the means of encryption but the degraded quality can be recovered by a decryption key. The nature of perceptual encryption leaves part of the data untouched and allows a decoder to recover unencrypted perceptual information without accessing to the decryption key. Ideally, a fully scalable perceptual encryption should allow the perceptual quality to be downgraded under the control of a continuous quality factor $q \in [0, 1]$. When $q = 1$, the maximum amount of perceptual information is encrypted and the perceptual quality becomes minimal; when $q = 0$, the minimum amount of perceptual information is encrypted and the perceptual quality is maximized; when $0 < q < 1$, part of the perceptual information is encrypted and the perceptual quality is downgraded accordingly. Figure 14.6 shows an illustrative view of a perceptual encryption scheme.

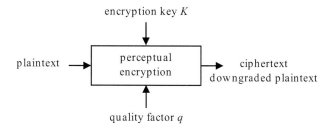

**FIGURE 14.6**

A perceptual encryption scheme.

Given the above definition of perceptual encryption, the following facts about the control factor $q$ should be noted:

- Despite extensive research on visual quality assessment (VQA), there still does not exist a well-accepted objective measure of visual quality that can replace an average human observer's subjective quality assessment. Therefore, the control factor $q$ is generally chosen to represent a rough measure of the quality degradation.

- While the relationship between the control factor $q$ and the quality degradation cannot be linear (due to the lack of a proper VQA metric), the monotonicity should be maintained, meaning that a larger $q$ should statistically correspond to a higher level of quality degradation. This implies that a selective format-compliant encryption scheme is not necessarily a perceptual encryption scheme unless the selective encryption can downgrade the perceptual quality in a monotonic manner under the control of a factor $q$.

- In most cases, the control factor $q$ is defined as a parameter of the encryption scheme that have a direct relationship with the visual quality of the encrypted image/video, for example, the percentage of syntax elements encrypted.

- For digital videos, the quality degradations of different frames may be different, so the control factor represents the average quality of all frames.

- Depending on applications, the control factor may have different levels of granularity, so there are may be only a limited number of possible values of $q$.

- Multiple control factors may be used to have a finer control of different aspects of visual quality degradation.

As in general multimedia encryption schemes, there are multiple performance factors to be considered when evaluating the performance of a perceptual encryption scheme. Among all the performance factors, main attention is focused on security since all the other factors are not unique for perceptual encryption. There are two unique features making the security of perceptual encryption schemes essentially different from other kinds of multimedia encryption schemes: i) the existence of two separate sets of data (one encrypted and one unencrypted) containing perceptual information; and ii) the unencrypted data are decodeable to an attacker who does not know the key. The need of maintaining some level of perceptual quality immediately leads to two different levels of security that should be considered:

- *cryptographical security* – measurement of security against cryptographical attacks that try to recovery the plaintext of the encrypted data;

- *perceptual security* – measurement of security against *perceptual attacks* that try to recover as much perceptual information as possible out of the unencrypted data *without* any attempt of breaking the underlying encryption algorithm.

Although most perceptual encryption schemes are based on cryptographically strong ciphers, the cryptographical security may not be guaranteed because the interface between the cipher and the image/video coding system may create new insecure points so that the cipher fails. Perceptual security is a unique concept for perceptual encryption and covers perceptual attacks that can recover perceptual information from the encrypted data in non-cryptographical means. Perceptual attacks are by definition indifferent of any details about how the encryption algorithm works, which means that they are more general and can also be generalized to applications where some data are missing, corrupted, or distorted due to other reasons (for example, transmission errors and noise, and/or intended or unintended data removal, truncation, and manipulation).

In the following, a brief survey of perceptual encryption schemes for digital images and videos is provided, then the two levels of security are separately discussed with greater detail.

### 14.4.2   Perceptual Encryption Schemes

This subsection does not attempt to give a complete coverage of all schemes in the literature; it aims at highlighting the main relevant techniques and some representative schemes that can better demonstrate interesting features of perceptual encryption schemes in general. Most perceptual encryption schemes are conceptually simple and straightforward, and the main technical question is what data should be selected for encryption given a specific image and coding standard as the carrier of the multimedia data. Despite the conceptual simplicity of many perceptual encryption, performance evaluation of them especially security analysis is non-trivial, as will be shown later. Throughout the subsection, it is assumed that perceptual encryption is implemented by adopting a cryptographically strong cipher like AES or RC4 unless otherwise stated.

#### 14.4.2.1   Scalability-Based Encryption

Some image and video coding standards, such as JPEG2000 [7], MPEG-2/4 [40], [41] and MPEG SVC [42], have inherent support on scalable coding, meaning that more than one quality layer can be included in a single image or video bitstream. As a consequence, it is natural to realize perceptual encryption by encrypting selected quality layers while leaving others untouched [43]. Perceptual encryption of this kind is also called scalable encryption or multilevel encryption due to the existence of several quality layers. The main problem with this perceptual encryption setting is the mandatory requirement on scalable coding, which may not be adopted by some image and video formats. The limited quality granularity caused by the small number of quality layers is another problem.

### 14.4.2.2 DCT-Based Image and Video Encryption

Discrete cosine transform (DCT) is the most widely used transform in image and video coding standards [9], [10], [40], [41], [44], [45],[3] so many researchers have proposed perceptual encryption schemes based on selective encryption of DCT coefficients and their bit planes. There are several approaches of achieving perceptual encryption in DCT domain [28], [46], [47], [48], [49], [50], [51]: DC encryption, AC encryption, sign bit encryption, bit-plane encryption, and hybrid encryption (for example, DC encryption plus sign bit encryption). Note that DCT encryption can implement scalable encryption because different DCT coefficients of a block represent visual details of different frequency components.

In most DCT-based image and video coding standards, DC coefficients and sign bits are coded differently from AC coefficients and other bit planes, so although conceptually DC encryption has no real difference from AC encryption, the overall performance of the perceptual encryption scheme can be very different. For instance, in JPEG and MPEG-1/2/4 standards, DC coefficients and sign bits are (partly or completely) coded as fixed-length codewords, which means that DC and sign bit encryption can achieve strict size preservation and maintain compression efficiency. AC encryption and bit-plane encryption are relatively complicated because they involve rule-length coding and variable-length coding (VLC). One simple way of circumventing the VLC problem is not to encrypt the value of each AC coefficient or bit plane, but to encrypt their positions by shuffling all encrypted elements. The shuffling process can be driven by a stream cipher. Another solution is to encrypt selected AC coefficients and bit planes before VLC happens, which unfortunately will influence compression efficiency in an uncontrollable manner.

In DCT-based perceptual encryption, the quality factor $q$ is usually set as the percentage of targeted syntax elements (DC/AC/sign bits/bit planes) encrypted. Perceptual encryption operations can be performed immediately after DCT transform, but only some can work at the bitstream level. For the former case, the compression efficiency will normally be compromised because encryption can increase the information entropy of the encrypted syntax elements. Therefore, the quality control factor also becomes a control factor of compromised compression efficiency. For the latter case, encryption cannot be directly applied to syntax elements in order to maintain format compliance, and often the only possible way of performing encryption is to randomly permute the syntax elements concerned. Depending on how the underlying image and video coding standard is designed, only some syntax elements can be permuted at the bitstream level. For instance, for JPEG and MPEG permutations of AC coefficients are possible by shuffling the VLC codewords representing a group of running zero AC coefficients and a nonzero AC coefficient.

### 14.4.2.3 DWT-Based Image and Video Encryption

As the second most widely used transform in image and video coding [7], [52], [53], discrete wavelet transform (DWT) has also attracted significant attention from researchers

---

[3]Note that some video coding standards use an integral approximation of DCT, so strictly speaking they are not based on DCT but DCT-like transforms. However, such a subtle difference does not influence how perceptual encryption and cryptanalysis work, and therefore it is ignored in this chapter.

working on perceptual encryption [54], [55], [56], [57], [58], [59]. The perceptual encryption settings for DCT can be largely generalized to DWT, but different DWT decomposition levels and high-/low-frequency subbands replace DC and AC coefficients in the DCT case.

Most DWT-based image and video coding schemes follow an incremental coding principle; that is, the most significant bits are encoded first and less significant bits come later. This means perceptual encryption can be naturally realized by gradually encrypting the encoded bitstream from the beginning of each independently coded unit. In addition, DWT has a pyramid structure that allows performing random permutations at different levels and on different subbands to achieve different degrees of quality degradation.

For perceptual encryption schemes working at the encoded bitstream level of JPEG2000 images/videos, a special issue has to be carefully handled to ensure format compliance of ciphertexts; the encrypted bitstream should not create fake termination markers that are words in the range 0xff90 to 0xffff [60], [61].

### 14.4.2.4 Motion Vector Scrambling

For digital videos, scrambling motion vectors can create lightweight perceptual encryption schemes controlling temporal visual quality only [28], [62]. Since intra-coded video frames do not contain any motion vectors, this approach cannot have any quality control over these frames. As a result, motion vector scrambling should be used as a supplemental option for further enhancing the performance of perceptual encryption schemes based on other techniques.

### 14.4.2.5 Fixed-Length Codeword Encryption

In most image and video coding standards, there are syntax elements coded as fixed-length codewords (FLCs). Taking MPEG-1/2/4 videos as an example, the following syntax elements are FLCs: sign bits of nonzero DCT coefficients, (differential) DC coefficients in intra blocks, ESCAPE DCT coefficients (DCT coefficients out of the scope of Huffman coding), and sign bits and residuals of motion vectors. By selectively encrypting some FLCs in an image/video, one can achieve strict size-preservation and format compliance at the same time. The size-preservation will further allow on-the-fly encryption and concurrent encryption, as pointed out in Section 14.3. A typical perceptual encryption based on FLC encryption is the so-called perceptual video encryption algorithm (PVEA) [28], which is designed for MPEG videos. A unique feature of PVEA is the use of three independent quality-control factors for the low-resolution rough (spatial) view, the high-resolution (spatial) details, and the (temporal) motions.

### 14.4.2.6 Fractal Coding-Based Encryption

Fractal coding has been developed as an image compression algorithm [63]. The basic idea behind is to represent an image as a set of transformations associated with an iterated function system, which can converge to a so-called fixed point (which is a normally a fractal set). In the case of fractal image coding, the fixed point will be an approximation of the original image. Since fractal coding represents an image as a set of transforms, it is possible to encrypt part of the parameters of the transforms to achieve perceptual encryption, as proposed in Reference [64].

### 14.4.3   Cryptographical Security

The following discusses cryptographical security from the perspectives of different attacks. If one can break the cipher used to encrypt the selected part of the image and video data, then the whole perceptual encryption scheme will collapse. While normally a cryptographically strong cipher like AES or RC4 is used to guarantee security, the way how such a cipher is applied to image and video data can make a difference. This becomes more obvious when the cipher has to be applied in a way different from its original design. For instance, some perceptual encryption schemes are based on random permutations of selected syntax elements of an image/video bitstream. A stream cipher is usually used to produce a pseudo-random permutation map, which is then applied to achieve the effect of encryption. Here, the use of the stream cipher differs from the normal usage in cryptography, where the keystream is used to mask the plaintext (normally via XOR or modular addition) rather than to shuffle their positions.

#### 14.4.3.1   Brute-Force Attacks

The simplest way of breaking a cryptosystem system is to exhaustively try all possible plaintexts/keys until the correct one is found. This attack can be estimated by the size of the plaintext/key space.

For the key space, the security against brute-force attack can be ensured by using a cryptographically strong cipher, which always uses a sufficiently large key space. The plaintext space is a bit more complicated because it depends on the number of bits of the encrypted part of the image/video bitstream. Here, one has to consider each independently encrypted unit rather than the whole encrypted data, because independent units can be guessed separately. Although it is possible to make all encrypted bits dependent on each other, for example, by using a mode with ciphertext feedback like CBC, those bits may still be separately guessed because of their semantic meanings. A typical example is DC encryption of DCT-transformed images and videos. One can make encryption results of all DC coefficients dependent on each other; however, the attacker only needs to break a small amount of them to reveal part of the image. Since part of an image may already contain sensitive perceptual information (for instance, a person's face), this partial brute-force attack can be considered successful although it is not in a cryptographical sense. This can be a serious problem when the quality control factor $q$ is small, meaning that the percentage of encrypted syntax elements is low and thus the average distance between consecutive syntax elements encrypted is large. In other words, this fact creates a lower bound of the quality control factor $q$. For a more detailed discussion of this subtle issue on an ad hoc design of perceptual encryption, readers are referred to Reference [28, Sec. III.B]. Note that this fact also blurs the boundary between cryptographical security and perceptual security because now one has to ask what is the consequence of revealing part of the encrypted data *without* considering what encryption algorithm is actually used.

Another related issue about brute-force attacks is how to judge a guessed key is correct. Apparently this should not be done by manually checking the perceptual quality of recovered plaintext unless this only requires a very small number of manual checks (say, hundreds). To automate the checking process, one needs some criteria. For image and video perceptual encryption, this can be naturally done by using a no-reference image/video qual-

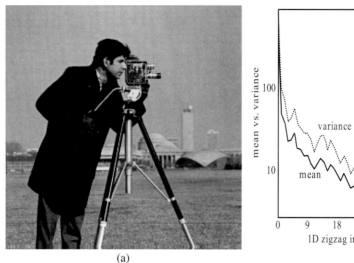
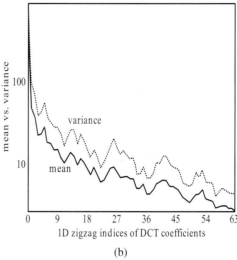

(a)                                                           (b)

**FIGURE 14.7**

The means and variances of coefficients of $8 \times 8$ DCT on a $256 \times 256$ test image *Cameraman*.

ity assessment tool that can measure the difference between the original plaintext and the ciphertext in terms of perceptual quality. For instance, DC encryption will create discontinuity along block boundaries, so the level of such blockiness can be used as a measurement of how good a guessed key is. The key giving the minimum level of blockiness can be selected as the correct key.

### 14.4.3.2 Ciphertext-Only Attacks

Since it is assumed here that a cryptographically strong cipher is used for encryption, the target of a ciphertext-only attack in this context is not the key, but the plaintext of the encrypted data. Being cryptographically strong, the underlying cipher should not allow an attacker to launch a successful ciphertext-only attack. However, in perceptual encryption, there is the context that is not encrypted, that is, what the encrypted data represent semantically. This may give the attacker a chance. This is the case when random permutations (driven by a stream cipher) are applied to DCT coefficients. Consider a perceptual encryption scheme random permutes the first eight most significant DCT coefficients, that is, DC coefficient and the first seven AC coefficients. The nature of DCT as a suboptimal decorrelation map implies that the both the mean and the variance of the DCT coefficients drop from the low-frequency to high-frequency bands (see Figure 14.7). As a consequence, if the permuted DCT coefficients are sorted according to their amplitudes, the original order of the eight encrypted DCT coefficients will likely be recovered with a non-negligible probability, thus recovering the image/video with a better overall quality. Figure 14.8 shows the result when such an attack is applied to a test image *Cameraman*. One can see that the visual quality of Figure 14.8b is obviously much better than that of the ciphertext shown in Figure 14.8a. A similar attack for DCT encryption of MPEG videos can be found in Reference [65, Sec. 3.4]. There are also ciphertext-only attacks that are completely independent of how encryption is implemented, but those attacks are about perceptual security and will be discussed in the next subsection.

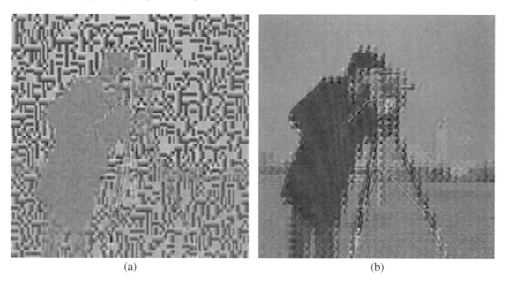

(a)    (b)

**FIGURE 14.8**

(a) The ciphertext of randomly permuting the first eight DCT coefficients in each block. (b) The result of the ciphertext-only attack by sorting the first eight DCT coefficients according to their amplitudes.

### 14.4.3.3   Known/Chosen-Plaintext Attacks and Chosen-Ciphertext Attacks

In these attacking scenarios, the target is the key since the plaintexts have been known. Under the assumption that the underlying cipher is cryptographically strong, such attacks are largely impossible. However, due to the special structure of stream ciphers, the key cannot be reused to avoid extraction of the keystream by comparing the plaintext and its ciphertext. The common practice is to change the key for each image/video frame or the minimum encryption unit (for example, a slice of an MPEG-1/2/4 video frame). This can be done by making the key dependent on the contents of the plaintext. One approach is to combine a long-term master key with the hash value of the image/video (or hash value of the concatenation of both). In this case, the hash value should also be transmitted with the image/video, but it does not need to be kept secret so it can be encoded in the user data field of the image/video. When the employed image and video coding format does not support user data, one can consider embedding the hash value into the image/video as a watermark. Since the watermark does not need to be secret but need to be lossless, it is also accepted to make the watermark visible at one corner of the encrypted image/video like a logo in a broadcasted TV program.

### 14.4.4   Perceptual Security

For perceptual security, one does not care about any detail of the encryption algorithm, but cares about how to recover a better quality out of the unencrypted data. This means that all attacks in this category fit into the class of ciphertext-only attacks.

From an attacker's point of view, the problem to be solved can be described as follows:

- *What is available* – the unencrypted visual data and the known semantics of such data (for example, the encrypted data are all DC coefficients or sign bits).

- *What is a successful attack* – if one can recover the plaintext image/video with a quality (statistically) better than that of the ciphertext (that is, the one defined by the quality control factor $q$).

- *What are the constraints* – both the time complexity and the space complexity should be sufficiently low to make the attack practical even for relatively large images and videos.

The above points are self-explanatory, but the last one needs a bit more discussion. How high the time complexity, that is, how slow an attack, is allowed depends on the application. If the attacker wants to crack a live broadcast program, the attack will have to be fast enough to catch the frame rate, meaning that the attack should be done in $O(10)$ milliseconds. If the attacker just wants to crack what is stored on a hard disk or a DVD disc, then it is possible to run the attack for hours or even days. When an attacker has access to a botnet or a supercomputer, the above requirement can be relaxed at an order of $O(10^6)$. Note that an attack running in polynomial time (in the so-called P complexity class) may not be practical at all if either the hidden constant factor or the fixed exponent is too large. The requirement on the space complexity is normally tighter because the storage available to an attacker is largely fixed and does not grow over time.

All attacks in the perceptual security category follow the same basic approach; namely, replace the encrypted data by an estimate from the available unencrypted data/information and then run the decoder to get an approximate edition of the plaintext image/video. The more accurate the estimate is, the better the recovered quality and the more successful the attack will be. Since this approach is very similar to error concealment in multimedia coding, these attacks are collectively called *error-concealment attacks* (ECAs) [34]. Some researchers also use the term *concealment attacks* or *replacement attacks* [59].

Error-concealment attacks have been so far the most successful attacks on perceptual encryption and some very promising results have been obtained in recent years on DCT-based encryption. It is also likely that these attacks can be easily generalized to other transforms, including DWT. Some error-concealment attacks also have more profound implications on image and video coding (compression) in general. Such attacks are discussed in more detail in the next section.

Since the performance of an error-concealment attack has to be judged against the quality defined by the ciphertext, measuring the perceptual information for the comparative purposes becomes an important task. The golden standard of perceptual information measurement is subjective evaluation by human experts. However, subjective quality assessment is very time-consuming and costly, so researchers have proposed various objective visual quality assessment (VQA) metrics for digital images and videos [66], [67]. Traditionally, image and video coding community has been using a simple metric called peak signal-to-noise ratio (PSNR). For an $M \times N$ image or video frame $I^* = \{I^*(i,j)\}$ and its distorted version $I = \{I(i,j)\}$, where $1 \leq i \leq M$ and $1 \leq j \leq N$, PSNR is defined as follows:

$$\text{PSNR} = 10 \cdot \log_{10} \frac{I_{\max}^2}{\sum_{i=1}^{M} \sum_{j=1}^{N} (I(i,j) - I^*(i,j))^2 / (MN)}, \qquad (14.5)$$

where the denominator, called mean square error (MSE), is also widely used in objective quality assessment. Unfortunately, neither PSNR nor MSE matches subjective quality very well [67, Figure 1.1], so more advanced VQA metrics, such as the structural similarity index (SSIM) [68], visual information fidelity (VIF) [69], and visual signal-to-noise ratio (VSNR) [70], have been proposed. Although VQA metrics are commonly tested on various test databases to demonstrate their good correlation with subjective data collected from human observers, a recent comprehensive comparative study has shown that none of existing VQA metrics consistently outperforms all others [71].

Some encryption-oriented VQA metrics have also been proposed to obtain more accurate measurements on encrypted images and videos. Such metrics include the luminance similarity score (LSS) and edge similarity score (ESS) [72], neighborhood similarity [73], similarity of local entropy [74], and a number of metrics based on image histogram and radial Fourier spectrum [75]. Basically speaking, these encryption-oriented VQA metrics focus more on a specific aspect of the overall visual quality that is supposed to be influenced by perceptual encryption in a more scalable way (meaning that no abrupt change will happen as the quality factor $q$ changes slightly). The main drawback of all the existing encryption-oriented VQA metrics is the lack of performance evaluation against subjective quality. This makes it difficult to judge whether there exist the best metric and if so, which one it is.

## 14.5 Error-Concealment Attacks

This section surveys all error-concealment attacks currently known. At first, it discusses the so-called naive ECA, which involves only replacement of encrypted data by a fixed value. Then, this section introduces different kinds of advanced ECAs that try to have a better estimate of the encrypted data by looking at the semantics/statistics of the encrypted data and its interplay with the unencrypted data. This section ends with the most promising ECA reported very recently, which represents a universal approach to breaking many perceptual encryption schemes with globally optimized results.

### 14.5.1 Naive Attacks

The naive ECA is the simplest ECA in the sense that it just tries to replace all encrypted syntax elements of the same kind by a fixed value according to their semantic meanings. This can be more easily explained by how such an attack works with DC encryption of DCT-transformed images. While one has no idea what the DC coefficients are, it is known that they have a range determined by the valid range of pixel values. For $N \times N$ DCT and eight-bit grayscale images, this range is $[0, 255N]$. By setting all DC coefficients to any value in this valid range, for example, 0 or $255N/2$, and then scale the dynamic range of the resultant image to $[0, 255]$, an image with an overall quality better than the ciphertext can be produced. Figure 14.9 shows the result on the test image *Cameraman*. One can see that the image obtained via the naive ECA obviously has a higher visual quality than the ciphertext image. Using PSNR as the objective VQA metric, the same conclusion can be reached: the ciphertext image has PSNR = 11.5883 dB, whereas the image recovered from

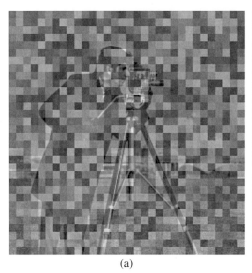 

(a)                                                                    (b)

**FIGURE 14.9**

(a) The ciphertext image of Figure 14.7a when all DC coefficients are encrypted. (b) The image recovered from the ciphertext image by setting all encrypted DC coefficients to zeros.

this ciphertext by the naive ECA has PSNR = 13.0428 dB. The naive ECA is the widely acknowledged attack. Most perceptual encryption schemes were tested against this attack to validate its perceptual security.

### 14.5.2 Local Constraints-Based Attacks

In the naive ECA, all encrypted syntax elements of the same kind are replaced by the same fixed value. Such a global replacement is not necessarily (generally not) the best choice. By considering a small local window around each encrypted syntax element, some local constraints with a tighter range than the global range may be found. Once again, this can be best demonstrated by the case of DC encryption. Since the two-dimensional DCT is performed blockwise, the local window can be naturally selected as the block each encrypted DC coefficient lies in. In this case, Property 14.1 gives an interesting local constraint of the unknown DC coefficients defined by the known AC coefficients.

**Property 14.1**
Given an $N \times N$ block $B$, the pixel values calculated from its DC-free edition $B^{(0)}$ (that is, the inverse DCT result by setting the DC coefficient to 0) define a constraint on the DC coefficient:

$$N(t_{\min} - \min(B^{(0)})) \leq \mathrm{DC}(B) \leq N(t_{\max} - \max(B^{(0)})), \tag{14.6}$$

where $[t_{\min}, t_{\max}]$ denotes the valid range of pixel values.                              $\square$

The local constraint defines a valid range of the DC coefficient of the block containing it, which is a proper subset of the global valid range with high probability. Therefore, if the midpoint of the tighter range is used to replace the encrypted DC coefficient, the ECA attack

**FIGURE 14.10**

The image recovered from the ciphertext image shown in Figure 14.9a by setting each encrypted DC coefficient to the midpoint of the valid range of that block. ©2011 IEEE

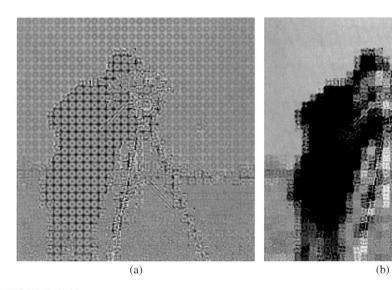

(a)        (b)

**FIGURE 14.11**

The results of attacks on the test image *Cameraman*. (a) The naive ECA by setting all encrypted DPSS co-efficients to zeros, resulting in PSNR = 12.7071 dB. (b) The advanced ECA as reported in Reference [76], resulting in PSNR = 18.2721 dB.

should give results closer to the plaintexts with a higher visual quality. Figure 14.10 shows the result of such an advanced ECA applied to the ciphertext in Figure 14.9a. Comparing Figure 14.10 with Figure 14.9b reveals that the visual quality is further improved, which is also reflected by higher PSNR, that is, 13.8837 dB.

This advanced ECA on DC coefficients can also be generalized to any encrypted DCT coefficients. When more than one DCT coefficient are encrypted in each block, the local constraints have to be calculated jointly by checking all possible values of the encrypted DCT coefficients and taking those values that do not produce invalid pixel values.

### 14.5.3   Prior Statistics-Based Attacks

If the attacker can have access to an image database that represents the statistics of the plaintext image of interest, then it will be more beneficial to replace the encrypted syntax elements by the means of the most likely values rather than the midpoints of the valid ranges. Such an advanced ECA was demonstrated in Reference [76] on an image encryption scheme based on two-dimensional discrete prolate spheroidal sequences (DPSS) [77]. This image encryption is not a real perceptual encryption scheme, but the basic idea behind the statistics-based ECA demonstrated in Reference [76] can be generalized to any perceptual encryption scheme. To be more exact, the advanced ECA exploits statistics of the ratios between the encrypted DPSS transform coefficients and the most significant unencrypted coefficient. Then, the statistics are used to derive a better estimate of each encrypted DPSS transform coefficient. Compared with the results of the naive ECA, the performance of the statistics-based ECA is significantly better, as shown in Figure 14.11.

### 14.5.4   DC Encryption Attacks Using Coefficient Recovery

All previously introduced ECAs are relatively simple and the recovered image can hardly reach a PSNR value larger than twenty. In Reference [38], the authors proposed a method that can boost the performance of an advanced ECA to a level far more than all previous ECAs. This attack was aimed at DC encryption of DCT-transformed images/videos. To facilitate the following discussion, this method is named here as USO by taking the initials of the three authors. The USO method is based on Property 14.1 of digital images and Property 14.2 of natural images.

**Property 14.2**
The difference between two neighboring pixels is a Laplacian variate with zero mean and small variance.                                                                                                    □

The two properties play two different roles and can work together to design an iterative ECA. Namely, Property 14.2 allows the attacker to estimate the DC coefficient of a block relative to its neighbor blocks by minimizing the pixel difference along the block boundaries, and Property 14.1 applies a tighter range of the DC coefficient on each block so that the intersection of all the relative DC coefficients can give a rough estimate of the global brightness of the image. After an attacker has the global brightness and all the relative DC coefficients, approximations of absolute values of the encrypted DC coefficients can immediately be obtained. Note that the estimation process of the relative DC coefficients involves a scanning process, which starts from a corner block of the image and then continues through the whole image toward the opposite corner to derive all the relative DC coefficients one after the other. It is obvious that such a scanning process suffers from error propagations. To mitigate this problem, Reference [38] proposed to have four different scans starting from the four different corners and then average the results to get a less noisy

**FIGURE 14.12**

The image recovered from the ciphertext image shown in Figure 14.9a by performing the USO attack, resulting in PSNR = 19.6975 dB.

result. After the whole process, the recovered plaintext image may still have pixel values out of the valid range, then a scaling or truncating operation is applied to remove those invalid pixel values. Figure 14.12 shows the result of the USO attack on the test image *Cameraman*. As it can be seen, the perceptual quality of the recovered image is boosted a lot. The PSNR value becomes 19.6975 dB, much higher compared to that of the naive ECA and the simple advanced ECA described in Sections 14.5.1 and 14.5.2. For some other images, such as *Lena*, the PSNR can go beyond 20 dB.

### 14.5.5 DC Encryption Attacks via Underflow/Overflow Rate Minimization

While the USO method works largely well, it sometimes suffers from a relatively low accuracy of the DC estimation when the error propagation is not under control so that many pixel values go out of the valid range. For the averaged result from the four scans shown in Figure 1 of Reference [78], the pixel values range in the interval $[-88.6, 303.0]$. By scaling the range to $[0, 255]$, the finally recovered image has a low PSNR value of 14.3 dB.

To further reduce the error propagation effect, Reference [78] proposed to apply Property 14.1 during the relative DC estimation process rather than after it. To be more exact, during the scanning process, the underflowed and overflowed pixel values are checked for each block. Once such underflows and/or overflows happens, the whole block is shifted toward the valid range of pixel values until all underflowed and overflowed pixel values disappear. The modified DC estimation process makes the final result more sensitive to the global brightness, or equivalently the DC coefficient of the first block in each scan. Reference [78] also proposed an iterative process to get a more accurate estimate of the first DC coefficient, which is based on an interesting observation that roughly holds for various natural images: the more accurate the estimate of the first DC coefficient, the less frequently underflowed and overflowed pixel values appear in each block. In other words, the underflow/overflow rate is a unimodal function (with only one global minimum) of the es-

**FIGURE 14.13**

The performance of the FRM method on recovering the test image *Cameraman* shown in Figure 14.11a, resulting in PSNR = 22.8142 dB.

timate of the first DC coefficient. Based on this observation, the authors proposed to search all possible values of the first DC coefficients and take the one with the minimum underflow/overflow rate as the estimate. Since this rate is a unimodal function, the searching process can be done effectively with a time complexity of $O\left(M_B \log_2(N(t_{\max} - t_{\min})/\Delta)\right)$. This method was named as underflow/overflow rate minimization (FRM).

Because the iterative process of searching for the optimal value of the first DC coefficient, the FRM method is generally slower than the USO method, but it outperforms the USO method significantly according to experimental results carried out on 200 test images. The authors tested the performance improvement measured in ten different VQA metrics, and the FRM method was proved to be consistently better than the USO method. Figure 14.13 shows an example of using the FRM and USO methods for recovering a test image *Cameraman*; it can be seen that the FRM produces a result with an obviously better visual quality. For a detailed comparison between the USO and FRM methods, the readers are referred to Reference [78, Sec. 3.3].

### 14.5.6 Optimization-Based Approach to DCT Encryption Attacks

Both the USO and FRM methods can break DC encryption only. They cannot be easily generalized to more general cases, such as AC encryption or combination of DC and AC encryption, because Properties 14.1 and 14.2 describe the relationship between *one* unknown coefficient and all other known coefficients. The question about where an generalized attack exists was answered by in Reference [79] affirmatively.

This time the whole ECA is modeled as a discrete optimization problem and solved as a linear program. For the ECA breaking DC encryption, the discrete optimization problem can be described as follows:

$$\text{minimize } f\left(\{x(i,j)\}_{0 \le i,j \le N-1}\right) \tag{14.7}$$

**FIGURE 14.14**

The image recovered from the ciphertext image shown in Figure 14.9a by performing the linear programming-based ECA, resulting in PSNR = 26.4866 dB. ©2011 IEEE

subject to $x = A \cdot y$, $x_{\min} \leq x(i,j) \leq x_{\max}$, and $y(k,l) = y^*(k,l)$ for all AC coefficients. The terms $x(i,j)$ and $y(k,l)$ are variables denoting the value of pixel $(i,j)$ and the DCT coefficient $(k,l)$, respectively. The term $A$ represents the linear relationship between pixel values and DCT coefficients defined by the blockwise two-dimensional DCT, and $y^*(k,l)$ represents the ground truth of $y(k,l)$. Note that the last constraint fixes all AC coefficients to their ground truths while the DC coefficient $y(0,0)$ remains as a variable. In principle, the objective function $f$ can be any convex function to allow a polynomial-time optimization algorithm. However, considering Property 14.2, the authors chose the objective function to be $\sum_{\{(i,j),(i',j')\}} |x(i,j) - x(i',j')|$, where the sum ranges over all pairs of neighboring pixels. This choice makes the optimization problem a linear one, which can be solved efficiently by using the linear programming (LP) technique [80]. The linearization is done by introducing variables $h_{i,j,i',j'}$ into the model as follows:

$$\text{minimize} \sum h_{i,j,i',j'} \tag{14.8}$$

subject to $x(i,j) - x(i',j') \leq h_{i,j,i',j'}$, $x(i',j') - x(i,j) \leq h_{i,j,i',j'}$, and the constraints related to Equation 14.7. In the above model, the variable $h_{i,j,i',j'}$ will be tight at $|x(i,j) - x(i',j')|$ for the optimum solution.

With the above optimization model, the generalization from DC encryption to DCT encryption is straightforward: one can simply fix $y(k,l) = y^*(k,l)$ for known DCT coefficients while keeping all unknown variables in their natural ranges. The generalized model maintains the linearity, so it can still be solved via linear programming. Given an $n \times m$ image and $U$ unknowns in each of $B$ blocks with dimensions $N \times N$, the optimization model contains $2nm - (n+m)$ variables $h$ and $UB$ variables $y$ while all other unknowns can be determined by these key variables in a presolve step. The average time complexity and space complexity of the optimization problem were experimentally verified to be $O(n^2 m^2 U)$ and $O(nmU)$, respectively.

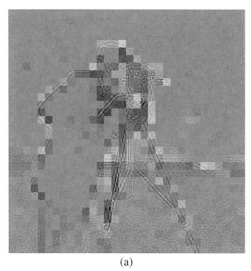

(a)                                                              (b)

**FIGURE 14.15**

The recovery results of the linear programming-based ECA when sixteen most significant DCT coefficients are
encrypted in each block of the image *Cameraman* shown in Figure 14.13a: (a) the image recovered by setting
all missing DCT coefficient to the midpoints of their valid ranges and (b) the image recovered by the linear
programming-based ECA.

The performance of the linear programming-based ECA works nearly perfectly for
breaking DC encryption. For the test image *Cameraman*, the recovered image reaches
a PSNR value of 26.4866 dB (see Figure 14.14). For the DC encryption case, the linear
programming-based method outperforms the FRM method statistically with a significant
margin (see Figure 2 of Reference [79]). What is more surprising, even when more than
ten most significant DCT coefficients are encrypted, the linear programming-based ECA
can still recover the plaintext with relatively high perceptual quality for some images. Figure 14.15 shows an example on the test image *Cameraman* when sixteen most significant
DCT coefficients are missing from each block.

The big success of the optimization-based ECA is rather unexpected, especially when it
turned out that a relatively simple optimization model can work so perfectly. The results
not only reveal that DCT encryption largely failed, but also lead to some more profound
implications on image and video coding. It raises the question about how much information
should be encoded for a DCT-based image and video coding scheme. In addition, some
digital watermarking and data hiding schemes can also be handled by the same optimization
model if they use some AC coefficients to hide the watermarks/hidden information.

## 14.6   Conclusion

Perceptual encryption of digital images and videos is useful for many applications and
can be easily implemented by selectively encrypting part of the whole image/video bit-

stream. However, recent research on error-concealment attacks has raised an alarm about the security of many perceptual encryption schemes.

Although the optimization-based ECA is very new, it has shown great potential of breaking probably almost all DCT-based perceptual encryption with high probability. Its generalization to DWT and other transforms is still to come, but it is highly likely that the generalized attacks will still work. Therefore, it is believed that more study on further extension of the optimization-based ECA should be the main research direction in the field. A combination of the optimization based method and the statistics-based approach could be another topic for further enhancing the results, especially when the plaintext image has unique statistics that are known to the attacker (for example, satellite images). In addition, applications of the optimization model to other fields, such as image and video coding, digital watermarking, and steganography, are also interesting directions that deserve further investigation.

It remains unclear whether the optimization-based attacks can also be generalized to non-perceptual multimedia encryption where the unencrypted data are not decodeable. It seems to be still possible that some new attacks can be developed by combining the current optimization model with the image and video codecs. For instance, although the unencrypted data are not decodeable, it is known that it will become decodeable if the encrypted data are correctly revealed. So the question becomes whether it is possible to build an optimization model that can find this optimal point which maximizes the degree of decodeablility of the non-decodeable data. This is certainly another big challenge to the multimedia encryption community.

## Acknowledgment

Shujun Li was supported by a fellowship from the Zukunftskolleg of the University of Konstanz, Germany, which is part of the "Excellence Initiative" Program of the DFG (German Research Foundation).

Figures 14.10 and 14.14 are reprinted from Reference [79] with the permission of IEEE.

## References

[1] Y.Q. Shi and H. Sun, *Image and Video Compression for Multimedia Engineering: Fundamentals, Algorithms, and Standards.* Boca Raton, FL, USA: CRC Press, 2nd edition, 2008.

[2] I.F. Akyildiz, T. Melodia, and K.R. Chowdhury, "Wireless multimedia sensor networks: Applications and testbeds," *Proceedings of the IEEE*, vol. 96, no. 10, pp. 1588–1605, October 2008.

[3] F. Dufaux, W. Gao, S. Tubaro, and A. Vetro, "Distributed video coding: Trends and perspectives," *EURASIP Journal on Image and Video Processing*, vol. 2009, pp. 508167:1–13, 2009.

[4] K.R. Rao and P.C. Yip, *Discrete Cosine Transform: Algorithms, Advantages, Applications.*

Boston, MA, USA: Academic Press, 1990.

[5] S. Mallat, *A Wavelet Tour of Signal Processing*. Boston, MA, USA: Academic Press, 2nd edition, 1999.

[6] T.M. Cover and J.A. Thomas, *Elements of Information Theory*. Hoboken, NJ, USA: John Wiley & Sons, 2nd edition, 2005.

[7] ISO/IEC, "Information technology – JPEG 2000 image coding system: Core coding system." ISO/IEC 15444-1, 2000.

[8] ISO/IEC, "Information technology – JPEG XR image coding system – Part 2: Image coding specification." ISO/IEC 29199-2, 2009.

[9] ISO/IEC, "Information technology – Coding of audio-visual objects – Part 10: Advanced Video Coding." ISO/IEC 14496-10, 2004.

[10] SMPTE (Society of Motion Picture and Television Engineers), "Standard for television – VC-1 compressed video bitstream format and decoding process." SMPTE 421M, 2006.

[11] A. Menezes, P. van Oorschot, and S. Vanstone, *Handbook of Applied Cryptography*. Boca Raton, FL, USA: CRC Press, 1996.

[12] B. Schneier, *Applied Cryptography – Protocols, Algorithms, and Source Code in C*. Hoboken, NJ, USA: John Wiley & Sons, 2nd edition, 1996.

[13] National Institute of Standards and Technology (US), "Specification for the advanced encryption standard (AES)." Federal Information Processing Standards Publication 197 (FIPS PUB 197), November 2001.

[14] K. Kaukonen and R.Thayer, "A stream cipher encryption algorithm Arcfour." Available online, http://www.mozilla.org/projects/security/pki/nss/draft-kaukonen-cipher-arcfour-03.txt.

[15] A. Kerckhoffs, "La cryptographie militaire," *Journal des sciences militaires*, vol. 9, pp. 5–38, January 1883.

[16] A. Kerckhoffs, "La cryptographie militaire," *Journal des sciences militaires*, vol. 9, pp. 161–191, February 1883.

[17] ISO/IEC, "Information technology – JPEG 2000 image coding system: Secure JPEG 2000." ISO/IEC 15444-8, 2007.

[18] ISO/IEC, "Information technology – Generic coding of moving pictures and associated audio information – Part 11: IPMP on MPEG-2 systems." ISO/IEC 13818-11, 2004.

[19] ISO/IEC, "Information technology – Coding of audio-visual objects – Part 13: Intellectual Property Management and Protection (IPMP) extensions." ISO/IEC 14496-13, 2004.

[20] "Advanced Access Content System (aacs) specifications." Available online, http://www.aacsla.com/specifications.

[21] M. Johnson, P. Ishwar, V. Prabhakaran, D. Schonberg, and K. Ramchandran, "On compressing encrypted data," *IEEE Transactions on Signal Processing*, vol. 52, no. 10, pp. 2992–3006, October 2004.

[22] L. Qiao and K. Nahrsted, "Comparison of MPEG encryption algorithms," *Computers & Graphics*, vol. 22, no. 4, pp. 437–448, August 1998.

[23] C. Fontaine and F. Galand, "A survey of homomorphic encryption for nonspecialists," *EURASIP Journal on Information Security*, vol. 2007, pp. 13801:1–10, 2007.

[24] C. Gentry, "Fully homomorphic encryption using ideal lattices," in *Proceedings of the 41st Annual ACM Symposium on Theory of Computing*, Bethesda, MD, USA, June 2009, pp. 169–178.

[25] B. Schneier, "Homomorphic encryption breakthrough." Available online, http://www.schneier.com/blog/archives/2009/07/homomorphic_enc.html.

[26] S. Li, Z. Li, and W.A. Halang, "Multimedia encryption," in *Encyclopedia of Multimedia Technology and Networking*, Hershey, PA, USA: IGI Global, August 2008, pp. 972–977.

[27] S. Li, C. Li, G. Chen, N.G. Bourbakis, and K.T. Lo, "A general quantitative cryptanalysis of permutation-only multimedia ciphers against plaintext attacks," *Signal Processing: Image Communication*, vol. 23, no. 3, pp. 212–223, March 2008.

[28] S. Li, G. Chen, A. Cheung, B. Bhargava, and K.T. Lo, "On the design of perceptual MPEG-video encryption algorithms," *IEEE Transactions on Circuits and Systems for Video Technology*, vol. 17, no. 2, pp. 214–223, February 2007.

[29] Y. Mao and M. Wu, "A joint signal processing and cryptographic approach to multimedia encryption," *IEEE Transactions on Image Processing*, vol. 15, no. 7, pp. 2061–2075, July 2006.

[30] T. Uehara and R. Safavi-Naini, "Attacking and mending arithmetic coding encryption schemes," in *Proceedings of the 22nd Australasian Computer Science Conference*, Auckland, New Zealand, January 1999, pp. 408–419.

[31] S. Li, G. Chen, A. Cheung, K.T. Lo, and M. Kankanhalli, "On the security of an MPEG-video encryption scheme based on secret Huffman tables," in *Proceedings of the Third Pacific Rim Symposium*, Tokyo, Japan, January 2009, pp. 898–909.

[32] K.P. Subbalakshmi and G. Jakimoski, "Cryptanalysis of some encryption schemes for multimedia," *IEEE Transactions on Multimedia*, vol. 10, no. 4, pp. 330–338, April 2008.

[33] I. Agi and L. Gong, "An empirical study of secure MPEG video transmission," in *Proceedings of ISOC Symposium on Network and Distributed Systems Security*, San Diego, CA, USA, February 1996, pp. 137–144.

[34] J. Wen, M. Severa, W. Zeng, M.H. Luttrell, and W. Jin, "A format-compliant configurable encryption framework for access control of video," *IEEE Transactions on Circuits and Systems for Video Technology*, vol. 12, no. 6, pp. 545–557, June 2002.

[35] T.D. Lookabaugh, D.C. Sicker, D.M. Keaton, W.Y. Guo, and I. Vedula, "Security analysis of selectively encrypted MPEG-2 streams," *Proceedings of SPIE*, vol. 5241, pp. 10–21, September 2003.

[36] C.P. Wu and C.C.J. Kuo, "Design of integrated multimedia compression and encryption systems," *IEEE Transactions on Multimedia*, vol. 7, no. 10, pp. 828–839, October 2005.

[37] A. Said, "Measuring the strength of partial encryption schemes," in *Proceedings of 2005 IEEE International Conference on Image Processing*, Genoa, Italy, September 2005, pp. 1126–1129.

[38] T. Uehara, R. Safavi-Naini, and P. Ogunbona, "Recovering DC coefficients in block-based DCT," *IEEE Transactions on Image Processing*, vol. 15, no. 11, pp. 3592–3596, November 2006.

[39] B.M. Macq and J.J. Quisquater, "Cryptology for digital TV broadcasting," *Proceedings of the IEEE*, vol. 83, no. 6, pp. 944–957, June 1995.

[40] ISO/IEC, "Information technology – Generic coding of moving pictures and associated audio information: Video." ISO/IEC 13818-2, 1996.

[41] ISO/IEC, "Information technology – Coding of audio-visual objects – Part 2: Visual." ISO/IEC 14496-2, 1999.

[42] H. Schwarz, D. Marpe, and T. Wiegand, "Overview of the scalable video coding extension of the H.264/AVC standard," *IEEE Transactions on Circuits and Systems for Video Technology*, vol. 17, no. 9, pp. 1103–1120, September 2007.

[43] J. Dittmann and A. Steinmetz, "Enabling technology for the trading of MPEG-encoded video," in *Proceedings of the Second Australasian Conference*, Sydney, NSW, Australia, July 1997, pp. 314–324.

[44] ISO/IEC, "Information technology – Digital compression and coding of continuous-tone still images: Requirements and guidelines." ISO/IEC 10918-1, 1994.

[45] ISO/IEC, "Information technology – Coding of moving pictures and associated audio for digital storage media at up to about 1,5 Mbit/s – Part 2: Video." ISO/IEC 11172-2, 1993.

[46] M.V. Droogenbroeck and R. Benedett, "Techniques for a selective encryption of uncompressed and compressed images," in *Proceedings of Advanced Concepts for Intelligent Vision Systems*, Ghent, Belgium, September 2002, pp. 90–97.

[47] A. Torrubia and F. Mora, "Perceptual cryptography of JPEG compressed images on the JFIF bit-stream domain," in *Proceedings of the IEEE International Conference on Consumer Electronics*, Los Angeles, CA, USA, June 2003, pp. 58–59.

[48] F. Dufaux and T. Ebrahimi, "Scrambling for privacy protection in video surveillance systems," *IEEE Transactions on Circuits and Systems for Video Technology*, vol. 18, no. 8, pp. 1168–1174, August 2008.

[49] L. Weng and B. Preneel, "On encryption and authentication of the DC DCT coefficient," in *Proceedings of the Second International Conference on Signal Processing and Multimedia Applications*, Barcelona, Spain, July 2007, pp. 375–379.

[50] G. Liu, T. Ikenaga, S. Goto, and T. Baba, "A selective video encryption scheme for MPEG compression standard," *IEICE Transactions on Fundamentals of Electronics, Communications and Computer Sciences*, vol. E89-A, no. 1, pp. 194–202, January 2006.

[51] M.I. Khan, J. Varun, and A.S. Malik, "On perceptual encryption: Variants of DCT block scrambling scheme for JPEG compressed images," in *Proceedings of the Signal Processing and Multimedia: International Conferences, SIP and MulGraB 2010*, Jeju Island, Korea, December 2010, pp. 212–223.

[52] BBC, "Dirac specification," Version 2.2.3. Available online, http://diracvideo.org/specifications.

[53] SMPTE (Society of Motion Picture and Television Engineers), "VC-2 video compression." ST 2042-1-2009, 2009.

[54] S. Lian, J. Sun, and Z. Wang, "Perceptual cryptography on SPIHT compressed images or videos," in *Proceedings of 2004 IEEE International Conference on Multimedia and Expo*, Taipei, Taiwan, July 2004, pp. 2195–2198.

[55] D. Engel and A. Uhl, "Parameterized biorthogonal wavelet lifting for lightweight JPEG 2000 transparent encryption," in *Proceedings of the 7th Workshop on Multimedia and Security*, New York, NY, USA, August 2005, pp. 63–70.

[56] J. Fang and J. Sun, "Compliant encryption scheme for JPEG 2000 image code streams," *Journal of Electronic Imaging*, vol. 15, no. 4, pp. 043013:1–4, November 2006.

[57] J.L. Liu, "Efficient selective encryption for JPEG 2000 images using private initial table," *Pattern Recognition*, vol. 39, no. 8, pp. 1509–1517, August 2006.

[58] T. Stütz and A. Uhl, "On efficient transparent JPEG2000 encryption," in *Proceedings of the 9th Workshop on Multimedia & Security*, Dallas, TX, USA, September 2007, pp. 97–108.

[59] D. Engel, T. Stütz, and A. Uhl, "A survey on JPEG2000 encryption," *Multimedia Systems*, vol. 15, no. 4, pp. 243–270, August 2009.

[60] H. Kiya, S. Imaizumi, and O. Watanabe, "Partial-scrambling of images encoded using JPEG2000 without generating marker codes," in *Proceedings of 2003 International Conference on Image Processing*, Barcelona, Spain, September 2003, pp. 205–208.

[61] H. Wu and D. Ma, "Efficient and secure encryption schemes for JPEG2000," in *Proceedings of the IEEE International Conference on Acoustics, Speech, and Signal Processing*, Montreal, QC, Canada, May 2004, pp. 869–972.

[62] Y. Bodo, N. Laurent, and J.L. Dugelay, "A scrambling method based on disturbance of motion vector," in *Proceedings of the 10th ACM International Conference on Multimedia*, Juan les Pins, France, December 2002, pp. 89–90.

[63] R. Hamzaoui and D. Saupe, "Fractal image compression," in *Document and Image Compression*, M. Barni (ed.), Boca Raton, FL, USA: CRC Press, May 2006, pp. 145–175.

[64] S. Roche, J.L. Dugelay, and R. Molva, "Multi-resolution access control algorithm based on fractal coding," in *Proceedings of the International Conference on Image Processing*, Lausanne, Switzerland, September 1996, pp. 235–238.

[65] L. Qiao, *Multimedia Security and Copyright Protection*. PhD thesis, University of Illinois at Urbana–Champaign, Urbana, IL, USA, 1998.

[66] S. Winkler, *Digital Video Quality: Vision Models and Metrics*. Hoboken, NJ, USA: John Wiley & Sons, 2005.

[67] Z. Wang and A.C. Bovik, *Modern Image Quality Assessment*. San Rafael, CA, USA: Morgan & Claypool Publishers, 2006.

[68] Z. Wang, A.C. Bovik, H.R. Sheikh, and E.P. Simoncelli, "Image quality assessment: From error visibility to structural similarity," *IEEE Transactions on Image Processing*, vol. 13, no. 4, pp. 600–612, April 2004.

[69] H.R. Sheikh and A.C. Bovik, "Image information and visual quality," *IEEE Transactions on Image Processing*, vol. 15, no. 2, pp. 430–444, February 2006.

[70] D.M. Chandler and S.S. Hemami, "VSNR: A wavelet-based visual signal-to-noise ratio for natural images," *IEEE Transactions on Image Processing*, vol. 16, no. 9, pp. 2284–2298, September 2007.

[71] W. Lin and C.C.J. Kuo, "Perceptual visual quality metrics: A survey," *Journal of Visual Communication and Image Representation*, vol. 22, no. 4, pp. 297–312, May 2011.

[72] Y. Mao and M. Wu, "Security evaluation for communication-friendly encryption of multimedia," in *Proceedings of the International Conference on Image Processing*, Singapore, October 2004, pp. 24–27.

[73] Y. Yao, Z. Xu, and J. Sun, "Visual security assessment for cipher-images based on neighborhood similarity," *Informatica*, vol. 33, pp. 69–76, March 2009.

[74] J. Sun, Z. Xu, J. Liu, and Y. Yao, "An objective visual security assessment for cipher-images based on local entropy," *Multimedia Tools and Applications*, vol. 53, pp. 75–95, May 2011.

[75] F. Ahmed and C.L. Resch, "Characterizing cryptographic primitives for lightweight digital image encryption," *Proceedings of SPIE*, vol. 7351, pp. 73510G:1–11, April 2009.

[76] S. Li, C. Li, K.T. Lo, and G. Chen, "Cryptanalysis of an image scrambling scheme without bandwidth expansion," *IEEE Transactions on Circuits and Systems for Video Technology*, vol. 18, no. 3, pp. 338–349, March 2008.

[77] D.V.D. Ville, W. Philips, R.V. de Walle, and I. Lemanhieu, "Image scrambling without bandwidth expansion," *IEEE Transactions on Circuits and Systems for Video Technology*, vol. 14, no. 6, pp. 892–897, June 2004.

[78] S. Li, J.J. Ahmad, D. Saupe, and C.C.J. Kuo, "An improved DC recovery method from AC coefficients of DCT-transformed images," in *Proceedings of the IEEE International Conference on Image Processing*, Hong Kong, September 2010, pp. 2085–2088.

[79] S. Li, A. Karrenbauer, D. Saupe, and C.C.J. Kuo, "Recovering missing coefficients in DCT-transformed images," in *Proceedings of the IEEE International Conference on Image Processing*, Brussels, Belgium, September 2011.

[80] D. Bertsimas and J.N. Tsitsiklis, *Introduction to Linear Optimization*. Nashua, NH, USA: Athena Scientific, 2nd edition, 1997.

# 15

# Exceeding Physical Limitations: Apparent Display Qualities

**Piotr Didyk, Karol Myszkowski, Elmar Eisemann, and Tobias Ritschel**

## 15.1  Introduction

Existing display devices introduce a number of physical constraints, which make real-world appearance difficult to realistically reproduce. For example, a direct reproduction of the luminance range of a moonless night to the intensity of the sun is technically out of reach. Similarly, the continuous nature of spatial and temporal information does not directly fit to the discrete notions of pixels and frames per second.

The human visual system (HVS) has its own limitations, which to certain extent reduce the requirements imposed on display devices. For example, through a luminance adaptation process (that can be extended in time) human eyes can operate both in dark-night and sunny-day conditions, however, simultaneously only four to five log10 units of luminance dynamic range can be perceived at once. Similarly, the limited density of photoreceptors in the retina (in the foveal region the size of cones amounts to 28 arcsec) as well as imperfections in the eye optics limit the spatial resolution of details that can be perceived. In the temporal domain, the critical flickering frequency (CFF) limits the ability to discern temporal signals over 60 Hz.

4

All such HVS-imposed limitations are taken into account, when designing display devices, but still a significant deficit of reproducible contrast, brightness, and spatial pixel resolution can be observed, which fall short with respect to the HVS capabilities. Moreover, unfortunate interactions between technological and biological aspects create new problems, which are unknown for real-world observation conditions. For example, the conflict between the eye accommodation adjusted to the display screen and the eye-ball vergence driven by depth (disparity) reproduced in three-dimensional (3D) stereo displays imposes limitations on the depth range that can be comfortably observed. Also, "frozen in time" discrete frames (for LCD displays, for instance) result in perceptual issues. While the entire sequence might appear smoothly animated, each frame is actually static for a short period of time. When the eye tracks dynamic objects (to keep their steady projection in the fovea), the static image is traversed smoothly and values crossed by the eye start to integrate on the retina, which results in a perceived hold-type blur. Note that such blur does not exist in the physical space (that is, in displayed images), but is created in a perceptual space. Nonetheless, hold-type blur can degrade the impression of perceived image quality in a similar way as physical blur introduced to images.

This chapter refers to *perceptual effects*, rather than to *physical effects*, which can be *experienced* but not *measured physically*. In particular, it aims at the exploitation of perceptual effects to help overcome physical limitations of display devices in order to enhance apparent image qualities. Section 15.2 shows how the perceived image contrast can be improved by exploiting the Cornsweet illusion. Section 15.3 introduces glare effect and shows how it can be used for brightness boosting. Section 15.4 presents techniques for reducing the negative impact of some perceptual effects, such as hold-type blur. It shows how high-quality frames can be interleaved with low-quality frames, still improving the overall appearance. Afterward, Section 15.5 demonstrates how the high quality of all frames can improve apparent spatial resolution. Section 15.6 discusses the role of perception in the context of stereovision and accommodation/vergence conflict reduction. Finally, this chapter concludes with Section 15.7.

## 15.2 Apparent Contrast Enhancement

Image contrast is an important factor for image perception. Nonetheless, there is usually a mismatch between the optical contrast of the displayed image and the real world due to the contrast limitations of typical display devices. Fortunately, humans, nonetheless, have the impression of a plausible real-world depiction when looking at images on a computer screen, despite the far lower luminance and contrast ranges in comparison to reality. So, the key issue for image reproduction is to allocate a sufficient contrast for a plausible reproduction of all important image features, while preserving the overall image structure [1]. The amount of allocated contrast influences the discrimination and identification of the objects depicted in the image, which are important factors for image quality judgments [2]. Numerous psychophysical studies show a clear preference toward images with enhanced contrast often by 10–20% with respect to their original counterparts [3], [4].

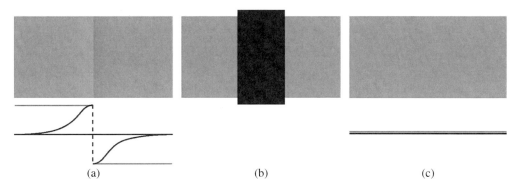

**FIGURE 15.1**

Cornsweet illusion. (a) By imposing the so-called Cornsweet luminance profile (as depicted by the black plot below), onto the center of isoluminant bar, a lightness illusion is created that the left bar side is brighter than the right one (as depicted by the gray plot). (b) By occluding the Cornsweet profile with the black rectangle the lightness of visible bar parts becomes the same, as one could expect for the isoluminant bar shown in (c).

Existing tone mapping and color grading operators are usually quite efficient in allocating an available dynamic range to reproduce image details and to preserve a certain overall image appearance. However, global contrast must often be traded off when texture details are faithfully reproduced. Conversely, preservation of contrast relations between major image components require a local contrast reduction that leads to a loss of small-scale information. Since physical contrast manipulations reach a dead end in such conditions, the question arises: Can the *perceived* contrast be enhanced?

Contrast perception is a far more complex phenomenon than accounting for physical contrast [5]. The eye's adaptation plays an important role, just like the display's luminance levels, ambient lighting, and spatial frequencies of the contrast signal [6]. While one can mostly reliably predict the influence of global adaptation, spatial-vision effects, which arise from intensity and color patterns created by neighboring pixels, are not easy to model [5]. This fact makes perceived contrast difficult to quantify in isolation from the actual image content because simultaneous contrast and Mach-band effects appear. While these effects are well known, their interactions can be complex and their controlled use to enhance apparent contrast has not yet been shown.

In this respect the Cornsweet perceptual illusion [7] is far more promising as it enables contrast enhancement along edges [8]. The illusion is created by introducing a pair of gradient profiles that are gradually darkening and, on the opposite side, lightening toward the common edge. At this edge they result in a sharp shading discontinuity, as shown in Figure 15.1. The lightness levels on both sides of the discontinuity are propagated through some filling-in mechanisms of human perception [9]. While this process is not fully understood, it is likely to be similar to image restoration processes that take place in the retinal blind spot where the optic nerve exits from the eyeball, and, for dark conditions, in the foveal blind spot due to the absence of foveal rods. The luminance propagation of the Cornsweet perceptual illusion creates the impression of a lightness step function, even though the gray levels on both sides match physically and do not actually correspond to a step function. Note that to trigger such an effect, one would usually need to introduce a physical contrast in form of a luminance step function, which would then require a certain

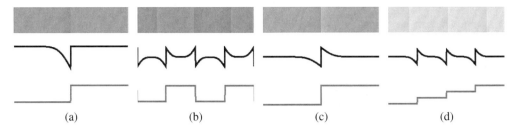

**FIGURE 15.2**

Different luminance profiles, which create the Craik-Cornsweet-O'Brien illusion: (a) Craik O'Brien, (b) missing fundamental, (c) Cornsweet, and (d) repeated Cornsweet.

dynamic range. For the Cornsweet perceptual illusion, the gradient profiles (called further the Cornsweet profiles) gradually tend to the original luminance levels on both sides of the edge, hence, the apparent contrast is produced without any extra use of dynamic range.

The illusion can also be achieved by inserting the Cornsweet profile into uniform luminance regions, as shown in Figure 15.1; it is even more interesting that the profile can be added to existing edges to boost their apparent contrast even further. In the latter scenario, skillfully inserted Cornsweet profiles affect the image appearance relatively little, as they contribute to a desirable apparent contrast enhancement at the edge, but do not produce additional sharp contrast patterns or ringing, as they gradually vanish. Obviously, when exaggerated, such profiles can create an infamous halo effect as discussed in the context of tone mapping, so an important issue is to control their strength. In art, adding such profiles along important lines and silhouettes is usually referred to as countershading [10].

In fact, a whole family of profiles similar to those shown in Figure 15.2, which are often called by the names of their inventors, can be used to trigger the apparent contrast illusion [7]. For example, the one-sided Craik-O'Brien profile is extremely useful to enhance image contrast in the proximity of saturated (clamped) image regions, where there is not enough dynamic range to use a fully symmetric Cornsweet profile.

Technically speaking, the Cornsweet profile can be generated by subtracting from the original image $Y$ its lowpass version $Y_\sigma$ obtained by means of Gaussian blurring. Then by adding back the difference $U = Y - Y_\sigma$ to $Y$ as:

$$Y' = Y + k \cdot U = Y + k \cdot (Y - Y_\sigma) \tag{15.1}$$

the resulting image $Y'$ has its contrast perceptually enhanced. This procedure is equivalent to the well-known image processing technique called *unsharp masking* [11], which is commonly used to sharpen image details (hereby, the apparent contrast increases as well [3], [12]). Unsharp masking produces Cornsweet profiles and, hence, its perceptual effect can be explained by the Cornsweet illusion [13].

While for naive unsharp masking the same amount of high frequencies is added to the whole image, one can imagine an adaptive procedure, where detail contrast is enhanced only in those regions, where the texture visibility is reduced. Reference [14] proposed a multi-resolution metric of local contrast, which detects feature loss in tone-mapped images with respect to a high-dynamic range reference and drives the spatial extent and the strength of Cornsweet profiles (Figure 15.3).

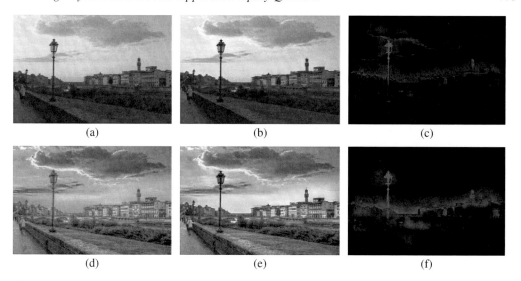

**FIGURE 15.3** (See color insert.)

Two examples of how the Cornsweet illusion can enhance the contrast of tone-mapped images are presented on the left. (a–c) Example processed using a global tone mapping operator, where countershading restores contrast in the heavily compressed dynamic range of the sky and highlights. (d–f) Example processed using a local tone mapping operator, which emphasizes on local details at expense of losing global contrast between the landscape and sky. In this case countershading restores brightness relations in this starkly detailed tone mapping result. Notice that the contrast restoration preserves the particular style of each tone mapping algorithm.

Although the Cornsweet illusion is the most effective for the luminance signal, it can also be applied to chromatic channels. For example, one can measure the loss of contrast in luminance, while the corrective signal $U$ affects chroma. Chroma enhancement strengthens the image colorfulness, and it turns out that increasing the original chroma values by 10% to 20% usually results in a preferred image appearance [15]. Reference [16] experimented with the introduction of the Cornsweet profile into the chroma signal to enhance the overall image contrast (Figure 15.4).

**FIGURE 15.4** (See color insert.)

The Cornsweet illusion used for color contrast enhancement: (a) original image and (b) its enhanced version using Cornsweet profiles in the chroma channel. In this example, the higher chroma contrast improves the sky and landscape separation and enhances the impression of scene depth.

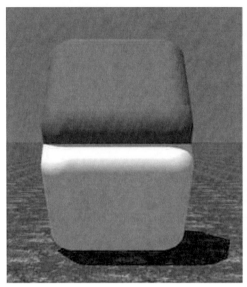

**FIGURE 15.5**

The Cornsweet effect strength in 3D scenes. Brightness of the bottom cube's wall looks different in these two images which differ only in the presence of black rectangle that occludes the Cornsweet profile [9].

Countershading has proved to be also useful in other applications, such as color-to-gray image conversion, where the dimensionality reduction from the usual three color channels into one achromatic channel may lead to an information loss. The lost chromatic contrast can be restored by adding the Cornsweet profiles into the achromatic channel [17].

In 3D graphics, other signals can be used to produce countershading profiles, for example, image depth [18]. Also, as discovered in Reference [9], the apparent contrast enhancement is significantly stronger when Cornsweet profiles are consistent with the scene lighting, undergo correct perspective foreshortening, and respect other cues resulting from 3D scene interpretation. Figure 15.5 shows an example where the apparent contrast enhancement was increased by 167% with respect to a standard two-dimensional Cornsweet profile, as reported in Reference [9]. Motivated by this observation, Reference [19] proposed a *3D unsharp-masking* technique, where the Cornsweet profiles are created directly in object space over the outgoing reflectance function. In this way, the depiction of various visual cues can be enhanced, including gradients from surface shading, surface reflectance, shadows, and highlights. These modifications lead to an improved overall image appearance and eases the interpretation of the 3D scene. Figure 15.6 shows an example of an image that was enhanced using 3D unsharp masking, where the Cornsweet profile was inserted at the shadow boundary and deepens the blackness impression. At the same time enough dynamic range is left (it is not completely black) in the shadow, so that the text on the opened book page remains legible.

In summary, contrast enhancement in displayed images is an important factor toward a better reproduction of real-world contrast ranges, and it leads to an overall preferred image appearance. By relying on apparent contrast enhancement by means of the Cornsweet illusion, the often contrast-limited displays can reach new levels. This effect was demonstrated for two-dimensional images or with an even stronger effect for 3D rendering. The Corn-

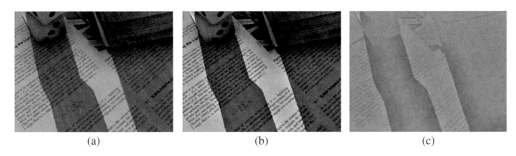

|  |  |  |
| :---: | :---: | :---: |
| (a) | (b) | (c) |

**FIGURE 15.6**

3D unsharp masking: (a) original image, (b) enhanced image, and (c) the corresponding correction signal $U'$ but computed and added to the surface shading in the 3D object space. ©ACM, 2008.

sweet illusion profiles can be introduced into achromatic (luminance) or chromatic channels to improve channel-related contrast deficits. Interestingly, this observation even opens up the road for other applications, such as color-to-gray conversion, where chromatic contrast in isoluminant regions is reproduced in the achromatic channel. Section 15.6 demonstrates another incarnation of the Cornsweet illusion that is used to enhance the impression of perceived depth by modifying the visual disparity signal.

## 15.3 Apparent Brightness Boost

This section discusses the problem of brightness enhancement by capitalizing on optical imperfections of the human eye, depicted in Figure 15.7. Every real-world optical system is affected by a certain amount of reflected and scattered stray light, which causes the contrast reduction due to a veil of luminance in the proximity of bright light sources and highlights.

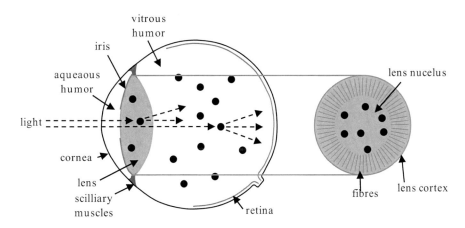

**FIGURE 15.7**

The human eye anatomy [22]. Notice opaque particles suspended in the lens and vitreous humor, which create the ciliary corona pattern due to light scattering, as shown in Figure 15.9. The right inset shows the lens structure with fiber membranes in the lens cortex, which act as a diffraction net and are responsible for the lenticular halo effect. ©The Eurographics Association 2009.

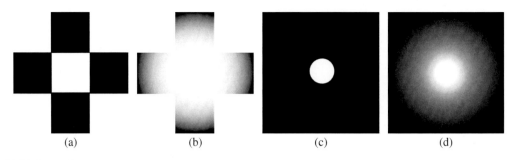

(a)                           (b)                           (c)                           (d)

**FIGURE 15.8**

It is not necessary to have physical light sources to evoke the impression that an object is self-luminous, as can be seen in (b) and (d). Painting halo (luminance gradients) around objects enhance their brightness or even creates an impression of glow without the actual light emission. Interestingly, the same surface representations without such luminance gradients cannot trigger the self-luminosity impression, and instead are perceived as reflective, as can be seen in (a) and (c). Redrawn from Reference [21].

This usually unwanted effect is called *veiling glare* or simply *glare*, and it occurs when bright objects are present in the field of view. In particular, in night scenes such scattered lighting may dominate at certain regions of the human eye retina. One may recall from the night driving experience how difficult is to read the registration plate of an approaching car while being exposed to its beam lights.

The human eyes are prone for glare effects due to structural shape imperfections of optical elements and the presence of various particles suspended in the lens and vitreous humor. Of particular interest is synthetic rendering of typical veiling glare patterns in images, which are somehow interpreted by the human brain as caused by the real presence of bright objects and lead to the so-called *glare illusion*. It turns out that when a veiling pattern is painted around nominally bright objects in the image, this illusion makes them appear brighter. Since the levels of luminance required to naturally trigger the glare effects in the eye cannot be achieved using traditional display technology, such glare patterns, painted directly in the images, improve the impression of realism. Moreover, even the effect of glowing caused by the introduction of smooth gradients around bright objects can be easily obtained [20], [21], as illustrated in Figure 15.8. Such gradients have been used by artists for centuries to improve the apparent dynamic range of their paintings, and it is just as attractive today in a digital imaging context.

A typical glare pattern for a small light source, as perceived by most subjects with normal eyes, is depicted in Figure 15.9. Two major component can be distinguished in the glare pattern [23], [24]: *bloom*, which refers to a general loss of contrast in the proximity of bright objects (veil), and *flare*, which comprises the *ciliary corona* (the sharp needles) and the *lenticular halo*. Some people report temporal changes in the glare appearance, such as a pulsation of the glare intensity and flickering of the fine needles in the ciliary corona [22].

The majority of existing approaches to computer-generated glare, while inspired by the knowledge about the human eye anatomy and physiology, are based on phenomenological results rather than explicit modeling of the underlying physical mechanisms. A common approach is to design convolution filters, which reduce image contrast in the proximity of glare sources by effectively adding luminance gradients around bright regions patterns similar as in Figure 15.8. In Reference [24], the authors base their filter on the point-spread

**FIGURE 15.9**

The glare appearance example [22]. ©The Eurographics Association 2009

function measured for the optics of the human eye [25], [26]. A set of Gaussian filters with different spatial extent, when skillfully applied, may lead to very convincing visual results as well. This approach is commonly used in computer games [27]. A recent perceptual study [28] has shown that the impression of displayed image brightness can be increased by more than 20% by convolving high-intensity pixels in the image with such simple filters.

Other glare effects, such as the ciliary corona and the lenticular halo, are often designed off-line and placed in the location of the brightest pixel for each glare source as a billboard (image sprite) [29], [24]. However, using billboards, it is difficult to realistically render glare for glare sources of arbitrary shape and non-negligible spatial extent.

Recently, References [30], [31], and [22] investigated the application of wave optics principles to glare modeling, considering various obstacles causing light scattering on its way from the pupil to the retina, as shown in Figure 15.7. Reference [31] investigated the role of small particles randomly distributed in the lens and floating in the vitreous humor in creating the ciliary corona pattern, as originally suggested in Reference [23]. Regularly spaced fiber membranes in the lens cortex act as the diffraction grating which produces lenticular halo [23], [29], [24]. Reference [22] considered the dynamic aspect of glare, observing that the pupil size, the shape of lens, and particle positions change in time.

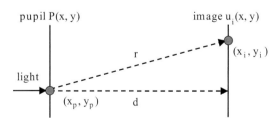

**FIGURE 15.10**

A simplified scheme of optical system in the eye used for wave optics simulation of light scattering in the eye [22]. ©The Eurographics Association 2009.

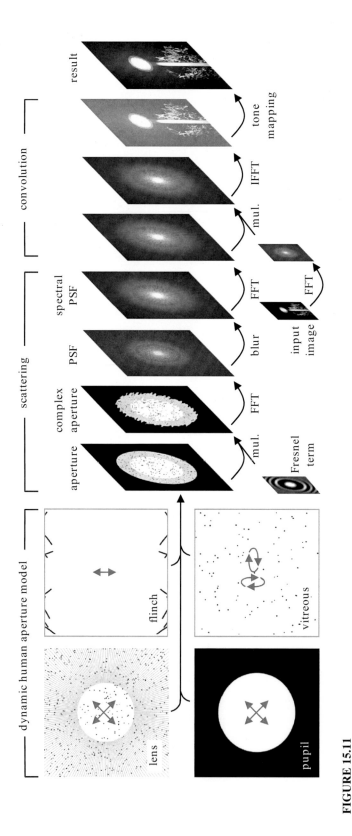

**FIGURE 15.11**

The glare computation pipeline, where the black arrows indicate GPU shaders that transform input texture(s) into output textures [22]. ©The Eurographics Association 2009.

It turns out that under a simplifying assumption that all these opaque light scattering obstacles are orthographically projected onto a single plane aperture Figure 15.10, a simple wave-optics model can be derived. Essentially, the model boils down to taking the Fourier transform $\mathscr{F}$ of the pupil aperture function $P(x_p, y_p)$, and computing its squared value $|\mathscr{F}\{P(x_p, y_p)\}|^2$ to derive the point-spread function (PSF) for this aperture. Then it is enough to convolve the input high-dynamic range image with the PSF to obtain a glare image. More formally, the Fresnel approximation to Huygen's principle [32] for the optical system depicted in Figure 15.10 under the assumption of homogeneous incident light of unit amplitude is given by

$$L_i(x_i, y_i) = K \left| \mathscr{F}\{P(x_p, y_p)E(x_p, y_p)\}_{p=\frac{x_i}{\lambda d}, q=\frac{y_i}{\lambda d}} \right|^2, \tag{15.2}$$

$$E(x_p, y_p) = e^{i\frac{\pi}{\lambda d}(x_p^2 + y_p^2)}, \quad K = 1/(\lambda d)^2, \tag{15.3}$$

where $(x_i, y_i)$ denotes the coordinate at the retina plane, $\lambda$ is the wavelength of the light, and $d$ is the distance between the pupil and the retina. The pupil aperture function $P(x_p, y_p)$ gives the opacity of each point in the pupil, with zero opacity values denoting transparent and ones denoting opaque attributes. The Fourier transform $\mathscr{F}$ is computed at the grid of points $(p, q) = \left(\frac{x_i}{\lambda d}, \frac{y_i}{\lambda d}\right)$.

Figure 15.11 summarizes the computation process that is performed in real time using recent graphics hardware. Namely, the high-resolution opacity image is composed from the images representing the pupil aperture, the lens diffraction grid (at the lens cortex) and particles, the eyelashes, and particles in the vitreous humor (obviously other light scattering elements can be considered as well). The resulting aperture function $P(x_p, y_p)$ is multiplied by the imaginary Fresnel term $E(x_p, y_p)$, as in Equation 15.2. Then, the fast Fourier transform (FFT) is performed over the resulting complex aperture, and the PSF for a single wavelength is computed. By rescaling the PSF for different wavelengths (refer to Reference [22] for more details) and summing up all resulting PSFs, the spectral PSF is computed. Then, the convolution of the input high-dynamic range image with the spectral PSF is performed in the Fourier domain, where it boils down to a simple multiplication. The inverse fast Fourier transform (IFFT) is performed to obtain the image with glare in the spatial domain, which is then tone mapped and displayed.

Figure 15.12a demonstrates a simple billboard approach on the candle image, where the spectral PSF is just placed at the flame center. Note the differences in the visibility of horizontal needles in the ciliary corona with respect to Figure 15.12b, which is computed using the processing pipeline depicted in Figure 15.11.

In summary, the glare illusion is a powerful tool to enhance the apparent image brightness and create an impression of light emission from the image. Similar observations have been made in the context of lens flares [33], and their simulation can be used to improve the overall perceived image appearance in the same way. The possibility to enhance brightness and to create the impression of light emission is highly desirable when aiming at a more realistic observer experience. As shown in this section, a direct glare simulation based on the principles of wave optics gives a convincing result and it is surprisingly inexpensive in terms of computations.

(a)                                                      (b)

**FIGURE 15.12 (See color insert.)**

The glare appearance for a luminous object of non-negligible spatial size [22]: (a) the light diffraction pattern (point-spread function) is just imposed on the background candle image, (b) the light diffraction pattern is convolved with the image. ©The Eurographics Association 2009.

## 15.4  Image Temporal Resolution Retargeting

The continuous demand for better-quality displays results in a huge development in the display industry. Attempts of creating mini-cinemas at home and/or visualization centers lead to the production of bigger screens. At the same time viewers tend to move closer to their displays in order to enjoy fine details of visualized content. Therefore, today's new designs – apart from improved contrast and brightness – need to ensure that the spatial display resolution is sufficient to convey a high-quality image appearance. This section will show that temporal aspects (that is, frame rate) can have a huge impact on the perception of spatial resolution.

Under standard conditions, the world is perceived as a crisp and sharp mental image. However, under certain conditions, the visual percept is blurred, which suggests that the human visual system indeed is a time-averaging sensor. This usually happens in the case of fast-moving objects, which cannot be easily tracked by the human eye (for example, car wheels or objects behind a fast train window). The blur, respectively, the temporal integration is because that quickly moving objects cannot be stabilized on the human eye's retina for certain velocities. This phenomena is usually called *motion blur* and is similar to the situation of taking a picture with a long exposure while objects are moving quickly in front of the camera. It is often assumed that the temporal integration of information by the HVS follows Bloch's law [34], which states that detectability of stimuli of similar characteristic depends only on their energy, which is a product of exposure time and luminance. Although this law holds only up to a short time duration ($\sim$40 ms), it becomes crucial for displays showing dynamic content, where the time of one frame, assuming 25 frames/s content, is

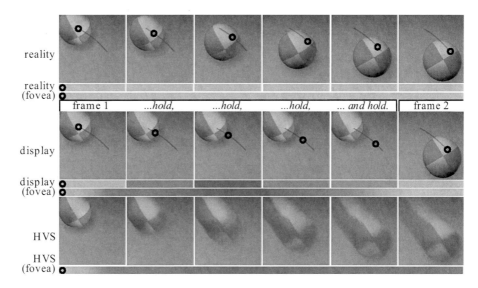

**FIGURE 15.13**

Illustration of hold-type blur for a ball under translational motion of constant velocity [35]. In the top row, physical reality denoted by six intermediate moments in time is shown. The open circle denote the eye fixation point. As the ball continuously moves and the human eyes perfectly track the ball (SPEM), its image is stabilized on the retina producing a crisp and sharp image. On a hold-type display shown in the middle row, the ball is displayed at the same position for the time period of one frame. However, at the same time, the fixation point will continuously move to smoothly track the ball. Therefore, as the ball's image on the retina is no longer stabilized, the signal registered on the retina will be blurred along the trajectory of motion, which is shown in the bottom row. ©The Eurographics Association 2010.

exactly 40 ms. This basically shows that an observer looking at a display does not only perceive what is currently shown but also what was shown and perceived before.

The eye integration has a huge influence on the appearance of moving objects on a display. Consider the continuously-moving real object shown in Figure 15.13 and assume that it is tracked perfectly by the HVS. Because of this tracking, a single receptor on the observer's retina will always receive the same signal over time. The percept is, hence, sharp and crisp. However, in all current display devices, objects are displayed for the time period of one frame at the same position (so-called *hold-type* effect), whereas the tracking performed by the HVS is continuous due to *smooth pursuit eye motion* (SPEM). Therefore, in contrast to what is found for real objects, the single receptor considered previously will not receive the same signal over time anymore. Instead, it will average values along the object trajectory resulting in a blurred percept. Note that this so-called *hold-type blur* is not due to any display limitations. It happens in human eyes, meaning that it is purely perceptual and cannot be captured easily by any camera.

A simple solution to reduce hold-type blur is to reduce the hold time, the period for which the same frame is shown on a display. This requires a high frame rate, which is, in case of broadcasting applications, usually not provided by the original content or too expensive in the case of rendered content. Therefore, many temporal upsampling techniques have been developed to increase the frame rate. Since this field is relevant to both the computer graphics and the display communities, solutions to reduce hold-type blur or to increase frame rate can be categorized in two according groups. The first considers the general problem of temporal upsampling, where solutions are usually implemented using a CPU or dedicated graphics hardware (GPUs). Here, the computation time can usually be traded in for higher quality. The second group is formed by solutions found in television screens, which have to be really efficient even on a low-cost computation units.

### 15.4.1   Computer Graphics Solutions

In the computer graphics community, a high frame rate is desired as it has a huge influence on animation smoothness, interactivity, and perceived image quality. Because high-quality image generation techniques are very often time consuming, a high frame rate is not always possible to achieve. Therefore, less expensive methods for improving the temporal resolution of the content without scarifying overall quality have been introduced.

One of the first attempts to increase the number of frames in the context of interactive applications was presented in Reference [36]. The method relied on depth information that allowed to reproject shaded pixels from one frame into another, thus creating many frames out of a single originally rendered frame. Similar ideas were exploited in later studies, where very expensive shaded samples were reused to speed up the image generation process. First such methods [37] scattered information from previously rendered frames into new frames by means of forward reprojections. Because of problems with occlusions and gaps that needed to be fixed explicitly, Reference [38] replaced forward reprojection by reverse reprojection, which also better fits common GPU architectures. Instead of pixel colors, one can also reuse computationally expensive intermediate values whose choice can be automated [39]. All these techniques can also been adapted to respect time-varying phenomena [40], which makes them particularly interesting for remote-rendering contexts as well [41]. More details can be found in Reference [42].

Another category of methods that allow for creating additional frames are morphing methods. An extended survey discussing various techniques was presented in Reference [43]. Recently, Reference [44] used content-preserving warps targeting the problem of video stabilization rather than temporal upsampling. However, similar techniques could be used also for temporal upsampling.

There are also methods that directly target the creation of intermediate frames. An example of such a method is presented in Reference [45], where the problem of temporal upsampling of video content has been addressed. The work is well-suited for a single dis-occlusion which allows for producing high-quality results for a standard content. However, it requires full knowledge of future frames and is computationally expensive; therefore, it is not really suitable for real-time applications. Although the quality of additionally created frames is important and can be produced using various methods, such as the one described above, in the case of temporal upsampling not all regions of the image are equally important. Reference [46] showed that high-quality moving edges are a key feature to the HVS. Therefore, any ghosting produced by temporal upsampling methods as well as hold-type blur can significantly lower perceived quality. The same work also proposed a method that addresses those perceptual effects and performs temporal upsampling, which preserves a high edge quality.

### 15.4.2 Industrial Solutions in Television Sets

Although the problem of temporal upsampling is well studied in computer graphics and many methods providing good quality have been developed, most of them are computationally expensive. Therefore, such solutions cannot be easily implemented as computational units manufactured for television screens. As the problem of hold-type blur is a widely known issue in the display industry, many off-the-shelf television sets nowadays offer solutions that are designed to reduce blur, but do not require much computational power. The key idea is, to increase the frame rate, usually to 100 or 200 Hz (respectively, 120 and 240 Hz for NTSC content), by introducing intermediate frames produced internally from a low frame rate signal. The following shortly summarizes existing solutions while an extended survey can be found in Reference [47].

The simplest solution is *black data insertion* (BDI), which reduces the hold-type effect by introducing new black frames interleaved with the original. This is similar to the way old cathode-ray tube (CRT) displays work; the light is emitted only for a small fraction of the frame time. However, this solution has some drawbacks and limitations. Similarly to classic CRT displays, it can significantly reduce brightness or introduce temporal flickering, especially in large bright areas, where the HVS temporal sensitivity is high.

Instead of inserting black frames other solutions turn the backlight of an LCD panel on and off, a procedure called *backlight flashing* (BF) [48], [47]. This is possible because in many available displays, LCD panels are illuminated using hundreds of LEDs, whose response is very fast. Therefore, it is easy to flash them at frequencies as high as 500 Hz. Note that this solution has similar drawbacks to that of BDI methods. Besides hold-type blur reduction, such techniques are also useful for reducing the effect of long LC response by turning on the backlight only at the moment when the desired state of the LCs is reached.

The problems of black data insertion methods can be overcome by using the original frames, that are duplicated and blurred, instead of black frames. On the other hand, this *blurred frame insertion* (BFI) solution [49] can cause visible ghosting since the blurred frames are not motion compensated.

Another category of methods reducing hold-type blur is *frame rate doubling* (FRD). For such solutions, additional frames are obtained by interpolating pairs of original frames along their optical flow trajectories [50]. Such techniques are commonly used in current television sets, where they can easily expand standard 24 Hz content to much higher frame rates (for example, 240 Hz) without reducing brightness or introducing flickering problems. The biggest limitation of frame interpolation comes from the optical flow estimation. While it is usually not a problem for objects that are moving slowly, optical flow estimation algorithms tend to fail when velocity gets bigger. As reported in Reference [35], in such situations, the optical flow is automatically deactivated and the original frames are simply duplicated, hereby, lowering the frame rate to the original, but avoiding visible errors. Such problems are because that the optical flow estimation problem is usually very complex and time consuming. Because all algorithms in television sets must be very efficient, they generally tend to show difficulties for challenging cases. Another drawback of using interpolation is the time lag introduced by the need of knowing the future frames in order to create intermediate frames. This is usually not a problem for broadcasting applications; however, in scenarios where a high interactivity is needed (for example, video games) the lag in interaction even beyond 90 Hz can be detected by the user and result in discomfort [35], [51].

Instead of increasing the frame rate, a software solution for reducing the hold-type blur is to apply an image filter to the content to be shown on the screen, that aims to invert hold-type blur. This technique is called *motion compensated inverse filtering* (MCIF). As hold-type blur can be modeled in image-space by a local one-dimensional convolution kernel oriented in the direction of the optical flow, a deconvolution technique can be used in order to inverse this process. In practice [52], it boils down to applying a local sharpening filter, which is computed according to the local optical flow. The effectiveness of such a technique is limited by the fact that hold-type blur is a lowpass filter, which removes certain frequencies. Therefore, none of those frequencies that are completely lost can be restored. Only those that are attenuated by hold-type blur can be recovered by amplifying them beforehand using linear filtering.

There exists a possibility to combine several methods described above; for example, the in-between frame derivation based on optical flow with the backlight flashing. However details on such custom solutions are not published.

### 15.4.3  Perceptually Motivated Temporal Upsampling for 3D Content

The solutions described in the previous section can be used for hold-type blur reduction and come either from the computer graphics community or television industry. Solutions from the first group are able to provide high-quality results at the expense of a high computational cost; therefore, depending on the complexity of the rendering, the achievable frame rate can be limited. Solutions from the second group are usually implemented as small chips in television sets, are well suited for real-time application (for example, broadcast content) but offer limited quality. The quality loss is because such solutions arc limited to the information found in the signal that is immediately displayed. Therefore, additional information, such as optical flow, needs to be recovered in order to create in-between frames. However, this step usually introduces visible artifacts since correct data cannot be obtained. Fortunately, in the case of synthetic content, such information is in fact available. The most common example is 3D rendering, where perfect optical flow, often needed for other purposes (for example, motion blur), can be computed very efficiently.

Reference [35] presented a method, which performs temporal upsampling for computer generated content by exploiting information available during 3D rendering and various aspects of the human perception. This method produces quality similar to that by solutions designed for computer graphics applications (Section 15.4.1), but at a low computational cost usually required in current industrial real-time solutions (Section 15.4.2). Another advantage of this method is that it performs extrapolation instead of interpolation, and hence, it does not produce any additional time lag. To this end, the method achieves trade-off between the single additional frame quality and the cost of its production by exploiting human perception.

In contrast to the previously described computer-graphics solutions, this method targets temporal upsampling from frame rates that are already relatively high. Namely, extrapolation was demonstrated from 40 Hz to 120 Hz signals, although the idea can be extended to even higher frame rates. The pipeline of the method is presented in Figure 15.14.

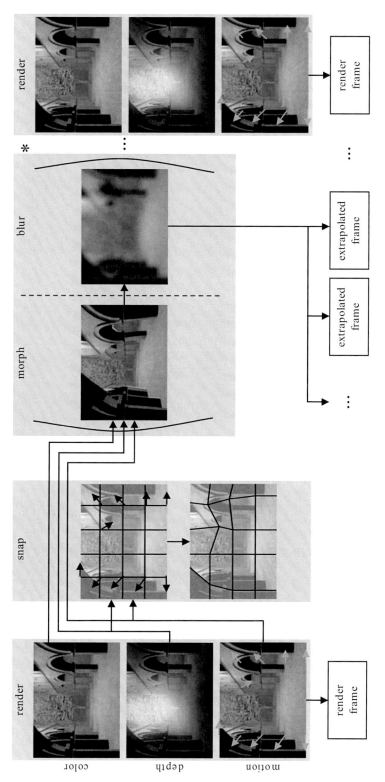

**FIGURE 15.14**

The pipeline of the temporal upsampling method from Reference [35]. From left to right; to extrapolate one or more additional frames, first a grid is created with vertices snapped to geometry discontinuities. Later the grid is used to morph the originally rendered frame into multiple in-between frames. These are later blurred in order to hide artifacts where morphing failed, for example, at disocclusions. Finally, the high-frequency losses are compensated for by frequency enhancement in the original frame.

(a)                                                                                    (b)

**FIGURE 15.15**

Simulation of the blur perceived when the same sequence is played at (a) 40 Hz and (b) 120 Hz. Notice how
the horizontal blur (hold-type effect), along the motion trajectory, was reduced. The simulation comes from a
user adjustment task where the level of blur in the presented images was matched with the blur perceived in
played animations [35]. ©The Eurographics Association 2010.

As the method is meant to work at a high frame rate, the time budget for the additional
frame computation is very low. The method first performs a fast and rough approximation
of the additional frames based on the knowledge of the past. This is achieved by extrapolat-
ing the current original frame using a mesh-based approach. First, a coarse grid is attached
to the original frame. This grid is then morphed, according to the optical flow that is obtained from the rendering pipeline, by translating each vertex of the grid by the vector of
the optical flow that corresponds to the location of the vertex. While the grid is deformed
also the underlying textures are morphed according to the optical flow. In the scenario of
upsampling 40 Hz content to 120 Hz, this technique was used to produce two predicted
frames. The low time consumption comes from the fact that morphing a coarse grid along
with a texture is relatively inexpensive using current GPUs. Unfortunately, this comes at
the expense of possible artifacts that can appear on discontinuities of optical flow (geometry). In order to suppress and hide them two solutions were proposed. First, to prevent
big drifts on geometry edges, vertices of the grid are snapped to geometry discontinuities
before the morphing. Although this aligns the grid with geometry well, small artifacts can
persist as the grid is too coarse to handle small geometric details. To improve the result further, the time-averaging property of the HVS is used. The extrapolated frames are blurred
in order to avoid such artifacts. The resulting high-frequency losses are compensated for in
the original frames that are, by definition, perfect. Because of the time averaging property
of the HVS, the perceived images are composed to the desired animation. However, the
compensation is not always feasible, particularly in high contrast regions. A black object
that is brightened by the blur would need a negative compensation, which is physically
impossible. In such places, the blur can be reduced, especially if artifacts are unlikely to be
visible. An example, how such a technique can reduce the hold-type effect, is shown in Figure 15.15. Table 15.1 compares the perceptually motivated temporal upsampling technique
to other techniques used in television sets described previously in Section 15.4.2.

This section showed that the temporal integration ability of the human visual system has
a significant impact on the perceived image quality. A low frame rate is the reason why

**TABLE 15.1**
Comparison of the various methods.

| Criterion | Method | | | | | |
|---|---|---|---|---|---|---|
| | BDI | BF | BFI | FRD | MCIF | Ref. [35] |
| LCD response required | high | moderate | high | high | no | high |
| backlight response required | no | high | no | no | no | no |
| optical flow quality | no | no | no | high | moderate | high |
| ghosting artifacts | possible | possible | yes | no | no | no |
| flickering artifacts | yes | yes | no | no | no | no |
| luminance reduction | yes | yes | no | no | no | no |
| limitation of blur reduction | flickering | flickering | no | no | frequency cut-off | no |
| other possible artifacts | no | no | no | fast motion | oversaturation | no |

objects moving on a hold-type screen are perceived as less sharp and crisp. The presented techniques reduce this problem and achieve a higher apparent visual quality using different temporal upsampling strategies or inverse filtering. It was also shown that the limitations of the HVS imply that not all frames need to be perfect. The latter can be exploited to make solutions more efficient. If performance is not an issue, one can even turn these HVS deficiencies to an advantage; as shown in the next section, carefully computed high-quality frames can even allow for the improvement of image quality (in terms of resolution and color depth) beyond the physical properties of a display.

## 15.5 Apparent Color and Resolution Enhancement

While hold-type blur leads to an objectionable reduction of perceived image quality, the fact that the eye integrates information over time can be turned into an advantage. The following describes two examples, that is, the increase of color and resolution, which allow enhancement of content appearance by exploiting the HVS.

The techniques presented here rely on the temporal effects. This idea can be illustrated on a prominent example of DLP projectors. These devices display the three color channels of an image independently, but in rapid succession. These mono-colored images are integrated in the eye and then lead to the impression of a full-color image.

Ideas similar to the DLP projector principle were also exploited in the early days of computer graphics and video games. The available color palettes were often limited and only a few could be used simultaneously. By making use of temporal effects, it is possible to virtually lift these constraints slightly. The idea is to flicker different colors that then integrate in the eye to a new mix that was not previously available.

For example, in a video game, one might want to add a shadow to the scene. To simulate such a shadow, one would usually rely on a darkened version of the affected pixels, yet, the necessary darker tints were not always available in the palette. To overcome this limitation, one could draw the shadow only in one out of two frames, resulting in a flickering shadow. If the refresh rate is high enough to exceed the critical flickering frequency (CFF) [53], the affected colors start to mix with the black of the shadow, hereby, leading to an apparent extension of the available colors.

Similar principles are also used in LCD screens under the name of *frame rate control*. In fact, many devices are limited to 6 bits per color channel, while the graphics card usually sends 8-bit color information per channel. Many screens exploit a simple technique to display the different nuances, which otherwise would be lost. Whenever a color is not representable by 6 bits, the screen displays its immediate color neighbors in quick succession over time [54]. Hereby, the apparent bit depth is effectively increased because the eye integrates the information and reports an average value to the brain.

The color impression is not the only feature that can be improved by exploiting the temporal domain. Resolution can also be enhanced to produce details that physically cannot be displayed on a screen. Such a possibility is of high interest because today's image content becomes increasingly detailed. There is an often striking difference between the resolution that a camera uses to record an image and the number of pixels that can actually be shown on a display device. Gigapixel images are no longer uncommon, but displaying them in full detail is a challenge.

In order to increase the apparent resolution of an image, it is again possible to make use of perceptual solutions. As explained in the previous section on hold-type blur, the eye integrates neighboring pixels when moving over the screen to track dynamic content. This effect would, for a static image, lead to an image blur that reduces the overall quality. On the other hand, if the display frequency is increased and the image does no longer appear static, the information perceived during the eye trajectory changes rapidly. Consequently, the information that is integrated in each photoreceptor of the eye can vary drastically. By choosing appropriate temporal color patterns, one can produce an illusion of a high-resolution image by stimulating each receptor in an appropriate way.

The precise setup, recently presented in Reference [55], is as follows. The goal is to display a single high-resolution image (animations will be discussed later) on a low-resolution screen. The idea is to move this high-resolution image slowly across the screen. The observer will track this image and the smooth pursuit eye motion will fixate image points at their corresponding position on the retina. As the movement is perfectly smooth, the eye photoreceptors will integrate the time-varying information in each pixel over a short period of time (Figure 15.16a). As explained before, this integration time is related to the CFF. For example, consider the CFF of 40 Hz; consequently, on a 120 Hz display, three frames are integrated together. In other words, if the eye tracks a point, it will constantly project to the same photoreceptor on the retina; hence, the receptor will perceive three temporal value changes besides the changes when moving from one pixel to the next.

Mathematically, if a receptor moves over an image $I$ (which varies in its pixels and over time) during a duration $T$ that conforms with the CFF along a path $p(t)$, the perceived result is

$$\int_0^T I(p(t),t)\,dt. \tag{15.4}$$

When tracking across the screen, retinal photoreceptors switch pixels of the low-resolution device at different time steps, which, in combination with the temporal changes of the image content, results in a sufficient variance of the perceived information to create the illusion of looking at a higher-resolution screen (Figure 15.16b).

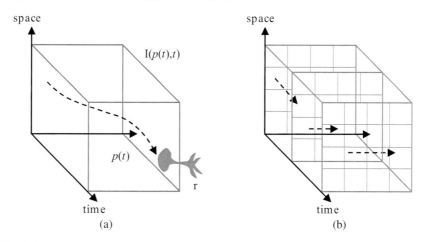

**FIGURE 15.16**

Perceived integration of an animated image: (a) the continuous case, where an animated image $I(p(t),t)$ is perceived by a receptor $r$ moving along a continuous path $p(t)$, and (b) the discrete case, where an animated image consists of discrete pixels at discrete frames of time with a discrete motion in every frame.

One important observation from Equation 15.4 is that the eye movement is crucial in order to enhance apparent resolution. The reason becomes clearer when looking at a given time $t_0$, as $I(p(t),t_0)$ is constant for all receptors inside of a pixel. Consequently, the same color sequence would be perceived and hence, the same integrated response will be reported. The image on the retina corresponds thus to a low-resolution image.

Interestingly, although Equation 15.4 looks continuous, the change of the frames (assuming a CFF of 40 Hz, three frames will be integrated) and the pixels are discrete. Thus, the integral becomes a weighted sum with the weights proportional to the time for which a pixel's color is seen by the photoreceptor:

$$\sum_{t=0}^{N} w_t I(p(t),t). \tag{15.5}$$

Formally, a weight $w_{i,j,t}$, where $i$ and $j$ refer to a discrete pixel position and $k$ the discrete time interval during which the pixel's color is constant, is computed as

$$w_{i,j,k} := \int \chi_{(i,j)}(p(t))\, \chi_k(t)\, dt, \tag{15.6}$$

where $\chi$ describes a characteristic function. Namely, $\chi_{(i,j)}(p(t))$ equals to one if $p(t)$ lies in pixel $(i,j)$; otherwise, it is zero. The term $\chi_k(t)$ denotes a characteristic function according to the time interval. In this step, one underlying assumption is that the temporal integration on the retina corresponds to a box filter in the temporal domain, further, time variation of the pixel colors is assumed to be produced instantaneously.

So far, only a single photoreceptor was considered. In Reference [55], the authors assumed that the these receptors are arranged on a uniform grid and hence, that each photoreceptor is actually located on a pixel of the high-resolution image. If the illusion of a higher-resolution image is to be reproduced, each photoreceptor should then perceive an in-

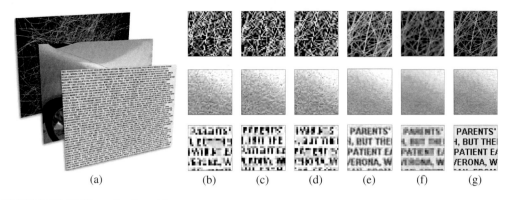

(a)            (b)       (c)       (d)       (e)       (f)       (g)

**FIGURE 15.17 (See color insert.)**

Resolution enhancement: (a) test images, (b–d) subimages obtained in the optimization process, (e) resolution enhancement via simulation of the images created on the retina once the three subimages are shown on a high frame rate display, (f) Lanczos downsampling as a standard method, and (g) original high-resolution images. Note, that even though the resolution was reduced three times, the presented method was able to recover fine details in contrast to the standard downsampling method.

tegrated response according to Equation 15.5 that is close to the value of its corresponding high-resolution image pixel.

For each receptor, Equation 15.5 describes a linear relationship. For all receptors, this results in a linear system that one can solve for the image sequence $I$ over time. The solution can be found in a least-square sense. A constrained optimization is often used to assure that the values of $I$ lie in the displayable range. Examples of the resolution enhancement are shown in Figure 15.17.

As the equation models the integration in the eye, one has to make sure that the CFF is respected and the eye properly integrates the information. Reference [55] proposed a postprocessing technique that blends the optimal solution with a low-resolution filtered version. The blending process is spatially varying, according to the predicted CFF, which is strongly scale and contrast dependent [56]. It ensures that the variance in the temporal signal is properly reduced and the integration assured. Consequently, flickering is avoided, but the resolution gain might be slightly reduced.

For general high-resolution animation, such as 3D rendering, an extension was introduced in Reference [57]. Here, the idea is to rely on optical flow (extracted from the movie or 3D scene) to predict the way the eye tracks general information. A novel GPU-based solver, which drastically reduces the computation time, was proposed. These measures lead to higher processing rates, that enable the treatment of general movies.

Overall, the exploitation of the temporal domain offers an interesting opportunity to produce a perceived resolution increase. This improvement has been proven in several user studies that show the superiority of the approach with respect to standard downsampling techniques. The gain in information has been illustrated by producing a font of $2 \times 3$ pixels that remains legible. This latter case illustrates that there is an actual gain in information and not just the illusion of it. In the future, the temporal domain is an important aspect that should be considered when producing image content.

## 15.6 Apparent Depth Manipulation

Recently, 3D stereo has received much attention in applications, such as computer games, feature films, and television production. Although 3D movies, 3D games, or even first 3D television channels are accessible to a wide range of customers, many challenges exist when aiming to produce stereo content that is perceptually convincing. The previous sections discussed how the knowledge of the HVS helps to improve luminance contrast, brightness, as well as spatial and temporal resolutions, which are important constituents of image quality. This section complements this argument by showing that taking into account the HVS in the context of 3D stereo improves the perceived quality and the viewing comfort of stereo content.

### 15.6.1 Depth Perception

In reality, the HVS relies on a number of different mechanisms that allow perceiving depth. They are known as *depth cues* and can be categorized [58] as pictorial (occlusion, perspective foreshortening, relative and familiar object size, texture and size gradients, shadows, aerial perspective), dynamic (motion parallax), ocular (accommodation, vergence), and stereoscopic (binocular disparity). The sensitivity of the HVS to different cues mostly depends on the distance between the observer and the observed objects [59]. The HVS is also able to combine the information coming from different cues even if they contradict each other [58, Chapter 5.5.10]. Recently, both the feature film and the computer game industries have been undertaking considerable effort to introduce binocular disparity (one of the strongest depth cues) into the content they produce. It is introduced artificially by showing different images to the left and the right eye. This strongly affects accommodation and vergence, which are closely coupled and due to the synthetic stimulation easily contradict each other. The problem is caused by the fact that accommodation tends to maintain the display screen within the depth of focus (DOF) that roughly falls into the range of $\pm 0.3$ diopters [60]. On the other hand, with increasing screen disparity, vergence drives the fixation point away from the screen plane. This results in a conflict between the fixation point and the focusing point, which can be tolerated to some degree by the accommodation-vergence mechanisms, but at the expense of possible visual discomfort [60] (Figure 15.18). However, even if the problem of depth conflicts would not exist, the HVS is limited in perceiving big disparities. Retinal images can be fused only in a region called *Panum's fusional area*, where disparities are not too big and otherwise double vision (*diplopia*) is experienced. It was also shown that the fusion depends on many other factors, such as individual differences, stimulus properties (better fusion for small, strongly textured, well-illuminated, static patterns), and exposure duration.

Binocular disparity, as one of the most appealing depth cues, can turn flat images shown on a screen into three-dimensional scenes with significantly enhanced realism. Therefore, it is crucial to understand how the HVS interprets this information, and possibly overcome the limitations stemming from using standard 3D equipment. Unfortunately, disparity perception has not yet received as much attention as brightness perception, which is already a well-studied topic, but it is interesting how many properties they share [61], [62], [63].

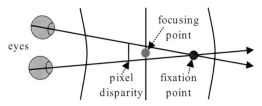

**FIGURE 15.18**

The problem of discomfort due to the disagreement between accommodation and vergence. When some none-zero disparity is introduced by showing different images to the left and the right eye, the fixation point is driven further from the screen plane, on which the focusing point is located. As the disparity gets bigger the distance between the focusing and the fixation point is increased. This disagreement can be tolerated only to some extent by the accommodation-vergence mechanism, causing discomfort when the fixation point gets too far the screen plane.

Sensitivity to disparity and brightness is defined by detection thresholds, which depend on the spatial frequency of depth corrugations and their amplitude. The "u"-shape of the disparity sensitivity function (DSF) is analogous to the contrast sensitivity function (CSF). However, it is shifted and has a peak around 0.3 to 0.5 cpd and a cut-off frequency around 3 to 5 cpd [63]. For brightness it was shown that CSF is an envelope of responses for a number of independent channels that are tuned to different spatial frequencies [64]. Similar conclusions were provided for disparity [65, Chapter 19.6.3]; however, it is not yet clear what the independent channel bandwidth for disparity modulation is, although some estimates suggest one to three octaves. Interestingly apparent depth, similarly to apparent brightness, is also dominated by the distribution of disparity contrasts rather than absolute disparities [61]. All the analogies listed above suggest that disparity and brightness perception undergo similar mechanisms. Therefore, new models, that potentially could handle disparities, can be inspired by already existing work for luminance.

### 15.6.2    Perceptual Disparity Model

There were many attempts to better understand how the HVS interprets depth and how sensitive it is to different cues [59], [65]. It is also interesting how disparity can be reduced in order to increase the viewing comfort [66]. Also many psychophysical measurements were performed to better understand the sensitivity of the HVS to disparities [67], [63], [65]. However, none of them allowed to build a model as complete as the ones available for luminance [64], [68].

Reference [69] recently developed a perceptual model for disparity based on a series of psychophysical measurements that allows for a better understanding of binocular disparity and predicts the response of the HVS to different stimuli. This work was mostly inspired by the existing similarities between brightness and depth perception described in the previous section. The key information needed to build such a model are measurements of disparity detection thresholds. The authors conducted a number of psycho-visual experiments to obtain the thresholds for sinusoidal patterns in depth with different frequencies and amplitudes. This allowed them to fit an analytic function to the obtained data, which describes detection thresholds for a whole range of possible sinusoidal depth corrugations. Such function can be later used for computing a so-called *transducer function* that maps disparity values to a perceptually linearized space. The transducer function is invertible, therefore changing from physical disparity values into a perceptual space and back becomes possible. This allows for performing all operations on disparity in a perceptually linearized space and take into account the nonlinearity of human disparity perception. As the disparity sensitivity depends on the spatial frequency of the depth pattern, before applying the transducer function, a disparity map needs to be decomposed into frequency bands and all operations have to be performed separately on each band.

Reference [69] showed a number of possible applications of the model. One of them is a disparity metric which allows for comparing two stereo images. Given an original disparity map and a distorted one, it computes a spatially varying map of perceived disparity distortion magnitudes. Both disparity maps are first decomposed into frequency bands and later transformed using the model to the perceptually uniform space, where they are expressed as just-noticeable difference units (JNDs). In order to compute the perceived difference, corresponding bands are first subtracted and per-band difference is calculated. Later, differences from all bands are combined to one map that illustrates the perceived differences. The metric was validated by the authors using scenes with different number of cues, where distortions were introduced as a postprocessing step using a warping method [70].

Another possible application is disparity compression. Since the model can transfer disparity maps into JND space, it is easy to decide which disparities are invisible to the human observer. Removing such information allows for good compression, preserving at the same time the high quality of images. As the response of the HVS can be predicted, it is also possible to enforce the same response on another person or different equipment, as those factors can influence the disparity sensitivity. Therefore, it is possible to display the same content on two different devices optimizing the disparity for both of them in such a way that the depth impression will be similar regardless of the device or the observer. The presented model allows also to adjust the disparity range to the viewing conditions. The usual scaling of disparities, such as the one presented in Reference [66], is performed without taking into account how the HVS perceives disparity. Using the presented model, the disparity can be scaled in the perceptual space and the nonlinear behavior of the HVS will be implicitly captured. Such a method will allow for bigger manipulations of disparities to which the HVS is not sensitive, whereas those places that are crucial for the HVS will be better preserved.

Reference [69] also showed *backward-compatible* stereo as a novel application of the proposed model. It allows for minimizing disparity in a way that the stereo content seen without glasses does not reveal any significant artifacts; however, once special equipment is used, the observer can experience a stereo effect. As an example application, one can consider the problem of printing a stereo image using anaglyph (red/green or red/cyan glasses) techniques. In such a situation, the stereo impression can be enjoyed only when a viewer wears anaglyph glasses, but when such equipment is not available, standard stereo images can be easily spoiled by color artifacts. In contrast, when the backward-compatible technique is used, the color artifacts can be minimized. The technique can also be useful for other 3D equipment. For example, when the content must be seen through shutter glasses, a person who does not have them will perceive blurred pictures, that is, an average of the left and right eye view.

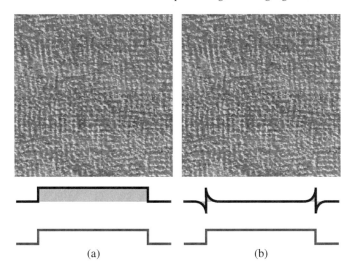

**FIGURE 15.19**

The Cornsweet illusion shown in anaglyph colors. A circle with (a) depth due to disparity and (b) apparent depth due to Cornsweet disparity profiles. The plots show the disparity profiles and their perceived shapes. The solid area depicts the total disparity, which is significantly smaller when using Cornsweet profiles.

The backward-compatible stereo technique exploits the Cornsweet illusion, which was shown (Section 15.2) to be able to enhance the contrast of luminance patterns. It turns out that the illusion holds for quite different signals, such as perceived line lengths or texture pattern density [71]. Reference [72] found that a depth Cornsweet profile adds to the perceived depth difference between real textured surfaces. In Reference [73], this effect was confirmed for random-dot stereograms. The authors observed that the typically induced depth difference over the whole surfaces amounted up to 40% of the depth difference at the discontinuity, which was roughly twice larger than in experiments conducted in Reference [72] for real surfaces. It was concluded that a gradual depth decay in the Cornsweet profile remains mostly invisible and the object dislocation at the discontinuity is propagated by the HVS over both surfaces [73]. This effect is illustrated in Figure 15.19.

In order to obtain a stereo image that does not exhibit disturbing artifacts while watched without special equipment, the backward-compatible technique first compresses (that is, flattens) the image disparity. At the same time, the disparity metric is employed to make sure that a specified minimum of perceived disparity remains. In this process, the *Craik-O'Brien-Cornsweet* illusion [72], [73] is created on disparity discontinuities by removing the low-frequency component. Since the HVS is less sensitive to low-frequency disparity corrugations, the obtained Cornsweet profiles are not visible. Instead, apparent depth is created by propagating the disparity discontinuities introduced by the profiles over surfaces [73] (Figure 15.20).

One additional advantage of exploiting the Cornsweet illusion comes from the fact that a cascade of the local Cornsweet profiles still conveys a consistent impression of discrete depth changes, while in the traditional approach disparity accumulation is required for proper stereoscopic effect. Thus, there is no need of accumulating global disparities, which further improves the backward compatibility.

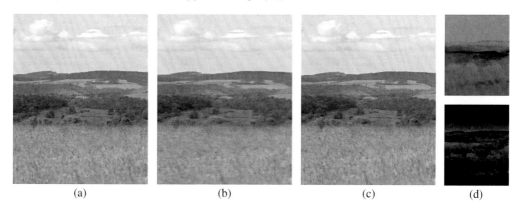

(a)  (b)  (c)  (d)

**FIGURE 15.20** **(See color insert.)**

Backward-compatible anaglyph stereo (c) offers good depth-quality reproduction with similar appearance to the standard stereo image. The small amount of disparity leads to relatively high-quality images when stereo glasses are not used. To achieve equivalent depth-quality reproduction in the traditional approach significantly more disparity is needed (b), in which case the appearance of anaglyph stereo image is significantly degraded once seen without special equipment.

Note, that the Cornsweet illusion can be also used for enhancement of 3D impression, similarly as it was done for luminance by introducing the unsharp masking profiles described in Section 15.2. Examples of such enhancement can be seen in Figure 15.21.

Because of the huge success of 3D equipment, the study of stereo perception plays an increasingly important role and is likely to still occupy researchers in the future. Especially, the nonlinear relationship between physical disparity and perceived depth is an interesting observation. It is important for a deeper understanding, but also crucial when creating stereo content. It has been even shown that the perceived depth can be augmented by perceptual means. Currently, experts are only scratching the surface of this topic and exciting avenues for future research remain.

## 15.7   Conclusion

This chapter demonstrated how perceptual effects can be employed to exceed physical limitations of display devices. By capitalizing on various aspects of the human visual system's insensibility, display qualities, at least perceptually, have been significantly enhanced in a stable and persistent way. Similar enhancement could often be achieved only by improving physical parameters of displays, which might be impossible without fundamental design changes in the existing display technology and clearly may lead to overall higher display costs.

The presented techniques achieve their goals by means of relatively simple image processing operations, which often rely on a skillful gradient introduction into the luminance and disparity domains. The Cornsweet gradient profile inserted across luminance and depth discontinuities (edges in the image) enhances apparent perceived contrast and disparity.

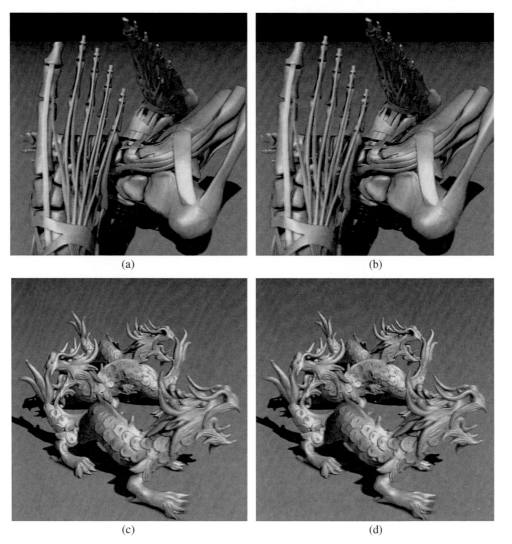

**FIGURE 15.21** (See color insert.)

Depth enhancement using the Cornsweet illusion for two different scenes with significant depth range: (a,c) original images and (b,d) enhanced anaglyph images. A better separation between foreground and background objects and a more detailed surface structure depiction can be perceived in enhanced images.

The halo effect surrounding bright image regions enhances perceived image brightness. Interleaving, in the temporal domain, high-quality sharpened and low-quality blurred frames on high-refresh rate displays, reduces perceptual hold-type blur and improves apparent motion smoothness. Similar in spirit, the principles of temporal signal integration in the retina can be used to improve color perception, reduce quantization errors on low bit-depth displays, and enhance apparent resolution.

In the research presented here, each quality dimension, such as image contrast, brightness, sharpness, temporal, and spatial resolution, have been handled separately. An interesting question remains: can the possible interactions between those dimensions also be exploited to even further improve apparent display qualities?

## Acknowledgment

Figure 15.3 is image courtesy of Grzegorz Krawczyk. Figure 15.4 is image courtesy of Kaleigh Smith and Grzegorz Krawczyk. Figure 15.5 is image courtesy of R. Beau Lotto (www.lottolab.org) and Dale Purves (www.purveslab.net). Figure 15.6 is image courtesy of Kaleigh Smith, and is reprinted from Reference [19] with the permission of ACM. Figure 15.8 is redrawn from Reference [21].

Figures 15.7, 15.9, 15.10, 15.11, and 15.12 are reprinted from Reference [22] and Figures 15.13 and 15.15 are reprinted from Reference [35], all with the permission of the Eurographics Association.

## References

[1] Z. Wang and A.C. Bovik, *Modern Image Quality Assessment*. San Rafael, CA, USA: Morgan & Claypool Publishers, 2006.

[2] R. Janssen, *Computational Image Quality*. Bellingham, WA, USA: SPIE Press, 2001.

[3] A.J. Calabria and M.D. Fairchild, "Perceived image contrast and observer preference: I. The effects of lightness, chroma, and sharpness manipulations on contrast perception," *The Journal of Imaging Science and Technology*, vol. 47, no. 6, pp. 479–493, 2003.

[4] A. Yoshida, R. Mantiuk, K. Myszkowski, and H.P. Seidel, "Analysis of reproducing real-world appearance on displays of varying dynamic range," *Computer Graphics Forum*, vol. 25, no. 3, pp. 415–426, March 2006.

[5] F. Ratliff, "Contour and contrast," *Proceedings of the American Philosophical Society*, vol. 115, no. 2, pp. 150–163, April 1971.

[6] P.G. Barten, *Contrast Sensitivity of the Human Eye and Its Effects on Image Quality*. Bellingham, WA, USA: SPIE Press, 1999.

[7] F. Kingdom and B. Moulden, "Border effects on brightness: A review of findings, models and issues," *Spatial Vision*, vol. 3, no. 4, pp. 225–262, January 1988.

[8] D. Burr, "Implications of the Craik-O'Brien illusion for brightness perception," *Vision Research*, vol. 27, no. 11, pp. 1903–1913, May 1987.

[9] D. Purves, A. Shimpi, and R. Lotto, "An empirical explanation of the Cornsweet effect," *The Journal of Neuroscience*, vol. 19, no. 19, pp. 8542–8551, October 1999.

[10] M. Livingstone, *Vision and Art: The Biology of Seeing*. New York, USA: Harry N. Abrams, 2002.

[11] W. K. Pratt, *Digital Image Processing*. New York, USA: John Wiley & Sons, Inc., 2nd edition, 1991.

[12] W. Lin, Y. Gai, and A. Kassim, "Perceptual impact of edge sharpness in images," *IEE Proceedings Vision, Image and Signal Processing*, vol. 152, no. 2, pp. 215–223, April 2006.

[13] K. Smith, *Contours and Contrast*. PhD thesis, MPI Informatik, Saarbruecken, Germany, 2008.

[14] G. Krawczyk, K. Myszkowski, and H.P. Seidel, "Contrast restoration by adaptive counter-shading," *Computer Graphics Forum*, vol. 26, no. 3, pp. 581–590, September 2007.

[15] E. Fedorovskaya, H. deRidder, and F. Blommaert, "Chroma variations and perceived quality of color images of natural scenes," *Color Research and Application*, vol. 22, no. 2, pp. 96–110, April 1997.

[16] K. Smith, G. Krawczyk, K. Myszkowski, and H.P. Seidel, "Beyond tone mapping: Enhanced depiction of tone mapped HDR images," *Computer Graphics Forum*, vol. 25, no. 3, pp. 427–438, September 2006.

[17] K. Smith, P.E. Landes, J. Thollot, and K. Myszkowski, "Apparent greyscale: A simple and fast conversion to perceptually accurate images and video," *Computer Graphics Forum*, vol. 27, no. 2, pp. 193–200, April 2008.

[18] T. Luft, C. Colditz, and O. Deussen, "Image enhancement by unsharp masking the depth buffer," *ACM Transactions on Graphics*, vol. 25, no. 3, pp. 1206–1213, July 2006.

[19] T. Ritschel, K. Smith, M. Ihrke, T. Grosch, K. Myszkowski, and H.P. Seidel, "3D unsharp masking for scene coherent enhancement," *ACM Transactions on Graphics*, vol. 27, no. 3, pp. 90:1–8, August 2008.

[20] D. Zavagno, "Some new luminance-gradient effects," *Perception*, vol. 28, no. 7, pp. 835–838, May 1999.

[21] D. Zavagno and G. Caputo, "The glare effect and the perception of luminosity," *Perception*, vol. 30, no. 2, pp. 209–222, September 2001.

[22] T. Ritschel, M. Ihrke, J.R. Frisvad, J. Coppens, K. Myszkowski, and H.P. Seidel, "Temporal glare: Real-time dynamic simulation of the scattering in the human eye," *Computer Graphics Forum*, vol. 28, no. 2, pp. 183–192, March 2009.

[23] G. Simpson, "Ocular haloes and coronas," *British Journal of Ophthalmology*, vol. 37, no. 8, pp. 450–486, August 1953.

[24] G. Spencer, P. Shirley, K. Zimmerman, and D.P. Greenberg, "Physically-based glare effects for digital images," in *Proceedings of the 22nd Annual Conference on Computer Graphics and Interactive Techniques*, Los Angeles, CA, USA, August 1995, pp. 325–334.

[25] G. Westheimer, "The eye as an optical instrument," in *Handbook of Perception and Human Performance: 1. Sensory Processes and Perception*, K. Boff, L. Kaufman, and J. Thomas (eds.), New York, USA: Wiley, 1986, pp. 4.1–20.

[26] R. Deeley, N. Drasdo, and W. Charman, "A simple parametric model of the human ocular modulation transfer function," *Ophthalmic and Physiological Optics*, vol. 11, no. 1, pp. 91–93, January 1991.

[27] M. Kawase, "Practical implementation of high dynamic range rendering," in *Proceedings of the Game Developers Conference*, 2005.

[28] A. Yoshida, M. Ihrke, R. Mantiuk, and H.-P. Seidel, "Brightness of the glare illusion," in *Proceedings of the ACM Symposium on Applied Perception in Graphics and Visualization*, Los Angeles, CA, USA, August 2008, pp. 83–90.

[29] P. Rokita, "A model for rendering high intensity lights," *Computers & Graphics*, vol. 17, no. 4, pp. 431–437, July 1993.

[30] M. Kakimoto, K. Matsuoka, T. Nishita, T. Naemura, and H. Harashima, "Glare generation based on wave optics," *Computer Graphics Forum*, vol. 24, no. 2, pp. 185–193, June 2005.

[31] T.J. van den Berg, M.P. Hagenouw, and J.E. Coppens, "The ciliary corona: Physical model and simulation of the fine needles radiating from point light sources," *Investigative Ophthalmology and Visual Science*, vol. 46, no. 7, pp. 2627–2632, July 2005.

[32] J. Goodman, *Introduction to Fourier Optics*. Englewood, CO, USA: Roberts & Co, 2005.

[33] M.B. Hullin, E. Eisemann, H.P. Seidel, and S. Lee, "Physically-based real-time lens flare rendering," *ACM Transactions on Graphics*, vol. 30, no. 4, pp. 108:1–9, August 2011.

[34] A. Gorea and C.W. Tyler, "New look at Bloch's law for contrast," *Journal of the Optical Society of America A*, vol. 3, no. 1, pp. 52–61, January 1986.

[35] P. Didyk, E. Eisemann, T. Ritschel, K. Myszkowski, and H.P. Seidel, "Perceptually-motivated real-time temporal upsampling of 3D content for high-refresh-rate displays," *Computer Graphics Forum*, vol. 29, no. 2, pp. 713–722, May 2010.

[36] W.R. Mark, L. McMillan, and G. Bishop, "Post-rendering 3D warping," in *Proceedings of the Symposium on Interactive 3D Graphics*, Providence, RI, USA, April 1997, pp. 7–16.

[37] B. Walter, G. Drettakis, and S. Parker, "Interactive rendering using the render cache," in *Proceedings of the 10th Eurographics Workshop on Rendering*, Granada, Spain, June 1999, vol. 10, pp. 235–246.

[38] D.F. Nehab, P.V. Sander, J. Lawrence, N. Tatarchuk, and J. Isidoro, "Accelerating real-time shading with reverse reprojection caching," in *Proceedings of the 22nd ACM Symposium on Graphics Hardware*, San Diego, CA, USA, 2007, pp. 25–35.

[39] P. Sitthi-Amorn, J. Lawrence, L. Yang, P.V. Sander, D. Nehab, and J. Xi, "Automated reprojection-based pixel shader optimization," *ACM Transactions on Graphics*, vol. 27, no. 5, December 2008.

[40] R. Herzog, E. Eisemann, K. Myszkowski, and H.P. Seidel, "Spatio-temporal upsampling on the GPU," in *Proceedings of ACM Symposium on Interactive 3D Graphics and Games*, Bethesda, MD, USA, February 2010, pp. 91–98.

[41] D. Pajak, R. Herzog, E. Eisemann, K. Myszkowski, and H.P. Seidel, "Scalable remote rendering with depth and motion-flow augmented streaming," *Computer Graphics Forum*, vol. 30, no. 2, pp. 415–424, April 2011.

[42] D. Scherzer, L. Yang, O. Mattausch, D. Nehab, P.V. Sander, M. Wimmer, and E. Eisemann, "A survey on temporal coherence methods in real-time rendering," in *In State of the Art Reports Eurographics*, Llandudno, UK, May 2010, pp. 1017–4656.

[43] G. Wolberg, "Image morphing: A survey," *The Visual Computer*, vol. 14, no. 8, pp. 360–372, December 1998.

[44] F. Liu, M. Gleicher, H. Jin, and A. Agarwala, "Content-preserving warps for 3D video stabilization," *ACM Transaction on Graphics*, vol. 28, no. 3, pp. 44:1–9, July 2009.

[45] D. Mahajan, F.C. Huang, W. Matusik, R. Ramamoorthi, and P. Belhumeur, "Moving gradients: A path-based method for plausible image interpolation," *ACM Transaction on Graphics*, vol. 28, no. 3, pp. 42:1–11, July 2009.

[46] T. Stich, C. Linz, C. Wallraven, D. Cunningham, and M. Magnor, "Perception-motivated interpolation of image sequences," *ACM Transactions on Applied Perception*, vol. 8, no. 2, pp. 11:1–25, January 2011.

[47] X.F. Feng, "LCD motion blur analysis, perception, and reduction using synchronized backlight flashing," in *Proceedings of SPIE*, vol. 6057, pp. M1–14, February 2006.

[48] H. Pan, X.F. Feng, and S. Daly, "LCD motion blur modeling and analysis," in *Proceedings of the IEEE International Conference on Image Processing*, Genoa, Italy, September 2005, pp. 21–24.

[49] H. Chen, S.S. Kim, S.H. Lee, O.J. Kwon, and J.H. Sung, "Nonlinearity compensated smooth frame insertion for motion-blur reduction in LCD," in *Proceedings of the IEEE 7th Workshop on Multimedia Signal Processing*, Shanghai, China, November 2005, pp. 1–4.

[50] T. Kurita, "Moving picture quality improvement for hold-type AM-LCDs," in *Proceedings of the Society for Information Display Conference*, San Jose, CA, USA, June 2001, pp. 986–989.

[51] D. Luebke, B. Watson, J.D. Cohen, M. Reddy, and A. Varshney, *Level of Detail for 3D Graphics*. New York, USA: Elsevier Science Inc., 2002.

[52] M.A. Klompenhouwer and L.J. Velthoven, "Motion blur reduction for liquid crystal displays: Motion-compensated inverse filtering," *Proceedings of SPIE*, vol. 5308, p. 690, January 2004.

[53] M. Kalloniatis and C. Luu, "Temporal resolution." Available online, http://webvision.med. utah.edu/temporal.html.

[54] O. Artamonov, "X-bit's guide: Contemporary LCD monitor parameters and characteristics. Page 11." Available online, http://www.xbitlabs.com/articles/monitors/display/lcd-guide\_11. html.

[55] P. Didyk, E. Eisemann, T. Ritschel, K. Myszkowski, and H.P. Seidel, "Apparent display resolution enhancement for moving images," *ACM Transactions on Graphics*, vol. 29, no. 4, pp. 113:1–8, July 2010.

[56] P. Mäkelä, J. Rovamo, and D. Whitaker, "Effects of luminance and external temporal noise on flicker sensitivity as a function of stimulus size at various eccentricities," *Vision Research*, vol. 34, no. 15, pp. 1981–91, August 1994.

[57] K. Templin, P. Didyk, T. Ritschel, E. Eisemann, K. Myszkowski, and H.P. Seidel, "Apparent resolution enhancement for animations," in *Proceedings of the 27th Spring Conference on Computer Graphics*, Vinicne, Slovak Republic, April 2011, pp. 85–92.

[58] S.E. Palmer, *Vision Science: Photons to Phenomenology*. Cambridge, MA, USA: The MIT Press, 1999.

[59] J. Cutting and P. Vishton, "Perceiving layout and knowing distances: The integration, relative potency, and contextual use of different information about depth," in *Perception of Space and Motion (Handbook Of Perception And Cognition)*, W. Epstein and S. Rogers (eds.), San Diego, CA, USA: Academic Press, 1995, pp. 69–117.

[60] D. Hoffman, A. Girshick, K. Akeley, and M. Banks, "Vergence-accommodation conflicts hinder visual performance and cause visual fatigue," *Journal of Vision*, vol. 8, no. 3, pp. 1–30, March 2008.

[61] A. Brookes and K. Stevens, "The analogy between stereo depth and brightness," *Perception*, vol. 18, no. 5, pp. 601–614, 1989.

[62] P. Lunn and M. Morgan, "The analogy between stereo depth and brightness: A reexamination," *Perception*, vol. 24, no. 8, pp. 901–904, 1995.

[63] M.F. Bradshaw and B.J. Rogers, "Sensitivity to horizontal and vertical corrugations defined by binocular disparity," *Vision Research*, vol. 39, no. 18, pp. 3049–3056, September 1999.

[64] S. Daly, "The visible differences predictor: An algorithm for the assessment of image fidelity," In *Digital Images and Human Vision*, A.B. Watson (ed.), Cambridge, MA, USA: MIT Press, 1993, vol. 4, pp. 179–206.

[65] I.P. Howard and B.J. Rogers, *Seeing in Depth: Depth Perception*. Toronto, ON, Canada: I. Porteous, vol. 2, 2002.

[66] M. Lang, A. Hornung, O. Wang, S. Poulakos, A. Smolic, and M. Gross, "Nonlinear disparity mapping for stereoscopic 3D," *ACM Transaction on Graphics*, vol. 29, no. 4, pp. 751–760, July 2010.

[67] C.W. Tyler, "Spatial organization of binocular disparity sensitivity," *Vision Research*, vol. 15, no. 5, pp. 583–590, May 1975.

[68] J. Lubin, "A visual discrimination model for imaging system design and development," in *Vision Models for Target Detection and Recognition*, A.R. Menendez and E. Peli (eds.), River Edge, NJ, USA: World Scientific, 1995, pp. 245–283.

[69] P. Didyk, T. Ritschel, E. Eisemann, K. Myszkowski, and H.P. Seidel, "A perceptual model for disparity," *ACM Transactions on Graphics*, vol. 30, no. 4, pp. 96:1–96:10, August 2011.

[70] P. Didyk, T. Ritschel, E. Eisemann, K. Myszkowski, and H.P. Seidel, "Adaptive image-space stereo view synthesis," in *Proceedings of the Vision, Modeling and Visualization Workshop*, Siegen, Germany, November 2010, pp. 299–306.

[71] D.M. Mackay, "Lateral interaction between neural channels sensitive to texture density?," *Nature*, vol. 245, no. 5421, pp. 159–161, September 1973.

[72] S. Anstis, I. Howard, and B. Rogers, "A Craik-O'Brien-Cornsweet illusion for visual depth," *Vision Research*, vol. 18, no. 2, pp. 213–217, 1978.

[73] B. Rogers and M. Graham, "Anisotropies in the perception of three-dimensional surfaces," *Science*, vol. 221, no. 4618, pp. 1409–11, September 1983.

# *Index*

Printed and bound by CPI Group (UK) Ltd, Croydon, CR0 4YY

18/10/2024

01776270-0019